T0212898

Communications
in Computer and Information Science　521

Editorial Board

More information about this series at http://www.springer.com/series/7899

Stanisław Kozielski · Dariusz Mrozek
Paweł Kasprowski · Bożena Małysiak-Mrozek
Daniel Kostrzewa (Eds.)

Beyond Databases, Architectures and Structures

11th International Conference, BDAS 2015
Ustroń, Poland, May 26–29, 2015
Proceedings

 Springer

Editors

Stanisław Kozielski
Silesian University of Technology
Gliwice
Poland

Bożena Małysiak-Mrozek
Silesian University of Technology
Gliwice
Poland

Dariusz Mrozek
Silesian University of Technology
Gliwice
Poland

Daniel Kostrzewa
Silesian University of Technology
Gliwice
Poland

Paweł Kasprowski
Silesian University of Technology
Gliwice
Poland

ISSN 1865-0929   ISSN 1865-0937   (electronic)
Communications in Computer and Information Science
ISBN 978-3-319-18421-0   ISBN 978-3-319-18422-7   (eBook)
DOI 10.1007/978-3-319-18422-7

Library of Congress Control Number: 2015937218

Springer Cham Heidelberg New York Dordrecht London

Printed on acid-free paper

Springer International Publishing AG Switzerland is part of Springer Science+Business Media
(www.springer.com)

Preface

Collecting, processing, and analyzing data became important branches of computer science. Many areas of our existence generate a wealth of information that must be stored in a structured manner and processed appropriately in order to gain the knowledge from the inside. Databases have become a ubiquitous way of collecting and storing data. They are used to hold data describing many areas of human life and activity, and as a consequence, they are also present in almost every IT system. Today's databases have to face the problem of data proliferation and growing variety. More efficient methods for data processing are needed more than ever. New areas of interests that deliver data require innovative algorithms for data analysis.

Beyond Databases, Architectures and Structures (BDAS) is a series of conferences that intends to give the state of the art of the research that satisfies the needs of modern, widely understood database systems, architectures, models, structures, and algorithms focused on processing various types of data. The aim of the conference is to reflect the most recent developments of databases and allied techniques used for solving problems in a variety of areas related to database systems, or even go one step forward – beyond the horizon of existing databases, architectures, and data structures. The 11th International BDAS Scientific Conference was a continuation of the highly successful BDAS conference series started in 2005 in Ustroń, Poland. For many years BDAS has been attracting hundreds or even thousands of researchers and professionals working in the field of databases. Among attendees of our conference were scientists and representatives of IT companies. Several editions of BDAS were supported by our commercial, world-renowned partners, developing solutions for the database domain, such as IBM, Microsoft, Sybase, Oracle, and others. BDAS annual meetings have become an arena for exchanging information on the widely understood database systems and data processing algorithms.

BDAS 2015 was the 11th edition of the conference, organized under the technical cosponsorship of the IEEE Poland Section. We also continued our successful cooperation with Springer, which resulted in the publication of this book. The conference attracted more than a hundred participants from 15 countries, who made this conference a successful and memorable event. There were five keynote talks given by leading scientists: Prof. Bora İ. Kumova from Department of Computer Engineering, İzmir Institute of Technology (İYTE), İzmir, Turkey spoke on 'Fuzzy syllogistic reasoning over relational data.' Prof. Dirk Labudde from Bioinformatics group Mittweida (bigM) and Forensic Science Investigation Lab (FoSIL), University of Applied Sciences, Mittweida, Germany gave an excellent talk entitled 'Bioinformatics and Forensics - How today's Life Science technologies can shape the Crime Sciences of tomorrow.' Prof. Jean-Charles Lamirel from SYNALP team, LORIA, France gave a very enlightening speech on 'New metrics and related statistical approaches for efficient mining in very large and highly multidimensional databases,' Prof. Mikhail Moshkov from King Abdullah University for Science and Technology (KAUST), Saudi Arabia honored us with a presentation on 'Extensions of dynamic programming for design and analysis of decision trees,' and Dr. Riccardo Rasconi from Institute of Cognitive Science and Technology, National Research Council, Rome, Italy spoke on 'Surveying the versatility of constraint-based large neighborhood search for scheduling problems.' The keynote speeches and plenary sessions gained insight into new areas.

BDAS is focused on all aspects of databases. It is intended to have a broad scope, including different kinds of data acquisition, processing, and storing, and this book reflects fairly well the large span of research presented at BDAS 2015. This volume consists of 53 carefully selected papers. The first three papers accompany the stunning keynote talks. The remainder of the papers are assigned to eight thematic groups:

- Database architectures and performance
- Data integration, storage, and data warehousing
- Ontologies and Semantic Web
- Artificial intelligence, data mining, and knowledge discovery
- Image analysis and multimedia mining
- Spatial data analysis
- Database systems development
- Applications of database systems

The first group is related to various database architectures, query optimization, and database performance. Papers gathered in this group discuss hot topics of query selectivity estimation, testing performance of various database systems, NoSQL and data consistency, temporal and probabilistic databases. The next group of papers concern issues related to data integration, data storage, and data warehousing. The group consists of seven papers presenting research devoted to the data mapping semantics while sharing and exchanging data, novel data integration architectures, efficiency of storage space configuration, new ETL concepts, and data warehouse modeling.

The third group consists of three papers devoted to ontologies and the Semantic Web. These papers discuss problems of automatic approaches for building ontology from relational data, data integration with ontology, and RDF graph partitioning. The research devoted to artificial intelligence and data mining is presented in eight papers gathered in the fourth group. These papers show a wide spectrum of applications of various exploration methods, like decision rules, knowledge-based systems, clustering, artificial immune systems and memetic algorithms, Dynamic Gaussian Bayesian Network models, to solve many real problems.

The next group of papers is focused on image analysis and multimedia mining. This group consists of six papers devoted to lossless compression of images, querying multimedia databases, real-time object detection from depth images, analysis of facial expressions, emotions, and medical images.

Some aspects of the spatial data collecting and processing are discussed in three successive papers. The next three papers show various aspects of database systems development. Finally, the last 10 papers present different usage of databases starting from mining and metallurgical industries, through different fuel and energy consumption related problems and ERP systems, ending with bioinformatics knowledgebase and databases storing affect-annotated data and faces.

We would like to thank all Program Committee members and additional reviewers for their effort in reviewing the papers. Special thanks to Piotr Kuźniacki - builder and for ten years administrator of our website www.bdas.pl. The conference organization would not have been possible without the technical staff: Dorota Huget and Jacek Pietraszuk.

We hope that the broad scope of topics related to databases covered in this proceedings volume will help the reader to understand that databases have become an important element of nearly every branch of computer science.

April 2015

Stanisław Kozielski
Dariusz Mrozek
Paweł Kasprowski
Bożena Małysiak-Mrozek
Daniel Kostrzewa

Organization

BDAS 2015 was organized by Institute of Informatics, Silesian University of Technology, Poland.

BDAS 2015 Program Committee

Honorary Member

Lotfi A. Zadeh   University of California, Berkeley, USA

Chair

Stanisław Kozielski   Silesian University of Technology, Poland

Members

Sansanee Auephanwiriyakul	Chiang Mai University, Thailand
Werner Backes	Sirrix AG Security Technologies, Bochum, Germany
Susmit Bagchi	Gyeongsang National University, South Korea
Patrick Bours	Gjøvik University College, Norway
George D.C. Cavalcanti	Universidade Federal de Pernambuco, Brazil
Po-Yuan Chen	China Medical University, Taichung, Taiwan, University of British Columbia, BC, Canada
Yixiang Chen	East China Normal University, Shanghai
Tadeusz Czachórski	IITiS, Polish Academy of Sciences, Poland
Andrzej Chydziński	Silesian University of Technology, Poland
Sebastian Deorowicz	Silesian University of Technology, Poland
Jack Dongarra	University of Tennessee, Knoxville, USA
Andrzej Drygajlo	École Polytechnique Fédérale de Lausanne, Switzerland
Moawia Elfaki Yahia	King Faisal University, Saudi Arabia
Rudolf Fleischer	German University of Technology, Oman
Hamido Fujita	Iwate Prefectural University, Japan
Krzysztof Goczyła	Gdańsk University of Technology, Poland
Marcin Gorawski	Silesian University of Technology, Poland
Jarek Gryz	York University, Ontario, Canada
Andrzej Grzywak	Silesian University of Technology, Poland
Brahim Hnich	Izmir University of Economics, Izmir, Turkey

Dominik Ślęzak	University of Warsaw and Infobright Inc., Poland
Andrzej Świerniak	Silesian University of Technology, Poland
Adam Świtoński	Silesian University of Technology, Poland
Karin Verspoor	University of Melbourne, Australia
Alicja Wakulicz-Deja	University of Silesia in Katowice, Poland
Sylwester Warecki	Intel Corporation, San Diego, California, USA
Tadeusz Wieczorek	Silesian University of Technology, Poland
Konrad Wojciechowski	Silesian University of Technology, Poland
Robert Wrembel	Poznań University of Technology, Poland
Stanisław Wrycza	University of Gdańsk, Poland
Mirosław Zaborowski	IITiS, Polish Academy of Sciences, Poland
Grzegorz Zaręba	University of Arizona, Tucson, USA
Krzysztof Zieliński	AGH University of Science and Technology, Poland
Quan Zou	Xiamen University, People's Republic of China

Organizing Committee

Bożena Małysiak-Mrozek	Silesian University of Technology, Poland
Dariusz Mrozek	Silesian University of Technology, Poland
Paweł Kasprowski	Silesian University of Technology, Poland
Daniel Kostrzewa	Silesian University of Technology, Poland
Piotr Kuźniacki	Silesian University of Technology, Poland

Additional Reviewers

Augustyn Dariusz Rafał
Bach Małgorzata
Bajerski Piotr
Brzeski Robert
Duszenko Adam
Frączek Jacek
Harężlak Katarzyna
Josiński Henryk
Kawulok Michał
Kozielski Michał
Michalak Marcin
Momot Alina
Niedbała Sławomir
Nowak-Brzezińska Agnieszka
Nurzyńska Karolina

Piórkowski Adam
Płuciennik-Psota Ewa
Respondek Jerzy
Romuk Ewa
Sitek Paweł
Świderski Michał
Szwoch Wioleta
Traczyk Tomasz
Tutajewicz Robert
Waloszek Wojciech
Werner Aleksandra
Wyciślik Łukasz
Zghidi Hafed
Zielosko Beata

Sponsoring Institutions

Technical cosponsorship of the IEEE Poland Section

Contents

Data Integration, Storage and Data Warehousing

Ontologies and Semantic Web

Artificial Intelligence, Data Mining and Knowledge Discovery

Image Analysis and Multimedia Mining

Spatial Data Analysis

Database Systems Development

Applications of Database Systems

Invited Papers

New Metrics and Related Statistical Approaches for Efficient Mining in Very Large and Highly Multidimensional Databases

Jean-Charles Lamirel[1,2(✉)]

[1] Department of Computer Science, University of Tartu,
J. Liivi 2, 50409 Tartu, Estonia
jean-charles.lamirel@ut.ee
[2] LORIA, Equipe Synalp, Bâtiment B,
54506, Vandoeuvre Cedex, France
lamirel@loria.fr

Abstract. As regard to the evolution of the concept of text and to the continuous growth of textual information of multiple nature which is available online, one of the important issues for linguists and information analysts for building up assumptions and validating models is to exploit efficient tools for textual analysis, able to adapt to large volumes of heterogeneous data, often changing and of distributed nature. We propose in this communication to look at new statistical methods that fit into this framework but that can also extent their application range to the more general context of dynamic numerical data.

For that purpose, we have recently proposed an alternative metric based on feature maximization. The principle of this metric is to define a measure of compromise between generality and discrimination based altogether on the properties of the data which are specific to each group of a partition and on those which are shared between groups. One of the key advantages of this method is that it is operational in an incremental mode both on clustering (i.e. unsupervised classification) and on traditional categorization. We have shown that it allowed to very efficiently solve complex multidimensional problems related to unsupervised analysis of textual or linguistic data, like topic tracking with data changing over time or automatic classification in natural language processing (NLP) context. It can also adapt to the traditional discriminant analysis, often exploited in text mining, or to automatic text indexing or summarization, with performance that are far superior to conventional methods. In a more general way, this approach that freed from the exploitation of parameters can be exploited as an accurate feature selection and data resampling method in any numerical or non numerical context.

We will present the general principles of feature maximization and we will especially return to its successful applications in the supervised framework, comparing its performance with those of the state of the art methods on reference databases.

Keywords: Classification · Feature selection · Resampling · Clustering · Big data

© Springer International Publishing Switzerland 2015
S. Kozielski et al. (Eds.): BDAS 2015, CCIS 521, pp. 3–20, 2015.
DOI: 10.1007/978-3-319-18422-7_1

1 Introduction

Since the 1990s, progress in computing, and in storage capacities, has allowed the handling of extremely large volumes of data: it is not rare to deal with space for the description of several thousand, or even tens of thousands, features. It could be thought that the classification algorithms are more effective with a large number of features, but the situation is not so simple. The first problem is the increase in the calculation time. Additionally, the fact that a large number of features are redundant, or irrelevant, for the classification task, considerably disrupts the functioning of the classifiers. Furthermore, most training algorithms use probabilities whose distributions may be difficult to estimate in the presence of a very large number of features. The integration of a process of feature selection in the frame of large dimension data classification has thus become a central issue. In the literature, essentially three types of approach are proposed for the selection of features: approaches directly incorporated into the classification methods, known as "embedded", methods based on techniques of optimization, or "wrapper", and approaches based on statistical tests, also named filter-based methods. Thorough states-of-the art have been described by numerous authors, such as Ladha et al. [21,3,13] ou [8]. Therefore, below we will simply give a brief overview of the existing approaches.

"Embedded" approaches integrate the selection of features in the learning process [5]. The most popular methods in this category are those based on SVM and the methods founded on neural networks. For example, RFE-SVM (Recursive Feature Elimination for Support Vector Machines) [14] is an integrated process, where the selection of features is carried out in an iterative manner using an SVM classifier and suppressing features that are the most distant from the decision boundary.

For their part, the "wrapper" methods use a performance criterion to seek out a pertinent sub-group of predictors [20]. Most often it is the error rate (but that can be a prediction cost, or the area under the ROC curve). As an example, the WrapperSubsetEval method begins with an empty set of features, and continues until the addition of new features no longer improves performance. It uses cross-validation to estimate learning for a given group of features [39]. Comparisons between methods, such as that of Forman [10] clearly demonstrate that, without taking their effectiveness into account, one of the principal drawbacks of these two classes of methods is that they require long calculation times. This prohibits their use in the case of strongly multidimensional data. In this context, a possible alternative is to exploit filter-based methods.

Filter-based approaches are selection methods that are used upstream and independently of the learning algorithm. Based on statistical tests, they require less calculation time than do other approaches. Most classical examples of filter-based methods are chi-squared method [21], mutual information-based methods, like MIFS [16], information gain-based methods, like CBS [7], correlation-based methods, like MODTREE [22], or, nearest-neighbour-based methods, like Relief or ReliefF [19].

As for all statistical tests, filter-based approaches are known to behave erratically in the case of very low frequency features, which are common in text classification [21]. In this article we show that, despite their diversity, all existing approaches are inoperative, or even detrimental, in the case of extremely unstable, multidimensional and noisy data, with a high degree of similitude between classes. As an alternative, we propose a new method of feature selection and contrast, based on the recently developed feature maximization metric. Furthermore, we compare the performance of this method to that of classical techniques in the context of help with patent validation. Then we extend the range of our study to habitually used textual reference data. The rest of this manuscript is structured as follows: section 2 presents our new approach for feature selection; section 3 details the data used; section 4 compares the results for the different data corpora of the classification with and without the use of the proposed approach; section 5 outlines the use of the method in unsupervised context; section 6 presents our conclusions and perspectives.

2   Feature Maximization for Feature Selection

Feature maximization (F-max) is an unbiased metric with whichto estimate the quality of an unsupervised classification, which uses the properties (i.e. the features) of data associated with each cluster without prior examination of the cluster profiles [24]. Its principal advantage is that it is totally independent of the classification method and of its operating mode. When it is used after learning, it can be exploited to establish global indices of clustering quality [26] or for cluster labelling [28].

Consider a group of clusters C which results from a method of clustering applied to a dataset D represented by a group of features F. The feature maximization metric favours clusters with a maximal feature F-measure. The feature F-measure $FF_c(f)$ of a feature f associated with a cluster c c is defined as the harmonic mean of the feature recall $FR_c(f)$ and of the feature precision $FP_c(f)$, themselves defined as follows:

$$FR_c(f) = \frac{\Sigma_{d \in c'} W_d^f}{\Sigma_{c' \in C} \Sigma_{d \in c'} W_d^f} \quad FP_c(f) = \frac{\Sigma_{d \in c} W_d^f}{\Sigma_{f' \in F_c, d \in c} W_d^{f'}} \tag{1}$$

with

$$FF_c(f) = 2 \left(\frac{FR_c(f) \times FP_c(f)}{FR_c(f) + FP_c(f)} \right) \tag{2}$$

where W_d^f represents the weight of the feature f for the data d and F_c represents all the features present in the dataset associated with the cluster C.

Taking into account the basic definition of the feature maximization metric, its use for the task of feature selection in the context of supervised learning becomes a simple process. Therefore, this generic metric can be applied to data associated with a class, as well as those associated with a cluster. The selection process can thus be defined as non-parametered, based on classes in which a class feature is characterised using both its capacity to discriminate between classes ($FP_c(f)$ index) and its ability to faithfully represent the class data ($FR_c(f)$ index). The Sc set of features that are characteristic of a given class c belonging to the group of classes C is translated by:

$$S_c = \{f \in F_c \mid FF_c(f) > \overline{FF}(f) \text{ and } FF_c(f) > \overline{FF}_D\} \text{ where} \tag{3}$$

$$\overline{FF}(f) = \Sigma_{c' \in C} \frac{FF_{c'}(f)}{|C_{/f}|} \text{ and } \overline{FF}_D = \Sigma_{f \in F} \frac{\overline{FF}(f)}{|F|} \tag{4}$$

where $C_{/f}$ represents the subset of C in which the f feature is represented.

Finally, the set of all selected features S_C is the subset of F defined by:

$$S_C = \cup_{c \in C} S_C. \tag{5}$$

In other words, the features that are judged relevant for a given class are those whose representations are better than average in this class, and better than the average representation of all the features in terms of feature F-measure.

In the specific context of the process of feature maximization, an improvement by contrast step can be exploited as a complement to the first step of selection. The role of this is to adapt the description of each single data to the specific characteristics of its associated class. This consists of modifying the data weighting schema in a distinct way for each class, taking into account the information gain supplied by feature F-measure of the features locally in this class.

The information gain is proportional to the relation between the F-measure value of a feature in the $FF_c(f)$ class and the average F-measure value of this feature for the whole partition. Given one single data and one single feature describing this data, the resulting gain acts as a contrast factor that adjusts the weight of this feature in the data profile, optionally taking into account its prior establishment. For a feature f belonging to the group of selected features S_c from a class C, the gain $G_c(f)$ is expressed as:

$$G_c(f) = (FF_c(f)/\overline{FF}(f))^k \tag{6}$$

where k is a magnification factor that can be optimized according to the resulting accuracy.

The active features of a class are those for which the information gain is higher than 1. Given that the proposed method is one of selection and of contrast based on the classes, the average number of active features per class is comparable to the total number of features selected in the case of habitual selection methods.

3 Validating the Approach on Real-World Data

One of the goals of the QUAERO project is to use bibliographic information to help experts to judge patent precedence. Thus, initially it was necessary to prove that it is possible to associate such information with the patent classes in a pertinent manner; or in other words, to classify it correctly within these classes. Main experimental data source comprised 6387 patents from the pharmacological domain in an XML format, grouped into 15 sub-classes of the A61K class (medical preparation). The bibliographic references in the patents were extracted from the Medline database[1]. 25887 citations were extracted from the 6387 patents. Interrogation of the Medline database with the extracted citations allowed bibliographic notices of 7501 references to be recovered. Each notice was then labelled with the first classment code of the citing patent [15]. Each notice's abstract was treated and transformed into a bag of words [36] using the TreeTagger tool [37]. To reduce the noise generated by this tool, a frequency threshold of 45 (i.e. an average threshold of 3 per class) was applied to the extracted descriptors. The result was a description space limited to the 1804 dimension. A last TF-IDF weighting step was applied [36]. The series of labelled notices, which were thus pre-treated, represented the final corpus on which training was carried out. This last corpus was highly unbalanced. The smallest class (A61K41) contained 22 articles, whereas the largest contained 2500 (A61K31 class). The inter-class similarity was calculated using a cosine correlation. This indicated that more than 70% of pairs of classes had a similarity of between 0.5 and 0.9. Thus, the ability of a classification model to precisely detect the correct class is strongly reduced. A solution commonly used to contend with an imbalance in classes' data is sub-sampling of the larger classes [12] and/or over-sampling of the smaller ones [6]. However, re-sampling, which introduced redundancy into the data, does not improve the performance of this dataset, as was shown by Hajlaoui et al. (2012). Therefore, we have proposed an alternative solution detailed below, namely to edit out the features that are judged irrelevant and to contrast those considered reliable [25].

As a complement, 4 other well-known reference text datasets have been exploited for validation of the method:

- The R8 and R52 corpora were obtained by Cardoso Cachopo[2] from the R10 and R90 datasets, which are derived from the Reuters 21578 collection[3]. The aim of these adjustments was to only retain data that had a single label. Considering only monothematic documents and classes that still had at least one example of training and one of test, R8 is a reduction of the R10 corpus (the 10 most frequent classes) to 8 classes and R52 is a reduction of the R90 corpus (90 classes) to 52 classes.
- The Amazon^{tm} corpus (AMZ) is a UCI dataset [2] derived from the recommendations of clients of the Amazon web site that are usable for author

---

[1] http://www.ncbi.nlm.nih.gov/pubmed/

[2] http://web.ist.utl.pt/~acardoso/datasets/

[3] http://www.research.att.com/~lewis/reuters21578.html

identification. To evaluate the robustness of the classification algorithms with respect to a large number of target classes, 50 of the most active users who have frequently posted comments in these newsgroups were identified. Thirty messages were collected for each of them. Each message included the authors' linguistic style, such as the use of figures, punctuation, frequent words and sentences.

- The 20Newsgroups dataset [19] is a collection of approximately 20,000 documents (almost) uniformly distributed among 20 different discussion groups. We consider two "bag of words" versions of this dataset in our experiments. In the (20N - AT) version, all words are preserved and non-alphabetic characters are converted into spaces. It resulted in a 11153 words description space. The (20N - ST) version is obtained after a additionnal step of stemming. The words of less than 2 characters, as well as stopwords (S24 SMART list [36]), are eliminated. The stemming is performed using Porter's algorithm [33]. The description space is thus reduced to 5473 words.
- The WebKB dataset (WKB) contains 8282 pages collected from the departments of computer science of various universities in January 1997 by the World WideKnowledge Base, a project of the CMU text learning group[4] (Carnegie Mellon University, Pittsburgh). The pages have been manually divided into 7 classes: student, faculty, department, course, personal, project, other. We operate on the Cardoso Cachopo's reduced version in which classes "department" and "staff" were rejected due to their low number of pages, and the class "other" has been deleted. Cleaning and stemming methods used for the 20Newsgroups dataset are then applied on the reduced dataset. It resulted in a 4158 items dataset described by a 1805 words description space.

4 Experiments and Results

4.1 Experiments

To carry out our experiments, we first took into consideration different classification algorithms that are implemented in the Weka tool box[5]: decision trees (J48) [35], random forests (RF)[4], KNN [1], habitual Bayesian algorithms, i.e. the Multinomial Nave Bayes (MNB) and Bayesian Network (BN), and finally, the SMO-SVM algorithm (SMO) [32]. The default parameters were used during the implementation of these algorithms, apart from KNN for which the number of neighbours was optimized based on the resulting precision. Secondly, we placed the accent more particularly on tests of the efficacy of feature selection approaches, including our new proposition (FMC). In our test, we included a panel of filter-based approaches applicable on large dimension data, using once again the Weka platform. The methods tested include: chi-squared [21], information gain [16], CBF [7], symmetric incertitude [40], ReliefF [19] (RLF), Principal

---

[4] http://www.cs.cmu.edu/afs/cs.cmu.edu/project/theo-20/www/data/
[5] http://www.cs.waikato.ac.nz/ml/weka/

Component Analysis [31] (PCA). Default parameters were used for most of these methods except for PCA, where the explained variance percentage is tuned with respect to the resulting accuracy. Initially we tested the methods separately. In a second phase, we combined the feature selection supplied by the different methods with the F-max contrast method that we have proposed (eq. 6). We used a 10-fold cross-validation in all our experiments.

4.2 Results

The different results are presented in tables 1 to 8. They are based on measurements of standard performance (level of true positives [TP] or recall [R], level of false positives [FP], Precision [P], F-measure [F] and ROC) weighted by class size, then averaged for all the classes. For each table and each combination of selection and classification methods, an indicator of performance gain/loss (TP Incr) is calculated using the TP of SMO level on original data as a reference. Finally, as the results for chi-squared, information gain and symmetric incertitude

Table 1. Classification results on initial data

	TP(R)	FP	P	F	ROC	TP Incr
J48	0.42	0.16	0.40	0.40	0.63	-23%
RandomForest	0.45	0.23	0.46	0.38	0.72	-17%
SMO	**0.54**	**0.14**	**0.53**	**0.52**	**0.80**	**0% (Ref)**
BN	0.48	0.14	0.47	0.47	0.78	-10%
MNB	0.53	0.18	0.54	0.47	0.85	-2%
KNN (k=3)	0.53	0.16	0.53	0.51	0.77	-2%

Table 2. Results of classification after the selection of features (BN classifier)

	TP(R)	FP	P	F	ROC	Nbr. var.	TP Incr
CHI+	0.52	0.17	0.51	0.47	0.80	282	-4%
CBF	0.47	0.21	0.44	0.41	0.75	37	-13%
PCA (50% vr.)	0.47	0.18	0.47	0.44	0.77	483	-13%
RLF	0.52	0.16	0.53	0.48	0.81	937	-4%
FMC	**0.99**	**0.003**	**0.99**	**0.99**	**1**	**262/cl**	**+90%**

Table 3. Results of classification after the selection of FMC features

	TP(R)	FP	P	F	ROC	TP Incr
J48	0.80	0.05	0.79	0.79	0.92	+48%
RandomForest	0.76	0.09	0.79	0.73	0.96	+40%
SMO	0.92	0.03	0.92	0.91	0.98	+70%
BN	**0.99**	**0.003**	**0.99**	**0.99**	**1**	**+90%**
MNB	0.92	0.03	0.92	0.92	0.99	+71%
KNN (k=3)	0.66	0.14	0.71	0.63	0.85	+22%

were identical, they only figure once in the tables, as results of the chi-squared type (and are noted CHI+).

For our main patent collection, table 1 shows that the performances of all classification methods are weak for the dataset considered, provided no feature selection process is carried out. In this context, this table also confirms the superiority of the SMO, KNN and Bayesian methods compared to the other two methods, based on decision trees. Additionally, SMO gave the best global performance in terms of discrimination, as demonstrated by its highest ROC value. However, this method is clearly not usable in an operational context of patent evaluation such as QUAERO, because of the major confusion between classes. This shows its intrinsic inability to cope with the attraction effect of the largest classes. Each time that a standard feature selection method is applied in our context, in association with the best classification methods, its use alters the quality of the results slightly, as indicated in table 2. Table 2 also underlines the fact that the reduction in the number of feature by the FMC method is similar to CHI+ (in terms of active features; see section 2 for more details), but that its use stimulates the performances of classification methods, particularly those of Bayesian methods (table 3), leading to impressive classification results in the context of highly complex classification: 0.987 accuracy i.e. only 94 misclassed data with the BN method, amongst a total of 7252.

The results presented in table 4 illustrate more precisely the efficiency of the F-max contrast method that acts on data description (eq. 6). In experiments relating to this table, the contrast is applied individually to the features extracted by each selection method, and in a second step a BN classifier is applied to the resulting contrasted data. The results show that, irrespective of the type of method used for feature selection, the performances of the resulting classification are re-enforced each time that the F-max contrast is applied downstream of the selection. The average performance increase is 44%. Finally, table 5 illustrates the ability of the FMC approach to efficiently confront the problems of imbalance and class similitude. The examination of TP level variations (especially in the small classes) seen in this Table shows that the attraction effect of data from the largest classes, produced at a high level in the case of the use of original data, is practically systematically overcome each time the FMC approach is exploited. The ability of this approach to correct class imbalance is equally clearly demonstrated by the homogeneous distribution of active features in the different classes, despite the extremely heterogeneous class size.

Table 4. Results of classification with different feature selection methods, and F-max contrast (BN classifier)

	TP(R)	FP	P	F	ROC	Nbr. var.	TP Incr
CHI+	0.79	0.08	0.82	0.78	0.98	282	+46%
CBF	0.63	0.15	0.69	0.59	0.90	37	+16%
PCA (50% vr.)	0.71	0.11	0.73	0.67	0.53	483	+31%
RLF	0.79	0.08	0.81	0.78	0.98	937	+46%
FMC	**0.99**	**0.003**	**0.99**	**0.99**	**1**	**262/cl**	**+90%**

```
=== Confusion Matrix ===

    a    b    c    d    e    f    g    h    i    j    k    l    m    n    o    <-- classified as
 2007    0   31   26  197  103    0   13   13    1    2    0    0    0  140 |   a = a61k31
   44    1    1    0    3    2    0    0    2    0    1    0    0    0    6 |   b = a61k33
  139    0  142    2   65   91    0    1    4    2    0    0    0    1   12 |   c = a61k35
  137    0    3   48    9    9    0    0    0    0    0    0    0    0    6 |   d = a61k36
  369    0   43    3  493  160    0    4    8    2    1    0    0    1   26 |   e = a61k38
  194    0   29    1  121  741    0    3   17    4    3    5    0    0   23 |   f = a61k39
   10    0    0    0    3    2    0    0    0    0    1    1    0    0    5 |   g = a61k41
  174    0    4    4   50   34    0   19    2    0    1    1    0    0    6 |   h = a61k45
   84    0    4    0   53   56    0    0   65    0    2    2    0    0   38 |   i = a61k47
   46    0    7    0   33   33    0    0    1   17    0    1    0    0    2 |   j = a61k48
   38    1    1    0    4    2    0    0    7    0   23    2    0    0   12 |   k = a61k49
   28    0    0    0   12    6    0    0    7    0    1   20    0    0    4 |   l = a61k51
   15    0    0    1   11    7    0    0    1    0    0    0    2    0   10 |   m = a61k6
   51    0    7    2    6    5    0    0    0    0    2    0    0    2   12 |   n = a61k8
  298    0    5    2   43   46    0    0   18    0    2    1    0    0  344 |   o = a61k9
```

```
=== Confusion Matrix ===

    a    b    c    d    e    f    g    h    i    j    k    l    m    n    o    <-- classified as
 2530    0    0    0    3    0    0    0    0    0    0    0    0    0    0 |   a = a61k31
    6   46    0    0    2    0    0    1    2    0    0    0    0    0    3 |   b = a61k33
    6    0  445    0    1    6    0    0    0    0    0    0    0    0    1 |   c = a61k35
   18    0    2  189    0    1    0    0    1    0    0    0    0    0    1 |   d = a61k36
   10    0    0    0 1097    3    0    0    0    0    0    0    0    0    0 |   e = a61k38
    4    0    0    0    2 1134    0    0    0    0    0    0    0    0    0 |   f = a61k39
    4    0    1    1    2    2    3    0    4    0    0    0    0    0    0 |   g = a61k41
   43    0    2    0    3    5    0  251    0    0    0    0    0    0    0 |   h = a61k45
   10    0    1    0    3   12    0    0  278    0    0    0    0    0    0 |   i = a61k47
    8    0    1    0    6   17    0    0    0  107    0    0    0    0    1 |   j = a61k48
    6    0    0    0    2    2    0    0    7    0   68    0    0    0    5 |   k = a61k49
    3    0    0    0    2    1    0    0    0    0    0   70    0    0    1 |   l = a61k51
    5    0    0    2    5    3    0    1    2    0    0    0   26    0    3 |   m = a61k6
   12    0    0    2    3    0    0    1    1    0    1    0    0   64    3 |   n = a61k8
   21    0    0    0    1    0    0    0    0    0    0    0    0    0  737 |   o = a61k9
```

Fig. 1. Confusion matrix of the optimal results before and after feature selection on PAT-QUAERO dataset (SMO classification)

Table 5. Characteristics/class before and after FMC selection (BN classifier)

Class Label	Size	Feat. Select.	% TP FMC	% TP before
a61k31	2533	223	**1**	0.79
a61k33	60	276	**0.95**	0.02
a61k35	459	262	**0.99**	0.31
a61k36	212	278	**0.95**	0.23
a61k38	1110	237	**1**	0.44
a61k39	1141	240	**0.99**	0.65
a61k41	22	225	**0.24**	0
a61k45	304	275	**0.98**	0.09
a61k47	304	278	**0.99**	0.21
a61k48	140	265	**0.98**	0.12
a61k49	90	302	**0.93**	0.26
a61k51	78	251	**0.98**	0.26
a61k6	47	270	**0.82**	0.04
a61k8	87	292	**0.98**	0.02
a61k9	759	250	**1**	0.45

The summary of the results of the four complementary datasets is presented in tables 6 to 8. These tables highlight the fact that the FMC method can very significantly improve the performance of the classifiers in different types of cases. As in the context of our previous experience (patents), the best performances are obtained with the use of the FMC method in combination with the MNB

Table 6. List of high contrast features (lemmes) for the 8 classes of the REUTERS8 corpus

Trade	Grain	Ship	Acq
6.35 tariff	5.60 agricultur	6.59 ship	5.11 common
5.49 trade	5.44 farmer	6.51 strike	4.97 complet
5.04 practic	5.33 winter	6.41 worker	4.83 file
4.86 impos	5.15 certif	5.79 handl	4.65 subject
4.78 sanction	4.99 land	5.16 flag	4.61 tender
Learn	**Money-fx**	**Interest**	**Crude**
7.57 net	6.13 currenc	5.95 rate	6.99 oil
7.24 loss	5.55 dollar	5.85 prime	5.20 ceil
6.78 profit	5.52 germani	5.12 point	4.94 post
6.19 prior	5.49 shortag	5.10 percentag	4.86 quota
5.97 split	5.16 stabil	4.95 surpris	4.83 crude

Table 7. Results of classifications after FMC feature selection (MNB/BN classifier)

		TP (R)	FP	P	F	ROC	TP Incr.
Reuters8 (R8)	-	0.937	0.02	0.942	0.938	0.984	
	FMC	**0.998**	**0.001**	**0.998**	**0.998**	**1**	**+6%**
Reuters52 (R52)	-	0.91	0.01	0.909	0.903	0.985	
	FMC	**0.99**	**0.001**	**0.99**	**0.99**	**0.999**	**+10%**
Amazon	-	0.748	0.05	0.782	0.748	0.981	
	FMC	**0.998**	**0.001**	**0.998**	**0.998**	**1**	**+33%**
20NewsGroup-AT (all terms)	-	0.882	0.006	0.884	0.881	0.988	
	FMC	**0.992**	**0**	**0.992**	**0.1**	**1**	**+13%**
20NewsGroup-ST (stemmed)	-	0.865	0.007	0.866	0.864	0.987	
	FMC	**0.991**	**0.001**	**0.991**	**1**	**1**	**+15%**
WebKB	-	0.842	0.068	0.841	0.841	0.946	
	FMC	**0.996**	**0.002**	**0.996**	**0.996**	**0.996**	**+18%**

Table 8. Dataset information an complementary results after FMC feature selection (5 reference datasets and MNB or BN classification)

	R8	R52	AMZ	20N-AT	20N-ST	WKB
Nb. class	8	52	50	20	20	4
Nb. data	7674	9100	1500	18820	18820	4158
Nb. feat.	3497	7369	10000	11153	5473	1805
Nb. sel. feat.	1186	2617	3318	3768	4372	725
Act. feat./class (av.)	268.5	156.05	761.32	616.15	525.95	261
Magnification factor	4	2	1	4	4	4
Misclassed (Std)	373	816	378	2230	2544	660
Misclassed (FMC)	**19**	**91**	**3**	**157**	**184**	**17**
Comp. time (s)	1	3	1.6	10.2	4.6	0.8

C1- 7(7) [315(315)]

Prevalent Label --- = Cause-Experiencer

0.273245 G-Cause-Experiencer
0.173498 C-SUJ:Ssub,OBJ:NP
0.138411 C-SUJ:NP,DEOBJ:PP
0.091732 C-SUJ:NP,DEOBJ:PP,DUMMY:REFL
. . .
****
****
0.013839 T-Asset
0.013200 C-SUJ:NP,DEOBJ:Ssub,POBJ:PP
0.009319 C-SUJ:Ssub,OBJ:NP,POBJ:PP
. . .
[flatter 0.907200 3(1)] [charmer 0.889490 3(0)] [ex-
ulter 0.889490 3(0)] [**frissonner 0.889490 3(0)]
[mortifier 0.889490 3(0)] [époustoufler 0.889490
3(0)] [pâtir 0.889490 3(0)] [ravir 0.889490 3(0)]
[**trembler 0.889490 3(0)] [**trembloter 0.889490
3(0)] [décourager 0.872350 2(2)]. . .

Fig. 2. Sample output for a French verb cluster produced with the IGNGF clustering method. The exploited features represent either verb subcategorization frames or
semantic labels.

and BN Bayesian classifiers. Table 7 presents the comparative results of such a
combination. It demonstrates that the FMC method is particularly effective in
increasing the performance of the classifiers when the complexity of the classification task becomes higher because of an increasing number of classes (AMZ
corpus). Table 8 supplies general information about the data and the behaviour
of the FMC selection method. They illustrate the significant reduction in the
classification complexity obtained with FMC because of the drop in the number
of features to manage, as well as the concomitant decrease of badly classed data.
It also stresses the calculation time, which is highly curbed for this method (the
calculation is carried out on Linux using a laptop computer equipped with an
Intel® Pentium® B970 2.3Ghz processor and 8Go of memory).

For these datasets, similar remarks to those mentioned for the patent dataset
can be made on the subject of the low efficiency of common feature selection
methods and the re-sampling methods. Table 8 also shows that the value of the
contrast magnification factor utilised to obtain the best performances can vary
throughout the experiments (from 1 to 4 in this last context). However, it can
be observed that by taking a fixed value for this factor, for example the highest
(here 4), the results are not down-graded. This choice thus represents a good
alternative to confront the problem of configuration.

The 5 most contrasted feature (lemmes) of the 8 classes issued from the
Reuter8 corpus are shown in table 6. The fact that the main lines of the themes
covered by the classes can be clearly demonstrated in this way illustrates the

Table 9. Classification results on UCI Wine dataset

	TPR	FP	P	F	ROC	TP Incr
J48	0.94	0.04	0.94	0.94	0.95	0% (Ref)
BN + FMC	**1**	**0**	**1**	**1**	**1**	**+6%**

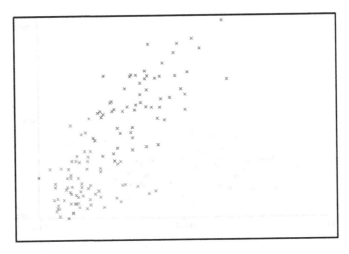

Fig. 3. WINE dataset: "Proline-Color intensity" decision plan generated by J48 - Proline is on Y axis on this and next figures

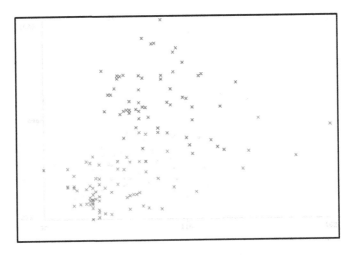

Fig. 4. WINE dataset: "Proline-Magnesium" decision plan generated by FMC (before data contrasting)

Fig. 5. WINE dataset: "Proline-Magnesium" decision plan generated by FMC (after data contrasting with a magnification factor k=1)

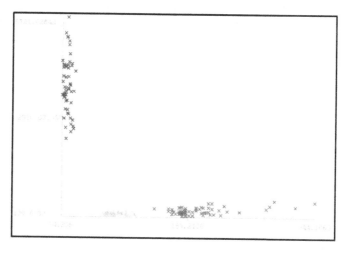

Fig. 6. WINE dataset: "Proline-Magnesium" decision plan generated by FMC (after data contrasting with a magnification factor k=4)

topic extraction capacities by the FMC method. Finally, the acquisition of very good performances by combining the FMC feature selection and constrast with a classification method such as MNB is a real advantage for large-scale usage, given that the MNB method has incremental abilities and that the two methods have low calculation times.

Complementary results obtained with the numerical UCI Wine dataset interestingly show that, with the help of FMC, NB/BN methods are able to exploit only two features (among 13) for classification as a decision tree classifier like J48 (i.e. C4.5 [27]) would do on standard data. The difference is that a perfect result is obtained with NB/BN and FMC whereas it is not the case with J48 (table 9). Some explanations are provided by looking up at the distribution of the class samples on the alternative decision plans of the two methods. In the "Proline-Color intensity" decision plan exploited by J48, the different classes are not clearly discriminable (Fig. 3). On its own side, the FMC method "apparently" generates an even more complex "Proline-Magnesium" decision plan, if contrast is not considered (Fig. 4). However, as shown in Fig. 5- 6, with the combined effect of contrast and magnification factor (4) on data features, the different classes become very clearly discriminable on that decision plan, especially when the magnification factor is increased sufficiently (Fig. 6).

5   Feature Maximization for Clustering

Like other neural free topology learning methods such as Neural Gas (NG) [30], Growing Neural Gas (GNG) [11], or Incremental Growing Neural Gas (IGNG) [34], the IGNGF method makes use of Hebbian learning [17] for dynamically structuring the learning space. Hebbian learning is inspired by a theory from neurosciences which explains how neurons connect to build neural networks. Whereas for NG the number of output prototypes is fixed, GNG adapts this number during the learning phase, guided by the characteristics of the data to be classified. Prototypes and connections between them can be created or removed depending on evolving characteristics of learning (as for example the "age" or "maturity" of connections and the cumulated error rate of each prototype). A drawback of this approach is that prototypes are created or removed after a fixed number of iterations yielding results which might not appropriately represent complex or sparse multidimensional data. With the IGNG clustering method this issue is addressed by allowing more flexibility when creating new prototypes: a prototype is added whenever the distance of a new data point to an existing prototype is above a predefined global threshold, the average distance of all the data points to the centre of the data set. The learning process thus becomes incremental: each incoming data point is considered as a potential prototype. For all the above-mentioned methods, at each iteration over all the data points, a data point is connected with the "closest" prototypes and at the same time interacts with the existing model by strengthening the connections between these "closest" prototypes and weakening those to other, less related prototypes. Because of these dynamically changing interactions between prototypes, these methods are "winner take most" methods in contrast to K-means (for example), which represents a "winner-take-all" method. The notion of "closeness" is based on a distance function computed from the features associated to the data points.

IGNGF uses the Hebbian learning process as IGNG, but the use of a standard distance measure as adopted in IGNG for determining the "closest" prototype is replaced in IGNGF by feature maximization.

With feature maximization, the clustering process is roughly the following. During learning, an incoming data point x is temporary added to every existing cluster, its feature profile constituted by its maximal features and its average feature F-measure are computed. Then the winning prototype is the prototype whose associated cluster maximises the κ criterion given in Equation (7),

$$\kappa(c) = \Delta(FF_c) * |F_c \cap F_x| - \frac{\|p_c, x\|}{weight} \qquad (7)$$

where $\Delta(FF_c)$ represents the gain in feature F-measure for the new cluster, $|F_c \cap F_x|$ represents the number of features shared by cluster c and the data point x and p_c is the codebook vector of the prototype associated to cluster c. This way, those clusters are preferred which share more features with the new data point and clusters which don't have any common feature with the data point are ignored. The gain in feature F-measure multiplied by the number of shared features is adjusted by the euclidean distance of the new data point x to the cluster's prototype codebook vector p_c. Thus, the smaller the euclidean distance to the cluster's prototype, less the κ value decreases. The influence of the euclidean distance can be parametrised with a *weight* factor ($\sqrt{2}$ for usual application). Clusters with negative κ score are ignored. The data point is then added to the cluster c with maximal $\kappa(c)$ and the connections between its associated prototype and the neighbour prototypes are updated. If κ value is negative for all clusters, a new prototype is created and an associated cluster is formed with the currently considered data point.

The IGNGF method was shown to outperform other usual neural and non neural methods for clustering tasks on sparse and/or highly multidimensionnal and/or noisy data [27]. Moreover, it can be fruitfully combined with unsupervised Bayesian reasoning for setting up the first parameter-free method capable of automatically tracking research topics evolving over time in a realistic multidimensionnal context [23]. It was also recently shown to outperform supervised classification methods in the context of websites classification task thanks to its capacity to highlight "latent classes" not initially planed by the analyst [29].

Another main advantage of the method is that maximized features used by IGNGF during learning can also be exploited in a final step for accurately labeling the resulting clusters. An example of such results is given in the case of French verb clustering [9,18]. In this specific context, the IGNGF clustering method does not only provides accurate verb clusters, outperforming state-of-the-art methods of the domain, like spectral clustering [38]. As a complementary result, it associates each verb cluster c with a profile containing syntactic and semantic features characteristic of that cluster. Features are displayed in decreasing order of feature F-measure given by Equation (2) and features whose feature F-measure is under the average feature F-measure of the overall clustering are clearly separated from others. In the sample cluster shown in Fig. 2 these are listed above the two star lines. In addition, for each verb in a cluster, a confidence score can be easily computed [9].

6   Conclusion

Our main aim was to develop an efficient method of feature selection and contrast, which would allow routine problems linked to the supervised classification of large volumes of textual data to be overcome. These problems are linked to class imbalance, with a high degree of similarity between them, as they house highly multidimensional and noisy data. To achieve our aim, we adapted a recently developed metric in the unsupervised framework to the context of supervised classification. By means of different experiments on large textual datasets, we illustrated numerous advantages of our approach, including its effectiveness to improve the performance of classifiers in such a context. Notably, this approach places the accent on the most flexible classifiers, and the least demanding in terms of calculation times, such as the Bayesian classifiers. Another advantage of this method is that it concerns an approach without parameters that depends on a simple feature extraction schema. The method can thus be used in numerous contexts, such as those of incremental or semi-supervised learning, and more generally, in large scale digital learning.

References

1. Aha, D., Kibler, D., Albert, M.: Instance-based learning algorithms. Machine Learning 6, 37–66 (1991)
2. Bache, K., Lichman, M.: Uci machine learning repository (http://archive.ics.uci.edu/ml): University of California, school of information and computer science, Irvine, CA, USA (2013)
3. Bolón-Canedo, V., Sánchez-Maroño, N., Alonso-Betanzos, A.: A review of feature selection methods on synthetic data. Knowledge and Information Systems 34(3), 483–519 (2013)
4. Breiman, L.: Random forests. Machine Learning 45(1), 5–32 (2001)
5. Breiman, L., Friedman, J.H., Olshen, R.A., Stone, C.J.: Classification and regression trees. Tech. rep., Wadsworth International Group, Belmont, CA, USA (1984)
6. Chawla, N.V., Bowyer, K.V., Hall, L.O., Kegelmeyer, W.P.: Synthetic minority oversampling technique. Journal of Artificial Intelligence Research 16, 321–357 (2002)
7. Dash, M., Liu, H.: Consistency-based search in feature selection. Artificial Intelligence 151(1), 155–176 (2003)
8. Daviet, H.: Class-Add, une procédure de sélection de variables basée sur une troncature k-additive de l'information mutuelle et sur une classification ascendante hiérarchique en pré-traitement. Thèse de doctorat, Université de Nantes (2009)
9. Falk, I., Gardent, C., Lamirel, J.-C.: Classifying French verbs using French and English lexical resources. In: Proceedings of ACL, Jeju Island, Korea (2012)
10. Forman, G.: An extensive empirical study of feature selection metrics for text classification. Journal of Machine Learning Research 3, 1289–1305 (2003)
11. Fritzke, B.: A growing neural gas network learns topologies. Advances in Neural Information Processing Systems 7, 625–632 (1995)
12. Good, P.: Resampling methods. Ed. Birkhauser (2006)

13. Guyon, I., Elisseeff, A.: An introduction to variable and feature selection. Journal of Machine Learning Research 3, 1157–1182 (2003)
14. Guyon, I., Weston, J., Barnhill, S., Vapnik, V.: Gene selection for cancer classification using support vector machines. Machine Learning 46(1), 389–422 (2002)
15. Hajlaoui, K., Cuxac, P., Lamirel, J.-C., François, C.: Enhancing patent expertise through automatic matching with scientific papers. In: Ganascia, J.-G., Lenca, P., Petit, J.-M. (eds.) DS 2012. LNCS, vol. 7569, pp. 299–312. Springer, Heidelberg (2012)
16. Hall, M.A., Smith, L.A.: Feature selection for machine learning: Comparing a correlation-based filter approach to the wrapper. In: Proceedings of the Twelfth International Florida Artificial Intelligence Research Society Conference, pp. 235–239 (1999)
17. Hebb, D.O.: The organization of behavior: a neuropsychological theory. John Wiley & Sons, New York (1949)
18. Lamirel, J.-C., Falk, I., Gardent, C.: Federating clustering and cluster labeling capabilities with a single approach based on feature maximization: French verb classes identification with igngf neural clustering. Neurocomputing, Special Issue on 9th Workshop on Self-Organizing Maps (WSOM 2012) 147, 136–146 (2014)
19. Lang, K.: Learning to filter netnews. In: Proceedings of the Twelfth International Conference on Machine Learning, pp. 331–339 (1995)
20. Kohavi, R., John, G.R.: Wrappers for feature subset selection. Artificial Intelligence 97(1-2), 273–324 (1997)
21. Ladha, L., Deepa, T.: Feature selection methods and algorithms. International Journal on Computer Science and Engineering 3(5), 1787–1797 (2011)
22. Lallich, S., Rakotomalala, R.: Fast feature selection using partial correlation for multi-valued attributes. In: Zighed, D.A., Komorowski, J., Żytkow, J.M. (eds.) PKDD 2000. LNCS (LNAI), vol. 1910, pp. 221–231. Springer, Heidelberg (2000)
23. Lamirel, J.C.: A new approach for automatizing the analysis of research topics dynamics: application to optoelectronics research. Scientometrics 93, 151–166 (2012)
24. Lamirel, J.C., Al Shehabi, S., François, C., Hoffmann, M.: New classification quality estimators for analysis of documentary information: application to patent analysis and web mapping. Scientometrics 60(3) (2004)
25. Lamirel, J.C., Cuxac, P., Chivukula, A.S., Hajlaoui, K.: Optimizing text classification through efficient feature selection based on quality metric. Journal of Intelligent Information Systems, Special Issue on PAKDD-QIMIE 2013, 1–18 (2014)
26. Lamirel, J.C., Ghribi, M., Cuxac, P.: Unsupervised recall and precision measures: a step towards new efficient clustering quality indexes. In: Proceedings of the 19th International Conference on Computational Statistics (COMPSTAT 2010), Paris, France (2010)
27. Lamirel, J.C., Mall, R., Cuxac, P., Safi, G.: Variations to incremental growing neural gas algorithm based on label maximization. In: Proceedings of IJCNN 2011, San Jose, CA, USA (2011)
28. Lamirel, J.C., Ta, A.P.: Combination of hyperbolic visualization and graph-based approach for organizing data analysis results: an application to social network analysis. In: Proceedings of the 4th International Conference on Webometrics, Informetrics and Scientometrics and 9th COLLNET Meetings, Berlin, Germany (2008)

29. Lamirel, J.-C., Reymond, D.: Automatic websites classification and retrieval using websites communication signatures. Journal of Scientometrics and Information Management: Special Issue on 8th International Conference on Webometrics, Informetrics and Scientometrics 8(2), 293–310 (2014)
30. Martinetz, T., Schulten, K.: A "neural-gas" network learns topologies. In: Artificial Neural Networks, pp. 397–402 (1991)
31. Pearson, K.: On lines an planes of closetst fit to systems of points in space. Philosophical Magazine 2(11), 559–572 (1901)
32. Platt, J.C.: Fast training of support vector machines using sequential minimal optimization. In: Schölkopf, B., Burges, C.J.C., Smola, A.J. (eds.) Advances in Kernel Methods, pp. 185–208. MIT Press, Cambridge (1999)
33. Porter, M.F.: An algorithm for suffix stripping. Program 14(3), 130–137 (1980)
34. Prudent, Y., Ennaji, A.: An incremental growing neural gas learns topologies. In: Proceedings of the 2005 IEEE International Joint Conference on Neural Networks, vol. 2, pp. 1211–1216 (2005)
35. Quinlan, J.R.: C4.5: programs for machine learning. Morgan Kaufmann Publishers Inc., San Francisco (1993)
36. Salton, G.: Automatic processing of foreign language documents. Prentice-Hall, Englewood Cliffs (1971)
37. Schmid, H.: Probabilistic part-of-speech tagging using decision trees. In: Proceedings of International Conference on New Methods in Language Processing (1994)
38. Sun, L., Korhonen, A., Poibeau, T., Messiant, C.: Investigating the cross-linguistic potential of verbnet-style classification. In: Proceedings of ACL, Beijing, China, pp. 1056–1064 (2010)
39. Witten, I.H., Frank, E.: Data Mining: Practical machine learning tools and techniques. Morgan Kaufmann (2005)
40. Yu, L., Liu, H.: Feature selection for high-dimensional data: a fast correlation-based filter solution. In: Proceedings of ICML 2003, Washington DC, USA, pp. 856–863 (2003)

Generating Ontologies from Relational Data with Fuzzy-Syllogistic Reasoning

Bora İ. Kumova[✉]

Department of Computer Engineering,
İzmir Institute of Technology, 35430 Turkey
borakumova@iyte.edu.tr

Abstract. Existing standards for crisp description logics facilitate information exchange between systems that reason with crisp ontologies. Applications with probabilistic or possibilistic extensions of ontologies and reasoners promise to capture more information, because they can deal with more uncertainties or vagueness of information. However, since there are no standards for either extension, information exchange between such applications is not generic. Fuzzy-syllogistic reasoning with the fuzzy-syllogistic system ^{4}S provides 2048 possible fuzzy inference schema for every possible triple concept relationship of an ontology. Since the inference schema are the result of all possible set-theoretic relationships between three sets with three out of 8 possible fuzzy-quantifiers, the whole set of 2048 possible fuzzy inferences can be used as one generic fuzzy reasoner for quantified ontologies. In that sense, a fuzzy syllogistic reasoner can be employed as a generic reasoner that combines possibilistic inferencing with probabilistic ontologies, thus facilitating knowledge exchange between ontology applications of different domains as well as information fusion over them.

Keywords: Relational data · Ontology learning · Syllogistic reasoning · Fuzzy logic

1 Introduction

Relational modelling facilitates maintaining data consistency, whereas ontological modelling facilitates logical reasoning with the data [30]. Since most data of every enterprise is usually maintain in relational models, there is increasing demand for automating reasoning with the relational data, in order to further utilise the information systems as decision support systems (Fig. 1). The objective of this paper is to review state-of-the-art in ontology learning and reasoning and to suggest a common possibilistic reasoner for probabilistic and quantified ontologies.

A variety of approaches for ontology learning have been proposed, ranging from unstructured data, like internet text search results, over semi structured data, like partially normalised data, to structured ones, like object-oriented data [29,51] or normalised relational data [37,20]. Here we will focus on relational data only.

© Springer International Publishing Switzerland 2015
S. Kozielski et al. (Eds.): BDAS 2015, CCIS 521, pp. 21–32, 2015.
DOI: 10.1007/978-3-319-18422-7_2

In most approaches, first an ontology is generated manually for a given relational data schema, thereafter the ontology is used in conjunction with the relational schema for reasoning with the data [18].

Ones an ontology is available for a particular relational data schema, an appropriate ontology reasoner can be chosen, which will then enable reasoning over the relational data, via the ontology. While an ontology is usually stored separately in a file or database, a reasoner is part of the semantic web application.

It is widely accepted that uncertainties of a domain that find reflections in the relational data, can be represented with related probabilities within ontologies.

Reasoners for probabilistic ontologies are mostly based on Bayesian networks. If data about such probabilities is unavailable, fuzzy ontologies [42,8,5] or possibilistic[1] ontologies [34] and reasoners may be preferred instead for processing vague information [28].

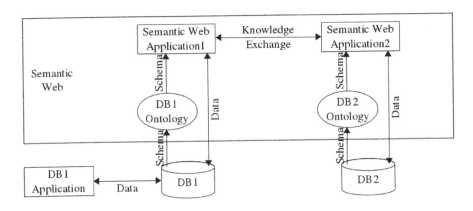

Fig. 1. Utilising relational data sources in the semantic web

The lack of standards for ontologies with probabilist or possibilistic extensions, prevent from efficient knowledge exchange between semantic web applications (Fig. 1) that have such extensions.

Here we propose the fuzzy-syllogistic (FS) reasoning system as a common logic for ontology reasoning. The 2048 syllogistic moods of the system cover any possible inference for any transitive concept relationship of the ontology [49]. Every such syllogistic inference has a fixed truth ratio [23], which can however change relative to the cardinalities of the probabilistic samples that make up the concepts and their relationships within the probabilistic ontology. A sample design for FS reasoning with an ontologies that was learned from text was presented elsewhere [49].

---

[1] In this work we will use the term "possibilistic" as a generic term that includes possibilistic logics as well as fuzzy logic, since fuzzy-syllogistic systems cover both.

Reasoning with fuzzy quantifiers or intermediate quantifiers is referred to as approximate reasoning [47]. Fuzzy syllogism refers to individual syllogistic moods that have fuzzy quantifiers [48]. Whereas fuzzy syllogistic system refers to the whole set of all 256 syllogistic moods, along with the truth ratio of every individual mood [24], where moods may have classical quantifiers or fuzzy quantifiers [22].

The paper is organised as follows. First the literature on ontology learning and reasoning with probabilistic and possibilistic approaches is reviewed, thereafter FS reasoning for ontology learning and reasoning is proposed as common reasoner for quantified ontologies with probabilist and possibilistic extensions.

2   Ontology Learning and Reasoning

Before we review approaches for learning ontologies, particularly database ontologies and approaches for reasoning with them, we briefly explain ontologies, database ontologies and how database ontologies can be compiled from databases.

2.1   Ontologies

An ontology is an object-oriented conceptualisation of a particular domain, such that it allows for logical reasoning about the domain. It is used to specify and share a domain in a common language. Ontologies are always attached with uncertainties, whether they emerge from probabilistic data or are created from possibilistic knowledge, but such extensions are not always reflected on the ontology.

An ontology consists of the following primitives [29]:

- Domain objects: Classes and instances.
- Object attributes: Aspects, properties, features, characteristics, parameters.
- Object relationships: Qualitative or quantitative relations between attributes.
- Processes & events involving objects: Functionalities modifying attribute values.
- Logic: Rules for valid reasoning information from attribute values.
- Uncertainties: Probabilities attached to anyone of the above primitives.
- Vagueness: Possibilities attached to anyone of the above primitives.

An ontology usually does not store or reference samples for any of its primitives, but could be extended with such a capability. Such an extension for database ontologies [27] is discussed below.

Database Ontologies

The relational database model is a derivation from first-order predicate logic [12]. However, there are no reasoners available for databases. Since ontologies are also based on first-order predicate logic, the primitives of the different models can be transformed into each other, based on common logical concepts. For instance, valid column values of a database table can be specified with predicate-logical quantifiers, which is analogous to valid values of an attribute of an ontology class (table 1).

Table 1. Mapping between relational data concepts and ontological concepts

Relational	Ontological
Entity Table	Class
Relation Table	Class Relationship
1-n Relationship[+]	Superclass-Subclass Relationship
n-n Relationship[+]	Multiple Superclass-Subclass Relationships
View	Process
Foreign Key	Attribute
Primary Key	Relational Instance
Attribute	Attribute
Attribute Value	Attribute Value
Row	Class Instance
Not Null; Unique	Cardinality Constraints

[+]Possible class-instance relationships instead of class hierarchy

2.2   Compiling Database Ontologies

Formal transformations between the models are one way for generating database ontologies.

Transformations could be performed principally bidirectional, if formal definitions for mapping the elements between the two models [26] are available. Whereby the time complexity of such mapping algorithms is usually linear [26]. However, since most solutions transform unidirectional, from database to ontology, we prefer the term compiling.

Two major approaches are distinguished for compiling an ontology from a database, reverse engineering and schema transformation. The former transform relational schema into ontological schema [18,44,21,1,26], whereas the latter transform entity-relationship (ER) schema into ontology schema [45,15,32]. The latter has the advantage that the generated ontology can be used for performing relational database operations via the ontology [27]. Even transformations of fuzzy extended ER into fuzzy ontologies are proposed [50].

In such systems the ontology can be used for strategic decisions using the data concept relationships, while the database is used for retrieving sample data that supports these decisions [38]. Ontological decision making is discussed below under reasoning with ontologies.

2.3   Learning Database Ontologies

Learning an ontology is achieved by conceptualising objects of the application domain, their attributes and relationships. This can be accomplished with probabilistic approaches that can identify regularities in the data. Therefore, probabilistic ontologies are the result of logical evaluations of domain statistics.

There are various approaches for learning database ontologies [20]. We will focus on approaches that work at least partially automated on the database and that learn either crisp, probabilistic or possibilistic database ontologies (table 2).

Table 2. Tools for ontology learning from relational models

Ontology Learning	Source Data Model	Ontology Logic
RDBToOnto [10]	ER$^\$$ Schema & Relational Schema	Crisp
MASTRO [14]	ER Schema	Crisp
PROGNOS [13]	Relational Schema	Probabilistic
SoftFacts [43]	Relational Schema	Fuzzy

$^\$$ER: Entity-Relationship

Some approaches do not generate an ontology, but aim at learning probabilistic models from large relational data [17]. Popular probabilistic approaches are mostly based on Bayesian updating and inferencing, like Multi-Entity Bayesian Networks (MEBN) [25], for learning from relational data [33].

Data mining is rich on techniques for learning regularity of a data set that may be structured or unstructured. Some of these techniques have been adapted to learning database ontologies too. RTAXON is such a learning technique that transforms statistically identified relational data into ontological relationships [11]. It takes both as input, the relational schema and the entity-relationship schema, and can discover data relations not specified in either schema, but found in the data [10].

Interesting is further a data mining example for learning fuzzy ontologies from unstructured data of a hyper-media database that can learn time-varying dynamics of the domain. The adaptive ontology is used in return, to adapt querying the dynamically changing database [7].

2.4   Reasoning with Ontologies

Formal systems have the disadvantage of having no learning capability. However they have the advantage that they can be axiomatised and formally checked for consistency, satisfiability, subsumption or redundancy. Such formal methods have to be part of any reasoner [28]. Any ontology, whether specified, compiled or learned, needs to be validated formally, in order to be acceptable as common knowledge.

Inferencing is rule execution in propositional logic. An inference engine is an implementation of the logic, with the objective to execute rules of the knowledge base according to given inference rules, ie resolving rules in forward or backward chaining. The engine may additionally use utility functions, like the Rete algorithm [16], for improving efficiency of rule searching (Fig. 2).

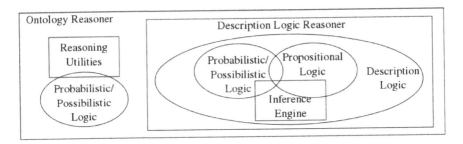

Fig. 2. Probabilistic/possibilistic extensions of logics used in ontology reasoners

Description Logic Reasoners

Most description logic reasoners use existing inference engines and utilise further techniques, such as decision tree, tableau algorithm or subsumption hierarchy, for handling specifics of description-logical (table 3).

Ontology Reasoners

An ontology reasoner is an implementation of a description logic reasoner for a particular domain ontology. Both reasoners may be extended separately with probabilistic or possibilistic information (Fig. 2).

For instance, the fuzzy ontology reasoner HyFOM extends the Fuzzy DL further with Mamdani inferencing [46]. Likewise SoftFacts extends Pellet and Fuzzy DL further with fuzzy database query features [43] (table 3).

Table 3. Ontology reasoners and underlying logics

Ontology Reasoner	Ontology Logic	Reasoning Logic[#]	DL Reasoner
MASTRO [14]	Crisp	–	Presto [36]
MEBN [25]	Probabilistic	Bayesian	MEBN
BUNDLE [35]	Probabilistic	Probabilistic	Pellet[+] [39]
HermiT [31]	Fuzzy	Hyper-Tableau [31]	HermiT [31]
FIRE [41]	Fuzzy		Fuzzy DL [40]
FuzzyDL [3]	Fuzzy	Fuzzy Rough Sets; Fuzzy; Łukasiewicz [5]	Pellet
DeLorean [2]	Fuzzy	Zadeh, Gödel Fuzzy Operators	Pellet
HyFOM [46]	Fuzzy	Mamdani	Fuzzy DL
SoftFacts [43]	Fuzzy DB	Fuzzy	Pellet
KAON [34]	Possibilistic	Possibilistic	Possibilistic

[#]Ontology reasoners extended DL reasoners to a specific reasoning logic
[+]Pellet implements tableau reasoning

3   Applications

Some of the above discussed ontology learning and reasoning approaches have been implemented as plug-in extensions for ontology development tools, such as Protégé [19] or KAON [34]. Fuzzy Protg is such a sample plug-in [19].

It turns out that only a few applications can learn database ontologies and reason with them probabilistically or possibilistically (table 4). We require these capabilities for our further extension of such systems with FS reasoning.

Table 4. Sample applications that use uncertain ontology learning or reasoning

Application	Originality	Ontology	
		Learning Tool	Reasoning Tool
PROGNOS [13]	Knowledge Fusion	MEBN [25]	MEBN [25]
UnBBayes [9]	Knowledge Fusion	MEBN [25]	MEBN [25]
FEER2FOnto [50]	Semantics-Preserving	Probabilistic	DeLorean [2]

Multi-Entity Bayesian Networks (MEBN) [25] extends first-order propositional ontologies with probabilistic information and infers within Bayesian networks of those probability distributions. Whereby the probability distributions are learned again Bayesian. Thus applications that use MEBN can learn and reason probabilistically. PROGNOS [13] and UnBBayes [9] use MEBN for knowledge fusion over relational databases.

FEER2FOnto [50] is interesting in that it combines probabilist learning and fuzzy reasoning with the reasoner DeLorean [2].

4   Fuzzy-Syllogistic Reasoning

Fuzzy-syllogistic (FS) reasoning is based on a fuzzy-logical extension of the syllogistic system that consists of all possible combinations of the well known categorical syllogisms. We interpret the system as one complex approximate reasoner that consists of all possible fuzzy-inferences for any given triple concept of an ontology. Here we introduce the fuzzy-syllogistic system ^{4}S that consists of four affirmative and four negative quantifiers.

4.1   Fuzzy-Syllogistic System

A categorical syllogism $\Psi_1\Psi_2\Psi_3F$ is an inference schema that concludes a quantified proposition $\Phi_3 = S\Psi_3P$ from the transitive relationship of two given quantified proportions $\Phi_1 = \{M\Psi_1P, P\Psi_1M\}$ and $\Phi_2 = \{S\Psi_2M, M\Psi_2S\}$:

$$\Psi_1\Psi_2\Psi_3F = (\Phi_1 = M\Psi_1P, P\Psi_1M, \Phi_2 = S\Psi_2M, M\Psi_2S, \Phi_3 = S\Psi_3P) \quad (1)$$

where F={1, 2, 3, 4} identifies the four possible combinations of Φ_1 with Φ_2, namely syllogistic figures and Ψ={A=all; ^{3}I=most; ^{2}I=half; ^{1}I=several; A=allNot; ^{3}I=mostNot; ^{2}I=halfNot; ^{1}I=severalNot} are the fuzzy quantifiers. Every syllogistic figure produces 8^3=512 permutations, which are called fuzzy-syllogistic moods. Thus the whole system ^{4}S has in total 2048 fuzzy-syllogistic moods.

Every mood has a structurally fixed truth ratio in [0,1], which is calculated algorithmically [24] by relating the number of its true cases to the number of false cases [23] (table 5 shows sample inferencing in the system S, since these are easier to follow manually). Moods of the FS system become inferences in FS reasoning.

Table 5. Sample fuzzy-syllogistic moods, their truth cases, truth ratios and sample interpretations of the fuzzy-syllogistic system S

Mood $\Psi_1\Psi_2\Psi_3F$	AAA1, AAI1	EEI1, 2, 3, 4
Cases $\Delta_i^{\#}$	t: 0100101	t: 0110010; t: 1010010 t: 1110010; f: 1110000
Truth Ratio τ	1t/(1t+0f)=1.0$^+$	3t/(3t+1f)=0.75
Interpretation of false cases*	–	At least $P \cap S \neq \Phi$ is missing
Example	All primates are mammals All humans are primates {All, Some} humans are mammals	Not All are {Turks, Muslim} Not All are {Orientals, Turks} Some Orientals are Muslim
Interpretation of Example	Concluding with All is true, probably without exception; concluding with Some is true only for the possible All case in Some	All four examples that can be loaded into the four moods are possibly more true than false, however possibly not fully true

$^+$t=true case; f=false case
*The conclusions of the examples assume that $P \cap S \neq \Phi$ equals the truth ratio τ of the mood
$^{\#}\Delta_i$ syllogistic case i=[1,96]; all possible distinct space permutations of the possible 7 spaces of three sets

4.2   Fuzzy-Syllogistic Learning and Reasoning

The objective of FS reasoning is to find the best matching fuzzy-syllogistic inference for a given triple concept with transitive relationships. The objective of FS reasoning within the learning process is to accumulate the samples for all possible 7 spaces of three sets, since their 96 possible distinct permutations constitute the universal set, from which the fuzzy-syllogistic moods match some as true syllogistic case and some as false [22]. The outcome of learning is an ontology with quantified relationships.

The FS reasoner calculates for every triple concept with transitive relationships a truth ratio using the very same algorithm for calculating the truth ratios of the individual moods/inferences [22]. In case of learning a fuzzy-syllogistic database ontology (Fig. 3) data quantities can easily be determined from the related tables.

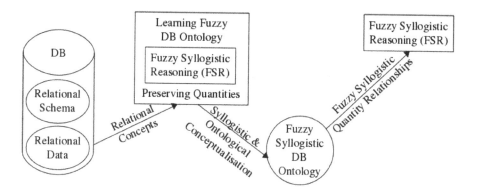

Fig. 3. Learning fuzzy syllogistic ontologies from relational databases using fuzzy syllogistic reasoning

5   Conclusion

We have reviewed state-of-the-art in ontology learning and reasoning, with an emphasis on database ontologies. We have pointed out that there are no standards for probabilist or possibilistic extensions of such system and therefore knowledge exchange between them is currently not efficient. Based on these observations, we have suggested fuzzy-syllogistic (FS) reasoning as a common logic for both, learning FS ontologies and reasoning with them. Principally any probabilist learning approach can be extended with FS reasoning, provided that all quantities of the data samples that lead to ontological concepts and their relationships can be calculated during the learning process and remain available along with the FS ontology after learning.

We have briefly discussed FS reasoning using the FS system ^{4}S that consists of 4 affirmative and 4 negative quantifiers.

Since FS reasoning is based on clear FS systems like, S or ^{4}S, it could be used as a common possibilistic reasoner for probabilistic ontologies, thus facilitate knowledge exchange in the semantic web. The reasoner can further adjust itself to changing quantities of the domain data, by applying the most suitable FS system ^{n}S to every triple concept relationship individually. With the FS reasoner individual optimisations are not required, like with fuzzy aggregation operators [6] or fuzzy integrals [4].

References

1. Albarrak, M.K., Sibley, E.H.: Translating relational and object-relational database models into owl model. In: IEEE IRI (2009)
2. Bobillo, F., Delgado, M., Gomez-Romero, J.: DeLorean: a reasoner for fuzzy OWL 1.1. In: Uncertainty Reasoning for the Semantic Web (URSW). LNAI. Springer (2008)
3. Bobillo, F., Straccia, U.: fuzzyDL: an expressive fuzzy description logic reasoner. In: Fuzzy Systems (FUZZIEEE). IEEE Computer Society (2008)
4. Bobillo, F., Straccia, U.: Fuzzy ontologies and fuzzy integrals. In: Intelligent Systems Design and Applications (ISDA). Springer (2011)
5. Bobillo, F., Straccia, U.: Reasoning with the finitely many-valued łukasiewicz fuzzy description logic sroiq. Information Sciences 181 (2011)
6. Bobillo, F., Straccia, U.: Aggregation operators for fuzzy ontologies. Applied Soft Computing 13 (2013)
7. Calegari, S., Loregian, M.: Using dynamic fuzzy ontologies to understand creative environments. In: Larsen, H.L., Pasi, G., Ortiz-Arroyo, D., Andreasen, T., Christiansen, H. (eds.) FQAS 2006. LNCS (LNAI), vol. 4027, pp. 404–415. Springer, Heidelberg (2006)
8. Calegari, S., Ciucci, D.: Fuzzy ontology, fuzzy description logics and fuzzy-OWL. In: Masulli, F., Mitra, S., Pasi, G. (eds.) WILF 2007. LNCS (LNAI), vol. 4578, pp. 118–126. Springer, Heidelberg (2007)
9. Carvalho, R., Laskey, K., da Costa, P.C.G., Ladeira, M., Santos, L., Matsumoto, S.: Unbbayes: Modeling uncertainty for plausible reasoning in the semantic web. In: Semantic Web. Intech (2010)
10. Cerbah, F.: Learning highly structured semantic repositories from relational databases: The RDBToOnto tool. In: Bechhofer, S., Hauswirth, M., Hoffmann, J., Koubarakis, M. (eds.) ESWC 2008. LNCS, vol. 5021, pp. 777–781. Springer, Heidelberg (2008)
11. Cerbah, F.: Mining the content of relational database to learn ontology with deeper taxonomies. In: Web Intelligence and Intelligent Agent Technology, IEEE, WIC (2008)
12. Codd, E.F.: Further normalization of the data base relational model. IBM Reaserch Report (1972)
13. da Costa, P.C.G., Laskey, K.B., Chang, K.C.: PROGNOS: Applying probabilistic ontologies to distributed predictive situation assessment in naval operations. In: International Command and Control Research and Technology Symposium (ICCRTS), C2 Journal (2009)
14. De Giacomo, G., Lembo, D., Lenzerini, M., Poggi, A., Rosati, R., Ruzzi, M., Savo, D.F.: Mastro: A reasoner for effective ontology based data access. In: OWL Reasoner Evaluation Workshop (ORE) (2012)
15. Fisher, M., Dean, M.: Automapper: Relational database semantic translation using OWL and SWRL. OWL experiences and Directions (OWLED), World Wide Web Consortium (w3c) (2008)
16. Forgy, C.: Rete: A fast algorithm for the many pattern/many object pattern match problem. Artificial Intelligence 19 (1982)
17. Getoor, L., Taskar, B.: Introduction to statistical relational learning. MIT Press (2007)
18. Ghawi, R., Cullot, N.: Database-to-ontology mapping generation for semantic interoperability. In: Very Large Databases (VLDB). ACM (2007)

19. Ghorbel, H., Bahri, A., Bouaziz, R.: Fuzzy protégé for fuzzy ontology models. In: International Protégé Conference (IPC), Stanford Medical Informatics (2009)
20. Hazman, M., El-Beltagy, S.R., Rafea, A.: A survey of ontology learning approaches. International Journal of Computer Applications 22(9) (2011)
21. He-ping, C., Lu, H., Bin, C.: Research and implementation of ontology automatic construction based on relational database. In: Computer Science and Software Engineering. IEEE Computer Society (2008)
22. Kumova, B.İ.: Symmetric properties of the syllogistic system inherited from the square of opposition (in review) (2015)
23. Kumova, B.İ., Çakır, H.: Algorithmic decision of syllogisms. In: García-Pedrajas, N., Herrera, F., Fyfe, C., Benítez, J.M., Ali, M. (eds.) IEA/AIE 2010, Part II. LNCS (LNAI), vol. 6097, pp. 28–38. Springer, Heidelberg (2010)
24. Kumova, B.İ., Çakir, H.: The fuzzy syllogistic system. In: Sidorov, G., Hernández Aguirre, A., Reyes García, C.A. (eds.) MICAI 2010, Part II. LNCS (LNAI), vol. 6438, pp. 418–427. Springer, Heidelberg (2010)
25. Laskey, K.B.: MEBN: A language for first-order bayesian knowledge bases. Artificial Intelligence 172 (2007)
26. Lin, L., Xu, Z., Ding, Y.: Owl ontology extraction from relational databases via database reverse engineering. Journal of Software 8(11) (2013)
27. Lubyte, L., Tessaris, S.: Automatic extraction of ontologies wrapping relational data sources. In: Bhowmick, S.S., Küng, J., Wagner, R. (eds.) DEXA 2009. LNCS, vol. 5690, pp. 128–142. Springer, Heidelberg (2009)
28. Lukasiewicz, T., Straccia, U.: Managing uncertainty and vagueness in description logics for the semanticweb. Web Semantics: Science, Services and Agents on the World Wide Web 6 (2008)
29. Maedche, A., Staab, S.: Ontology learning for the semantic web. IEEE Intelligent Systems and Their Applications 16(2) (2005)
30. Martinez-Cruz, C., Blanco, I.J., Vila, M.A.: Ontologies versus relational databases: are they so different? A comparison. Artificial Intelligence Review 38 (2011)
31. Motik, B., Shearer, R., Horrocks, I.: Hypertableau reasoning for description logics. Journal of Artificial Intelligence Research 36 (2009)
32. Myroshnichenko, I., Murphy, M.C.: Mapping er schemas to owl ontology. Semantic Computing, Berkeley (2009)
33. Park, C.Y., Laskey, K.B., Costa, P., Matsumoto, S.: Multi-entity bayesian networks learning in predictive situation awareness. In: International Command and Control Research and Technology Symposium (ICCRTS), US DoD (2013)
34. Qi, G., Pan, J.Z., Ji, Q.: A possibilistic extension of description logics. Description Logics (DL), Sun SITE Central Europe (CEUR) (2007)
35. Riguzzi, F.: Probabilistic description logics under the distribution semantics. Semantic Web Journal, SWJ (2013)
36. Rosati, R., Almatelli, A.: Improving query answering over DLLite ontologies. In: Principles of Knowledge Representation and Reasoning (KR). AAAI (2010)
37. Sahoo, S.S., Halb, W., Hellmann, S., Idehen, K., Thibodeau Jr., T., Auer, S., Sequeda, J., Ezzat, A.: A survey of current approaches for mapping of relational databases to rdf. W3C RDB2RDF Incubator Group (2009)
38. Santoso, H.A., Haw, S.C., Abdul-Mehdi, Z.T.: Ontology extraction from relational database: Concept hierarchy as background knowledge. Knowledge-Based Sys. (2011)
39. Sirin, E., Parsia, B., Cuenca-Grau, B., Kalyanpur, A., Katz, Y.: Pellet: A practical owldl reasoner. Journal of Web Semantics 5(2) (2007)

40. Stoilos, G., Simou, N., Stamou, G., Kollias, S.: The fuzzy description logic fshin. In: Uncertainty Reasoning for the Semantic Web, CEUR Electronic Workshop (2005)
41. Stoilos, G., Simou, N., Stamou, G., Kollias, S.: Uncertainty and the semantic web. IEEE Intelligent Systems 21, 5 (2006)
42. Straccia, U.: Reasoning within fuzzy description logics. Journal of Artificial Intelligence Research 14 (2001)
43. Straccia, U.: SoftFacts: A top-k retrieval engine for ontology mediated access to relational databases. In: Systems, Man and Cybernetics (SMC). IEEE (2010)
44. Trinkunas, J., Vasilecas, O.: Building ontologies from relational databases using reverse engineering methods. In: Computer Systems and Technologies. ACM (2007)
45. Xu, J., Li, W.: Using relational database to build owl ontology from xml data sources. In: Computational Intelligence and Security Workshops. IEEE Computer Society (2007)
46. Yaguinuma, C.A., Magalhães Jr., W.C.P., Santos, M.T.P., Camargo, H.A., Reformat, M.: Combining fuzzy ontology reasoning and mamdani fuzzy inference system with hyFOM reasoner. In: Hammoudi, S., Cordeiro, J., Maciaszek, L.A., Filipe, J. (eds.) ICEIS 2013. LNBIP, vol. 190, pp. 174–189. Springer, Heidelberg (2014)
47. Zadeh, L.A.: Fuzzy logic and approximate reasoning. Syntheses 30 (1975)
48. Zadeh, L.A.: Syllogistic reasoning in fuzzy logic and its application to usuality and reasoning with dispositions. IEEE Transactions on Systems, Man and Cybernetics 15(6) (1985)
49. Zarechnev, M., Kumova, B.I.: Ontology-based fuzzy-syllogistic reasoning. In: Industrial, Engineering and Other Applications of Applied Intelligent Systems (IEA-AIE). LNCS. Springer (2015)
50. Zhang, F., Ma, Z.M., Yan, L., Cheng, J.: Construction of fuzzy OWL ontologies from fuzzy EER models: A semantics-preserving approach. Fuzzy Sets and Sys. 229 (2013)
51. Zhang, F., Ma, Z.M., Yan, L., Wang, Y.: A description logic approach for representing and reasoning on fuzzy object-oriented database models. Fuzzy Sets and Systems 186 (2012)

Surveying the Versatility of Constraint-Based Large Neighborhood Search for Scheduling Problems

Riccardo Rasconi[✉], Angelo Oddi, and Amedeo Cesta

Institute of Cognitive Sciences and Technologies, CNR, Rome, Italy
{riccardo.rasconi,angelo.oddi,amedeo.cesta}@istc.cnr.it

Abstract. Constraint-based search techniques have gained increasing attention in recent years as a basis for scheduling procedures that are capable of accommodating a wide range of constraints. Among these, the Large Neighborhood Search (LNS) has largely proven to be a very effective heuristic-based methodology. Its basic optimization cycle consists of a continuous iteration of two steps where the solution is first relaxed and then re-constructed. In Constraint Programming terms, relaxing entails the retraction of some previously imposed constraints, while re-constructing entails imposing new constraints, searching for a better solution. Each iteration of constraint removal and re-insertion can be considered as the examination of a large neighborhood move, hence the procedure's name. Over the years, LNS has been successfully employed over a wide range of different problems; this paper intends to provide an overview of some utilization examples that demonstrate both the versatility and the effectiveness of the procedure against significantly difficult scheduling benchmarks known in literature.

Keywords: Scheduling · Metaheuristics · Constraint Programming · Databases

1 Introduction

Among the existing constraint-based search techniques, the Large Neighborhood Search (LNS) has proven to be a remarkably effective and versatile problem solving methodology. Initially proposed to tackle Vehicle Routing problems ([24]), the LNS has been henceforth successfully used with many other kind of difficult combinatorial problems. This paper intends to present a survey of utilization examples of a LNS-based procedure called Iterative Flattening Search (IFS) applied to the scheduling problem domain. IFS represents a family of stochastic local search ([11]) techniques that was originally introduced in [5] as a non-systematic approach to solve difficult time and resource-constrained scheduling problem instances, and it has been shown to have very good scaling capabilities. The IFS procedure basically iterates two solving steps: (1) a *relaxation step*, where a subset of solving decisions made at iteration $(i - 1)$ are randomly retracted at iteration i, and (2) a *flattening step*, where a new solution is re-computed after

S. Kozielski et al. (Eds.): BDAS 2015, CCIS 521, pp. 33–43, 2015.
DOI: 10.1007/978-3-319-18422-7_3

each previous relaxation. The choice of the term "flattening" stems from the fact that finding a solution equates to pushing down the resource usage profiles below the maximum capacity threshold of each resource involved in the scheduling problem.

Besides representing a lively area in research, scheduling problems notoriously find a wide range of different applications in the real world, ranging from the manufacturing processes field, to space related domains. Among the many applications of scheduling, DataBase management is of no secondary importance. Data Service Providers need effective query scheduling algorithm in order to guarantee efficient quality of service in high workload situations ([16]); more and more efficient scheduling algorithms are required to guarantee the timing constraints associated with transactions by means of dynamic serialization adjustments (concurrency control) in real-time databases ([25]); multi-objective scheduling algorithms are typically required in the cases where the stringent timing requirements associated with real-time querying meet the need to guarantee a high quality of the data to be retrieved ([27]).

This paper has the following structure: section 2 provides some technical details about the IFS schema, while section 3 describes the Constraint Satisfaction Problem (CSP), i.e., the constraint-based problem representation used throughout this survey. Section 4 is devoted to the description of some results in the literature obtained utilizing the IFS against two difficult scheduling problem benchmarks. Finally, section 5 presents some conclusions.

2   The Iterative Flattening Search Schema

Figure 1 introduces the generic IFS procedure. The algorithm basically implements an optimization loop, alternating relaxation and flattening steps as long as improved solutions are found, and terminates when a maximum number of non-improving iterations is reached. The procedure takes three parameters as input: (1) an initial solution S; (2) a positive integer $MaxFail$, which specifies the maximum number of non objective-improving moves that the algorithm will tolerate before terminating; (3) a parameter γ which controls the solution relaxation extent. After the initialization (Steps 1-2), a solution is repeatedly modified within the while loop (Steps 3-10) by applying the RELAX procedure, and subsequently rebuilt by means of the SOLVE procedure. At each iteration, the RELAX step reintroduces the possibility of resource contention, and the SOLVE procedure (i.e., the IFS's flattening step) is called again to restore resource feasibility. In the case a better makespan solution is found (Step 6), the new solution is saved in S_{best} and the *counter* is reset to 0. If no improvement is found within $MaxFail$ moves, the algorithm terminates and returns the best solution found.

Following the approach described in section 3, the SOLVE step in Fig. 1 relies on a *core* constraint-based search procedure which returns a scheduling solution (further detailed in the next section), while in the RELAX step of the IFS cycle, a feasible schedule is relaxed into a possibly resource infeasible but precedence feasible schedule, by retracting some number of the precedence constraints

IFS(S,$MaxFail$, γ)
1. $S_{best} \leftarrow S$
2. $counter \leftarrow 0$
3.   **while** ($counter \leq MaxFail$) **do**
4.   RELAX(S, γ)
5. $S \leftarrow$ SOLVE(S)
6.   **if** $C(S) < C(S_{best})$ **then**
7. $S_{best} \leftarrow S$
8.   counter $\leftarrow 0$
9.   **else**
10.   counter $\leftarrow$ counter $+ 1$
11. **return** (S_{best})

Fig. 1. The IFS schema

previously imposed in any of the SOLVE steps. In general, the relaxation step can be executed according to different strategies.

In the next section we first provide a formal definition of the Constraint Satisfaction Problem, as well as a description of how this paradigm has been used to represent and solve scheduling problems.

3   A CSP-Based Scheduling Problems Representation

Constraint Programming (see Rossi et al. 2006) is an approach to solving combinatorial search problems based on the Constraint Satisfaction Problem (CSP) paradigm (Montanari 1974, Kumar 1992, Tsang 1993). Constraints are just relations and a CSP states which relations should hold among the given problem decision variables. This framework is based on the combination of search techniques and constraint propagation. Constraint propagation consists of using constraints actively to prune the search space.

Constraint satisfaction and propagation rules have been successfully used to model, reason and solve about many classes of problems in such diverse areas as scheduling, temporal reasoning, resource allocation, network optimization and graphical interfaces. In particular, CSP approaches have proven to be an effective way to model and solve complex scheduling problems (see, for instance, [8,23,26,1,2,6]).

A CSP is defined by a set of *variables*, $X_1, X_2, ..., X_n$, and a set of *constraints*, $C_1, C_2, ..., C_m$. Each variable X_i has a nonempty domain D_i of possible values. Each constraint C_i involves some subset of the variables and specifies the allowable combinations of values for that subset. A solution of a CSP is defined as a *consistent* assignment of values to all of the variables $X_1, X_2, ..., X_n$, i.e., an assignment that satisfies all the constraints.

There are different ways to formulate this problem as a *Constraint Satisfaction Problem* (CSP) [15]. Analogously to [7,20], we treat the problem as the one of

CSP-Solver(P)

1.   **if** CheckConsistency(P) **then**
2.   **if** IsSolution(P) **then**
3. $S \leftarrow P$
4.   **return**(S)
5.   **else**
6. $v_i \leftarrow$ SelectVariable(P)
7. $\lambda_i \leftarrow$ ChooseValue(P, v_i)
8.   CSP-Solver($P \cup \{\lambda_i\}$)
9.   **else**
10.   **return**($\varnothing$)

Fig. 2. A general CSP-based algorithm

establishing *precedence constraints* between pairs of activities that require the same resource, so as to eliminate all possible conflicts in the resource use.

The general form of a complete CSP solving procedure is presented in Fig. 2, and consists of the following three steps: (a) the current problem P is checked for consistency (CheckConsistency(P)) by the application of a constraint propagation procedure; if the propagation fails (i.e., at least one constraint is violated) the algorithm exits with failure. Otherwise, the following two steps are executed; (b) a CSP variable v_i is selected by means of a *variable ordering* heuristic; (c) a value λ_i is chosen by a *value ordering* heuristic and added to P. Lastly, the solver is recursively called on the updated problem $P \cup \{\lambda_i\}$.

When applied to scheduling problems, the procedure generates a consistent ordering of activities that require the same resource by incrementally adding *precedence constraints* between activity pairs belonging to a temporally feasible solution. Resource constraints are super-imposed on the problem's activities by projecting "resource demand profiles" over time. The time intervals characterized by resource oversubscription are detected as *resource conflicts*, and are then resolved by iteratively posting simple precedence constraints between pairs of activities competing for the resource involved in the conflict ([6]). The procedure iteratively propagates the current temporal constraints, and then proceeds to select another resource conflict until no more conflicts are detected (i.e., a solution is found), or an *unresolvable conflict* is encountered, in which case the procedure terminates with no solution found.

4   Putting the IFS to the Test

In this section we provide evidence of the efficacy and of the versatility of the IFS approach. In particular, we describe the utilization of the approach against two classes of particularly interesting scheduling problems, as their structure underlies a number of real-world scheduling applications: the Resource Constrained Scheduling Problem with Time Windows (RCPSP/max), and the

Flexible Job Shop Scheduling Problem (FJSSP). The RCPSP/max ([17]) is an extended formulation of the Resource Constrained Project Scheduling Problem (RCPSP) frequently used to model manufacturing, logistics and project management scheduling problems, and is considered particularly difficult due to the presence of temporal separation constraints (in particular maximum time lags) between project activities (even the search of a feasible solution is *NP-hard*). The FJSSP ([22]) is an extension of the Job Shop Scheduling Problem (JSSP), of particular interest in the field of Flexible Manufacturing Systems development, whose resolution integrates a *routing policy* search to the ordinary sequencing of activities that concurrently use the same resource; being an extension of the JSSP, also the FJSSP is *NP-hard*.

4.1 Tackling the Resource Constrained Scheduling Problem with Time Windows (RCPSP/max)

The RCPSP/max can be formalized in terms of the following three sets: (1) a set V of n non-preemptive activities where each activity a_i has a fixed duration d_i. Each activity has a start-time s_i and a completion-time e_i that satisfies the constraint $s_i + d_i = e_i$; (2) a set E of temporal constraints that may exist between any two activities $\langle a_i, a_j \rangle$ of the form $s_j - s_i \in [T_{ij}^{min}, T_{ij}^{max}]$, called start-to-start constraints (time lags or *generalized precedence relations* between activities); (3) a set R of renewable resources, where each resource r_k is characterized by a maximum integer capacity $c_k \geq 1$. The execution of an activity a_i requires some capacity from one or more resources. For each resource r_k the integer $rc_{i,k}$ represents the capacity required by the activity a_i. A schedule S is an assignment of values to the start-times of all activities in V ($S = (s_1, \ldots, s_n)$). A schedule is said to be *feasible* if it is both *time* and *resource-feasible*. Solving the RCPSP/max optimization problem equates to finding a feasible schedule with *minimum makespan* MK, where $MK(S) = max_{a_i \in V}\{e_i\}$.

From section 3 it is clear that in order to provide a CSP-based representation of the RCPSP/max, it is sufficient to determine the decision variables of the problem, as well as the value domains associated to each such variables. In our case, the project scheduling problem is formulated as a CSP as follows: the *decision variables* of the CSP are identified as the so-called *Forbidden Sets* also known as *Minimal Critical Sets*, or MCS (see [12]). As anticipated in section 2, given a generic resource r_k, a *conflict* is a set of activities requiring r_k, which can mutually overlap and whose combined resource requirement is in excess of the resource capacity c_k. A Minimal Critical Set $MCS = \{a_1, a_2, \ldots, a_k\}$ represents a resource conflict of minimal size (each subsets is not a resource conflict), which can be *resolved* by posting a single precedence constraint between two of the competing activities in the conflict set. Hence, in CSP terms, a decision variable is defined for each $MCS = \{a_1, a_2, \ldots, a_s\}$ and the domain of possible values is the set of all possible feasible precedence constraints $a_i \preceq a_j$ which can be imposed between each pair of activities in the MCS. A *solution* of the scheduling problem is a set of precedence constraints that, when added to the original problem, removes all the MCSs from the problem.

The general SOLVE step of the IFS procedure (whose CSP-based schema is presented in Fig. 2) therefore proceeds by iteratively selecting MCS variables (i.e., resource conflicts), and resolving them by posting simple precedence constraints between pairs of competing activities ([6]) that belong to the selected MCS. The procedure iteratively propagates the current temporal constraints, and then proceeds to select a new resource conflict until no more MCSs exist in the problem, in which case a solution is found.

Table 1. Experimental results on the UBO200 set (RCPSP/max)

Metrics	cp-based	chain-based	best
No. improved MKs (Impr.)	14	21	**27**
Avg. Makespan Gap (Δ_{mk})	1.47	0.02	**-0.44**
Avg. Cpu time (Cpu)	6867.08	13395.42	-
Avg. IFS Cycles	580.04	693.38	-
No. new optimal solutions	2	1	**3**

Relatively to the RELAX step, two different versions of the procedure have been compared in [18]: the first strategy, used in [5,13], removes precedence constraints between pair of activities on the *critical path* of a solution, and hence is called *cp-based relaxation*. Such relaxation is more targeted to directly reducing the makespan of a solution, as the latter equates to the length of the critical path, by definition, even though it seems more prone to becoming trapped in *local minima*. The second strategy, from [10], starts from a POS-form solution (see [21]) and randomly *breaks* some chains in the input POS schedule, and hence is given the name *chain-based relaxation*; this strategy promotes a search with an higher degree of *diversification*. Both RELAX strategies are compared against two of the most challenging and widely used RCPSP/max benchmarks, namely the J30 set (medium size, 30 activities per instance) and the UBO-200sets (large size, 200 activities per instance), both available at http://www.om-db.wi.tum.de/psplib/otherlibs.html. In particular, the *cp-based relaxation* procedure is controlled by two parameters: n_r determines the number of individual relaxation attempts performed at each relaxation cycle, and for each attempt, p_r determines the percentage of decision constraints that will be randomly removed from the critical path. On the other hand, the relaxation procedure within the *chain-based relaxation* uses the p_r parameter only, which takes a slightly different meaning, determining the percentage of activities to be randomly selected and eliminated from the solution.

In [18], the experimental analysis has proceeded in two steps. First, a number of tests has been executed on the J30 set, establishing a significant superiority of the *chain-based relaxation* w.r.t. the *cp-based relaxation*; secondly, both RELAX methods have been used against the UBO-200 set. The experiment results of the second run are summarized in table 1. The table is organized as follows: the first two columns, with names *cp-based* and *chain-based*, represent the performance of the proposed IFS algorithm. The last column, named *best*, represents the results obtained merging the best performances of the previous two columns.

These experiments demonstrate the efficacy of the IFS, as 27 instances have been improved over the best results as known at the time of the analysis, and 3 new optimal solutions have been discovered (of these improvements, 21 have been found using the *chain-based relaxation* method).

4.2 Adding Flexibility on Resource Availability: The Flexible Job Shop Scheduling Problem (FJSSP)

In this section we evaluate the IFS procedure for solving a scheduling problem with a different structure than the multi-capacity job-shop problem. We focus specifically on the Flexible Job Shop Scheduling Problem (FJSSP), a generalization of the classical JSSP where a given activity may be processed on any one of a designated set of available machines.

The Flexible Job Shop Scheduling Problem (FJSSP) entails synchronizing the use of a set of machines (or resources) $R = \{r_1, \ldots, r_m\}$ to perform a set of n activities $A = \{a_1, \ldots, a_n\}$ over time. The set of activities is partitioned into a set of nj jobs $\mathcal{J} = \{J_1, \ldots, J_{nj}\}$. The processing of a job J_k requires the execution of a strict sequence of n_k activities $a_i \in J_k$ and cannot be modified. All jobs are released at time 0. Each activity a_i requires the exclusive use of a *single resource* r_i for its entire duration chosen among a *set of available resources* $R_i \subseteq R$. No *preemption* is allowed. Each machine is available at time 0 and can process more than one operation of a given job J_k (i.e., *recirculation* is allowed). The processing time p_{ir} of each activity a_i depends on the selected machine $r \in R_i$, such that $e_i - s_i = p_{ir}$, where the variables s_i and e_i represent the start and end time of a_i. A *solution* $S = \{(\overline{s_1}, \overline{r_1}), (\overline{s_2}, \overline{r_2}), \ldots, (\overline{s_n}, \overline{r_n})\}$ is a set of pairs $(\overline{s_i}, \overline{r_i})$, where $\overline{s_i}$ is the assigned start-time of a_i, $\overline{r_i}$ is the selected resource for a_i and all the above constraints are satisfied. An *optimal* solution S^* is a solution S with the minimum value of the makespan C_{max}.

The FJSSP is more difficult than the classical Job Shop Scheduling Problem (which is itself *NP-hard*), since it is not just a sequencing problem; in addition to deciding how to sequence activities that require the same machine, it is also necessary to choose a *routing policy*, that is which machine will process each activity. There are different ways to model the problem as a CSP; in [19] an approach has been used that focuses on *assigning resources to activities* (a distinguishing aspect of the FJSSP) and on *establishing precedence constraints* between pairs of activities that require the same resource, so as to eliminate all possible conflicts in the resource usage. In order to correctly represent both previous aspects of the FJSSP in CSP terms, once again it is necessary to determine the problem decision variables as well as the associated value domains. In particular, *two sets of decision variables* are identified: (1) a variable x_i is defined for each activity a_i to select one resource for its execution, the domain of x_i is the set of available resources R_i: (2) A variable o_{ijr} is defined for each pair of activities a_i and a_j requiring the same resource r $(x_i = x_j = r)$, which can take one of two values $a_i \preceq a_j$ or $a_j \preceq a_i$.

Table 2. Results on the BCdata Benchmark (FJSSP)

inst.	best	γ					
		0.2	0.3	0.4	0.5	0.6	0.7
mt10x	**918**	980	936	936	934	**918**	**918**
mt10xx	**918**	936	929	936	933	**918**	926
mt10xxx	**918**	936	929	936	926	926	926
mt10xy	**905**	922	923	923	915	**905**	909
mt10xyz	**847**	878	858	851	862	**847**	851
mt10c1	**927**	943	937	986	934	934	**927**
mt10cc	**910**	926	923	919	919	**910**	911
setb4x	**925**	967	945	930	**925**	937	937
setb4xx	**925**	966	931	933	**925**	937	929
setb4xxx	**925**	941	930	950	950	942	935
setb4xy	916	<u>**910**</u>	941	936	936	916	914
setb4xyz	**905**	928	909	**905**	**905**	**905**	**905**
setb4c9	**914**	926	937	926	926	920	920
setb4cc	909	929	917	<u>**907**</u>	914	<u>**907**</u>	909
seti5x	1201	1210	<u>**1199**</u>	<u>**1199**</u>	1205	1207	1209
seti5xx	**1199**	1216	**1199**	1205	1211	1207	1206
seti5xxx	**1197**	1205	1206	1206	1199	1206	1206
seti5xy	**1136**	1175	1171	1175	1166	1156	1148
seti5xyz	**1125**	1165	1149	1130	1134	1144	1131
seti5c12	**1174**	1196	1209	1200	1198	1198	1175
seti5cc	**1136**	1177	1155	1162	1166	1138	1150

The SOLVE procedure used in [19] to tackle the FJSSP instances is an extension of the SP-PCP procedure proposed in [20], in which both the *variable* and *value* ordering heuristics guiding the search have been extended to incorporate the additional set of x_i decision variables, related to the resource selection. The procedure (again following the general schema of Fig. 2) selects the decision variables x_i and o_{ijr}, and respectively decides their *values* in terms of imposing a duration constraint on a selected activity or a precedence constraint (i.e., $a_i \preceq a_j$ or $a_j \preceq a_i$) between two activities assigned to the same resource. The temporal constraints deriving from each decision are iteratively propagated, until all decision variables are assigned and no conflict is detected.

The RELAX step adopted in [19] is the *chain-based relaxation*. Based on the fact that in the FJSSP each scheduling decision is either a *precedence constraint* between a pair of activities that are competing for the same resource capacity and/or a *resource assignment* to one activity, the *chain-based relaxation* starts from a POS-form solution S and randomly *breaks* some total orders (or *chains*) imposed on the subset of activities requiring the same resource r. In more details, it proceeds by randomly selecting a subset of activities a_i from the input solution S, with each activity having an uniform probability $\gamma \in (0, 1)$ to be selected (γ is called the *relaxing factor*). For each selected activity, the resource assignment is removed and the original set of available options R_i is re-established.

In [19], the IFS framework is tested against one of the most challenging FJSSP benchmark composed of 21 instances initially provided by Barnes and Chambers (*BCdata*), where the objective of minimizing the *makespan*. Table 2 presents a summary of the results obtained in [19]. The table lists all the BCdata instances, and shows the makespan value obtained by the IFS varying the value of the *relaxing factor* γ from 0.2 to 0.7. The **best** column contains the best results known in current literature at the time of the analysis (obtained with at least one of the approaches described in [3,14,9,4]). Table 2 confirms the competitiveness of the IFS approach against the FJSSP. The efficacy of the IFS approach is proven by the fact that 10 previously known bests on a total of 21 instances are confirmed (in bold). More importantly, the three bold underlined instances in the table represent newly improved instances w.r.t. the current bests. For the record, the same improvements had also been obtained in [4] by means of a *specialized* tabu search algorithm using a high performance GPU with 128 processors. The fact that the IFS succeeded in finding the same results running on a single processor machine represents a remarkable confirmation of the efficacy of the approach.

5   Conclusions

Iterative Flattening Search (IFS) is a meta-heuristic strategy that has been largely and successfully used in literature to solve many types of scheduling problems. Prior research has shown IFS to be an effective and scalable heuristic procedure for minimizing schedule makespan in both multi-capacity and flexible resource settings.

In this paper, we have surveyed the performances of the Iterative Flattening Search (IFS) model obtained in previous research against two significant and notoriously difficult scheduling problem benchmarks. The scheduling problem types selected for this analysis are the Resource Constrained Scheduling Problem with Time Windows (RCPSP/max) and the Flexible Job Shop Scheduling Problem (FJSSP). The RCPSP/max is a notoriously hard and general scheduling problem, such that even the search of a feasible solution is *NP-hard*, while the FJSSP an extension of the Job Shop Scheduling Problem (JSSP) whose resolution integrates a *routing policy* search to the ordinary sequencing of activities that concurrently use the same resource. Being an extension of the JSSP, also the FJSSP is *NP-hard*.

The analysis carried on in this survey confirms the versatility and the efficacy of the IFS methodology; the broad sampling of the search space produced by the relax-solve loop, integrated with the local optimization guaranteed by the inner solving procedure within each iteration, provide an overall mechanism that inherently promotes both diversification and intensification. Versatility is demonstrated by the simplicity of the optimization loop and by the high adaptability of the procedure; the efficacy is demonstrated by comparing the approach's performances against different solutions, for a range of differently structured scheduling problems of medium/large size.

References

1. Beck, J.C., Davenport, A.J., Davis, E.D., Fox, M.S.: The ODO project: toward a unified basis for constraint-directed scheduling. Journal of Scheduling 1(2), 89–125 (1998)
2. Baptiste, P., Pape, C.L., Nuijten, W.: Constraint-Based Scheduling, 1st edn. Springer US (2001), doi:10.1007/978-1-4615-1479-4
3. Ben Hmida, A., Haouari, M., Huguet, M.J., Lopez, P.: Discrepancy search for the flexible job shop scheduling problem. Computers & Operations Research 37, 2192–2201 (2010)
4. Bożejko, W., Uchroński, M., Wodecki, M.: Parallel meta^{2}heuristics for the flexible job shop problem. In: Rutkowski, L., Scherer, R., Tadeusiewicz, R., Zadeh, L.A., Zurada, J.M. (eds.) ICAISC 2010, Part II. LNCS (LNAI), vol. 6114, pp. 395–402. Springer, Heidelberg (2010)
5. Cesta, A., Oddi, A., Smith, S.F.: Iterative Flattening: A Scalable Method for Solving Multi-Capacity Scheduling Problems. In: 17th National Conference on Artificial Intelligence, AAAI/IAAI, pp. 742–747 (2000)
6. Cesta, A., Oddi, A., Smith, S.F.: A constraint-based method for project scheduling with time windows. J. Heuristics 8(1), 109–136 (2002)
7. Cheng, C., Smith, S.F.: Generating Feasible Schedules under Complex Metric Constraints. In: Proceedings of the 12th National Conference on AI, AAAI 1994 (1994)
8. Fox, M.S.: Constraint Guided Scheduling: A Short History of Scheduling Research at CMU. Computers and Industry 14(1-3), 79–88 (1990)
9. Gao, J., Sun, L., Gen, M.: A hybrid genetic and variable neighborhood descent algorithm for flexible job shop scheduling problems. Computers & Operations Research 35, 2892–2907 (2008)
10. Godard, D., Laborie, P., Nuitjen, W.: Randomized Large Neighborhood Search for Cumulative Scheduling. In: Proceedings of the 15th International Conference on Automated Planning & Scheduling, ICAPS 2005, pp. 81–89 (2005)
11. Hoos, H.H., Stützle, T.: Stochastic Local Search. Foundations and Applications. Morgan Kaufmann (2005)
12. Laborie, P., Ghallab, M.: Planning with Sharable Resource Constraints. In: Proceedings of the 14th Int. Joint Conference on Artificial Intelligence, IJCAI 1995 (1995)
13. Laurent, M., Van Hentenryck, P.: Iterative Relaxations for Iterative Flattening in Cumulative Scheduling. In: Proceedings of the 14th International Conference on Automated Planning & Scheduling, ICAPS 2004, pp. 200–208 (2004)
14. Mastrolilli, M., Gambardella, L.M.: Effective neighbourhood functions for the flexible job shop problem. Journal of Scheduling 3, 3–20 (2000)
15. Montanari, U.: Networks of Constraints: Fundamental Properties and Applications to Picture Processing. Information Sciences 7, 95–132 (1974)
16. Moon, H.J., Chi, Y., Hacigümüş, H.: Performance evaluation of scheduling algorithms for database services with soft and hard slas. In: Proceedings of the Second International Workshop on Data Intensive Computing in the Clouds, DataCloud-SC 2011, pp. 81–90. ACM, New York (2011), http://doi.acm.org/10.1145/2087522.2087536
17. Neumann, K., Schwindt, C.: Activity-on-Node Networks with Minimal and Maximal Time Lags and Their Application to Make-to-Order Production. Operation Research Spektrum 19, 205–217 (1997)

18. Oddi, A., Rasconi, R.: Solving resource-constrained project scheduling problems with time-windows using iterative improvement algorithms (2009), http://aaai.org/ocs/index.php/ICAPS/ICAPS09/paper/view/736/1142

19. Oddi, A., Rasconi, R., Cesta, A., Smith, S.F.: Iterative flattening search for the flexible job shop scheduling problem. In: Proceedings of the 22nd International Joint Conference on Artificial Intelligence, IJCAI 2011, Barcelona, Catalonia, Spain, July 16-22, pp. 1991–1996 (2011), http://ijcai.org/papers11/Papers/IJCAI11-332.pdf

20. Oddi, A., Smith, S.F.: Stochastic Procedures for Generating Feasible Schedules. In: Proceedings 14th National Conference on AI (AAAI 1997) (1997)

21. Policella, N., Cesta, A., Oddi, A., Smith, S.F.: From Precedence Constraint Posting to Partial Order Schedules. AI Communications 20(3), 163–180 (2007)

22. Rossi, A., Dini, G.: Flexible job-shop scheduling with routing flexibility and separable setup times using ant colony optimisation method. Robotics and Computer-Integrated Manufacturing 23(5), 503–516 (2007)

23. Sadeh, N.: Look-ahead Techniques for Micro-opportunistic Job-Shop Scheduling. School of Computer Science, Carnegie Mellon University (1991)

24. Shaw, P.: Using constraint programming and local search methods to solve vehicle routing problems. In: Maher, M., Puget, J.-F. (eds.) CP 1998. LNCS, vol. 1520, pp. 417–431. Springer, Heidelberg (1998), http://dx.doi.org/10.1007/3-540-49481-2_30

25. Son, S.H., Park, S.: A Priority Based Scheduling Algorithm for Real-Time Databases. Journal of Information Science and Engg. 11(2), 233–248 (1995)

26. Smith, S.: OPIS: A Methodology and Architecture for Reactive Scheduling. In: Zweben, M., Fox, M. (eds.) Intelligent Scheduling. Morgan Kaufmann (1994)

27. Thiele, M., Bader, A., Lehner, W.: Multi-objective scheduling for real-time data warehouses. Computer Science - Research and Development 24(3), 137–151 (2009), http://dx.doi.org/10.1007/s00450-009-0062-z

Database Architectures
and Performance

Query Workload Aware Multi-histogram
Based on Equi-width Sub-histograms
for Selectivity Estimations of Range Queries

Dariusz Rafał Augustyn[✉]

Institute of Informatics, Silesian University of Technology,
16 Akademicka St., 44-100 Gliwice, Poland
draugustyn@polsl.pl

Abstract. Query optimizer uses a selectivity parameter for estimating the size of data that satisfies a query condition. Selectivity value calculations are based on some representation of attribute values distribution e.g. a histogram. In the paper we propose a query workload aware multi-histogram which contains a set of equi-width sub-histograms. The multi-histogram is designated for single-attribute-based range query selectivity estimating. Its structure is adapted to a 2-dimensional distribution of conditions of last recently processed range queries. The structure is obtained by clustering values of boundaries of query ranges. Sub-histograms' resolutions are adapted to a variability of a 1-dimensional distribution of attribute values.

Keywords: Selectivity estimation · Range query · Multi-histogram · embedded sub-histogram · Query workload · Data clustering · Bucket boundaries distribution · Variability metrics

1 Introduction

Selectivity factor is used by a database query optimizer to choose the best query execution plan. It is needed for an early estimation of size of the data that satisfying a query condition. For a simple single-table selection condition the selectivity is the number of rows satisfying the condition divided by the number of all table rows. For a simple range condition based on single attribute x with continuous domain, it may be defined as follows:

$$sel(Q(a < x < b)) = \int_a^b f(x)dx. \tag{1}$$

where x – a table attribute, a and b – range boundaries, $f(x)$ – a probability density function (PDF) that describes x attribute values distribution.

There are many approaches to representing an attribute values distribution using different types of histogram [8]. Most of them use only an information

© Springer International Publishing Switzerland 2015
S. Kozielski et al. (Eds.): BDAS 2015, CCIS 521, pp. 47–59, 2015.
DOI: 10.1007/978-3-319-18422-7_4

about x values distribution but also there are some that take into account an information about query workload [6, 5, 9, 11, 1–3].

The proposed method also uses information about processed queries, but it only collects data about the range conditions (values of range boundaries), not about their real selectivity values obtained just after a query execution (like the approaches presented in [6, 5, 9, 11]. [7]. Some of those approaches to query-workload-based selectivity estimation (so-called feedback driven) are dedicated for multi-dimensional queries (m-D range queries with conditions based on many attributes), e.g. the approaches that use: self-tunning histogram and STHoles [5, 9, 10], ISOMER – the maximum entropy based algorithm for feedback-driven m-D histogram creation [11], proactive and reactive m-D histogram [7].

In this paper we introduce a new representation of attribute values distribution – a multi-histogram – which consists on non-overlapping equi-width sub-histograms. Domains of sub-histograms depend on 2-D distribution of pairs (a, b) that describe range boundaries of last recently processed queries. Such 2-D representation is more detailed than 1-D one given by the including function proposed in [1, 2]. In the proposed approach we use clustering of query range boundary values (like in [3]) for adapting the multi-histogram to historical data about conditions of processed range queries. This allows to divide whole domain of multi-histogram and to use simple equi-width histograms (called sub-histograms) in obtained sub-domains (there is no usage of sub-histogram in [3]).

The contributions of the paper are as follows:

- query workload aware multi-histogram representation of an attribute values distribution,
- methods of improvement of sub-histograms' resolutions (also partially adapted to query workload) by increasing them in domain regions of high variability of $\text{PDF}(x)$.

2 Description of the Proposed Method and the Example of Usage

2.1 Exemplary Attribute Values Distribution

The proposed method will be presented by using a sample distribution of x attribute [3]. To build an exemplary distribution representation we use a pseudorandom generator based on superposition of $G = 4$ Gaussian clusters with bounded support (limited to $[0, 1]$), where parameters of used univariate truncated normal distributions are shown in table 1. The relevant PDF is defined as follows:

$$\text{PDF}(x) = \sum_{i=1}^{G} p_i \, \text{PDF}_{\text{TN}}(x, m_i, \sigma_i, 0, 1) \tag{2}$$

The distribution consists of two narrow clusters (no 3, 4 with small sigmas) and two wide ones (no 1, 2). Of course, we may use here any type of 1-D distribution, based not only on Gaussian clusters.

Table 1. Parameters of clusters used in the definition of exemplary PDF of x attribute [3]

Cluster no	1	2	3	4
p_i		0.25	0.25	0.3	0.2
m_i		0.2	0.8	0.6	0.8
σ_i		0.12	0.12	0.001	0.01

In the further consideration we will use a high resolution equi-width histogram which is based on $N = 100$ buckets. We assume that it will describe $\mathrm{PDF}(x)$ with enough accuracy. To build this histogram we used 10000 samples of x values that were generated using $\mathrm{PDF}(x)$. The histogram uses a series of obtained f_i values (series of frequencies of falling a x values in the i-th bucket) where $i = 1, \ldots, N$. It defines a probability density function:

$$f(x) = \frac{1}{h} f_i I_i(x) \wedge I_i(x) = \begin{cases} 1 \text{ if } x \text{ belongs to the } i\text{-th bucket} \\ 0 \text{ otherwise} \end{cases} \tag{3}$$

where $h = (\max(x) - \min(x))/N$ is a width of buckets of the histogram. The probability density function $f(x)$ is presented in Fig. 1. It will be called a high resolution referential histogram.

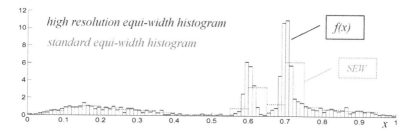

Fig. 1. Referential representation of x attribute values distribution – the high resolution equi-width histogram with $N = 100$ buckets (solid lines); SEW – standard equi-width histogram with $B = 20$ buckets (dotted lines) further defined in section 2.2

This histogram is named the high resolution one because other considered-below low resolution histograms will have significantly less number of buckets ($B \ll N$). The histogram presented in Fig. 1 (solid lines) will be used as temporary referential accurate distribution representation. It will be used for obtaining exact values of selectivities for any query ranges during creating standard equi-width histograms or multi-histograms.

2.2   Standard Equi-width Histogram

Let us use standard equi-width histogram (with B buckets) as a well-known typical representation of the distribution of x attribute values. It will be called *SEW*. Bucket's boundaries of *SEW* are uniformly distributed along the x domain. $B = 20$ is the assumed number of buckets in our exemplary histogram shown in Fig. 1 (dotted lines).

In the paper, we will try to find a better distribution representation (more accurate for selectivity estimations) subject to the assumed value of number of buckets (B) and taking into account an additional information about a distribution of previously processed range queries.

2.3   Exemplary of Distribution of Range Query Condition Bounds

Let us assume that we have information about a distribution of boundaries (a_j, b_j) of previously processed range queries $Q_j(a_j < x < b_j)$, where $0 \leq a_j \leq b_j \leq 1$. We assume that we have a sample – a set named *Qset* – which consists of pairs (a_j, b_j) for $j = 1, \ldots, M$ that come from conditions of M last processed range queries. Our exemplary *Qset* presented in Fig. 2 has $M = 20$ pairs [3]. 16 of them (circles) are highly clustered in so-called hot regions A, B, C). Zoomed parts of domain $a \times b$ were shown in Fig. 2 for presenting hot regions.

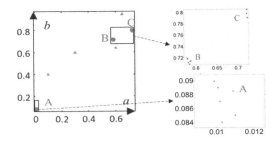

Fig. 2. *Qset* – the set of boundaries of recently processed range query – the exemplary set of pairs (a_j, b_j) for $j = 1 \ldots M(M = 20)$; A, B, C – hot regions [3]

2.4   Clustering Range Query Boundaries from Learning Set

To take into account a distribution of boundaries of query ranges we will use centers of some clusters that were built from values of a_j and b_j as some boundaries of buckets in a new histogram.

To obtain the error-optimal number of clusters we use K-fold cross validation method. In the k-th step of K-fold procedure (where $k = 1, \ldots, K$) we divide *Qset* into a learning set $Qset\_learn_k$ and a testing one $Qset\_test_k$ [3]. $Qset\_learn_k$ will be used for obtaining some boundaries of new histogram's buckets. $Qset\_test_k$ will be used for validation of the new histogram using some selectivity estimation error metrics.

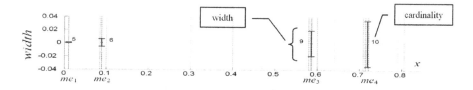

Fig. 3. Distribution of values from S_k which consists of either a or b values from $Qset\_learn_k$ – dashed lines; Medians of four accepted clusters (with their cardinality and width)

Let us assume that K equals 5 in our example. Thus $Qset\_learn_k$ consists of 16 boundaries pairs. Using $Qset\_learn_k$ we build a vector S_k. S_k has a 32 values (either a_j or b_j) that all come from $Qset\_learn_k$. Elements of S_k was presented in Fig. 3 (dashed lines).

By applying clustering procedure for S_k we may obtain a few clusters. In the example we use Fuzzy C-means algorithm (FCM) [4]. After that we eliminate some so-called weak clusters (clusters with relatively low cardinality or too wide clusters i.e. with relatively high values of standard deviation) [3]. For our example we get $C_{acc} = 4$ accepted clusters (Fig. 3).

We will use centers of accepted clusters i.e. medians $me_1, \ldots, me_{acc} = 0.0103$, 0.0888, 0.586, 0.7190 as some buckets of the new histogram. Those steps are described more detailed in [3].

2.5   Creating Equi-width-Based Multi-histogram

A new type of histogram – a multi-histogram denoted by MH – is constructed as a series of equi-width sub histograms (sH) embedded in intervals defined by the centers of clusters obtained from historical data about the distribution of boundaries of query ranges.

Due to C_{acc} clusters, we have $C_{acc} + 1$ sub-histograms. They are located between C_{acc} cluster centers, i.e. $\min(x), me_1, \ldots, me_{acc}, \max(x)$. Each equi-width sub histogram sHr is described by: s_r – a start value, e_r – an end value, B_r – a number of buckets where $r = 1, \ldots, C_{acc} + 1$ and $e_r = s_r + 1$ for $r \leq C_{acc}$, $s_1 = \min(x), e_{C_{acc}+1} = \max(x)$.

Let us assume that B is a given total number of bucket in the multi-histogram (i.e. in all sub-histograms). Thus:

$$B = \sum_{r=1}^{C_{acc}+1} B_r. \tag{4}$$

The multi-histogram has $B + 1$ buckets boundaries. $C_{acc} + 2$ boundaries are already defined by domains of sub-histograms, i.e. set of pairs (s_r, e_r), and $\min(x)$, and $\max(x)$. Remaining $B + 1 - (C_{acc} + 2)$ boundaries should be distributed among sub-histograms.

To obtain a final multi-histogram definition we should propose values $B_r - 1$, i.e. numbers of internal boundaries in each sub-histogram sH_r.

In this first approach to distributing locations of $B - C_{acc} - 1$ boundaries we assume that those locations should be almost uniformly distributed. Let us denote $L_r = e_r - s_r$ as a width of the sub-histogram sH_r, and $L = \max(x) - \min(x) = \sum_{r=1}^{C_{acc}+1} L_r$ as a width of the whole multi-histogram. We assume here that $B_r - 1$ should by approximately proportional to a relative width of the r-th sub-histogram:

$$B_r - 1 \approx A \frac{L_r}{L} \tag{5}$$

where A is some unknown constant.

Using (4) and (5) we may obtain A and B_r:

$$A \approx B - C_{acc} - 1, \tag{6}$$

$$B_r - 1 \approx (B - C_{acc} - 1) \frac{L_r}{L}. \tag{7}$$

In (5)–(7) we used symbol $\approx$ because B_r is a natural number so, in fact, we numerically find the optimal series of $B_1, \ldots, B_r, \ldots, B_{Cacc+1}$ and $B_r \in \mathcal{N}$ that minimizes some square evaluation function $F(B_1, \ldots, B_r, \ldots, B_{Cacc+1}) = \sum_{r=1}^{C_{acc}+1} ((B - C_{acc} - 1) \frac{L_r}{L} + 1 - B_r)^2$ subject to (4).

Having B_r we may construct all sub-histograms i.e. the final multi-histogram, using values of the high-resolution referential histogram (which is shown in Fig. 1).

To evaluate an accuracy of any histogram H we use the following error metrics:

– a relative selectivity estimation error for Q (a given condition range query):

$$RelErrSel_H(a,b){=}RelErrSel_H(Q(a < x < b)){=}\frac{|\widehat{sel_H}(Q){-}sel(Q)|}{sel(Q)} \cdot 100\%, \tag{8}$$

– a mean relative selectivity estimation error for QS (a given set of conditions):

$$MeanRelErrSel_H(QS) = \text{mean}_{(a,b)\in QS} \ RelErrSel_H(a,b). \tag{9}$$

$\widehat{sel_H}$ denotes an approximated selectivity value calculated with a H histogram. H is SEW (standard equi-width histogram) or MH (multi-histogram). sel denotes an exact value of selectivity calculated with the high-resolution referential histogram from Fig. 1.

Using (9) and the testing set $Qset\_test_k$ (see section 2.3) as QS we obtain: $MeanRelErrSel_{MH}(Qset\_test_k) \approx 15.8 < MeanRelErrSel_{SEW}(Qset\_test_k) \approx 31.7$. Thus, in our example the multi-histogram (MH shown in Fig. 4a) gives better accuracy (than SEW) in selectivity estimations for range query conditions from $Qset\_test_k$.

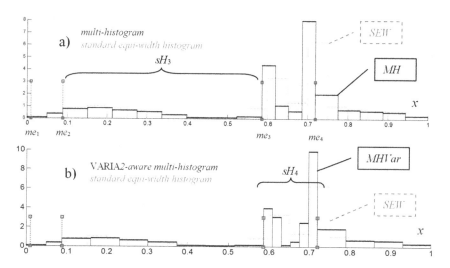

Fig. 4. *SEW* – standard equi-width histogram (dashed lines); locations of borders between sub histograms (vertical dotted lines) – series of medians; multi-histograms: a) *MH* – simple multi-histogram (solid lines), b) *MHVar* (alternatively called VARIA2-aware) – frequencies variability aware multi-histogram (solid lines) further defined in section 2.7

2.6 Improving Multi-histogram by Eliminating Boundaries from those Sub-histograms that Describe almost Uniform Distributions

To improve a multi-histogram we propose to take into account variability of frequencies (describing x distribution) in process of obtaining a distribution of numbers of internal buckets (B_r) in sub-histograms (sH_r).

In this second approach to built a multi-histogram we will assume that after finding C_{acc} we do not introduce any internal boundaries into such sub-histogram where there exist no significant changes of frequencies f_i (eq. (3) and Fig. 1) that belong to the domain of this sub-histogram. After the proposed step, such sub-histogram will have only a one bucket. So there will be more boundaries to distribute among remaining sub-histograms.

To select such sub-histogram sH_r we propose to use such condition:

$$\text{VARIA1}_r = \frac{\text{std}(f(x))|_{x \in [s_r, e_r]}}{\text{MV}_r} \leq \text{THR} \tag{10}$$

where THR is some threshold value (e.g. from $1\% \sim 10\%$), and

$$\text{MV}_r = \text{mean}(f(x))|_{x \in [s_r, e_r]} = \int_{s_r}^{e_r} f(x)dx, \tag{11}$$

and $\text{std}(f(x))|_{x \in [s_r, e_r]} = \left(\int_{s_r}^{e_r} (f(x) - \text{MV}_r)^2 dx \right)^{\frac{1}{2}}$.

For the exemplary sub-histograms we obtain $(\text{VARIA1}_r)_{r=1}^{C_{acc}+1} = (0.01, 72.3,$ 89.3, 97.6, 114) what allows to select only the first sub-histogram ($r = 1$) according to the assumed threshold value (THR = 10%). This sub-histogram i.e. sH_1 will not be taken into account in further distributing of $B - C_{acc} - 1$ locations of boundaries (described in section 2.7). It will have only a one bucket.

In fact, the above-considered try of improvement of this MH does not change the distribution of $B - C_{acc} - 1$ locations of boundaries in our example (because the selected sub-histogram sH_1 already has only a one bucket) but it may have an impact on the distribution of boundaries locations for other distribution of x values or values in $Qset$'s elements.

2.7   Improving Multi-histogram by Increasing Resolution of Selected Sub-histograms

In this section we propose more advanced metrics of variability of distribution of frequencies f_i within a domain $[s_r, e_r]$ of some sub-histogram sH_r. Using it we may increase a resolution of some sub-histograms for high variability of frequencies.

Till now we have assumed that we take into account a distribution of query conditions i.e. a 2-dimensional distribution of pairs (a, b) (samples from this distribution are given by $Qset\_learn_k$) to find extreme boundaries of sub-histograms (series of cluster's medians). Now let us also take into account a distribution of widths or query ranges, i.e. a 1-dimensional distribution of $z = b - a$. This distribution of z will have an impact on a distribution of number of internal boundaries of sub-histograms. Such approach may allows to partially adapt a multi-histogram to a (possible in future) shifted distribution of query ranges. The exemplary discrete z values distribution (obtained from $Qset\_learn_k$) is shown in Fig. 5.

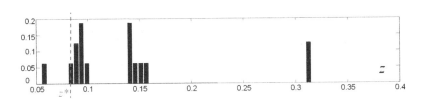

Fig. 5. The distribution of $z = b - a$, i.e. the distribution of widths of query ranges obtained from $Qset\_learn_k$

z distribution allows to obtain a window size for further analysis i.e. we chose as the window size such maximal z^* that:

$$P(z \geq z^*) = p \tag{12}$$

where p is an assumed value of confidence level.

Let us assume $p = 0.9$. Then for the considered example (i.e. the z distribution from Fig. 5) we obtain $z^* = 0.082$ (dashed line in Fig. 5).

We will use the value of z^* to determine a maximal resolution with which we will evaluate a variability of frequencies f_i (see section 2.1) within a domain of a sub-histogram.

Let us define a window function $w(t)$ with width equals z^*:

$$w(t) = 1(t) - 1(t - z^*) \tag{13}$$

where $1(t)$ is the step function.

We will consider a series of queries with ranges that are included in a domain of a sub-histogram sH_r and with range lengths equal z^*. This means that we will consider window queries $Q_w(t < x < t + z^*)$ for all $t \in [s_r, e_r - z^*]$.

If we assume for a little that a sub-histogram sH_r has only a one bucket then such histogram contains only one single value equals MV_r (given by (11)). Let us find a selectivity of Q_w for given t using such one-bucket sH_r:

$$sel_{1bck\text{-}sHr}(Q_w(t < x < t + z^*)) = \int_t^{t+z^*} MV_r d\tau = MV_r \, z^* = \text{const.} \tag{14}$$

We may find a selectivity of Q_w for given t using $f(x)$:

$$sel(Q_w(t < x < t + z^*)) = \int_t^{t+z^*} f(\tau)w(\tau - t)d\tau \tag{15}$$

as a function convolution of f and w on interval $[t, t + z^*]$.

Let us define a scaled selectivity formula as follows:

$$\frac{sel(Q_w(t < x < t + z^*))}{z^*} = \frac{1}{z^*} \int_t^{t+z^*} f(\tau)w(\tau - t)d\tau. \tag{16}$$

We may consider the scaled $sel(Q_w)$ as a moving average filter. The result of applying the filter for the exemplary frequencies f_i (see section 2.1) and z^* equals 0.08 is shown in Fig. 6.

We introduce a new metrics of variability of frequencies within a sub-histogram sH_r (using a filter based on value of z^*) as mean selectivity estimation absolute error:

$$\begin{aligned}
&VARIA2_r = \int_{s_r}^{e_r - z^*} |sel_{1bck\text{-}sHr}(Q_w(t<x<t+z^*)) - sel(Q_w(t<x<t+z^*))| dt, \\
&VARIA2_r = \int_{s_r}^{e_r - z^*} |MV_r \, z^* - sel(Q_w(t < x < t + z^*))| dt, \\
&VARIA2_r = frac1z^* \int_{s_r}^{e_r - z^*} |MV_r - \frac{sel(Q_w(t<x<t+z^*))}{z^*}| dt.
\end{aligned} \tag{17}$$

The presented above definition of $VARIA2_r$ ratio (17) is valid for $e_r - s_r \geq z^*$. For a narrow sub-histogram sH_r where $e_r - s_r < z^*$, a value of $VARIA2_r$ equals 0.

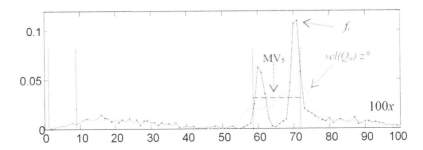

Fig. 6. Applying scaled selectivity $sel(Q_w)/z^*$ as a moving average filter ($z^* = 0.08$); frequencies f_i (connected by solid lines); averaged frequencies after applying the scaled selectivity filter (dotted lines)

Let us denote VARIA2 defined as follows:

$$\text{VARIA2} = \sum_{r=1}^{C_{acc}+1} \text{VARIA2}_r. \tag{18}$$

Using $\text{VARIA2}_r/\text{VARIA2}$ ratios we may refine the formula (5) for obtaining $B_r - 1$ as follows:

$$B_r - 1 \approx A \left(\alpha \frac{L_r}{L} + \beta \frac{\text{VARIA2}_r}{\text{VARIA2}} \right) \tag{19}$$

where α – a weight of an importance of a sub-histogram domain width and β – a weight of an importance of variability of frequencies within this sub histogram ($\alpha, \beta \geq 0$, $\alpha + \beta = 1$).
Using (4) and (19) we may obtain:

$$A \approx \frac{B - C_{acc} - 1}{\sum_{r=1}^{C_{acc}+1} \left(\alpha \frac{L_r}{L} + \beta \frac{\text{VARIA2}_r}{\text{VARIA2}} \right)}. \tag{20}$$

A multi-histogram which numbers of SHr's boundaries (B_r) depend either on L_r and VARIA2_r will be called $MHVar$ (or VARIA2-aware histogram).
Here, in the considered example, we propose equal impacts of the both ratios i.e. $\alpha = \beta = 1/2$. Thus using (19) and (20) and $B_r \in \mathcal{N}$ we obtain ($B_1, \ldots, B_r$, $\ldots, B_{Cacc} + 1) = (1, 2, 7, 6, 4)$. The new $MHVar$ based on (1, 2, 7, 6, 4) is presented in Fig. 4b (solid lines). The 4-th sub-histogram of $MHVar$ has 6 buckets (domain of sH_4 is a region of higher variability of frequencies f_i). It is the greater value than 4 – the number of bucket of the 4-th sub-histogram of MH (shown in Fig. 4a).
To evaluate $MHVar$ we again use (9) and the testing set $Qset\_test_k$ (section 2.3): $MeanRelErrSel_{MHVar}(Qset\_test_k) \approx 10.1 < MeanRelErrSel_{MH}$ $(Qset\_test_k)$. For $Qset\_test_k$ applying the VARIA2-aware histogram ($MHVar$) gives a little better selectivity estimation accuracy than applying the previously obtained multi-histogram (MH).

2.8   Selecting Number of Clusters of Range Boundaries to Obtain Error-Optimal Multi-histogram

K-fold cross validation method allows to find a value of the error-optimal C_{opt} amnong C values – numbers of clusters in S_k, where $C \in \mathcal{N}$ and $1 \leq C \leq B-1$ ($B-1$ because two of boundaries from all $B+1$ boundaries are already defined by $\min(x)$ and $\max(x)$). For each value of C we obtain a value of accepted clusters C_{acc} ($0 \leq Cacc \leq C$) after eliminating weak clusters.

Averaged MeanRelErrSel (which uses (9)) allows finding the optimal C_{opt} for our example as we can see in Fig. 7. We show here only values $3 \leq C \leq 9$ (not $1 \leq C \leq 19$) because for the other C values we have C_{acc} values are equal 0 (there are no accepted clusters).

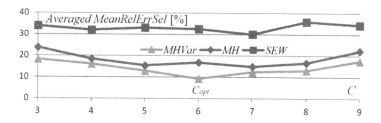

Fig. 7. K-fold cross validation results: the dependency between Averaged MeanRelErrSel (calculated for histograms: MHVar, MH, SEW) and the number of clusters equals C

For our example we obtain $C = 6$ (with corresponding $C_{acc} = 4$) as the error-optimal value, i.e. with the smallest Averaged MeanRelErrSel equals 10.9%.

MHVar based on $C_{opt} = 6$ clusters proved to be the most accurate representation for the considered example i.e. for the exemplary attribute distribution (given by $f(x)$ in Fig. 1) and for the exemplary distribution of boundaries of query ranges (given by Qset in Fig. 2).

3   The Algorithm for Obtaining Multi-histogram

The proposed method allows obtaining the error-optimal multi-histogram MHVar for an arbitrary given value of B i.e. the number of MHVar's buckets.

We assume that we have available Qset i.e. M pairs of boundaries of last recently processed range query conditions.

The proposed method consists on the following steps:

1. Create a temporary referential representation of x attribute values distribution i.e. build a high resolution equi-width histogram which describes $f(x)$.
2. Create SEW – a low resolution standard equi-width histogram using the high resolution histogram.

3. For each C from $1, \ldots, B-1$ (using K-fold cross validation method) obtain an error metrics value i.e. an *Average MeanRelErrSel* which evaluates *MHVar* histogram based on C clusters:

   3.1. Cluster a learning set *Qset_learn* (formed from values of *Qset*) using FCM.

   3.2. Eliminate weak clusters, i.e. obtain C_{acc} accepted clusters (this determines $C_{acc} + 1$ sub-histograms sH_r where $r = 1, \ldots, C_{acc} + 1$).

   3.3. Assume an only one bucket in those sub-histograms sH_r where VARIA1$_r$ $\leq$ THR (eq. (10)).

   3.4. Distribute $B - C_{acc} - 1$ locations of bucket boundaries among sub-histograms according to (19) and (20) using VARIA2$_r$ (but omitting the sub-histograms with one bucket that were selected in 3.3).

   3.5. Having bucket's boundaries, create *MHVar* using the high resolution histogram.

   3.6. Obtain *MeanRelErrSel* for *MHVar* and *SEW* using query ranges from a testing set (i.e.: *Qset_test = Qset \ Qset_learn*).

   3.7. Aggregate values of *MeanRelErrSel*.

4. Choose this *MHVar* multi-histogram which has the smallest *Averaged Mean-RelErrSel* (if *Averaged MeanRelErrSel* for *MHVar* is less than the one for *SEW*, else choose *SEW*).

The proposed method of obtaining *MHVar* is designated to be invoked during update statistics (not during on-line query processing) so it is rather not a time-critical operation (in opposite to selectivity calculation).

4 Conclusions

The proposed method of range query selectivity estimation is based on a multi-histogram – *MH*. A *MH* is such representation of an attribute values distribution which additionally takes into account a distribution of range boundaries of recently processed queries. Query workload is reflected in a division of *MH* into sub-histograms by using centers of clusters of range boundaries values. These results in creating equi-width sub-histograms embedded in *MH*. In the paper we also propose an improved multi-histogram – *MHVar* – which additionally supports an increased resolution in sub-histograms based on regions of high variability of an attribute values distribution. In *MHVar* we do not distribute buckets among the sub-histogram where attribute distribution is almost uniform. Additionally, having a knowledge of past query workload we may set a window size in some moving average filter and measure the variability of frequencies in sub-histograms. This allows refining the distribution of boundaries among selected sub-histograms embedded in *MHVar*.

In future we plan to confirm advantages of *MHVar* (in accuracy of range query selectivity estimations) against different query workload profiles in more experiments.

The future work may concentrate on improving of handling historical data about query workload. In this approach we store M last processed queries. This may be too short-time description of a past query workload. Thus in future, we plan to use the micro clustering technique [12] which allows taking into account some impact of older queries.

Another direction of research may be considering the other type of histogram as a sub-histogram (i.e. not necessary equi-width one) like equi-high one or V-optimal one.

References

1. Augustyn, D.R.: Query-condition-aware histograms in selectivity estimation method. In: Czachórski, T., Kozielski, S., Stańczyk, U. (eds.) Man-Machine Interactions 2. AISC, vol. 103, pp. 437–446. Springer, Heidelberg (2011), http://dx.doi.org/10.1007/978-3-642-23169-8_47

2. Augustyn, D.R.: Query-condition-aware v-optimal histogram in range query selectivity estimation. Bulletin of the Polish Academy of Sciences. Technical Sciences 62(2), 287–303 (2014), http://dx.doi.org/10.2478/bpasts-2014-0029

3. Augustyn, D.R.: Query selectivity estimation based on improved V-optimal histogram by introducing information about distribution of boundaries of range query conditions. In: Saeed, K., Snášel, V. (eds.) CISIM 2014. LNCS, vol. 8838, pp. 151–164. Springer, Heidelberg (2014), http://dx.doi.org/10.1007/978-3-662-45237-0_16

4. Bezdek, J.C.: Pattern Recognition with Fuzzy Objective Function Algorithms. Kluwer Academic Publishers, Norwell (1981)

5. Bruno, N., Chaudhuri, S., Gravano, L.: Stholes: A multidimensional workload-aware histogram. SIGMOD Rec. 30(2), 211–222 (2001), http://doi.acm.org/10.1145/376284.375686

6. Chen, C.M., Roussopoulos, N.: Adaptive selectivity estimation using query feedback. SIGMOD Rec. 23(2), 161–172 (1994), http://doi.acm.org/10.1145/191843.191874

7. He, Z., Lee, B.S., Wang, X.S.: Proactive and reactive multi-dimensional histogram maintenance for selectivity estimation. J. Syst. Softw. 81(3), 414–430 (2008), http://dx.doi.org/10.1016/j.jss.2007.03.088

8. Ioannidis, Y.: The history of histograms (abridged). In: Proc. of VLDB Conference (2003)

9. Khachatryan, A., Müller, E., Stier, C., Böhm, K.: Sensitivity of self-tuning histograms: Query order affecting accuracy and robustness. In: Ailamaki, A., Bowers, S. (eds.) SSDBM 2012. LNCS, vol. 7338, pp. 334–342. Springer, Heidelberg (2012), http://dx.doi.org/10.1007/978-3-642-31235-9_22

10. Luo, J., Zhou, X., Zhang, Y., Shen, H.T., Li, J.: Selectivity estimation by batch-query based histogram and parametric method. In: Proceedings of the Eighteenth Conference on Australasian Database, ADC 2007, vol. 63, pp. 93–102. Australian Computer Society, Inc., Darlinghurst (2007), http://dl.acm.org/citation.cfm?id=1273730.1273741

11. Srivastava, U., Haas, P.J., Markl, V., Kutsch, M., Tran, T.M.: Isomer: Consistent histogram construction using query feedback. In: Proceedings of the 22nd International Conference on Data Engineering, ICDE 2006, pp. 39–51. IEEE Computer Society, Washington, DC (2006), http://dx.doi.org/10.1109/ICDE.2006.84

12. Zhang, T., Ramakrishnan, R., Livny, M.: Birch: An efficient data clustering method for very large databases. SIGMOD Rec. 25(2), 103–114 (1996), http://doi.acm.org/10.1145/235968.233324

Analysis of the Effect of Chosen Initialization Parameters on Database Performance

Wanda Gryglewicz-Kacerka[1] and Jarosław Kacerka[2(✉)]

[1] Państwowa Wyższa Szkoła Zawodowa we Włocławku,
3 Maja 17, 87-800 Włocławek, Poland
wkacerka@gmail.com
[2] Instytut Automatyki Politechniki Łódzkiej,
Stefanowskiego 18/22, 90-924 Łódź, Poland
jaroslaw.kacerka@p.lodz.pl

Abstract. The paper presents an analysis of the influence of chosen initialization parameters on database efficiency and processing time. Experimental research is performed on a real banking system model which includes data structures characteristic to banking systems - structures supporting bank accounts, and reflecting a typical bank organization. Experimental results are compared with results of a real banking system.

Keywords: Database · Database efficiency

1 Introduction

Obtaining maximal performance of a complex system like Oracle database is a complicated task. It requires database tuning, optimization of SQL queries, as well as appropriate configuration of the hardware running the DBMS.

System performance is indicated by response time and throughput. Response time is a time measured between user request and system reply, while throughput shows the number of tasks (business transactions) processed in a time unit. For the overall performance of a system based on object-relational Oracle database performance of the database is critical. In order to achieve desired performance one has to consider two aspects [8]:

- code optimization and database design reflecting a relational or object-relational model,
- selection of system parameters optimal for certain installation.

Oracle has numerous initialization parameters [7]. Part of them is responsible for database performance. Selection of these parameters values provides a way of system tuning for actual application, user and operating system requirements.

Oracle DBMS is characterized by high functionality including mechanisms assisting the administrator in the process of adjusting parameters important for database operation [1–4, 6, 9]. It can be assumed that only some of the available publications contain experimental results. In [3] Boronski examined the

© Springer International Publishing Switzerland 2015
S. Kozielski et al. (Eds.): BDAS 2015, CCIS 521, pp. 60–68, 2015.
DOI: 10.1007/978-3-319-18422-7_5

effect of db_file_multiblock_read_count parameter (which is responsible for setting the maximal number of blocks read in a single operation) on the SQL query processing time. experiments start with a default value of parameter which is later modified. Described results show clearly that the default setting of db_file_multiblock_read_count parameter is suboptimal in terms of time required to read the complete table. Optimal values of the parameter are proposed for several cases, based on experimental results.

Oracle database optimization is a complicated process and requires a wide range of adjustments - starting from hardware running the DBMS to optimization of SQL queries. Li at al. [4] divided the process into four levels of optimization: operation system, database configuration, database design, SQL application level. Presented in [4] are specific methods and measures aimed at Oracle database performance adjustment. Proposed is also a method of calculating a metric of shared pool size optimality based on cache hit ratio. Qingfeng and Weishan inspected [9] various ways of optimization of a sample application and their effect on database response time. Some adjustments on database design and SQL level may be performed automatically. Belknap at al. [1] present a self-tuning mechanism introduced in Oracle 11g which - by analyzing performance statistics - may automatically correct database designers errors such as missing indexes and propose corrections to SQL queries. Suggested optimizations comply with general rules given e.g. in [5]. Berkovic at al. inspected the effect of using bulk operations on performance in connection to allocated memory, i.e. system and process global area sizes (PGA and SGA) [2]. Ling at al. also inspected the effect of PGA and SGA size on performance [6].

2   Subject of Study

The file containing database parameters is automatically generated and filled with starting values of required initialization parameters. Values of initialization parameters are automatically set to defaults (depending on database and operating system versions) which are sufficient to start up the database, however they require changes in order to optimize database operation. Database administrator may change individual parameters in order to:

- optimize processing performance by adjusting memory structure to database requirements, e.g. selecting a certain number of memory buffers, assigning memory for sorting, hash-join, etc.
- change some of the default values,
- set database constraints, e.g. maximal number of database users, number of simultaneous transactions, public rollback segments, etc.

Depending on the characteristics of processing required by an application various criteria may be applied:

- maximal processing power,
- maximal processing performance,
- transaction count per time unit,
- minimal transaction time.

For an application which generates millions of small transactions every day most important is the criterion of maximal processing power allowing for performance increase. For decision support systems performance equals to short response time. At the stages of designing and construction of the system application designers have a possibility of selecting such combination of system resources and Oracle mechanisms, which will result in the best realization of system tasks and requirements. Reactive tuning regards avoiding situations which could severely impact system operation performance. It is important that Oracle server and application be tuned to operate best on specific platform and exploitation conditions. Full use of Oracle server capabilities should result in achieving top system operation performance.

System processing performance may decrease during operation because of expansion of data volume, which will degrade the whole system. Therefore it is important to tune the system during normal operation.

3   Test Database Model

Database chosen for tests is a real commercial system for complex centralized banking. A typical bank structure is implemented in the system. In such model database various tests were performed in order to show what is the effect of chosen configuration parameters on processing performance.

The following assumptions were made: known is the total data size stored in real database as well as the relation between row counts in particular tables.

Only the following elements of database storage structure were included in the basic database model:

- tablespace set,
- about 500 tables storing all the data,
- indexes,
- primary, foreign and unique keys,
- set of constraints regarding certain table fields.

Also included were certain descriptional values:

- set of R_{Pi} parameters describing the structure of every i-th tablespace,
- rowcount for i-th table W_i,
- total database rowcount W,
- set of rowcount ratios between relation-connected tables Z_{nm} where $Z_{nm}=W_n/W_m$.

4   Simplified Model

A set of eight tables comprising database core were taken from the real database and appended by 5 tables for storing simple client registration data in order to automatically reduce the set of indexes, keys and constraints. The rowcount for individual tables chosen for the simplified model is limited as well. Assumed is a new set of tablespaces and parameters to characterize their structure.

The simplified model has the following structure:

- data tablespace storing all database tables with all the contained data,
- indexes tablespace storing all database indexes,
- swycof tablespace storing rollback segments,
- pstym tablespace storing temporary segments.

In such model database various benchmarks were run in order to show the influence of chosen configuration parameters and their storage on processing performance.

In order to perform the research a special tool was designed which among others comprises modules used to:

- analyze the results of performed experiments,
- display plots of all measured parameters as a function of time.

5   Measurements

The primary measured value was the time needed to complete a single test. To eliminate random errors every test was performed multiple times until repeatable results were obtained. In the same way the influence of initial database buffer filling was compensated; for large SGA it could noticeably affect the results.

To correctly interpret the time measurement results additional information was required: thorough operation statistics for the database and operating system. This information was obtained from:

- Performance monitor for the Windows system running the database
- Oracle Statspack package collecting data from database system perspective

Monitored operating system performance included:

- Bytes written to disk,
- Bytes read from disk,
- Virtual memory file activity.

Database operation statistics were obtained from statspack package, which may generate a database activity report for a chosen period of time. For the analyzed parameters the following statistics were most useful:

- buffer cache hit ratio,
- library cache hit ratio,
- memory sorts ratio,
- number of checkpoint calls,
- number of block reads,
- number of consistent block gets.

The above parameters were useful because of their impact on disk activity and memory caching level.

6 Test Research

The following parameters were examined during the tests: DB_BLOCK_SIZE, DB_BLOCK_BUFFERS, SHARED_POOL_SIZE. Due to the high importance of block size for Oracle database operation the following tests were conducted:

- TestA contained large number of queries requesting or modifying little amount of data; in such queries access to specific row is usually accomplished by direct addressing with ROWID value obtained from appropriate index leaf,
- TestB contained mostly queries asking for large amount of data; database access is then realized by way of FULL SCAN, i.e. reading whole tables.

The above tests were to investigate the influence of DB_BLOCK_SIZE parameter changes on processing performance. Anticipated test results are connected to their characteristics.

TestA requested little amounts of data contained in rows spread across the whole tablespace. As a result of a single query to a database with *large block* a number of blocks containing rows with data irrelevant to the query will be inserted into SGA. Thus the SGA memory will be partially filled with unnecessary data. For a *smaller block* the amount of data not relevant to the query is smaller, thus the advantage of using of smaller blocks.

TestB requires reading a large number of adjacent rows in every query. The advantage of large block size results from getting larger amount of data in one read. The time effect may be less visible here and dependent on buffer size and method of buffering employed in the file system.

Fig. 1 and Fig. 2 present the results of experimental research including:

- Test completion time depending on DB_BLOCK_SIZE,
- Number of write operations during the test as a function of DB_BLOCK_SIZE,
- Number of read operations during the test as a function of DB_BLOCK_SIZE.

Average number of read and write operations per second are characteristics of the operating system performance from which a conclusion may be drawn that in a database whose operation is dominated by a large number of little queries it is advantageous to use the smallest data block possible. Plots of blocks written and read by database should be interpreted as showing the number of disk access operations.

The value of DB_BLOCK_BUFFERS parameter determines the number of database blocks cached in SGA memory by Oracle. Basing on the assumption that processing performance is proportional to the amount of in-memory buffering (because of shorter access time of RAM than that of a hard drive) one should try to set this parameter to maximal value. The restriction is the amount of physical memory which can be assigned to one Oracle instance without paging and swapping being used, i.e. secondary caching of operational memory in a swap-file. Value of this parameter has a visible effect on processing time. The following tests were run:

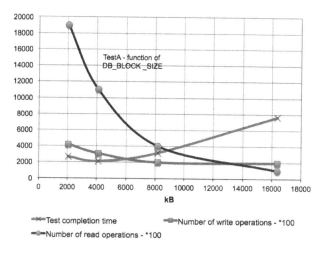

Fig. 1. Measured parameters as a function of DB_BLOCK_SIZE (TestA)

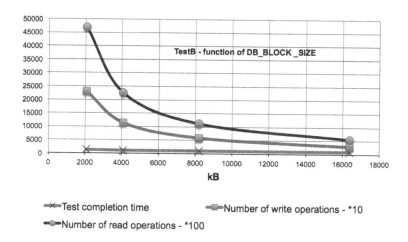

Fig. 2. Measured parameters as a function of DB_BLOCK_SIZE (TestB)

- TestC - many operations processing small number of rows in a single query,
- TestD - only queries modifying a large number of rows without committing,
- TestE - only queries modifying a large number of rows with commit.

The influence of DB_BLOCK_BUFFERS parameter value on processing performance between SGA and disk memory is shown in Fig. 4.

Analysis of the influence of DB_BLOCK_BUFFERS value on data processing time shows that increase of database blocks cached in memory (higher value of the DB_BLOCK_BUFFERS parameter) results in slightly increased processing performance for all the above tests.

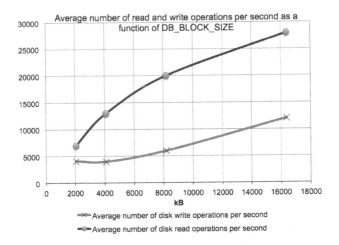

Fig. 3. Average number of read and write operations per second as a function of DB_BLOCK_SIZE (TestA)

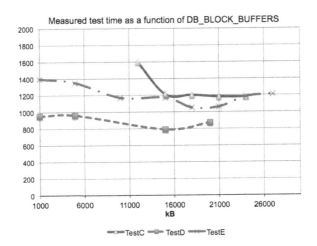

Fig. 4. Measured test time as a function of DB_BLOCK_BUFFERS

Measured data shown in Fig. 4 allows to estimate an optimal value od DB_BLOCK_BUFFERS parameter above which a drop of processing performance occurs. The decrease in performance is caused mostly by the limits of hardware configuration and the necessity to use disk swap file (additional read and write operations).

The value of SHARED_POOL_SIZE parameter determines the size of shared area containing shared cursors, stored procedures, control structures, etc. Shared pool allows for reduction of processing time for same SQL queries issued by one or multiple users by buffering realization schemes for those queries.

TestF contained a large number of different queries. Queries were simple, thus required short processing time. The test was constructed to fill the whole shared pool of certain close to maximal value. In this way operation of an application consisting of N shared queries executed by M users was simulated. Such operation is common to properly designed OLTP applications based on Oracle database. In such applications queries from the outside of set N occur sporadically or do not occur at all. The aim of this test was to measure the influence of shared pool on processing performance, in particular: the influence of shortage of shared pool. The influence of SHARED_POOL_SIZE parameter value on processing time, database processing and processor load is shown in Fig. 5.

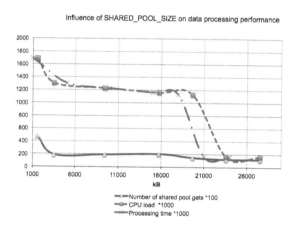

Fig. 5. Influence of shared memory on data processing performance

Test results show a distinct effect of processing performance drop - in case of shortage of shared memory the processing time increases. Performed test shows that for an application utilizing a shared queries model there exists an optimal amount of shared pool above witch no gain in performance is obtained. This value should be estimated for every production application.

7   Conclusions

Experimental results show a great influence of parameter value changes on Oracle database operation. It should be stated here, that only a thorough analysis of numerous measured values and statistics indicated a possibility of undesirable effects impeding the performance of certain processing type caused by a change of value of certain parameter. It has been shown that characteristics of the database application requires contradictory parameters tuning. For example a little data block is appropriate for OLTP applications, but not optimal for DSS applications. Additional factors that complicate tuning is the pursue of balance between minimization of database volume and reduction in I/O operations, obtaining a maximal number of concurrent transactions and efficiently balancing index trees. Some of the above problems are solved by the maker in newer database versions, e.g. introduction of variable block size, others require application designers and database administrators to have thorough experience and knowledge. After analyzing the obtained results it may be concluded that tuning of Oracle database in a purely automated manner is not possible.

References

1. Belknap, P., Dageville, B., Dias, K., Yagoub, K.: Self-tuning for sql performance in oracle database 11g. In: IEEE 25th International Conference on Data Engineering, ICDE 2009, pp. 1694–1700 (March 2009)
2. Berkovic, I., Ivankovic, Z., Markoski, B., Radosav, D., Ivkovic, M.: Optimization of bulk operation performances within oracle database. In: 2010 8th International Symposium on Intelligent Systems and Informatics (SISY), pp. 163–167 (September 2010)
3. Boronski, R.: Wplyw ustawien parametru wieloblokowego sekwencyjnego czytania danych na czas wykonywania zapytania sql w bazie danych oracle 11g. In: Modele Inzynierii Teleinformatyki. Wyd. Politechniki Koszalinskiej, Koszalin (2012)
4. Li, Q., Honglin, X., Yan, G.: Research on performance optimization and implementation of oracle database. In: Third International Symposium on Intelligent Information Technology Application, IITA 2009, vol. 3, pp. 520–523 (November 2009)
5. Lightstone, S., Teorey, T., Nadeau, T.: Physical Database Design: The Database Professional's Guide to Exploiting Indexes, Views, Storage, and More. Morgan Kaufmann Publishers Inc., San Francisco (2007)
6. Ling, Z., Qi, L., Qianyuan, Z., Wei, C.: The study of adjustment and optimization of oracle database in information system. In: 2013 Fifth International Conference on Computational and Information Sciences (ICCIS), pp. 442–445 (June 2013)
7. Oracle: Oracle database online documentation 12c, release 1 (12.1), http://docs.oracle.com/database/121/index.htm
8. Oracle Tips by Burleson Consulting: Oracle instance tuning: Examining the internals, http://www.dba-oracle.com/art_tuning_instance.htm
9. Qingfeng, D., Weishan, Z.: Performance optimization analysis based on oracle database application system. Computer Engineering 14, 91–93 (2005)

Database Under Pressure - Scaling Database Performance Tests in Microsoft Azure Public Cloud

Dariusz Mrozek[✉], Anna Paliga, Bożena Małysiak-Mrozek, and Stanisław Kozielski

Institute of Informatics, Akademicka 16, 44-100 Gliwice, Poland
dariusz.mrozek@polsl.pl

Abstract. Making changes in production database or changes in database system configuration often requires these changes to be priorly tested in a test system. This also requires to replay the original workload in the test environment by simulating client's activity on many workstations. In the paper, we show how this task can be realized with the use of many Workload Replay Agents working in Microsoft Azure public cloud. We present model and architecture of widely scalable, cloud-based stress testing environment, called CloudDBMonitor, which allows controlled execution of captured SQL scripts against a specified database in Microsoft SQL Server database management system. The stress testing environment provides the possibility to investigate how the tested database works under a large pressure generated by many simulated clients.

Keywords: Databases · Stress testing · Performance testing · Efficiency · Cloud computing · Workload replay · Scalability · Microsoft Azure

1 Introduction

Changes in an existing database system may affect performance of the system and efficiency while processing client transactions. The main risk when implementing different changes lies in the question how these changes affect the system performance, whether they will have positive or negative effect on the system, and how they influence the user experience of the database application.

Companies may want to improve their database systems in response to a number of factors and feedbacks from customers. For example, the company may notice a growing number of customers that connect to their database from client applications. Or it can notice that due to the growing amount of data the database system becomes inefficient. Or the company extends the database system toward other areas. In all these and other situations, the company may respond by introducing several types of changes.

This project was supported by Microsoft Research in USA within Microsoft Azure for Research Award.

© Springer International Publishing Switzerland 2015
S. Kozielski et al. (Eds.): BDAS 2015, CCIS 521, pp. 69–81, 2015.
DOI: 10.1007/978-3-319-18422-7_6

1.1 Use Cases

The company may decide to change the architecture of the system. Typical client-server architecture with one database server and many clients may turn out to be inefficient. A growing number of clients may put too much pressure on the single server. Single server may not cope with the number of client transactions. The company may want to switch to the architecture, which consists of a farm of servers connected by a network that work in various topologies, but is still visible to clients as one logical database server. Or due to availability reasons, the company may want to switch to the high-availability cluster that contains many nodes.

The company may also change the physical structure of the database. For example, due to maintenance requirements the database can be splitted into several data files and each file can be placed in a different location. Transaction logs can be divided to different files. The space for temporal operations can be separated from data files and located on different hard drive with the expectations to get some improvements. New indexing strategy or data partitioning schema can be adopted and the company may want check how it works for a real workload.

The company may decide to make some improvements in the code of existing stored procedures or user defined functions. Some code units recognized as inefficient can be rewritten, for example, to benefit from new indexes created in the database or as a part of the code optimization process. Transaction isolation levels can be elevated or lowered in some particular transactions.

The company may decide to migrate to a different hardware or increase computational resources available for the database server. This may include changing the number of processors or CPU cores, increasing the amount of RAM, increasing the number of hard drives available or the use of RAID storage systems, changing components of the network infrastructure, e.g. switches. Migration can be also done between various versions or editions of the database management system (DBMS). The company may migrate to new version of the DBMS or elevate the edition from Standard to Enterprise in order to get access to some advanced features.

Finally, the configuration of the DBMS or the database can be changed in order to make use of new resources. For example, database administrators of the company may change the number of CPU cores utilized by the DBMS, the minimum and maximum of RAM used, the number of concurrent batches that can be executed against the DBMS, data page size, fill factor for existing indexes, and others.

Changes in all these areas of the database environment may affect performance much. However, there is always a risk associated with such changes that some of them may not be beneficial in terms of efficiency. Therefore, they should be verified in a test environment before they are implemented in the production system. The verification requires the test environment to be established, real workload to be captured and replayed in the test database system (Fig. 1). It also requires the stress testing environment that will be able to replay the workload, monitor the database and provide appropriate statistics when the testing phase is completed. Establishing the stress testing environment that simulates the real activity of clients can be sometimes difficult due to the lack of appropriate computer resources. Cloud computing is a computing model that allows to lease the resources in minutes on a pay-as-you-go basis giving the possibility to scale the testing process on many computers.

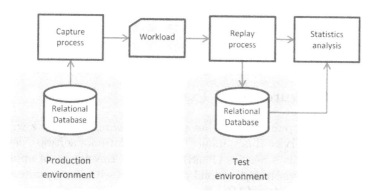

Fig. 1. Overview of workload capturing, replaying, and analysis

1.2   Related Works

Testing performance of database systems or testing application compatibility for databases after reconfigurations became one of the important problems in recent years. In terms of performance, various approaches and benchmark sets has been designed and deployed. The Transaction Processing Council (TPC) is an institution that develops standard workloads that can be used to test performance of various types of database systems, including OLTP, OLAP, cloud and Big Data systems [8,12]. On the other hand, there are also various commercial tools, like Oracle Database Replay [1] and SQL Server Distributed Replay [11], and scientific frameworks, like AgentTeam [4] and DBPerfTester@UMAP [6], created for database management and for testing workloads on various DBMSs. A commercial Oracle Database Replay enables users to replay a full production workload on a test system to assess the overall impact of system changes [1]. Similarly, Microsoft SQL Server Distributed Replay is also a commercial tool used for application compatibility testing, database performance testing, or capacity planning. It allows replaying a workload from multiple computers and better simulation of a mission-critical workload [11]. On the other hand, more sophisticated database architectures require dedicated approaches, like the one presented in [4] which uses the AgentTeam framework for dynamic database management (DDBM) and CourseMan environment for testing different DDBM protocols on heterogeneous DBMSs. In [3] authors present how to test multi-tier database applications by the use of a number of agents. And in [2] the same authors present an agent-based approach for the maintenance of database applications in terms of changes on the schema of the database underlying the application. Finally, in [6] we show application of the UMAP [7,14] multi-agent system in testing performance of DBMSs.

In this paper, we present model and architecture of the CloudDBMonitor, the stress testing and SQL workload replay environment for databases. The system has been developed in Microsoft Azure public cloud, which assures its high scalability. Before we discuss details of the system in chapter 3 and 4, we provide basic definitions regarding Microsoft Azure public cloud and Microsoft Azure programming model in chapter 2.

2   Microsoft Azure Cloud Platform

Microsoft Azure is a public cloud that provides compute resources within the Infrastructure as a Service (IaaS) model [5] and platform for software developers within the Platform as a Service (PaaS) model [5]. Developers of applications that work in the Microsoft Azure cloud are equipped with appropriate Application Programming Interface (API) that allows a programmatic creation of the application components. Among many components of the Microsoft Azure API the following elements are used in the construction of CloudDBMonitor [15]:

- **Compute** - refers to compute capabilities of the Microsoft Azure platform providing separate services for particular needs:
  - **Cloud Services** - represent an application that is designed to run in the cloud service and XML configuration files that define how the cloud service should run; the application is defined in terms of component roles that implement the logic of the application; configuration files define the roles and resources for an application; the following roles can be used to implement the logic of the application:
    * **Web role** - is a virtual machine instance used for providing a web based front-end for the cloud service;
    * **Worker role** - is a virtual machine instance used for generalized development that performs background processing and scalable computations, accepts and responds to requests, and performs long running or intermittent tasks;
  - **Virtual Machines** - represent instances of virtual machines with pre-installed operating systems (Windows Server or Linux);
- **App Services** - provide multiple services related to security, performance, workflow management, and finally, messaging including **Storage Queues** and Service Bus providing efficient communication between application tiers running in Microsoft Azure.

Basic tier of the Microsoft Azure platform provides five classes of virtual machines (compute units): ExtraSmall, Small, Medium, Large, ExtraLarge. They differ with the number of cores possessed, CPU/core speed and amount of memory delivered, and efficiency of I/O channel. These compute units can be used while building custom cloud-based applications. Detailed features of available compute units are listed in [13].

3   Model of the Stress Testing System

Model of the stress testing system that we propose is based on the Microsoft Azure programming model. Microsoft Azure programming model introduces the concept of role as a working unit: Web role that provide web-based Graphical User Interface (GUI), and Worker role for long running computational tasks. During the design phase we defined the stress testing environment STE as a set of roles R_{STE} and queues Q_{STE}:

$$STE = \langle R_{STE}; Q_{STE} \rangle. \tag{1}$$

The set of roles R_{STE} is defined as follows:

$$R_{STE} = \{r_{Web}\} \cup R_{Wrk}, \tag{2}$$

where: r_{Web} is a Web role responsible for the interaction with the STE users, and R_{Wrk} is a set of worker role instances that serve as Workload Replay Agents that replay in parallel the specified SQL workload.

Queues are very important for the functioning of the entire testing environment, since they provide asynchronism in gathering test requests by the Web role and processing these requests by Worker roles. By introducing queues to the system, the Web role does not have to wait for any reply from Worker roles and can accept user's requests as they come. There are three queues in the STE stress testing environment:

$$Q_{STE} = \{q_{CS}, q_{SQL}, q_{res}\}, \tag{3}$$

where: q_{CS} is the connection string queue that broadcasts connection strings of tested database(s) to Workload Replay Agents (Worker role instances), q_{SQL} is the SQL workload queue for transferring the same workload to Workload Replay Agents, q_{res} is the result queue for transferring results of stress tests back to the Web role.

The set of instances of the Worker role R_{Wrk} is defined as follows:

$$R_{Wrk} = \{r_{Wrk(i)} | i = 1, ..., n\} \tag{4}$$

where $r_{Wrk(i)}$ is a single instance of the Worker role (single Workload Replay Agent), and n is the maximum number of Workload Replay Agents that work in the STE ($n \in \mathbb{N}_+$).

Scalability is a key property of the system and one of the main reasons why the system is designed to work in the cloud. We assumed that the system will be scaled mainly horizontally (scaling out) by increasing the number of Workload Replay Agents n of the same size (i.e. possessing the same compute capabilities).

Communication between Web role and Worker roles is achieved by sending and receiving messages of various types t. The set of messages M is defined as follows:

$$M_{STE} = \bigcup_{t \in T} M_t, \tag{5}$$

where T defines types of messages:

- CS - messages for transferring connection strings to tested databases;
- SQL - messages for transferring SQL workload;
- res - messages for transferring results of performance tests.

For the communication between components of the system we also define the following operations:

- $a(m_t) \succ b$ means that component a sends message m_t to component b, and
- $a(m_t) \prec b$ means that component a gets message m_t from component b.

Having such defined communication operations, we assumed the following rules for acts of communication that may arise in the system:

$$\begin{aligned}
r_{Web}(m_{CS}) &\succ q_{CS} & r_{Wrk}(m_{CS}) &\prec q_{CS} \\
r_{Web}(m_{SQL}) &\succ q_{SQL} & r_{Wrk}(m_{SQL}) &\prec q_{SQL} \\
r_{Wrk}(m_{res}) &\succ q_{res} & r_{Web}(m_{res}) &\prec q_{res}
\end{aligned} \tag{6}$$

4   Architecture of the Stress Testing Environment

On the basis of the presented model we have developed the architecture of the stress testing environment. Fig. 2 shows the architecture of the entire environment, including stress testing system that works in the Microsof Azure cloud and tested databases. These databases may be located in different places, e.g. local hardware infrastructure of the company or remote hardware infrastructure leased from a cloud provider.

Stress testing system implements the model presented in sections 3. Users interact with the system through the Web role. The Web role provides a kind of wizard that helps a user to start testing the specified database. Passing through the steps of the wizard the Web role implements steps of Algorithm 1. It gathers the data that are needed to create the connection string to the tested database (line 1). Then, it creates a message m_{CS} with the connection string (line 2) and places in the queue q_{CS} one copy of the message m_{CS} for each Worker role $r_{Wrk(i)}$ employed for tests (lines 3-5). The SQL workload that is provided by the user in one of the wizard's steps is then encoded as message m_{SQL} (line 6). The Web role then sends $\#iter$ copies of the message to the q_{SQL} queue (lines

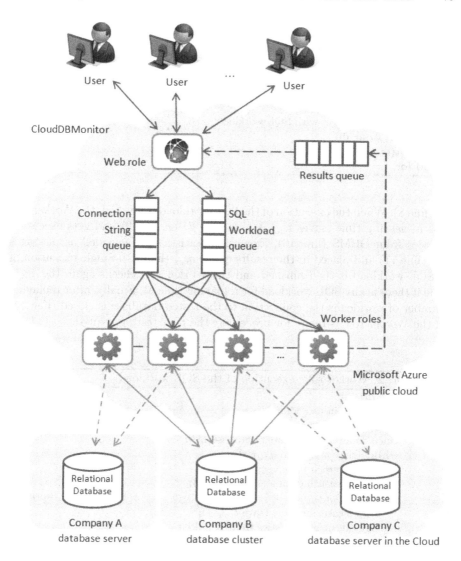

Fig. 2. General architecture of the CloudDBMonitor

7-9). The SQL workload is replayed #iter times against the tested database by instances of the Worker role.

Worker role instances implement the logic of Algorithm 2. An instance of the Worker role works continuously and accepts messages from the connection string queue and SQL workload queue. When it gets message m_{CS} containing a connection string to the database (line 2-3), it opens connection to the database (line 4). While there is a SQL workload prepared for execution (line 5), it gets the workload from the queue q_{SQL} (line 6), starts measuring time (saves $t_{start(j)}$, $j = 1, .., \#iter$, line 7), and executes the workload against the tested database

Algorithm 1. Web role: processing a testing request

1: Gather connection string data from the web page;
2: Create message m_{CS};
3: **for each** $r_{Wrk(i)}$ **do**
4:    $r_{Web}(m_{CS}) \succ q_{CS}$;
5: **end for**
6: Create message m_{SQL} with SQL workload;
7: **for** $i \leftarrow 1, \#iter$ **do**
8:    $r_{Web}(m_{SQL}) \succ q_{SQL}$;
9: **end for**

DB (line 8). When the execution of the SQL workload is finished, the Worker role stops measuring time (saves $t_{end(j)}$, $j = 1, .., \#iter$, line 9) and gets execution statistics from DBMS (line 10). Execution statistics are encoded as messages m_{res} (line 11) and placed in the results queue q_{res} (line 12). Single execution of the SQL workload is then finished and the Worker role checks again the q_{SQL} queue if there is any SQL workload for replaying (line 5). Finally, after replaying a number of workloads the connection to the tested database is closed (line 14) and the Worker role is ready for processing the next testing request.

Algorithm 2. Worker role: execution of the SQL workload

1: **while** true **do**
2:    **if** exists(m_{CS}) in q_{CS} **then**
3:        $r_{Wrk(i)}(m_{CS}) \prec q_{CS}$;
4:        Open connection to tested database DB;
5:        **while** exists(m_{SQL}) in q_{SQL} **do**
6:            $r_{Wrk(i)}(m_{SQL}) \prec q_{SQL}$;
7:            Start measuring time (save $t_{start(j)}$);   ▷ $j = 1, .., \#iter$
8:            $DB \leftarrow m_{SQL}$ ▷ Execute SQL workload
9:            Stop measuring time (save $t_{end(j)}$);
10:            Get execution statistics from $DBMS$;
11:            Create message m_{res} with execution statistics;
12:            $r_{Wrk(i)}(m_{res}) \succ q_{res}$;
13:        **end while**
14:        Close connection to DB;
15:    **end if**
16: **end while**

When instances of the Worker role process the SQL workload, the Web role passes to the monitoring state and behaves according to Algorithm 3. The role connects to the tested database (line 1) and continuously gets values of specified counters and displays them to the user as a chart (lines 2-6), like the one presented in Fig. 3. Counter values are collected periodically every d seconds (d is configured internally).

Algorithm 3. Web role: online monitoring of the database

1: Connect to tested database DB;
2: **while** monitoring in progress **do**
3:     Get counter values;
4:     Display counter values to the user;
5:     Wait d seconds;
6: **end while**

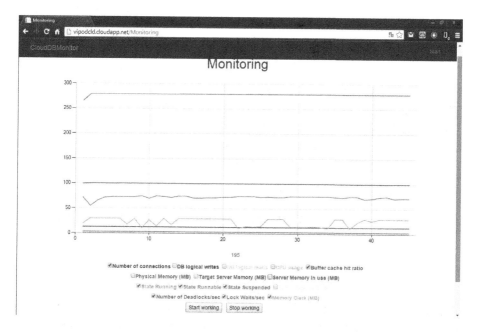

Fig. 3. Online monitoring of the specified database with CloudDBMonitor

There are different counter values that can be collected. Counters depend on the database management system. For example, for Microsoft SQL Server we can monitor:

- The number of connections to the tested database;
- CPU usage of the server hosting the database;
- Buffer cache hit ratio;
- Physical memory (MB);
- Target server memory (MB);
- Server memory in use (MB);
- Number of DBMS threads in particular states: running, runnable (waiting for CPU), suspended (waiting for resources);
- SQL compilations/sec;

– Number of deadlocks/sec;
– Lock waits/sec;
– and others.

After the whole SQL workload has been replayed $\#iter$ times and testing phase has been completed the Web role collects and displays results of the whole testing process. It implements the Algorithm 4 for this purpose. The Web role displays the Summary web page with aggregated counter values as a table (line 1) and periodical counter values in particular moments of time as charts (line 2). Then, it collects execution statistics from all messages $m_{res(j)}$ $(j = 1, ..., \#iter)$ stored in the results queue q_{res} and calculates the total execution time (lines 3-6). Finally, the role shows on a chart execution times for $\#iter$ successive executions of the SQL workload. Sample charts for selected counters are presented in Fig. 4.

Algorithm 4. Web role: gathering and displaying monitoring results

1: Show aggregated counter values;
2: Display distribution of counter values in time;
3: **for each** $m_{res(j)}$ in q_{res}, $j = 1, .., \#iter$ **do**
4:     $r_{Web}(m_{res(j)}) \prec q_{res}$;
5:     $t_{total} \leftarrow max(t_{end(j)}) - min(t_{start(j)})$;
6: **end for**
7: Display execution times for successive iterations;

---

5   Discussion and Concluding Remarks

Replaying the captured SQL workload is very important for newly established database systems or production database systems after reconfigurations. It allows to check how the new architecture or new configuration affects the performance of the database system. CloudDBMonitor stress testing environment allows the simulation and replay of an increased SQL workload in the chosen database management system by executing the SQL workload from many virtual workstations in the cloud. In such a way, it allows to test how the specified database works under a large pressure from concurrent transactions executed by many simulated client applications.

The main goal of the CloudDBMonitor is then the same as commercial systems, like Oracle Database Replay and Microsoft SQL Server Distributed Replay, and DBPerfTester@UMAP scientific prototype. The purpose of other mentioned frameworks and tools is slightly different - a dynamic management of various DBMSs with AgentTeam [4], and maintenance of complex database applications [2,3]. The core functionality of CloudDBMonitor is very similar to compared commercial tools and DBPerfTester@UMAP and all systems are based on the

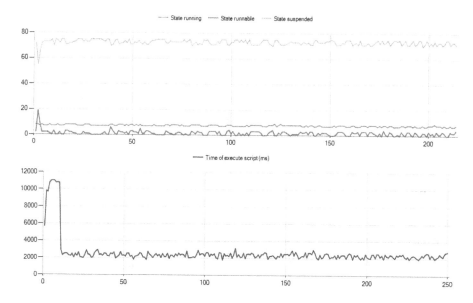

Fig. 4. Selected counter values: (top) the number of DBMS threads in particular states (*running, runnable, suspended*), (bottom) script execution time (ms), displayed by CloudDBMonitor for successive iterations (*x*-axe) of a sample SQL workload

same idea of employing many workload replay agents. However, CloudDBMonitor introduces queues into the system architecture, which provides temporal decoupling and independence of the system components. Commercial tools provide more functionality in terms of workload capturing, preparation, analysis and reporting. Scientific DBPerfTester@UMAP collects only basic statistics of the transaction execution time. CloudDBMonitor collects more statistics than its predecessor, DBPerfTester@UMAP. CloudDBMonitor also has a unique feature of possible infinite scalability as it is the system designed to work in the cloud.

Likewise its predecessor presented in [6], CloudDBMonitor stress testing system can be also used to verify database performance in the following situations:

- when migrating a database to work under control of different DBMS, e.g. when thinking of changing a database provider,
- when providing changes in database schema, business logic implemented in stored procedures, functions and triggers,
- when changing transaction isolation levels,
- when changing the SQL code in user's database application,
- when scaling hardware resources,
- when making changes in DBMS configuration.

At the moment, capabilities of the CloudDBMonitor are limited only to testing and monitoring Microsoft SQL Server DBMS. Tested databases can reside on stand-alone servers, clusters owned by a company, or in the cloud. The internal structure of the system is prepared for further development toward testing and monitoring other database management systems. This remains one of the directions for further development of the system in the future. Future works will also cover more sophisticated methods for controlling the activity of workload replay agents, including e.g., artificial immune systems [9,10]. This bio-inspired technique is widely used to solve many optimization problems. We believe that it will be a correct step toward the coordination of distributed systems containing a large number of agents.

References

1. Galanis, L., et al.: Oracle Database Replay. In: Proceedings of the 2008 ACM SIGMOD International Conference on Management of Data, pp. 1159–1170 (2008)
2. Gardikiotis, S.K., Lazarou, V.S., Malevris, N.: An agent-based approach for the maintenance of database applications. In: Proceedings of the 5th ACIS International Conference on Software Engineering Research, Management & Applications (SERA 2007), pp. 558–568 (2007)
3. Gardikiotis, S.K., Lazarou, V.S., Malevris, N.: Employing agents towards database applications testing. In: Proceedings of the 19th IEEE International Conference on Tools with Artificial Intelligence (ICTAI 2007), pp. 173–180 (2007)
4. Kumova, B.I.: Dynamic re-configurable transaction management in AgentTeam. In: Proceedings of the Ninth Euromicro Workshop on Parallel and Distributed Processing, pp. 258–264 (2001)
5. Mell, P., Grance, T.: The NIST definition of Cloud Computing. Special Publication 800-145 (accessed on November 25, 2014), http://csrc.nist.gov/publications/nistpubs/800-145/SP800-145.pdf
6. Mrozek, D., Małysiak-Mrozek, B., Mikołajczyk, J., Kozielski, S.: Database under pressure - Testing performance of database systems using Universal Multi-Agent Platform. In: Gruca, A., Czachórski, T., Kozielski, S. (eds.) Man-Machine Interactions 3. AISC, vol. 242, pp. 631–641. Springer, Heidelberg (2014)
7. Mrozek, D., Małysiak-Mrozek, B., Waligóra, I.: UMAP - A Universal Multi-Agent Platform for .NET Developers. In: Kozielski, S., Mrozek, D., Kasprowski, P., Małysiak-Mrozek, B., Kostrzewa, D. (eds.) BDAS 2014. CCIS, vol. 424, pp. 300–311. Springer, Heidelberg (2014)
8. Nambiar, R., Poess, M., Masland, A., Taheri, H.R., Emmerton, M., Carman, F., Majdalany, M.: TPC benchmark roadmap 2012. In: Nambiar, R., Poess, M. (eds.) TPCTC 2012. LNCS, vol. 7755, pp. 1–20. Springer, Heidelberg (2013)
9. Poteralski, A.: Optimization of mechanical structures using artificial immune algorithm. In: Kozielski, S., Mrozek, D., Kasprowski, P., Małysiak-Mrozek, B., Kostrzewa, D. (eds.) BDAS 2014. CCIS, vol. 424, pp. 280–289. Springer, Heidelberg (2014)
10. Poteralski, A., Szczepanik, M., Ptaszny, J., Kuś, W., Burczyński, T.: Hybrid artificial immune system in identification of room acoustic properties. Inverse Problems in Science and Engineering 21(6), 957–967 (2013)
11. SQL Server Distributed Replay, SQL Server 2012 Books Online, Quick Reference (June 2012), http://msdn.microsoft.com/en-us/library/ff878183.aspx (accessed on November 7, 2014)

12. Transaction Processing Performance Council, TPC-C/App/E BENCHMARKTM Standard Specification, http://www.tpc.org (accessed on November 7, 2014)
13. Virtual Machine and Cloud Service Sizes for Azure, http://msdn.microsoft.com/ library/azure/dn197896.aspx (accessed on November 25, 2014)
14. Waligóra, I., Małysiak-Mrozek, B., Mrozek, D.: UMAP - Universal Multi-Agent Platform. Studia Informatica 31(2A(89)), 85–100 (2010)
15. What is Azure? http://msdn.microsoft.com/library/azure/dd163896.aspx (accessed on November 25, 2014)

Comparison Between Performance of Various Database Systems for Implementing a Language Corpus

Dimuthu Upeksha, Chamila Wijayarathna[✉], Maduranga Siriwardena, Lahiru Lasandun, Chinthana Wimalasuriya, N.H.N.D. de Silva, and Gihan Dias

Department of Computer Science and Engineering,
University of Moratuwa, Sri Lanka
{upeksha.10,cdwijayarathna.10,maduranga.10,lasandun.10,
chinthana,nisansadds,gihan}@cse.mrt.ac.lk
http://www.cse.mrt.ac.lk

Abstract. Data storage and information retrieval are some of the most important aspects when it comes to the development of a language corpus. Currently most corpora use either relational databases or indexed file systems. When selecting a data storage system, most important facts to consider are the speeds of data insertion and information retrieval. Other than the aforementioned two approaches, currently there are various database systems which have different strengths that can be more useful. This paper compares the performance of data storage and retrieval mechanisms which use relational databases, graph databases, column store databases and indexed file systems for various steps such as inserting data into corpus and retrieving information from it, and tries to suggest an optimal storage architecture for a language corpus.

Keywords: Corpus · Relational Databases · NoSQL · Graph Database · Neo4j · Index file system · Apache Solr · Column stores · Cassandra

1 Introduction

Today, corpus based approach can be identified as the state of the art methodology in language studying for both prominent and inconspicuous languages in the world. Corpus based approach mines new knowledge on a language by answering two main questions:

1. What particular patterns are associated with lexical or grammatical features of the language?
2. How do these patterns differ within varieties and registers?

According to Bennet, et al. [2]: A corpus is a principled collection of authentic texts stored electronically that can be used to discover information about language that may not have been noticed through intuition alone.

From the top view architecture, a corpus may have three main components. And those are a mechanism to insert new data into the corpus, a data storage

S. Kozielski et al. (Eds.): BDAS 2015, CCIS 521, pp. 82–91, 2015.
DOI: 10.1007/978-3-319-18422-7_7

and information retrieval system, and data visualization tools. Among these components, data storage system is the most vital part, because the performance of data insertion and retrieval mainly depends on it. In this study we have carried out a comprehensive comparison on a set of widely used database systems and technologies in their role of being a data storage component for a corpus in order to find an optimal solution with optimal performance.

The remaining five sections of this paper describe our study. First in section 2, we briefly look at the storage mechanisms that have been used in existing corpora and the data storage technologies we are going to analyse with relevant architectures. Section 3 describes the method we are using to compare those storage mechanisms. Then in section 4 we present the results and observations of our study and finally in section 5, we present the drawn conclusions.

2   Suggested Solutions for Comparison

According to the literature there are two main storage systems used in the existing corpora. They are relational databases and indexed file systems.

Popular corpora like Corpus of Contemporary American English (COCA) and Corpus del Espaol use relational databases. Davies [3,4] has described the relational database model used in COCA and the details of database model used in Corpus del Espaol.

On the other hand British National Corpus (BNC), uses indexed file systems where data is stored as XML like files which follows a scheme known as the Corpus Data Interchange Format (CDIF).Aston, et al. [1] describes about the storage mechanism of BNC.

The idea of this study is to compare existing database systems and to come up with strengths and weaknesses of each system as a candidate for a storage mechanism of a language corpus.

In this study we are using H2 database as a relational database system because its licensing permits us to publish benchmarks and according to H2 benchmark tests [5], H2 has performed better than other relational databases such as MySQL, PostgreSQL, Derby and HSQLDB that allow publishing benchmarks. We did not do these performance testings on relational databases such as Oracle, MS SQL Server and DB2 because their licenses are not compatible with benchmark publishing.

We used Apache Solr which is powered by Apache Lucene as an indexed file system. We selected Solr because Lucene is a widely used text search engine and Solr is licensed under Apache License which allows us to do benchmark testing on that.

When storing details about words, bigrams, trigrams and sentences, one of the biggest challenges faced is storing the relationships between each of these entities. One way to represent relationships between entities is to use a graph. Hence, we also considered graph databases in our study as a candidate for the optimal storage system. Currently there are several implementations of graph databases such as Neo4j, OrientDB, Titan and DEX. In our study we used

Neo4j as our graph database system because, it has been identified by Jouili and Vansteenberghe [6] that Neo4j has performed better than the other graph databases.

Column databases improves the performance of information retrieval at the cost of higher insertion time and lack of consistency. Since one of our main goals is fast information retrieval, we considered column databases as a candidate too. We used Cassandra since it has proven to give higher throughput than other widely used column databases [7].

3   Method of Comparison

This section describes the method that we used for the comparison and how we set up each database system. All codes and dataset used for this analysis is available on a Github repository (goo.gl/JyoHRa).

3.1   Benchmarking Conditions

Dataset - The dataset we used in this study contains data which were crawled from online sources that are written in the Sinhala language. The final dataset consisted of 5 million word tokens, 400000 sentences and 20000 posts.

Testing Environment - All tests were run in a 2.4 GHz core i7 machine with 12GB of physical memory and a standard hard disk (spinning disk). Operating system was Ubuntu 14.04 with java version 1.8.0_05.

Data Insertion Mechanism - For every database system mentioned in the previous section, words were added in 50 iterations in which each iteration added 100,000 words with relevant sentences and posts.

Data Retrieval Mechanism - At the end of each iteration mentioned above, queries to retrieve information for following scenarios were passed to each database. Same query was executed 6 times in each iteration and selected the median of recorded values.

1. Get frequency of a given word in corpus using same word for every database system
2. Get list of frequencies of a given word in different time periods and different categories
3. Get most frequently used 10 words in a given time period or a given category
4. Get latest 10 posts that include a given word
5. Get latest 10 posts that include a given word in a given time period or a given category
6. Get 10 words which are most frequent as the last word of a sentence
7. Get the frequency of a given bigram in given time period and a given category
8. Get the frequency of a given trigram in given time period and a given category

9. Get most frequent bigrams
10. Get most frequent trigrams
11. Get most frequent word after a given word
12. Get most frequent word after a given bigram

3.2   Method of Evaluating H2 DB

We used H2 version 1.3.176 for this performance analysis. Evaluation was done using JDBC driver version 1.4.182. Cache was set to 0 Mb. We used relational schema shown in Fig. 1.

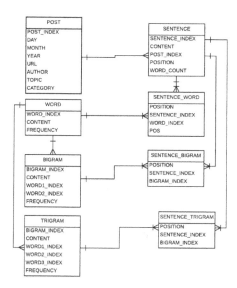

Fig. 1. Schema used in H2

We used below indexes for H2 database.

1. Index on CONTENT column of the WORD table
2. Index on FREQUENCY column of the WORD table
3. Index on YEAR column of the POST table
4. Index on CATEGORY column of the POST table
5. Index on CONTENT column of the BIGRAM table
6. Index on WORD1_INDEX column of the BIGRAM table
7. Index on WORD2_INDEX column of the BIGRAM table
8. Index on CONTENT column of the TRIGRAM table
9. Index on WORD1_INDEX column of the TRIGRAM table
10. Index on WORD2_INDEX column of the TRIGRAM table
11. Index on WORD3_INDEX column of the TRIGRAM table
12. Index on POSITION column of the SENTENCE_WORD table

When inserting the data to the database we dropped all the indexes except the one on CONTENT column of the WORD table. The indexes were recreated when retrieving information.

3.3   Method of Evaluating Neo4j

We used Neo4j community distribution 2.1.2 licensed under GPLv3. Evaluation was done using java embedded Neo4j database mode with heap size of 4048 MB which can pass database queries through a java API. We measured performance using both warm cache mode and cold cache mode. In warm cache mode, all caches were cleared and a set of warm up queries were run before running each actual query. In cold cache mode, queries were run with an empty cache. We did batch insertion through csv import function with pre-generated csv files.

We used graph structure in Fig. 2 to evaluate the performance.

Fig. 2. Graph structure used in Neo4j

Properties of Nodes

1. Post - post_id, topic, category, year, month, day, author and url
2. Sentence - sentence_id, word_count
3. Word - word_id, content, frequency

Properties of Relationships

1. Contain - position
2. Has - position

Procedure of Data Insertion. We used csv file formatted database dumps of relational database in-order to insert data into Neo4j using its csv import function.

1. Word.csv includes word_id, content and frequency
2. Post.csv includes post_id, topic, category, year, month, day, author and url
3. Sentence.csv includes sentence_id, post_id, position and word_count
4. Word_Sentence.csv includes word_id, sentence-id and position

Evaluations were done for both data insertion times and data retrieval times. Data retrieval time was calculated by measuring the execution time for each cypher query. For each cypher query, two time measurements were recorded for warm cache mode and cold cache mode.

3.4   Method of Evaluating Apache Solr

We used Apache Solr version 4.10.2. Cache sizes of LRUCache, FastLRUCache and LFUCache were set to 0 in solrconfig.xml and autoWarmCount for each cache was set to 0.

The following field types were defined to store data in Solr.

1. text_general - This was implemented using solr TextField data typeand the standard tokenizer was used.
2. text_shingle_2 - This was implemented using solr TextField data type and ShingleFilterFactory with minShingleSize="2" and maxShingleSize="2".
3. text_shingle_3 - This was implemented using solr TextField data type and ShingleFilterFactory with minShingleSize="2" and maxShingleSize="2".

To evaluate the performance, fields shown in table 1 were added to documents.

Table 1. Data types definition of Solr search engine

Fieldname	Fieldtype	Indexed	Sorted
id	string	true	true
date	date	true	true
topic	text_general	true	true
author	text_general	true	true
link	text_general	false	true
context	text_general	true	true
content_shingled_2	text_shingle_2	true	true
content_shingled_3	text_shingle_3	true	true

For each doc, an unique id of 7 characters was generated. All fields except link were indexed and all the fields were sorted.

3.5   Method of Evaluating Cassandra

We used Apache Cassandra version 2.1.1. Both Key Caching and Row Caching were disabled when retrieving information from the database.

Cassandra uses a query based data modeling where it maintains different column families to address querying needs. Hence, when designing our database we also maintained different column families for different querying needs and consistency among them was maintained at the application level that we used to insert data.

Following are the column families we used, with the indexes used in each one of them.

1. word_frequency (id bigint, content varchar, frequency int, PRIMARY KEY (content))

2. word_time_frequency (id bigint, content varchar, year int, frequency int, PRIMARY KEY (year, content))

3. word_time_inv_frequency (id bigint, content varchar, year int, frequency int, PRIMARY KEY((year) , frequency, content))

4. word_usage (id bigint, content varchar, sentence varchar, date timestamp, PRIMARY KEY (content, date, id))

5. word_yearly_usage (id bigint, content varchar, sentence varchar, position int, postname text, year int, day int, month int, date timestamp, url varchar, author varchar, topic varchar, category int, PRIMARY KEY ((content, year, category) ,date,id))

6. word_pos_frequency (id bigint, content varchar, position int, frequency int, PRIMARY KEY ((position), frequency, content))

7. word_pos_id (id bigint, content varchar, position int, frequency int, PRIMARY KEY (position, content))

8. bigram_with_word_frequency (id bigint, word1 varchar, word2 varchar, frequency int, PRIMARY KEY (word1, frequency, word2))

9. bigram_with_word_id (id bigint, word1 varchar, word2 varchar, frequency int, PRIMARY KEY (word1, word2))

10. bigram_time_frequency (id bigint, bigram varchar, year int, frequency int, PRIMARY KEY (year, bigram))

11. trigram_time_frequency (id bigint, trigram varchar, year int, frequency int, PRIMARY KEY (year, trigram))

12. bigram_frequency (id bigint, content varchar, frequency int, category int, PRIMARY KEY (category, frequency, content))

13. bigram_id (id bigint, content varchar, frequency int, PRIMARY KEY (content))

14. trigram_frequency (id bigint, content varchar, frequency int, category int, PRIMARY KEY (category, frequency, content))

15. trigram_id (id bigint, content varchar, frequency int, PRIMARY KEY (content))

16. trigram_with_word_frequency (id bigint, word1 varchar, word2 varchar, word3 varchar, frequency int, PRIMARY KEY ((word1, word2), frequency, word3))

17. trigram_with_word_id (id bigint, word1 varchar, word2 varchar, word3 varchar, frequency int, PRIMARY KEY (word1, word2, word3))

4   Observations and Results

Fig. 3 shows the comparison between the data insertion time for each data storage mechanism. Fig. 4 and Fig. 5 illustrate the comparison between times taken for the each information retrieval scenario for each data storage mechanism. When plotting graphs, we have ignored relatively higher values to increase the clarity of graphs.

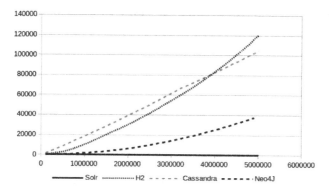

Fig. 3. Data insertion time in each data storage system

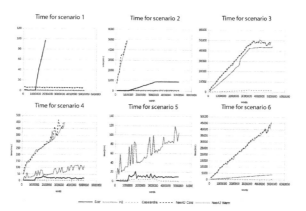

Fig. 4. Data retrieval time for each scenario in each data storage system - part 1

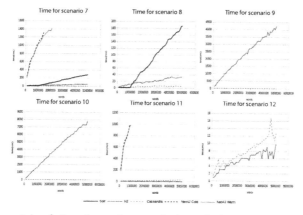

Fig. 5. Data retrieval time for each scenario in each data storage system - part 2

Normalized Time Measurements for Data Insertion. All data insertion time measurements of each database were normalized with respect to data insertion time measurements of H2 database.

Normalized time for data insertion in database

$$DB_i = \frac{1}{number\_of\_iterations} \sum \frac{time\_to\_insert\_data\_to\_DB_i}{time\_to\_insert\_data\_t\_H2} \tag{1}$$

Table 2. Normalized Time for Data Insertion in Each Data Storage System

H2	Solr	Neo4j	Cassandra
1	0.00047	0.2333	1.5306

Normalization of Time Measurements for Data Retrieval for Each Scenario. All time measurements were normalized with respect to time measurements of H2 database.

Normalized time for query q in database

$$DB_i = \frac{1}{number\_of\_iterations} \sum \frac{measured\_time\_of\_DB_i\_iteration\_k\_for\_query\_q}{measured\_time\_of\_H2\_iteration\_k\_for\_query\_q} \tag{2}$$

Table 3. Normalized data retrieval time for each scenario in each data storage system

Normalized time	Q1	Q2	Q3	Q4	Q5	Q6	Q7	Q8	Q9	Q10	Q11	Q12
H2	1	1	1	1	1	1	1	1	1	1	1	1
Solr	192	8.34	N/S	0.26	0.13	N/S	3.25	3.23	N/S	N/S	N/S	N/S
Neo4j Cold Cache	5.3	224	1.35	5.84	197	12.4	60	586	T/E	T/E	1228	2464
Neo4j Warm Cache	6	221	1.32	5.52	192	12	56	566	T/E	T/E	1141	2286
Cassandra	0.7	0.02	0.04	0.04	0.06	0.03	0.1	0.23	0.002	0.001	11.0	1.25

Reasons for having N/S and T/E values for the entries in table 3 are:

1. N/S (Not Support) - Storage system or the schema does not support data retrieval of such scenarios.
2. T/E (Time Exceeds) - Data retrieval time takes more than 120 seconds in such scenarios

5   Conclusion

Based on our observations, in data insertion point of view, Solr performed better than the other three databases with a significant amount of time gap (Solr : H2 : Neo4j : Cassandra = 1 : 2127 : 495 : 3256). That means Solr is approximately

2127 times faster than H2, 495 times faster than Neo4j and 3256 times faster than Cassandra with respect to data insertion time.

From data retrieval point of view Cassandra performed better than other databases. In couple of scenarios H2 outperformed Cassandra. But in those scenarios, Cassandra also showed a considerably good speed. Neo4j did not perform well in any scenario and hence it is not suitable to create a structured dataset such as a language corpus. Solr also marked a decent performance in some scenarios but there were issues in implementing some scenarios because its underlying indexing mechanism did not provide support to do those.

Only issue with Cassandra was its less flexibility in supporting new queries. If we had another information need other than the above mentioned scenarios, we would have to create new column families in Cassandra that support given information need and insert data from the beginning which is a very expensive process.

In conclusion Cassandra can be named as a good candidate for a data storage system of a language corpus because

1. Data insertion time of Cassandra is linear which can be very effective in a long term data insertion process
2. Performed well in 10 out of all 12 scenarios for information retrieval.

References

1. Aston, G., Burnard, L.: The BNC Handbook:Exploring the British National Corpus with SARA, http://corpus.leeds.ac.uk/teaching/aston-burnard-bnc.pdf
2. Bennet, G.R.: Using Corpora in the Language Learning Classroom, http://www.international.ucla.edu/media/files/Using-corpora-in-the-language-learning-classroom-Corpus-linguistics-for-teachers-my-atc.pdf
3. Davies, M.: The advantage of using relational databases for large corpora. International Journal of Corpus Linguistics 10(3), 307–335 (2005)
4. Davies, M.: The 385+ million word corpus of contemporary American English (1990–2008+) design, architecture, and linguistic insights. International Journal of Corpus Linguistics 14(2), 159–191 (2009)
5. H2: Performance, http://www.h2database.com/html/performance.html
6. Jouili, S., Vansteenberghe, V.: The advantage of using relational databases for large corpora. In: International Conference on Social Computing (SocialCom), pp. 708–715 (2013)
7. Rabl, T.M., et al.: Solving big data challenges for enterprise application performance management. PVLDB, 1724–1735 (2010)

A Comparison of Different Forms of Temporal Data Management

Florian Künzner and Dušan Petković[✉]

University of Applied Sciences
Rosenheim, Hochschulstr. 1, 83024, Germany
`flo.kuenzner@gmx.net, petkovic@fh-rosenheim.de`
`http://www.fh-rosenheim.de`

Abstract. Recently, the ANSI committee for the standardization of the SQL language has published the specification for temporal data support. This new ability allows users to create and manipulate temporal data in a significantly simpler way instead of implementing the same features using triggers and database applications. In this article we examine the creation and manipulation of temporal data using built-in temporal logic and compare its performance with the performance of equivalent hand-coded applications. For this study, we use an existing commercial database system, which supports the standardized temporal data model.

Keywords: Temporal data · SQL:2011 · Performance · Trigger

1 Introduction

A database represents a model of objects of the real world. During the lifetime of an object stored in a database, its properties can change. For such objects it is necessary to consider their time-variant aspects. For this reason, it is an important task of database systems to support management of temporal data, i.e. that current, past and future values of time-variant attributes can be persistently stored.

Timestamps can be represented using time points, time intervals and set of time intervals. Time points are an infinite but countable ordered set, which is used to specify a time domain [16]. Time intervals specify a time domain of an entity as the continuous set of time instants. Interval-based temporal data models are introduced in TSQL2 [14] and in the SQL:2011 standard [6]. Sets of time intervals represent time domain as a finite union of time intervals [15]. Almost all existing temporal data models today are based upon time intervals.

Also, there are three fundamental and orthogonal kinds of time: user-defined, valid and transaction time. User-defined time is a time representation designed to meet the specific needs of users. Valid time specifies when certain conditions in the real world are, were or will be valid. Several models supporting valid time are described in [15]. Transaction time automatically captures changes made to the state of time-variant data in a database. This time dimension represents the time period during which an instance is recorded in the database. Taxonomy

© Springer International Publishing Switzerland 2015
S. Kozielski et al. (Eds.): BDAS 2015, CCIS 521, pp. 92–106, 2015.
DOI: 10.1007/978-3-319-18422-7_8

for classifying databases in terms of valid time and transaction time has been developed in [13].

The first proposal for standardization of temporal data for database systems has been submitted in 1995 and was based on [14]. This proposal showed some shortcomings and was criticized in an article [3], thus failed to get a support of the SQL standardization committee. Another proposal [9] has been submitted, but the committee did not accept it. The next attempt to standardize temporal data as a part of the SQL specification started in 2006 and ended successfully with the publishing of the SQL:2011 standard [7]. The standard has adopted a model, where a time period represents all time granules starting from and including the start time, and ending with the last time granule before the end time.

The standard introduces three different forms of tables [11]: application-time period tables, system-versioned tables and bitemporal tables. An application-time period table captures time periods during which data are valid in the real world. For this reason, the user is responsible for setting the start and end times of each time-variant attribute. Also, the user modifies the validity periods of rows, when an error is detected. Generally, the valid time support is provided by these tables.

One of main requirements for system-versioning table is that any modification of rows immediately preserves the old state of these rows before executing the UPDATE or DELETE statement. By specifying a table with system-versioning, the user tells the system to immediately capture changes made to the state of tables rows and, at the same time to memorize the old state of the same rows. Therefore, the time period of a row in a system-versioning table begins at the time in which the row was inserted into the table and ends when the row was deleted or updated.

Bitemporal tables store and manage data with valid as well as with transaction time. Therefore, a bitemporal table is a union of data from a corresponding application-time period table and system-versioned table, and rows in such tables are associated with both application-time period as well as with system-time period. (Note that SQL:2011 does not use any particular name for these tables. In this article, the phrase bitemporal is used in accordance to its use in the literature.)

The aim of this paper is to examine whether the built-in support for temporal logic reduces execution time in relation to hand-coded database applications. Our tests showed that in the most cases execution times of built-in code is ca. 1.5 times quicker than execution time of the corresponding triggers and stored procedures.

For this study, we use an existing commercial database system, IBM DB2, which supports the standardized temporal logic [12]. Besides DB2, Oracle Version 12c as well as Teradata support temporal logic. Teradatas temporal support does not correspond to the specification in the SQL:2011 standard, while Oracles implementation is incomplete in relation to the DB2 support of this specification.

1.1 Related Work

Temporal databases have been a research topic since more than 25 years. During this time, numerous temporal models and query languages have been proposed. An annotated bibliography on temporal aspects can be found in [5]. The glossary of temporal database concepts is given in [8]. Taxonomy for the classification of temporal databases according to time dimensions has been developed in [13]. According to this taxonomy, Gadias work, described in [15] is a temporal data model concerning valid time. Ben-Zvi proposed the first data model for bitemporal databases, a temporal query language, storage architecture, indexing, recovery, concurrency, synchronization, and its implementation [4]. Snodgrass attached four implicit attributes to each time-varying relation, and presented a corresponding temporal query language [15]. The bitemporal data model, BCDM [1] forms the basis for the Temporal Structured Query Language. The article [17] discusses the standardization of database systems from another angle of view.

Performance issues in relation to temporal data have been investigated in several papers. In her work [1], Attay compared attribute and tuple timestamping in relation to the BCDM data model. Our work is similar to the work at IBM described in [12]. In contrast to our tests, the authors in [12] compare implementation of a subset of valid time capabilities with built-in temporal data management versus hand-coded implementation of equivalent logic. The built-in implementation of temporal logic reduced coding requirements by ca. 90% in relation to the implementation of the same in corresponding applications. In relation to lines of code, triggers and Java applications required 16 and 45 times more LOCs, respectively.

1.2 Roadmap

The rest of this article is organized in the following way: Section 2 deals with issues in relation to application-time period tables. Different cases concerning integrity constraints as well as DML statements on such tables are analyzed and compared. Section 3 discusses two different cases in relation to DML statements of system-versioned tables. In Section 4 we show cases in relation to integrity constraints and DML statements concerning bitemporal tables. Section 5 summarizes all experimental results from Chapters 2-4. The last section gives conclusions and discusses future work. The article has an appendix (Supplement), which contains the code of all triggers and stored procedures used for testing.

Each of the following three sections (Sections 2-4) has the same structure. First, we give basic specifications of all tables used in the following subsections. After that, the issues in relation to key integrity are discussed. In the final subsection, the discussion of DML statements is given.

2 Application-Time Period Tables

The creation of application-time period tables requires that the user explicitly defines two time-variant columns, which specify the start and end times of the period of each row. Both columns must be defined with the NOT NULL property and their type can be any date type (DATE or TIMESTAMP). The PERIOD clause, which is a new extension to SQL, instructs the system to use these time-variant columns to track the start and end points of time values for each row. (This clause implicitly enforces the constraint that start_time < end_time.)

2.1 Basic Specifications

The **employees** table is an SQL table used for testing of a logic implemented in hand-coded applications. The corresponding CREATE TABLE statement is given in Example 1. Similarly, the **T_employees** table is the corresponding table with the built-in temporal logic and handles valid time. (IBM calls such a table "table with business time".) The corresponding statement is given in Example 2.

Example 1

```
CREATE TABLE employees (
    E_Id INT NOT NULL, E_Name CHAR(20) NOT NULL,
    E_Start DATE NOT NULL, E_End DATE NOT NULL,
    E_Dept INT,
    PRIMARY KEY (E_Id, E_Start));
-- Start_Time <= End_Time ->   must be explicitly checked
ALTER TABLE employees ADD CONSTRAINT emp_check
    CHECK(E_Start <= E_End);
```

Example 2

```
CREATE TABLE T_employees (
    E_Id INT NOT NULL, E_Name CHAR(20) NOT NULL,
    E_Start DATE NOT NULL, E_End DATE NOT NULL,
    E_Dept INT,
    PERIOD BUSINESS_TIME (E_Start, E_End),
    PRIMARY KEY (E_Id, BUSINESS_TIME WITHOUT OVERLAPS));
```

For our tests, each of these tables is consecutively loaded with 100, 1000, 10.000, and 100.000 rows. Examples for loading data in both tables are given in Examples 3 and 4, respectively. As can be seen from these examples, both INSERT statement are equivalent. The reason is that in both CREATE TABLE statements the user has to specify start and end time of the corresponding time period.

Example 3

```
INSERT INTO employees
    VALUES (115, 'Dante', DATE '2012-01-01', DATE '2012-12-01', 10);
```

Example 4

```
INSERT INTO T_employees
    VALUES (115, 'Dante', DATE '2012-01-01', DATE '2012-12-01', 10);
```

2.2 Key Integrity Constraints on Application-Time Period Tables

Primary Key Integrity. There are two key integrity constraints: one concerning primary key (PK) and the other concerning foreign key i.e. referential integrity (RI). The first consequence of adding time support to an application-time period table is that the "convenient" primary key for the relational model is not sufficient for temporal logic and tables with time-variant columns have to consider the timestamp. In other words, the **E_Id** column in Example 2 does not suffice to guarantee uniqueness of rows in the **T_employees** table and the primary key specification has to include the **E_start** and **E_End** columns. However, this is still not enough for the uniqueness of rows, while we wish to specify that there can be only one **E_Id** value at any given time. In other words, we want that an employee belongs to exactly one department at the same time. For this reason, the standard specifies the WITHOUT OVERLAPS clause, which forbids overlapping of time periods. The first test we applied concerns the problem of primary key integrity. Example 5 tries to insert a row, which violates the constraint specified with the WITHOUT OVERLAPS clause in the CREATE TABLE statement of Example 2 and the already existing row inserted in Example 3.

Example 5

```
INSERT INTO T_employees
    VALUES (115, 'Dante', DATE '2012-01-01', DATE '2012-03-01', 10);
```

The same constraint for the convenient SQL table (**employees**) is shown in Example 6. The trigger in this example rejects the execution of INSERT statements that violate the PK constraint on the **employees** table. Note that this trigger is the only one, which is presented in the text of this article to demonstrate implementation of hand-coded applications. All other applications that will be discussed below can be found in the Supplement.

Example 6

```
-- the '!' sign is used as a delimiter
CREATE OR REPLACE TRIGGER EMPLOYEES_TRIGGER_INSERT
    BEFORE INSERT ON employees REFERENCING NEW AS newRow
    FOR EACH ROW
    BEGIN
        DECLARE i INTEGER;
        SELECT COUNT(*) INTO i
            FROM employees
            WHERE newRow.E_Id = E_Id
                AND newRow.E_End > E_Start
                AND newRow.E_Start < E_End;

        IF i > 0 THEN CALL
            RAISE_APPLICATION_ERROR(-20001, 'PK Times overlaps');
        END IF;
    END!
```

Referential Integrity. If we suppose that there is the **T_Departments** table, which is a parent table of the **T_employees** table, the convenient referential integrity involving these two tables cannot be satisfied, if the similar requirements from the last subsection are valid, i.e. that every value of the **E_Dept** column of the **T_employees** table corresponds to the department number column of the **T_Departments** table at every point in time. The SQL:2011 standard specifies the corresponding FOREIGN KEY clause (with the REFERENCES option), but the implementation of such a clause does not yet exist in IBM DB2 [10]. For this reason, a corresponding case cannot be tested at this moment.

2.3 DML Statements and Application-Time Period Tables

The semantics for DML statements differ significantly for application-time period tables in relation to the conventional SQL. Table 1 shows the differences for all three of these statements.

Table 1. The semantics of DML statements for application-time tables

	Conventional SQL	SQL Temporal
INSERT (one row)	Inserting the new row	Inserting a new row or prolonging the valid time of an existing row.
UPDATE (one row)	Updating the row	Shortening the valid time of the row and inserting a new row or prolonging valid time of an existing row.
DELETE (one row)	Deleting the row	Shortening the valid time of the row or deleting the row.

The case we examine concerns the DELETE statement. As can be seen from Table 1, the deletion of a row with a time-variant attribute means that its valid time is shortened or deleted. Example 7 shows such a DELETE statement for the **T_employees** table. After execution of this statement, the result contains two rows with **E_Id** = 115, one with the time period ending on 15.7.2012, and the other beginning on 15.8.2012.

Example 7

```
DELETE FROM T_employees FOR PORTION OF BUSINESS_TIME
    FROM DATE '2012-07-15' TO DATE '2012-08-15' WHERE E_Id = 115;
```

The example above uses the new clause - FOR PORTION -, which is the temporal extension of the DELETE (and UPDATE) statement. The Supplement shows the implementation of the corresponding stored procedure. (In this case, we used a stored procedure to implement temporal logic, because that way the explicit values of start and end times can be passed to the system using two parameters.)

3 System-Versioned Tables

System-versioned tables comprise system times, which are always updated, when the data modification statements are executed. These tables contain two additional columns, for system start time and system end time. In contrast to application-time period tables, the system maintains the start and end times of the periods of rows.

3.1 Basic Specifications

The **departments** table, created in Example 8 is the parent table of the **employees** table from Example 1. This table and the corresponding history table are used for testing a logic implemented in hand-coded applications and for transaction time. Similarly, the **S_departments** table is the corresponding system-versioned table with the built-in temporal logic. (IBM calls such a table "table with system time".) The corresponding CREATE TABLE statement is given in Example 9.

The two time-variant columns of the **S_departments** table (**D_Start** and **D_End**) are specified by the user, but their values are provided by the system. The third column, **TRANS_Start**, is used to track when the transaction first executed a statement that changes the tables data. The PERIOD clause has the same meaning as the clause with the same name for application-time period tables.

As can be seen from Example 9, IBM DB2 uses two tables, one for current data, (**S_departments**) and one for historical data (**Dept_history**). This is in contrast to the SQL:2011 specification, which defines only one table for current and historical data. Both DB2 tables are compatible to each other, and the system automatically moves the non-current data in the history table. Additionally, the ALTER TABLE statement in Example 9 alters the current table to enable versioning and identify the corresponding history table.

Example 8

```
CREATE TABLE departments(
    D_id INT NOT NULL, D_Name CHAR(20),
    D_Start TIMESTAMP(12), D_End TIMESTAMP(12),
    PRIMARY KEY (D_Id));
CREATE TABLE departments_History(
    D_id INT NOT NULL, D_Name CHAR(20),
    D_Start TIMESTAMP(12) NOT NULL, D_End TIMESTAMP(12) NOT NULL,
    PRIMARY KEY (D_Id, D_Start));
```

Example 9

```
CREATE TABLE S_departments(
    D_id INT NOT NULL, D_Name CHAR(20),
    D_Start TIMESTAMP(12) GENERATED ALWAYS AS ROW BEGIN NOT NULL,
    D_End TIMESTAMP(12) GENERATED ALWAYS AS ROW END NOT NULL,
    TRANS_Start GENERATED ALWAYS AS TRANSACTION
                START ID IMPLICITLY HIDDEN,
    PERIOD SYSTEM_TIME (D_Start, D_End), PRIMARY KEY (D_id));
CREATE TABLE Dept_History LIKE S_departments;
ALTER TABLE S_departments ADD VERSIONING USE HISTORY TABLE Dept_History;
```

The following examples show how rows can be inserted in both tables created in Example 8 and Example 9.

Example 10

```
INSERT INTO departments
    VALUES (1, 'D1', CURRENT TIMESTAMP, DATE '9999-12-31');
```

Example 11

```
INSERT INTO S_departments (D_Id, D_Name) VALUES ('1', 'D1');
```

The INSERT statements in Examples 10 and 11 are different, because in temporal logic, values for transaction start and end times are assigned by the system.

3.2 Key Integrity Constraints on System-Versioned Tables

The specification of key integrity constraints on system-versioned tables is significantly simpler than the specification of the same integrity constraints on application-time period tables. The reason is that these constraints (primary key constraint and referential integrity) must be enforced only on the current rows, because the same constraints have been already checked at the time where historical rows were current ones. For this reason, there is no need to include time-variant columns (begin system time and end system time) in the primary key and referential integrity definitions. In other words, the key integrity constraints on system-versioned tables are analogous to the same constraints in the conventional SQL.

3.3 DML Statements and System-Versioned Tables

The semantics for the INSERT, UPDATE and DELETE statements differ for system-versioned tables in relation to the conventional SQL. Table 2 shows the differences for all three of these statements.

Table 2. The semantics of DML statements in system-versioned tables

	Conventional SQL	SQL Temporal
INSERT	Inserting a new row	Inserting a new row
UPDATE (one row)	Modifying the row	Ending the currentness of that row and inserting a new row as a substitution for the ex-current row.
DELETE (one row)	Deleting the row	Ending the currentness of the row.

For system-versioned tables, our tests for DML concern the UPDATE (Example 12) and DELETE (Example 13) statements.

Example 12

```
UPDATE S_departments SET D_Name = 'D3' WHERE D_id = 1;
```

As can be seen from Table 2, the UPDATE statement in Example 12 will be replaced by an UPDATE and an INSERT statement: the former concerning the table with current values and the latter concerning the history table. The existing row in the **S_departments** table will be modified, with the new value for the **D_Name** column, and the start system time of the row will be set to the time, when the UPDATE statement has been performed. The historical row will have the end transaction time set to the time, when the UPDATE statement has been performed. (All other values of this row will be unchanged.)

Example 13

```
DELETE FROM S_departments WHERE D_id = 1;
```

The DELETE statement in Example 13 ends the currentness of the row with **D_id**=1. This means that two statements will be executed: a DELETE and an INSERT. In other words, the row will be deleted from the **S_departments** table and the new row will be inserted in the history table with the value of end transaction time set to the current time.

4 Bitemporal Tables

As already stated, valid time and transaction time represent two different kinds of time, which are orthogonal to each other. In the case that an application needs both of these dimensions, bitemporal tables are used. For this reason, the main requirement in relation to bitemporal tables is to present start and end of the valid time as well as its currentness in relation to transaction i.e. system time.

4.1 Basic Specifications

The **B_departments** table is an SQL table used for testing of bitemporal logic implemented in hand-coded applications. The CREATE TABLE statement for this table is given in Example 14. (The second CREATE TABLE statement in this example creates the corresponding history table.) Similarly, the **BITemp_departments** table is a table with the built-in temporal logic and it handles bitemporal time. As in the case of system-versioned tables, the user has to create a corresponding history table and to activate versioning for it. The corresponding CREATE TABLE statements are given in Example 15.

Example 14

```
CREATE TABLE B_departments(
    D_id INT NOT NULL, D_Name CHAR(20), D_Dept INT,
    V_Start DATE NOT NULL, V_End DATE NOT NULL, -- BUSINESS_TIME
    D_Start TIMESTAMP(12), D_End TIMESTAMP(12), -- SYSTEM_TIME
    PRIMARY KEY (D_id, V_Start));
-- Start_Time <= End_Time ->  must be explicitly checked
ALTER TABLE B_departments
    ADD CONSTRAINT B_departments_check CHECK(V_Start <= V_End);
-- History Table
CREATE TABLE B_departments_History(
    D_id INT NOT NULL, D_Name CHAR(20), D_Dept INT,
    V_Start DATE NOT NULL, V_End DATE NOT NULL,
    D_Start TIMESTAMP(12) NOT NULL, D_End TIMESTAMP(12) NOT NULL,
    PRIMARY KEY (D_id, D_Start, V_Start));
```

Example 15

```
CREATE TABLE BITemp_departments(
    D_id INT NOT NULL, D_Name CHAR(20), D_Dept INT,
    -- BUSINESS_TIME
    V_Start DATE NOT NULL, V_End DATE NOT NULL,
```

```
    PERIOD BUSINESS_TIME (V_Start, V_End),
    -- SYSTEM_TIME
    D_Start TIMESTAMP(12) GENERATED ALWAYS AS ROW BEGIN NOT NULL,
    D_End TIMESTAMP(12) GENERATED ALWAYS AS ROW END NOT NULL,
    TRANS_Start GENERATED ALWAYS AS TRANSACTION
                START ID IMPLICITLY HIDDEN,
    PERIOD SYSTEM_TIME (D_Start, D_End),
    PRIMARY KEY (D_Id, BUSINESS_TIME WITHOUT OVERLAPS));
CREATE TABLE BITemp_Dept_history LIKE BITemp_departments;
ALTER TABLE BITemp_departments
    ADD VERSIONING USE HISTORY TABLE BITemp_Dept_history;
```

The following two examples show how rows can be inserted in the
B_departments and **BITemp_departments** tables, respectively.

Example 16

```
INSERT INTO B_Departments (D_Id, D_Name, V_Start, V_End, D_Start, D_End)
    VALUES (1, 'D2', DATE '2011-01-01', DATE '2012-01-01',
            CURRENT TIMESTAMP, DATE '9999-12-31');
```

Example 17

```
INSERT INTO BITemp_Departments (D_Id, D_Name, V_Start, V_End)
    VALUES (1, 'D1', DATE '2011-01-01', DATE '2012-01-01');
```

The INSERT statements in Examples 16 and 17 are different, because in the
case of built-in temporal logic, the start system time and end system time are
assigned by the system.

4.2   Key Integrity Constraints on Bitemporal Tables

The integrity constraint for primary key on bitemporal tables is similar to the
corresponding constraint for application-time period tables. The INSERT statement in Example 18 violates this constraint.

Example 18

```
INSERT INTO BITemp_Departments (D_Id, D_Name, V_Start, V_End)
    VALUES (115, 'D1', DATE '2011-02-01', DATE '2012-03-01');
```

4.3   DML Statements and Bitemporal Tables

Temporal modifications for bitemporal tables are similar to temporal modifications for application-time period tables. The main difference is that updates
(instead of deletions) are performed on transaction end time.

Example 19

```
DELETE FROM BITemp_departments FOR PORTION OF BUSINESS_TIME
    FROM DATE '2011-01-01' TO DATE '2011-02-01' WHERE D_id = 115;
```

For the DELETE statement in Example 19 above, the system performs two
statements, an UPDATE and an INSERT. The UPDATE statement will modify
the existing row. The start valid time of the row will be updated to the new
value (2011-02-01), while the start system time and end system time will be
the time, when the DELETE statement above has been executed and forever,
respectively. (The end valid time will be unchanged.) The INSERT statement

inserts a new row in the history table and modifies the system end time of the original row from forever to the time, when the row has been deleted. Both valid time values as well as the start system time will be unchanged.

5   Test Results

To evaluate performance tests described above, we use a host system with 8 GB RAM and the Intel Q6600 processor with 2.4 GHz. The software was installed on a virtual machine (VMware) with 3 GB RAM and four kernels. The operating system was Windows 7 Professional 64 Bit.

The database system used for testing is IBM DB2 V 10.5 (64 bit edition), with IBM Data Studio V 3.2 as interface to the database server. Although Data Studio displays execution times for each executed statement, our tests showed that these times are unreliable i.e. vary significantly from each other and therefore this tool could not be used for performance test. For this reason, we used a command script, which starts the **db2batch** command. Each table is consecutively loaded with 100, 1000, 10000, and 100000 rows and tests with each load have been executed ten times. The tables below contain the following columns: number of rows, the tables name, average of each execution time and corresponding standard deviation. The last column shows the ratio between execution times of built-in code and execution times of corresponding triggers. All measures are given in milliseconds.

Table 3. Application-time period table, test of PK integrity

# Rows	Table	Mean (ms)	Standard devation (ms)	Speedup
100	Employees	72	9	
100	T_employees	59	9	1.21
1000	Employees	824	68	
1000	T_employees	620	58	1.33
10000	Employees	9135	202	
10000	T_employees	6878	95	1.33
100000	Employees	90229	1397	
100000	T_employees	69661	469	1.30

Table 3 shows execution times for violation of PK integrity with the INSERT statement in Examples 5 and 6. For all loads, the ratio shows that execution time of temporal logic is ca. 1.3 times faster than corresponding hand-coded applications.

Table 4 shows execution times for the DELETE statement in Example 7 and the corresponding stored procedure. For all but first load, the ratio shows the similar results as for the previous test, i.e. that execution time of temporal logic is ca. 1.45 times faster than corresponding hand-coded applications. (The anomalous result is due to the very small amount of rows in the first load.)

Table 4. Application-time period table: Execution of a DML statement (DELETE)

# Rows	Table	Mean (ms)	Standard devation (ms)	Speedup
100	Employees	109	12	
100	T_employees	514	54	0.21
1000	Employees	1007	11	
1000	T_employees	694	92	1.45
10000	Employees	9004	273	
10000	T_employees	6089	247	1.48
100000	Employees	88388	613	
100000	T_employees	58956	223	1.50

Table 5. System-versioned table: Execution of a DML statement (UPDATE)

# Rows	Table	Mean (ms)	Standard devation (ms)	Speedup
100	Departments	79	16	
100	S_departments	63	9	1.24
1000	Departments	813	100	
1000	S_departments	616	70	1.32
10000	Departments	8033	94	
10000	S_departments	6936	178	1.16
100000	Departments	82026	2444	
100000	S_departments	70803	723	1.16

Table 6. System-versioned table: Execution of a DML statement (DELETE)

# Rows	Table	Mean (ms)	Standard devation (ms)	Speedup
100	Departments	69	10	
100	S_departments	59	8	1.18
1000	Departments	658	85	
1000	S_departments	544	78	1.21
10000	Departments	7885	164	
10000	S_departments	6597	524	1.20
100000	Departments	82987	2302	
100000	S_departments	71040	696	1.17

Table 5 shows execution times for UPDATE in Example 12 and the corresponding trigger. The execution times differ non-significantly from the times in Table 3 and 4. The slightly better ratio for hand-coded applications can be explained with the minor complexity of system-versioned tables (see Section 3.3). The same is true for execution times of the DELETE statement in Table 6.

Table 7 shows execution times for violation of PK integrity with the INSERT statement for bitemporal tables. The corresponding statement is given in Example 18 and the corresponding trigger. The semantics of this constraint

Table 7. Bitemporal table: Test of PK integrity

# Rows	Table	Mean (ms)	Standard devation (ms)	Speedup
100	B_departments	75	17	
100	BITemp_departments	59	10	1.27
1000	B_departments	895	76	
1000	BITemp_departments	668	50	1.34
10000	B_departments	10546	200	
10000	BITemp_departments	7713	146	1.37

Table 8. Bitemporal table: Execution of a DML statement (DELETE)

# Rows	Table	Mean (ms)	Standard devation (ms)	Speedup
100	B_departments	297	25	
100	BITemp_departments	58	11	5.13
1000	B_departments	3175	262	
1000	BITemp_departments	573	50	5.54
10000	B_departments	45078	1813	
10000	BITemp_departments	7277	225	6.19

for bitemporal tables is equivalent to the semantics of the same constraint for application-time period tables. Therefore, the ratio of execution times is similar.

Table 8 shows execution times for a DML statement (Example 19) for bitemporal tables. As can be seen in Section 4.3, the semantics of UPDATE and DELETE statements on bitemporal tables is very complex. (The most complex form of an UPDATE statement requires an UPDATE and three INSERT statements.) Therefore, execution times of triggers and stored procedures for DML statements of bitemporal tables are several times more slowly than the corresponding implementation of temporal logic.

6   Conclusions and Future Work

In database systems, which do not support built-in temporal logic, users have to implement this functionality in their application code. The SQL:2011 standard with its specification of temporal data and IBM DB2 with the implementation of the same dramatically simplify the code that has to be written.

In order to show execution times of built-in temporal logic vs. corresponding hand-coded applications, we performed experiments to measure the performance of both groups using the same data. Our performance tests showed three important results. First, execution times of built-in code is ca. 1.5 times quicker than execution time of the corresponding triggers and stored procedures. Second, standard deviation of hand-coded applications is significantly greater than corresponding deviation for applications implemented using built-in temporal logic. Another conclusion is: the more complex the semantics of a statement, the bigger the difference in their execution times.

It is well known that the first specification of temporal data model in SQL:2011 is non-satisfying [11]. The main defect is that the standard, at this time, does not specify two very important temporal features: the PERIOD data type and coalescing. The PERIOD data type is defined as the time interval, which contains a set of consecutive time units. This data type has the lower and upper limit, which are both of type DATE or TIMESTAMP. The most important property of this data type is that it can be used in the natural way to represent time intervals. Additionally, it supports operations, such as CONTAINS, EQUALS, PRECEDES and OVERLAPS as methods of the data type. We expect that the support of this data type will come soon, and the study of performance advantages of the PERIOD data type is one of our main goals in the future. Also, the current standard lacks the support for coalescing. Coalescing means that the system automatically merges the rows of a table, which overlaps [2]. This problem appears very often when an INSERT (UPDATE) statement is performed, and time-invariant attributes of the rows are equal and at the same time their timestamps overlaps or adjoin. It can be expected that performance gains of the support for coalescing will bring significant benefits in relation to hand-coded solution. For this reason, this is also one of the goals in our work.

References

1. Atay, C.: A Comparison of Attribute and Tuple Time Stamped Bitemporal Relational Data Models. In: Int. Conf. on Applied Computer Science (2010)
2. Boehlen, M., Snodgrass, R.: Coalescing in Temporal Databases, VLDB (1996)
3. Darwen, H., Date, C.: An Overview and Analysis of Proposals Based on the TSQL2 Approach (1996), http://www.dcs.warwick.ac.uk/~hugh/TTM/OnTSQL2.pdf (last visit: February 14, 2014)
4. Gadia, S.: Ben-Zvi's Pioneering Work in Relational Temporal Databases. In: Tansel, A., et al. (eds.) Temporal Databases. Benjamin/Cummings (1993)
5. Grandi, F.: Introducing an Annotated Bibliography on Temporal and Evolution Aspects in the Semantic Web. SIGMOD Records 41(4) (2012)
6. Kulkarni, K., Michels, J.: Temporal Features in SQL:2011. SIGMOD Records 41(3) (2012)
7. ISO/IEC 9075-2:2011: Database languages: SQL, Part 2 (2011)
8. Jensen, C.S., et al.: The consensus glossary of temporal database concepts - February 1998 version. In: Etzion, O., Jajodia, S., Sripada, S. (eds.) Temporal Databases - Research and Practice. LNCS, vol. 1399, pp. 367–405. Springer, Heidelberg (1998)
9. Lorentzos, N.: The Interval-extended Relational Model and Its Applications to Valid-time. In: Temporal Databases (1993)
10. Nicola, M., Sommerlandt, M.: Managing time in DB2 with temporal consistency. IBM Developers Works (2011)
11. Petković, D.: Was lange währt, wird endlich gut: Temporale Daten im SQL-Standard. Datenbank-Spektrum 13(2), 131–138 (2013) (in German)
12. Saracco, C., Nicola, M., Gandhi, L.: A matter of time: Temporal data management in DB2 (2012), http://www.ibm.com/developerworks/data/library/techarticle/dm-1204db2temporaldata/dm-1204db2temporaldata-pdf.pdf (last visit: February 14, 2014)

13. Snodgrass, R., Ahn, I.: Performance Evaluation of a Temporal Database Management System. Communications of ACM (1986)
14. Snodgrass, R.: The TSQL2 Temporal Query Language. Kluwer (1995)
15. Tansel, A., Clifford, J., Gadia, S., Jajodia, S., Segev, A., Snodgrass, R.: Temporal Databases (1993)
16. Toman, D.: A Point-based Temporal Extension of SQL. In: Bry, F., Ramakrishnan, R., Ramamohanarao, K. (eds.) DOOD 1997. LNCS, vol. 1341, pp. 103–121. Springer, Heidelberg (1997)
17. Bach, M., Werner, A.: Standardization of NoSQL Database Languages. In: Kozielski, S., Mrozek, D., Kasprowski, P., Małysiak-Mrozek, B., Kostrzewa, D. (eds.) BDAS 2014. CCIS, vol. 424, pp. 50–60. Springer, Heidelberg (2014)

Performance Aspects of Migrating a Web Application from a Relational to a NoSQL Database

Katarzyna Harezlak[(✉)] and Robert Skowron

Silesian University of Technology,
Gliwice, Poland
katarzyna.harezlak@polsl.pl

Abstract. There are many studies which discuss the problem of using various NoSQL databases and compare their efficiency thus confirming their usefulness and performance quality. However, there are very few studies dealing with the problem of replacing data storage for systems currently working. This lack has become a motivating factor to examine how difficult and laborious it is to move an existing, regularly used application, based on the relational environment to a non-relational data structure. The difficulty of carrying out a data migration process, the scope of changes which would have to be done in the existing environment and the efficiency of an application while using new data structures were considered in the presented research. As an example one of on–line games, being a good representative for popular web applications, was chosen.

Keywords: Web application · Relational database · NoSQL · Efficiency · Data migration

1 Introduction

In the field of the database science NoSQL databases has become more and more popular. Many computer users mistakenly interpret this term as system, which does not use SQL language, while NoSQL stands for "not only SQL". Data stored in such databases does not use strictly defined schemas, data access is often realized without joins, which make NoSQL databases more scalable and efficient. Sometimes these systems are identified with one technology while there are many various solutions being an alternative for relational databases.

This appearance of new data models has led some applications developers to question whether it is more beneficial to transfer existing data from relational databases to these new structures. Going further the next questions regarding the possibility, feasibility and reasonableness of undertaking such tasks should be asked. To answer these questions the difficulty of carrying out a data migration process, the scope of changes which would have to be done in the existing environment and the efficiency of an application while using new data structures were considered in the presented research.

© Springer International Publishing Switzerland 2015
S. Kozielski et al. (Eds.): BDAS 2015, CCIS 521, pp. 107–115, 2015.
DOI: 10.1007/978-3-319-18422-7_9

There are many studies which discuss the problem of using various NoSQL databases and compare their efficiency thus confirming their usefulness and performance quality, yet they focused their attention mainly on comparing features or a performance of particular databases [1–4, 10, 12]. Some research compared NoSQL and relational databases but taking another point of view or SQL databases differing from that used in this research [7, 13, 14]. However, there are very few studies dealing with the problem of replacing data storage for systems currently working. This lack has become a motivating factor to examine how difficult and laborious it is to move an existing, regularly used application, based on the relational environment to a non-relational data structure. As an example one of on–line games, being a good representative for popular web applications, was chosen.

2   The Research Environment

The on-line web game **www.wygrajmecz.pl**, uniting many thousands of users, was chosen as a test environment. It combines features of managerial and role playing games (RPG) in a football area. The users impersonate football club owners and compete with each other by playing games, which, together with their results, are the basis for building the game ranking. The game is very similar to many other games on the market. Its main module is a kernel, with which cooperate all the elements of the application. The module is developed using PHP language. The operation of this game is based on the MySQL relational database; in which the way of data storing and modelling can be treated as universal and commonly used. There is a lack of untypical solutions with one exception. There was a new text type defined, in which the structure depends on a record type. From that point of view it may be compared to XML documents. The database structures have their equivalents in the application classes described by three terms:

- MetaClass – an equivalent to a database table,
- MetaList – a set of rows of a particular table,
- MetaAttributes – a column cell including one or a few values.

MetaClass is a base class, after which inherits all classes defined by developers.

Each of the previously–mentioned classes plays an important role in setting the communication between the application and the database using the functionality included in the *TConnectionDB.php* file. The contribution of a particular element in that cooperation is presented in Fig. 1.

The modification of those files was required to achieve one of aims of the research – provide the on–line game with connection possibilities for both relational database and the analysed structures minimizing the necessity of changing the application code and development patterns used. As a new database structures two representatives of NoSQL databases; MongoDB - a document database and Cassandra – referred as a column database were chosen. Usage of a graph database were also considered [11]. However an analysis of its characteristic

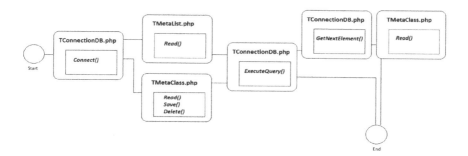

Fig. 1. Schema of the connection process to the relational database

features indicated that such type of database is thought to be used in other application types than that studied in the research.

3   Tests of Data Migration

The basis for conducting the experiments planned was developing a connection between the application and databases. The changes, which had to be done in the application kernel turned to be simple and easy to implement. Drivers of both selected database types for PHP language (*php_mongo.dll* and *phpCassa* respectively) are equipped with appropriate functions to facilitate opening and closing connection to a database [8].

Establishing this connection opened possibilities to carry out a data migration process from the relational database to NoSQL tools [6]. It was based on scripts defined especially for this purpose. This process was repeated for various number of records to check scalability of the solution. The outcomes of the experiments are presented in table 1.

Table 1. Duration of data migration [s]

Database	100 records	500 records	1000 records	10000 records
MongoDB	1.571	8.281	16.652	174.986
Cassandra	4.694	23.908	49.912	512.164

MongoDB, in this process, turned out to be meaningfully more efficient than Cassandra – the duration of the same task for Cassandra was three times longer than for MongoDB. Nevertheless, it should be emphasized that such operation is realized only once thus its efficiency should not be a key problem if a performance of execution of other statement types is satisfactory. Dealing with these issues was the direction of subsequent studies.

However, the problem may arise when such relational and NoSQL databases will be used in federated database systems, in which process of synchronizing data is repeated on a regular basis.

4 Mapping Database Statements

Fulfilling requirements of minimizing an influence of changes made on developers work, as it was mentioned above, was a simple task in case of opening and closing a connection to databases. The more complicated problem to be faced was query processing. The original solution assumed using SQL language to formulate operations for the relational database. For this reason all the application functions cooperating with a data source expect a parameter of this form. The variety of NoSQL database languages in conjunction with requirements defined earlier has led to working out a mechanism translating SQL operation to a text formatted in a way ensuring a possibility for representing SQL constructions in a notation typical for a particular NoSQL language. The description of these symbols is presented in table 2.

Table 2. Description of formatted text symbols

Symbol	Description
|	Dividing a string into parts describing a table and filtering conditions
/	The separation of a table name from an operation to be performed
^	The separation of WHERE and ORDER BY clauses into parts including filtering conditions
#	The separation of subsequent WHERE and ORDER BY clauses terms
&	Defining a key and its value in a single term

Each of SQL statements defined by MetaClass and MetaList methods is translated to elements being subsequent parts of a formatted string, returned as results of these method execution (Fig. 2). The example of two operation in MongoDB and Cassandra databases written using proposed convention is presented in subsequent listings.

– MongoDB

```
$ls_sql = "$temp->TableName/find|$ls_key."&".$ls_value.
"#".$ls_key."&".$ls_value."^".$ls_key."&".$ls_value
"#".$ls_key."&".$ls_value;
```

The first section of the string includes data on a collection (TableName) and an operation to be performed (find). The filtering conditions being criterions for MongoDB *find* method constitutes the second part of the string. There can be WHERE (|) and ORDER BY ("^") clause distinguished, both with two elements of *key–value* type.

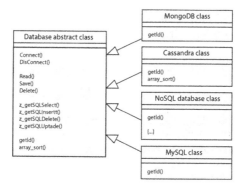

Fig. 2. Dependency of the abstract database class and classes cooperating with databases of different types

– Cassandra

```
$ls_sql = "".$this->TableName."/update|".$this->PmFieldName."&"
.$this->Pm."^".$lo_attr->Name."&".$lo_attr->Value."#"
.$lo_attr->Name."&".$lo_attr->Value."";
```

In this string the symbols separate table name, Cassandra function to be performed (update), primary key name (PmFieldName) and its value (Pm), and attributes to be changed (Name and Value)

5   The Application Efficiency Comparison

Developing one general solution for coding database operations allowed for extending the scope of the experiments to compare an efficiency of the application tasks realization while using different data storages. Tests were built taking two points of view into account. First of them assumed checking the application performance in terms of its functionality. The second one regarded the analysis of the efficiency of various database statement types. As an environment for the experiments the youth club module was chosen, in which task is to give an application user an opportunity to search for new football players. The analysed scope of the functionality included:

– sending a scout to search a new football junior,
– cancelling search activities,
– accepting a scout suggestion,
– rejecting a scout suggestion.

The listed functionalities consisted of *select*, *insert*, *update* and *delete* operations and were studied using small databases – filled with several thousand records as well as bigger sets – with thirty thousand records. The latter configuration corresponds to a situation, in which ten thousand players are registered in a database. At the beginning of the experiments small and big databases for each system studied were filled with the same numbers of records, which was changing during the experiments.

5.1   The Application Performance Discussion

The analysis of the results was conducted for small and big databases independently (tables 3 and 4 respectively), however in both cases similar outcomes were obtained. The performance of tasks realized on the basis of MongoDB turned out to be almost the same as in MySQL instance. Furthermore in some cases the application cooperation with MongoDB database was more efficient than with MySQL one (*Sending scout* and *Accepting junior*). Comparing the results for different database sizes it may be observed that both of discussed systems are highly scalable. Duration of particular tasks realization was almost the same for both small and big databases. The outcomes for the third system, Cassandra, differ significantly from those obtained for MongoDB and MySQL databases. Operating on data in Cassandra databases was several times less efficient (in case of bigger database even hundreds times) and less scalable than in other databases.

Table 3. Small databases - tasks duration [s]

Small database	Presenting database content	Sending scout	Cancelling search	Rejecting junior	Accepting junior
MySQL	0.015	0.2	0.03	0.03	0.25
MongoDB	0.015	0.075	0.04	0.04	0.05
Cassandra	0.4	1.0	0.8	0.8	1.1

Table 4. Big databases - task duration [s]

Big database	Presenting database content	Sending scout	Cancelling search	Rejecting junior	Accepting junior
MySQL	0.015	0.2	0.03	0.03	0.45
MongoDB	0.018	0.09	0.03	0.03	0.05
Cassandra	6.0	5.0	5.26	5.26	5.7

Table 5. MySQL: various type operation duration [s]

MySQL: Small database	Presenting database content	Sending scout	Cancelling search	Rejecting junior	Accepting junior
SELECT	0.015	0.02	0.026	0.026	0.065
INSERT	-	0.0175	-	-	0.18
UPDATE	-	0.005	0.002	0.002	0.002
DELETE	-	-	0.002	0.002	0.003

Table 6. MongoDB: various type operation duration [s]

MongoDB: Small database	Presenting database content	Sending scout	Cancelling search	Rejecting junior	Accepting junior
SELECT	0.015	0.01	0.025	0.025	0.015
INSERT	-	0.05	-	-	0.025
UPDATE	-	0.0015	0.01	0.01	0.005
DELETE	-	-	0.005	0.005	0.005

5.2   The Analysis of Statement Efficiency

To explore the reasons of the previously obtained results, the subsequent step of the research was carried out. It included the studies on the application performance in terms of an executed operation type. To achieve better readability they were presented for each of database servers independently. Times of SQL statements realization in MySQL database were shown in table 5. The results indicate that a task performance was mostly influenced by INSERT operations (in case a task included such statements). The second most time consuming task was SELECT one, whereas both DELETE and UPDATE statements had a little impact on the application efficiency. In case of MongoDB databases, although the general results were similar to outcome obtained for MySQL one, a contribution of particular statements were slightly different. Once again inserting records proved to be the most time consuming operation, however meaningful difference is visible only in one task (*Accepting Junior* – table 6). The bigger impact of the UPDATE and DELETE operations is also worth noting. As it was discussed earlier summarized results shown low efficiency of the application and Cassandra database cooperation. The studies of particular operation types revealed huge influence of SELECT duration on the final outcome (between 40–75% of the total execution time). It results from the fact that while reading data from a Cassandra database redundant records, filtered locally in the application, had to be taken, lengthening a statement execution time. The INSERT statement which requires unordered data scanning to find the maximal value of a record identifier also turned out to be time–consuming operation (between 20 and 40%

Table 7. Cassandra: various type operation duration [s]

Cassandra: Small database	Presenting database content	Sending scout	Cancelling search	Rejecting junior	Accepting junior
SELECT	0.4	0.43	0.6	0.6	0.7
INSERT	-	0.42	-	-	0.3
UPDATE	-	0.15	0.05	0.05	0.05
DELETE	-	-	0.15	0.15	0.05

of the execution time). In case of the most of the other statements durations differences are on the 8–15% level.

The results obtained for bigger databases created in MySQL and MongoDB system were almost identical as for smaller databases. All proportion was kept as well. The outcomes related to Cassandra usage confirmed earlier insights, however the influence of SELECT statement increased to 75-99% of the total time of the task execution. It maked duration of other operations negligible.

6   Conclusions

The research presented in the paper revealed that the MongoDB database, in most of the analysed cases (during the application adaptation process and data migration), turned out to be the better of the studied solutions. The performance tests strengthened this impression – Cassandra seemed to be a less efficient system and from the developers point of view using it in the analysed system would be useless and pointless. One of the main reasons of the weaker performance is the shortcomings of the Cassandra driver for PHP language. It may be expected that the development of such software, more adjusted to the environment used, could significantly improve the quality of the Cassandra system utilisation.

Nevertheless, the analysis conducted in the research requires another point of view to be taken into consideration as well. The application modification to work with new types of data sources and the later experiments had to be organized in a precisely defined and strictly imposed way. These limitations probably made Cassandra not fit to the analysed environment. It may be anticipated that developing an application from scratch to cooperate with this kind of database would provide a more efficient solution. In fact there are studies ([1, 5, 9]), in which the results show that Cassandra is a system of high availability and scalability, yet they were obtained based on different types of experiments.

The most important findings reveal a very good interoperability of the application and MongoDB databases, in a few cases it was even better than using the original data source. Although these results were obtained for a subset of the game functionality, they allow having an expectation of a performance recurrence while using other its modules. Such a prospect is reasonable because the chosen part of the application included representatives for all types of data and operations available in the application. The confirmation of these assumptions as well as involving other types of NoSQL databases is planned as a future work.

References

1. Abramova, V., Bernardino, J.: NoSQL Databases: MongoDB vs Cassandra. In: Proceedings of the International C* Conference on Computer Science and Software Engineering, C3S2E 2013, pp. 14–22. ACM, New York (2013), http://doi.acm.org/10.1145/2494444.2494447
2. Bushik, S.: Evaluating NoSQL Performance: Time for Benchmarking (2014), http://www.slideshare.net/tazija/evaluating-nosql-performance-time-for-benchmarking
3. Cattell, R.: Scalable SQL and NoSQL Data Stores. SIGMOD Rec. 39(4), 12–27 (2011), http://doi.acm.org/10.1145/1978915.1978919
4. Lungu, I., Tudorica, B.G.: The Development of a Benchmark Tool for NoSQL Databases. Open PDF Journal – Database Systems Journal IV(2), 13–20 (2012), http://www.dbjournal.ro/archive/12/12_2.pdf
5. Manoj, V.: Comparative study of NoSQL Document, Column Store Databases And Evaluation Of Cassandra. International Journal of Database Management Systems (IJDMS) 6(4), 11–26 (2014)
6. MongoDB Manual 2.6: Import and Export MongoDB Data (2014), http://docs.mongodb.org/manual/core/import-export/
7. Parker, Z., Poe, S., Vrbsky, S.V.: Comparing NoSQL MongoDB to an SQL DB. In: Proceedings of the 51st ACM Southeast Conference, ACMSE 2013, pp. 5:1–5:6. ACM, New York (2013), http://doi.acm.org/10.1145/2498328.2500047
8. PhP Manual: Database Extensions (2014), http://php.net/manual/en/refs.database.php
9. Planet Cassandra: Apache Cassandra NoSQL Performance Benchmarks (2014), http://planetcassandra.org/nosql-performance-benchmarks/
10. Stonebraker, M.: SQL Databases V. NoSQL Databases. Commun. ACM 53(4), 10–11 (2010), http://doi.acm.org/10.1145/1721654.1721659
11. Tezer, O.S.: A Comparison Of NoSQL Database Management Systems And Models (2014), https://www.digitalocean.com/community/tutorials/a-comparison-of-nosql-database-management-systems-and-models
12. Tudorica, B.G., Bucur, C.: A comparison between several NoSQL databases with comments and notes. In: 2011 10th Roedunet International Conference (RoEduNet), pp. 1–5 (June 2011)
13. van der Veen, J.S., van der Waaij, B., Meijer, R.J.: Sensor Data Storage Performance: SQL or NoSQL, Physical or Virtual. In: Proceedings of the 2012 IEEE Fifth International Conference on Cloud Computing, CLOUD 2012, pp. 431–438. IEEE Computer Society, Washington, DC (2012), http://dx.doi.org/10.1109/CLOUD.2012.18
14. Wei-ping, Z., Ming-xin, L., Huan, C.: Using MongoDB to implement textbook management system instead of MySQL. In: 2011 IEEE 3rd International Conference on Communication Software and Networks (ICCSN), pp. 303–305 (May 2011)

A Consensus Quorum Algorithm
for Replicated NoSQL Data

Tadeusz Pankowski[✉]

Institute of Control and Information Engineering,
Poznań University of Technology, Poland
tadeusz.pankowski@put.poznan.pl

Abstract. We propose an algorithm, called Lorq, for managing NoSQL
data replication. Lorq is based on consensus quorum approach and is
focused on replicating logs storing update operations. Read operations
can be performed on different levels of consistency (from strong to eventual consistency), realizing so-called service level agreements (SLA). In
this way the trade-off among availability/latency, partition tolerance and
consistency is considered. We discuss correctness of Lorq and its importance in developing modern information systems based on geo-replication
and cloud computing.

1 Introduction

New classes of information systems such as Web services, Web search [8,7], e-
commerce [9], and social networks [15], demand data stores using paradigms
different from those of centralized conventional databases. They need more scalable distributed solutions for managing huge data repositories based on NoSQL
key-value data models [6,4], and guaranteeing very high availability and low
latency [9,15,19]. An important way in which such data stores differ from conventional databases is that they have to deal with the trade-off between consistency, availability/latency and partition tolerance. This issue was formulated by
so-called CAP theorem [5]. The theorem states that these three features cannot
be achieved simultaneously, and a lot of research have been done to propose
solutions considering this trade-off [1,12,14].

In this paper, we propose a new data replication algorithm called Lorq (*Log
Replication based on Quorum consensus*). The algorithm is based on a *consensus quorum* approach [11,16], and on *eventual consistency* [21,20,13], which is a
weak variant of strong consistency. In Lorq, the replication is realized through
replication of logs (storing update operations) instead of data, like in Raft [18]
managing data in a replicated state machine. To solve the trade-off between consistency and high availability, read operations in Lorq may be performed using
different levels of guaranteed consistency – from strong consistency to one of
four kinds of eventual consistency.

The outline of this paper is as follows. The next sections reviews ideas of
consensus quorum algorithms. Section 3 presents the Lorq algorithm and some
ways for achieving strong and eventual consistency in Lorq. The correctness of
Lorq is discussed in section 4. Finally, section 5 concludes the paper.

S. Kozielski et al. (Eds.): BDAS 2015, CCIS 521, pp. 116–125, 2015.
DOI: 10.1007/978-3-319-18422-7_10

2   Basic Idea of Quorum Consensus Algorithms

In a quorum-based data replication, it is required that an execution of an operation (i.e., a propagation of an update operation or a read operation) is committed if and only if a sufficiently large number of servers acknowledge the successful termination of this operation [11]. Let us denote: N – a number of servers storing copies of data (replicas); R – an integer called *read quorum*, meaning that at least R copies were successfully read; W – an integer called *write quorum*, meaning that propagations to at least W servers have been successfully terminated. The following relationships hold between N, R and W:

$$W > N/2, \tag{1}$$
$$R + W > N. \tag{2}$$

To commit a read operation, a server must collect the read quorum, and to commit a write operation must collect the write quorum. Condition (1) guarantees that the majority of copies is updated, and (2) that among read copies at least one is up-to-date.

The aim of consensus algorithms is to allow a collection of servers to process users' commands (updates and reads) as long as the number of active servers is not less than $max\{W, R\}$. It means that the system is able to survive failures of some of its servers.

3   Lorq – Log Replication Based on Quorum Consensus Algorithm

During last decade, the research on consensus algorithms is dominated by Paxos algorithms [10,16,17]. Lately, a variant of Paxos, named Raft [18], was presented as a consensus algorithm for managing a replicated log. Lorq is based on ideas underlying Paxos and Raft, and includes such steps as: (1) leader election; (2) log replication, execution and commitment of update operations; (3) realization of read operations on different consistency levels.

3.1   Architecture

The architecture of Lorq (Fig. 1), like in the case of Raft [18], is organized having in mind: operations, clients, and servers occurring in the system managing data replication.

Operations. We distinguish three *update* operations: *set*, *insert*, and *delete*, and one *read* operation. In order to informally define syntax and semantics of operations, we assume that there is a NoSQL database $DB = \{(x, \{A : a\})\}$ storing a key-value data object $(x, \{A : a\})$. Additionally, any data object in DB has a timestamp of the last operation updating this object. Operations are specified as follows:

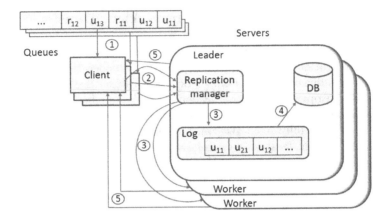

Fig. 1. Architecture of Lorq system. Thera are update and read operations in queues. Update operations are delivered (1) by clients from queues to one leader (2). A replication module delivers them to leader's log and to logs at all workers (3). Sequences of operations in all logs tend to be identical and are applied in the same order to databases (4). States of all databases also tend to be identical (eventually consistent). A client may read (5) data from any server.

- $set(dataId, dataVal)$ – the value of an existing data with identifier $dataId$ must be set to $dataVal$, e.g., $set(x, \{A : b\})$ and $set(x, \{B : b\})$ executed in DB changes its state to $\{(x, \{A : b\})\}$ and $\{(x, \{B : b\})\}$, respectively.
- $ins(dataId, dataVal)$ – the new key-value data $(dataId, dataVal)$ is inserted, or the value of the existing data object $dataId$ is modified accordingly, e.g.: successive execution of $ins(x, \{B : b\})$ and $ins(y, \{C : c\})$ against DB, changes its state to $DB = \{(x, \{A : a, B : b\}), (y, \{C : c\})\}$.
- $del(dataId, dataVal)$ – the existing key-value data $(dataId, dataVal)$ is deleted, or the value of the existing data object $dataId$ is modified accordingly, e.g.: successive execution of $del(x, \{B : b\})$ and $del(y, \{C : c\})$ against the DB state considered above, gives $DB = \{(x, \{A : a\})\}$.
- $read(dataId)$ – reads the key-value data with identifier $dataId$.

Queues. Operations from users are serialized and put in queues. An operation is represented as follows:

- any update operation, $opType(dataId, dataVal)$, $opType \in \{set, ins, del\}$, is represented as a tuple $(opId, clId, stTime, opType, dataId, dataVal)$;
- any read operation, $read(dataId)$ – as a tuple $(opId, clId, stTime, dataId)$,

operation identifier $(opId)$, *client identifier* $(clId)$, and *start timestamp* $(stTime)$ are determined, when the operation is delivered by a client to the leader.

Clients. Each queue is served by one *client*.

Servers. A server maintains one replica of a database along with the log related to this replica, and runs the software implementing Lorq protocols. The number of servers depends on the assumed efficiency of the system and the intended number of tolerated failures. For example, to be able to tolerate one failure, we need three servers, but to tolerate two failures we need five servers (that follows from conditions (1) and (2)). One server plays the role of *leader* and the rest roles of *workers*. The state of each server is characterized by the server's log and database state. The following variables relate to the state of a server:

- $lastIdx$ – last log index (the highest position storing an operation),
- $lastOpTime$ – the $stTime$ stored at $lastIdx$ position,
- $lastCommit$ – the highest log index storing a committed operation,
- $currentLeader$ – identifier of the current leader, 0 – the leader is not elected,
- $lastActivity$ – the observed latest activity time of the leader.

3.2   States and Roles in Lorq System

The Lorq system can be in one of the following three states: (1) UnknownLeader – no leader is established in the system (when the system starts and a short time after a leader failure). (2) LeaderElection – election of a leader is in progress. (3) KnownLeader – a leader is known in the system. In these states, a client plays, respectively, the following roles: Asking, Waiting, and Querying, whiles roles of a server are, respectively: Server, Elector, and Leader or Worker.

Client in Asking role: Initially, and when the system is in UnknownLeader state, a client asynchronously sends the request *GetLeader* to all servers. In response, each available server can return:

- 0 – if a leader can not be elected because the majority quorum cannot be achieved (the system is unavailable);
- $currentLeader$ – identifier of the leader that is either actually known to the server or has been elected in reaction to the *GetLeader* request.

Client in Waiting role: After issuing *GetLeader* request, the client plays the Waiting role. Next, depending on the reply, changes its role to Asking or Querying.

Client in Querying role: A client reads from a queue the next operation, or reads again a waiting one, if necessary, and proceeds as follows:

- determines values for $opId$, $clId$, and $stTime$;
- any update operation is sent to the leader using an *Update* command

$$Update(opId, clId, stTime, opType, dataId, dataVal); \qquad (3)$$

and the operation is treated as a *"waiting"* operation;
- any *read* operation is handled according to the required consistency guarantees (see section 3.4), and is sent using a *Read* command

$$Read(opId, clId, stTime, dataId). \qquad (4)$$

If the *Update* command replays committed, the corresponding operation is removed out from the queue. Otherwise, the client changes its role to Asking.

Server in Server role: A server plays the Server role when the system starts, and when a worker detects that election timeout elapses without receiving any message from the leader (that means the leader failure). A server in Server role starts an election issuing the command *StartElection* to all servers. Next, the systems goes to the LeaderElection state, and all servers change their roles to Elector.

Server in Elector role: The community of servers attempt to chose a leader among them. The leader should be this server that has the highest value of *lastCommit*, and by equal *lastCommit*, the one with the highest identifier (or satisfying another priority criterion). Next, the system goes to the KnownLeader state, the server chosen as the leader changes its role to Leader, and the remainder servers to Worker.

Server in Leader role: The leader acts as follows:

1. After receiving *Update* command (3), a leader issues asynchronously to each server (including itself) the command appending the operation to logs

$$Append(opId, clId, stTime, opType, dataId, dataVal, \qquad (5)$$
$$lastIdx, lastOpTime, lastCommit).$$

   The process can return with an exit code indicating:
   - success – the entry was successfully appended to the server's log,
   - inconsistency – the server's log is inconsistent with the leader's one. Then a procedure recovering consistency is carried out. After this, the leader retries sending the *Append* command. It is guaranteed, that after a finite number of repetition, the command returns with success (unless the worker fails).

   If the number of servers returning with success is at least equal to the write quorum (W), then the leader sends asynchronously to all these servers (also to itself) the request to execute the appended operation and to commit it. Next, replays committed to the client. If a worker does not respond to *Append* (because of crash, delay, or lost of network packet), the leader retries *Append* indefinitely (even after it has responded to the client). Eventually, all workers store the appended entry or a new election is started.

2. If the leader activity timeout elapses the leader sends

$$Append(lastIdx, lastOpTime, lastCommit)$$

   to each worker, i.e., *Append* command with empty data-part (so-called *heartbeat*), to confirm its role and prevent starting a new election. The response to this message is the same as to regular *Append*. In particular, this is important for checking log consistency, especially for restarting workers.

3. When a new leader starts, then:

- There can be some uncommitted entries in the top of the leader's log ($lastCommit < lastIdx$). These entries must be propagated to workers by *Append* command in increasing order. If a delivered log entry is already present in worker's log, it is ignored.
- Some *"waiting"* operations in a queue, i.e., denoted as already sent to a leader, could not occur in the leader's log (the reason is that they have been sent to a previous leader and that leader crashed before reaching the write quorum). Then these operations must be again sent by the client from the queue to the newly elected leader.
- After the aforementioned two operations have been done, the client starts delivering the next update operation from the queue.

A server plays the Leader role until it fails. After recovery, it plays a role of Worker.

Server in Worker role: Let a worker receive an *Append* command (possibly with empty data-part) of the form (5), where leader's parameters are denoted as: $lastIdxL$, $lastOpTimeL$, $lastCommitL$. Then:

1. If $lastIdx = lastIdxL$ and $lastOpTime = lastOpTimeL$, then: if the data-part is not empty, then append the new entry; execute and commit all uncommitted entries at positions less than or equal to $lastCommitL$ in increasing order; reply success.
2. If $lastIdx < lastIdxL$, then reply inconsistency. The worker expects that the leader will decide to send all missing entries.
3. If $lastIdx = lastIdxL$ and $lastOpTime < lastOpTimeL$, then delete $lastIdx$ entry and reply inconsistency.
4. If $lastIdx > lastIdxL$ then delete entries at positions greater than or equal to $lastIdxL$ and reply inconsistency.

A server plays the Worker role until its failure or failure of the leader – then it returns to the Server role.

3.3 Example

Now, we identify and discuss some steps in a process of receiving, replicating, executing and committing operations in Lorq system. We assume that there are five servers in the system ($N = 5$) (Fig. 2), and the write quorum is three ($W = 3$). By "+" we denote committed entries, and by "?" the waiting ones.

Step 1. Operations a and b are already committed, but c and d are waiting, i.e., their execution in the system is not completed. S_1 is the leader that propagated c and d to S_3, and after this failed. We assume that also S_5 failed. There are two uncommitted operations, c and d, in the top of S_3, these operations are also denoted as waiting.

Step 2. In the next election, S_3 has been elected the leader. The client receives information about the new leader and sends the waiting operations c and d to him. Since c and d are already in the S_3's log, they are ignored. S_3 propagates

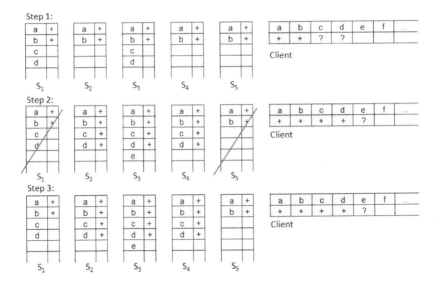

Fig. 2. Illustration to a scenario of managing data replication in Lorq

c and d to available workers S_2, and S_4. The write quorum is reached, so they are executed and committed. Next, S_3 receives the operation e, and fails.

Step 3. Now, assume that all servers are available and S_4 is elected as the leader. The client receives information about the new leader and sends the waiting operation e to S_4. Next, the leader propagates e as follows: (1) Propagation to S_1: e is appended at fifth position. Since leader's *lastCommit* is 4, thus c and d are executed in the S_1's database and committed. (2) Propagation to S_2: e is successfully appended at fifth position. (3) The leader (S_4) recognizes that the write quorum is already reached. Thus: asynchronously sends to S_1, S_2 and to itself the command to execute and commit e; replies committed to the client; continue propagations of e to S_3 and S_5. Note that e is already in S_3's log, so the append is ignored. There is an inconsistency between logs in S_4 and S_5. Thus, all entries at positions 3, 4 and 5 are propagated to S_5. Moreover, operations c, d and e are executed in S_5's database and committed. In the result, all logs will be eventually identical.

3.4   Consistency Models

A strong consistency can be provided by Lorq when reading operations are executed in the set of servers constituting the read quorum (R). Then the answer with the highest timestamp is chosen, and it is guaranteed, in force of (2), that the chosen value is up-to-date. However, providing the strong consistency is costly. For example, Amazon charges twice as much for strongly consistent reads as for eventually consistent reads in DynamoDB [2]. Thus, applications are often interested in possibility to declare their consistency and latency priorities [20].

Except of strong consistency, a user may expect a weaker kind of consistencies, which are referred to as *eventual consistencies*. This is similar to declaring isolation levels in conventional SQL databases. Some kinds of weak consistencies was discussed in [20]. In Lorq, some of them can be realized as follows:

1. *Consistent prefix:* A reader is guaranteed to see a state of data that existed at some time in the past, but not necessarily the latest state. In Lorq, this consistency is realized by reading from any server. It is guaranteed that databases in all servers store past or current states, and that these states were up-to-date sometimes in the past.
2. *Bounded staleness:* A reader is guaranteed that read results are not too out-of-data. To guarantee this, each server provides *lastUpdate* timestamp indicating when the last update of the server's database took place. The read operation is done against this server which guarantees acceptable staleness.
3. *Monotonic read:* A reader is guaranteed to see data states that is the same or increasingly up-to-date over time. In Lorq, the monotonic read is guaranteed by so-called *session guarantee* meaning that all read operations of a client are always addressed to the same worker.
4. *Read-my-writes:* It is guaranteed that effects of all updates performed by the client, are visible to the client's subsequent reads. In Lorq, it is guaranteed by reading from the leader database with considering the contents of its log.

4 Correctness

In this section we will discuss correctness of Lorq. To this order we will consider some invariable *correctness assertions*, which are always satisfied.

1. *Delivery guarantee:* Operations sent be a client are not lost. If the timeout elapses and the sent operation is not committed, it means that the leader fails. Then the client asks for a new leader and re-sends the operation.
2. *Superprioryty of leader's lastCommit:* The highest position of a committed operation in the leader's log is not less than this position in logs of any worker, i.e., $l.lastCommit \geq w.lastCommit$, where l denotes the leader and w any worker.
3. *Uniqueness of logs:* Operations on logs cannot be duplicated. Before appending an operation to a log, it is checked whether it already is stored in this log. So, any operation is stored only once.
4. *Monotonicity of leader's log:* Any operation that is sent to the leader's log can be either appended or ignored (if it is already in the log). No position in the leader's log can be removed or overwritten.
5. *Monotonicity of database states:* Changes in database states are caused by operations with increasing timestamps. It means, that there are no rollback (abort) of some previously executed operation.
6. *Prefix matching of logs:* If logs of two servers agree at a position k, then they agree at any position $i, i \leq k$. This follows from the precondition, which must be satisfied while an entry is appended to the log.

The following lemmas are consequences of the Lorq properties. They prove some aspects of correctness of Lorq but of course they do not form any complete proof.

Lemma 1. *Operations are executed by all servers in the same order.*

Proof. By contradiction, let us assume that a server srv_1 has executed an operation u_1 before u_2, and a server srv_2 executed these operations in the reverse order, denoted $srv_1.u_1 < srv_1.u_2$ and $srv_2.u_2 < srv_2.u_1$. Then there are indexes $j_1 < j_2$ and $k_2 < k_1$, such that $log_1[j_1] = u_1$, $log_1[j_2] = u_2$, $log_2[k_1] = u_2$, $log_2[k_2] = u_1$, where log_1 and log_2 are logs of servers, respectively, srv_1 and srv_2. Let $j_1 < k_1 < j_2 < k_2$. Then, for the log log_0 of the leader, we have $log_0[j_1] = u_1, log_0[k_1] = u_2, log_0[j_2] = u_2, log_0[k_2] = u_1$. We see that operations are duplicated in the log. The same will be implied by another ordering of j_1, j_2, k_1, k_2. Thus, the log cannot be under control of Lorq algorithms.

Now let us assume that u_1 from a client cl_1 has been stored in $log_0[j_1]$ and propagated to all workers. Before ending the propagation, an operation u_2 has been sent by a client cl_2, stored at $log_0[j_1 + 1]$ and propagated. Assume that the propagation to srv_1 succeeded, i.e., $log_1[j_1] = u_1$, and the propagation of u_2 and arrived to srv_2 before u_1. However, u_2 cannot be appended since $leader.lastIdx = j_1$, but $srv_2.lastIdx = j_1 - 1$. According to Lorq protocol, the leader must send all missing entries preceding position j_1. Thus, either the leader will sent u_1 or the propagation of u_1 will arrive (the later of these two will be ignored). □

By $srv.read(x)(r) = (x, v, t)$, we denote that a read operation $read(x)$ has been issued in a real time r against a server srv and its result is (x, v, t).

Lemma 2. *Let $srv_1.read(x)(r_1) = (x, v_1, t_1)$, $srv_2.read(x)(r_1) = (x, v_2, t_2)$, and $t_1 < t_2$ (i.e., srv_1 is delayed with respect to srv_2). Then there is $r_2 > r_1$ such that $srv_1.read(x)(r_2) = (x, v_2, t_2)$.*

Proof. Let database states of srv_1 and srv_2 (in real time r_2) were determined by execution of all operations from logs, respectively, $log_1[1..j]$ and $log_2[1..k]$, where $log_1[1..j] \subset log_2[1..k]$. It follows from Lemma 1 that after $(k - j)$ appends log_1 will have k-entries and $log_1[1..k] = log_2[1..k]$. This equality will have place after a time interval Δ necessary to realize these $(k - j)$ appends. So, $r_2 = r_1 + \Delta$.

Note that in force of Lemma 2, the eventual consistency is guaranteed.

5   Conclusions

We discussed an algorithm, called Lorq, managing replicated NoSQL data using a consensus quorum approach and based on replication of logs storing update operations by respecting the write quorum. A system controlled by Lorq protocol consists of a set of autonomous servers, among them a leader is chosen. Applications have possibilities to declare their consistency and latency priorities, from strong consistency to one kind of eventual consistencies. In this way the server level agreement (SLA) is provided. We briefly discussed different kinds of consistency, and some aspects of correctness of Lorq. The Lorq system have been designed and implemented using tools oriented to asynchronous and parallel programing [3]. This research has been supported by Polish Ministry of Science and Higher Education under grant 04/45/DSPB/0136.

References

1. Abadi, D.: Consistency tradeoffs in modern distributed database system design. IEEE Computer 45(2), 37–42 (2012)
2. Amazon DynamoDB Pricing (2014), http://aws.amazon.com/dynamodb/pricing
3. Asynchronous Programming with Async and Await (2014), http://msdn.microsoft.com/en-us/library/hh191443.aspx
4. Bach, M., Werner, A.: Standardization of noSQL database languages. In: Kozielski, S., Mrozek, D., Kasprowski, P., Małysiak-Mrozek, B.z. (eds.) BDAS 2014. CCIS, vol. 424, pp. 50–60. Springer, Heidelberg (2014)
5. Brewer, E.A.: Towards robust distributed systems (Invited Talk). In: ACM Principles of Distributed Computing, Portland, Oregon, p. 7 (2000)
6. Cattell, R.: Scalable SQL and NoSQL data stores. SIGMOD Record 39(4), 12–27 (2010)
7. Cooper, B.F., Ramakrishnan, R., Srivastava, U., Silberstein, A., et al.: PNUTS: Yahoo!'s hosted data serving platform. PVLDB 1(2), 1277–1288 (2008)
8. Corbett, J.C., Dean, J., Epstein, M., et al.: Spanner: Google's globally distributed database. ACM Trans. Comput. Syst. 31(3), 8 (2013)
9. DeCandia, G., Hastorun, D., Jampani, M., et al.: Dynamo: Amazon's highly available key-value store. SIGOPS Oper. Syst. Rev. 41(6), 205–220 (2007)
10. Gafni, E., Lamport, L.: Disk Paxos. In: Herlihy, M.P. (ed.) DISC 2000. LNCS, vol. 1914, pp. 330–344. Springer, Heidelberg (2000)
11. Gifford, D.K.: Weighted Voting for Replicated Data. In: ACM SIGOPS 7th Sym. on Op. Systems Principles, SOSP 1979, pp. 150–162 (1979)
12. Gilbert, S., Lynch, N.A.: Perspectives on the CAP Theorem. IEEE Computer 45(2), 30–36 (2012)
13. Golab, W., Rahman, M.R., AuYoung, A., Keeton, K., Li, X.: Eventually consistent: Not what you were expecting? Commun. ACM 57(3), 38–44 (2014)
14. Google: Balancing Strong and Eventual Consistency with Google Cloud Datastore (2014), https://cloud.google.com/developers/articles
15. Lakshman, A., Malik, P.: Cassandra: A decentralized structured storage system. SIGOPS Oper. Syst. Rev. 44(2), 35–40 (2010)
16. Lamport, L.: Generalized Consensus and Paxos. In: Technical Report MSR-TR-2005-33, pp. 1–63. Microsoft Research (2005)
17. Lamport, L.: Fast Paxos. Distributed Computing 19(2), 79–103 (2006)
18. Ongaro, D., Ousterhout, J.: In Search of an Understandable Consensus Algorithm. In: USENIX Annual Technical Conference, pp. 305–319 (2014)
19. Stonebraker, M., Cattell, R.: 10 rules for scalable performance in 'simple operation' datastores. Commun. ACM 54(6), 72–80 (2011)
20. Terry, D.: Replicated data consistency explained through baseball. Commun. ACM 56(12), 82–89 (2013)
21. Vogels, W.: Eventually consistent. Commun. ACM 52(1), 40–44 (2009)

Preserving Data Consistency in Scalable Distributed Two Layer Data Structures

Adam Krechowicz[(✉)], Stanisław Deniziak, Grzegorz Łukawski,
and Mariusz Bedla

Kielce University of Technology, Poland
{a.krechowicz,s.deniziak,g.lukawski,m.bedla}@tu.kielce.pl

Abstract. The scalable NoSQL systems are often the best solutions to store huge amount of data. Despite that they in vast majority do not provide some features known from database systems (like transactions) they are suitable for many applications. However, in systems that require data consistency, e.g. payments or Internet booking, the lack of transactions is still noticeable. Recently the need of transactions in such data store systems can be observed more and more often. The Scalable Distributed Two Layer Data Structures (SD2DS) are very efficient implementation of NoSQL system. This paper exposes its basic inconsistency problems. We propose simple mechanisms which will be used to introduce consistency in SD2DS[1].

Keywords: NoSQL · SD2DS · Consistency

1 Introduction

For fulfilling modern needs for data storing and processing, distributed systems are frequently used. Unfortunately, classic relational database model is inefficient for this type of systems. Splitting the data into multiple tables stored on many servers combined with locking usually results in a poor throughput. All the problems led to the new movement called NoSQL and beginning an era of "polyglot persistence" [14]. Many various solutions tailored to the specific needs were worked out. Depending on the field of application some requirements concerning consistency, persistence and method of data accessing should be redefined. The data does not have to be normalized and can be stored in some form of a schemaless aggregate. As a result the consistency restrictions are relaxed and the whole system offers better scalability.

NoSQL datastores offer outstanding scalability and throughput, but in contrast to databases may not entirely support consistency, transactions and sometimes even persistence. Assuring transactions and keeping consistency in a distributed system is a well known and widely discussed problem [13,16,2,5,17]. There are many worked out solutions addressing these problems, but none of

---

[1] The research used equipment funded by the European Union in the Innovative Economy Programme, MOLAB - Kielce University of Technology.

S. Kozielski et al. (Eds.): BDAS 2015, CCIS 521, pp. 126–135, 2015.
DOI: 10.1007/978-3-319-18422-7_11

them solves them completely and for all the cases. In practice, each distributed data structure should be considered individually in terms of preserving data consistency depending of the usage. On the other hand, transactions are not required in many modern applications while lack of them can be catastrophic in others. The banking, flight booking or any other systems that require payments will always need full transactions mechanisms [17].

2 Related Work

The vast majority of distributed stores do not use classical transactions to preserve data consistency. They base on weaker mechanisms that proved to be sufficient in many applications [1]. For example the BigTable systems can only support consistency which concerns only single data portion [4]. Many others, like CloudTPS, offer only *eventual consistency* which guarantees that eventually data will propagate to all its replicas if there are no further data modifications [17]. In that case the newest version of data could not be visible to all of the clients. The vast majority of NoSQL systems uses data versioning to preserve integrity of the data. NoSQL that provide strong consistency, like Neo4J, offer very weak scalability features.

In general, systems that do not support transactions are good solutions only if data can be well partitioned [16]. Because of that in the last few years there is a noticeable trend of returning to the classic ACID (Atomic, Consistent, Independent and Durable) transactions in NoSQL systems. The examples of such systems with transactions supports are Google Percolator [13], Google Spanner [5], Google Megastore [2], CloudTPS [17] or Calvin [16].

The vast majority of such systems were developed on top of the classic Big Data systems and introduced the transactions. For example Calvin can use any NoSQL system that provides basic data manipulation operations CRUD to provide full ACID transactions [16]. Some of them support transaction only in a limited manner. Google Megastore partitions the data into so called entity group. The full ACID transactions are provided only within the entity group [2].

CloudTPS is one of such systems that supports the transactions in full extent. It was built on the top of the open source implementation of BigTable called HBase and Amazon SimpleDB system. It provides Atomicity by using of two-phase commit protocol which allows to confirm the changes in a distributed system. The system assumes that if the transactions are performed correctly, the Consistency is preserved. It uses timestamps to order and sequentially perform the operations in conflicted transactions to fulfill the Isolation property. The Durability property is preserved by ensuring that all of the changes will be written to the store when the transaction is completed [17].

Percolator is another example of Big Data systems that provides transactions. It was originally developed to handle web search indexing [13]. It was built on the top of the Google BigTable system and utilize its versioning system to preserve consistency. It also introduce locking mechanisms and timestamps for stronger data consistency [13].

There are numerous software solutions that present different approach to a processes coordination in distributed applications. The main feature of these systems is the elimination of locking and waiting for the resource. They use strong consistency model by utilizing asynchronous version of *linearizability* [8] called *A-linearizability*. One of the most recognizable example of such a system is ZooKeeper [11]. It provides kernel for building more complicated coordination mechanism at the client. ZooKeeper has two basic ordering guarantees: *Linearizable writes* and *FIFO client order*. A client can use ZooKeeper API to build mechanisms such as *Rendezvous*, *Read/Write Locks*, *Double barrier*, and others.

3   Scalable Distributed Data Structures

Main purpose of Scalable Distributed Data Structures (SDDS) is storing huge sets of data using distributed RAM in a multicomputer [6]. The data is divided into records identified by the unique keys. The records with the keys are stored in so called buckets which are distributed among multicomputer nodes. All the buckets in a single addressing space form a file. There is a number of proposed architectures using different algorithms for addressing records [6]. An SDDS file consists of at least three different components: buckets storing the data, clients manipulating the data and additional control components. The file scales through splitting of the overloaded buckets.

Main feature of SDDS is that there is no central directory for addressing records. The distributed directories, so called file images, are used instead. As a result, a client may commit an addressing error while trying to access the data using an outdated file image. In such a case, the incorrectly addressed message is forwarded to the correct recipient and the client receives so called Image Adjustment Message (IAM). The IAM updates the client's file image, so it will never make the same addressing error again.

The SDDS offers outstanding scalability and throughput (as the data is stored in the distributed RAM). There is a number of extended SDDS architectures introducing storage of objects [3] and fault tolerance [12]. The transactions in SDDS were introduced in [9,10]. Ensuring consistency in SDDS is simplified, because each record with its key is always stored completely in one place (bucket).

Scalable Distributed Two Layer Data Structures (SD2DS) were developed on a basis of SDDS [15]. Main difference of SD2DS, in contrast to the SDDS, is that the structure is build of two layers, managed and addressed independently. The data is divided into so called components. Each component consists of two parts:

- The header, including the unique key of the component and so called locator pointing a place where the second part of the component is stored. Headers may contain also other metadata describing the components.
- The body, consisting of the component data supplemented with the key.

Headers are stored in the first layer of SD2DS, called the file, while bodies are stored in the second layer, called the storage. Both layers consist of buckets.

To manipulate the data, indirect addressing is used, so a client must firstly communicate with the first layer to obtain the locator and then gets access to the desired component body.

Main advantage of SD2DS over SDDS is that the component bodies are stored always in the same bucket, thus they are not moved during evolution of the file. Only the first layer must be split during the expansion of the file. Because the first layer consists of small headers, the process is fast and efficient.

Many different strategies may be applied for management of the first and the second layer. Moreover, the component header may be expanded with additional metadata such as checksums, additional locators, counters, etc. This allows for ensuring fault tolerance, load balancing and additional capabilities [15]. Summarizing, the SD2DS is a very efficient, scalable and flexible data structure, which may be easily adopted to many specific needs.

On the other hand, the separation of a header and a body in SD2DS may lead to serious problems concerning consistency. During the indirect access to the data there is a possibility of seeing not consistent state. It may happen when a client's operation is interrupted between access to the first and the second layer by another client. From the first client's point of view the data may become incomplete or partially correct.

4 SD2DS Architectures Providing Consistency

We propose two extended SD2DS architectures which significantly decrease the number of consistency faults in real-world applications, like non-critical social web applications. The following well known mechanisms are used as a basis: locking of the data and versioning of the data. Although these two methods are quite simple, the results proved that they provide not only efficiency but also safety when combined with SD2DS.

4.1 Component Locking

The component locking is one of the basic methods used to cope with the lack of consistency [7]. We propose a new SD2DS with Component Locking (SD2DS-CL). The locking mechanism involves applying the lock on a component so only one client can update it sequentially. The locking mechanism is managed by the first layer of SD2DS-CL and all necessary information is stored as component metadata. If a component is locked, any other clients are unable to access it, until the client that puts the lock will remove it. In that scenario we assume that all clients work without faults and all clients will eventually commit their changes. Otherwise, each server must periodically check not committed changes and rollback the expired ones.

Algorithm 1 presents the procedure of inserting a component into SD2DS-CL using locking. The lock is released after the client completes the operation and commits the changes to the store.

Algorithm 1. Inserting a component into a SD2DS-CL store

1: The client C sends a message to the first layer of the store with request to add a component CP

2: The corresponding server SH from the first layer creates a header H for the component and the component locator CL. The header H is marked as locked with a current timestamp T.

3: The server SH sends the CL and the timestamp T to the client C.

4: The client C inserts the component body in place indicated by the CL.

5: The client C commits the changes of the desired component (indicated by H) by sending a message with the previously received timestamp T to the server SH.

6: The server SH compares the received timestamp with the timestamp stored in the header H and if they are equal the component lock is removed.

The timestamps are used to prevent from removing the lock put by other clients. This may occur when a single component is modified by many clients simultaneously. Such situation leads to many consistency faults. However it may be neglected in some applications by replacing the old version of the component body with the new one.

In the environment vulnerable to faults, the locking mechanism may cause serious delays in accessing the data. Moreover, it can lead to starvation of clients. In the worst case a client can wait forever for desired component while other clients continuously lock the component. Consider a situation when a client C_1 locks a component, but does not commit the changes. Afterwards, a client C_2 wants to access the same component. The client C_2 requests the component header from the first layer. The first layer responds with information that the desired component is currently locked and cannot be accessed. The client C_2 waits until the component is accessible again.

A limited number of attempts eliminates the possibility of locking the entire client application when trying to access the continually locked component. In the optimistic case that scenario never occurs but can not be neglected. After a specified (predefined, e.g. 10) number of attempts the client may decide whether to abort the process, repeat the operation again or access different component instead.

The length of the delay between attempts must be carefully considered. If the time is longer the risk of starvation is reduced but the clients' latency is greatly increased. We have analysed different strategies to find the best answer. In the simplest solution there is no delay at all. We have also considered strategies with constant, increasing and randomly chosen delays. The experimental results are presented in the chapter 5.

4.2 Component Versioning

As it was presented above, simple locking is not the best and the most effective solution. Thus, we propose the second variant of the $SD2DS$ with Component Versioning (SD2DS-CV).

Algorithm 2 presents the outline of retrieval of a component with versioning. At any moment there can be two component locators which point to two different versions of the associated body. One of the locator points to the current version L_1 and the another one to the backup L_2 (read-only). All the time the backup version is accessible, and current version can be locked and then modified.

Algorithm 2. Retrieving a component with versioning in SD2DS-CV

1: Request the component header (H)
2: **if** H.state $\neq$ locked **then**
3:     Send locator L_1 to client
4: **end if**
5: **if** H.state $=$ locked **then**
6:     Send locator L_2 to client
7: **end if**
8: Client retrieves the appropriate version of component body

---

Algorithm 3 presents outline of an algorithm for performing modification. Before the next modification if the current version is not locked it becomes the backup version and the current version can be modified. It ensures that the second version of the component under L_2 is always consistent and clients can access it at any time even when another client is currently modifying the data. The backup version is also useful if clients' faults are frequent, in such case it is used to revert the incorrect data.

Algorithm 3. Modification of a component in SD2DS-CV

1: Request the component header (H)
2: **if** H.state $\neq$ inuse **then**
3:     Swap locators L_1 and L_2
4: **end if**
5: H.state:$=$ inuse
6: Retrieve the header H
7: Retrieve component body from L_2
8: Process component
9: Send body to L_1
10: Commit changes (H.state $=$ available)

Two versions of components are enough to provide consistency. In such a case the total size of a store is doubled but all consistency faults are eliminated. If the limitation of data space is critical, the backup version can be removed after successful modification. The space for that version can be recycled for another component. This simple versioning system eliminates the need of waiting for a locked component in almost every situation. The client will be forced to wait only when the desired component was not yet added completely.

5   Experimental Results

For the evaluation purposes a prototype multicomputer implementation of the proposed architectures was developed. The first layer consisted of 4 buckets with the capacity of 50 headers and the second layer was built of 5 buckets with the capacity of 50 bodies.

During the tests a variable number of clients retrieving (C_r) and updating (C_u) components was used. Each client operated on approximately 2000 randomly chosen components.

5.1   Performance Analysis

During the experiment, the time of retrieving and updating operations for the basic and proposed architectures was measured. During the experiment, 2000 components of 10MiB each were randomly retrieved and updated. In the case of the SD2DS-CL architecture, clients are waiting between the consecutive attempts in order to gain access to an unlocked component. The following strategies for waiting were evaluated:

- CL_N – no delays between attempts,
- CL_C – constant delays of 2.5s between attempts,
- CL_P – delays increased by 0.5s proportionally to the number of attempts (e.g. 5; 5.5; 6, . . .),
- CL_R – delays randomly chosen from range ¡1.5s; 3.5s¿.

In the figures, the SD2DS-CV is denoted as 'CV' and the basic SD2DS as 'Basic'.

Average times of execution for different ratio of updating/retrieving clients are similar (Fig. 1 and 2). The results show that in case of SD2DS-CL, the time for processing the data was longer, in comparison to SD2DS-CV and the basic SD2DS. The SD2DS-CV got the best results among all tested architectures, because none of the clients was forced to wait for an access to an unlocked component. Moreover, the second copy of a component improves the average time, as both copies may be accessed simultaneously by different clients.

5.2   Starvation Measurement

During the experiment, the number of starved requests was measured. If a component is still locked after 10 consecutive retries, the request is considered as starved. The Fig. 3 shows the percentage of starved requests during retrieval.

For all tested architectures, the possibility of starvation during retrieving and updating is increasing, it increases along with the number of updating clients. However, the results for CL_N are significantly worse. In the case of SD2DS-CL, introducing the delays between attempts significantly reduces the number of starved requests. An interesting fact is that for the constant delays (CL_C) the number of starved requests was lower than for the increasing delays (CL_P). There was no starvation in the case of SD2DS-CV, because this architecture never requires locking of the components.

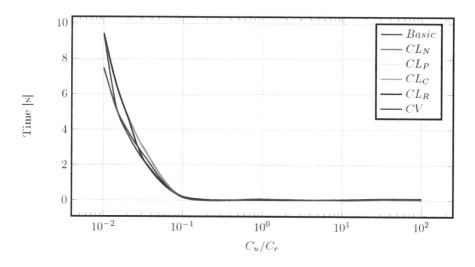

Fig. 1. Time comparison of retrieving operations for the basic and proposed architectures

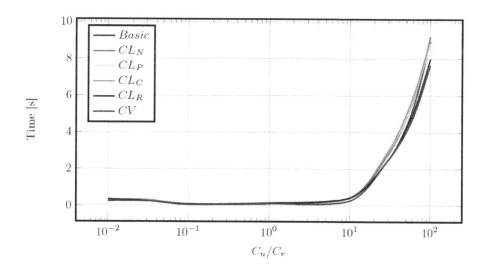

Fig. 2. Time comparison of updating operations for the basic and proposed architectures

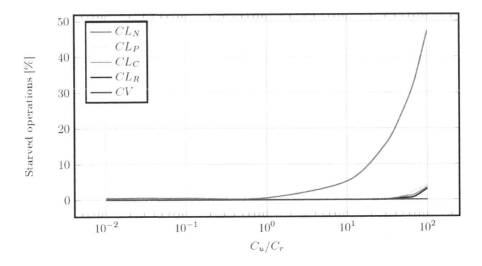

Fig. 3. Starvation during retrieving for the basic and proposed architectures

6   Conclusions and Future Work

In the paper the problem of preserving consistency of the data in SD2DS was discussed. As a result two new SD2DS architectures, utilizing component locking and versioning (denoted SD2DS-CL and SD2DS-CV respectively), were developed. Next, a prototype, multicomputer implementation of both SD2DS-CL and SD2DS-CV were prepared and then used in the experiments.

The results of our experiments show that the component locking proved to be good enough to prevent from consistency problems but can lead to poor latency. Thus it may be unacceptable in some real-life applications. Moreover, the locking mechanism may lead to starvation of clients. To reduce the problem of starvation many different strategies may be applied. We found that, despite worse client effectiveness, the delays between accesses and limited number of attempts are the best solutions.

The versioning mechanism used in SD2DS-CV provides the remedy for the client starvation and eliminates the problem completely. Moreover, the efficiency of the store is greatly improved by storing two versions of each modified component. Both versions may be processed by clients simultaneously thus the actual load of components is balanced. This method proved to be useful but may lead to doubled size of the store.

In our future work we are going to introduce to the SD2DS more advanced methods of components versioning and other mechanisms known from transactions. Moreover, architectures which can provide consistent data even in the case of system faults will be developed. We believe that the future research in that area can have significant influence on improving the consistency in all Big Data and NoSQL systems.

References

1. Augustyn, D.R., Bajerski, P., Brzeski, R.: Zachowanie spójności danych w wybranych systemach nosql. Studia Informatica 33(2A), 27–48 (2012)
2. Baker, J., Bond, C., Corbett, J.C., Furman, J.J., Khorlin, A., Larson, J., Léon, J.M., Li, Y., Lloyd, A., Yushprakh, V.: Megastore: Providing scalable, highly available storage for interactive services. CIDR 11, 223–234 (2011)
3. Bedla, M., Sapiecha, K.: Scalable store of java objects using range partitioning. In: Szmuc, T., Szpyrka, M., Zendulka, J. (eds.) CEE-SET 2009. LNCS, vol. 7054, pp. 84–93. Springer, Heidelberg (2012)
4. Chang, F., Dean, J., Ghemawat, S., Hsieh, W.C., Wallach, D.A., Burrows, M., Chandra, T., Fikes, A., Gruber, R.E.: Bigtable: A distributed storage system for structured data. ACM Transactions on Computer Systems (TOCS) 26(2), 1–14 (2008)
5. Corbett, J.C., Dean, J., Epstein, M., Fikes, A., Frost, C., Furman, J.J., Ghemawat, S., Gubarev, A., Heiser, C., Hochschild, P., et al.: Spanner: Google's globally distributed database. ACM Transactions on Computer Systems (TOCS) 31(3), 1–22 (2013)
6. Di Pasquale, A., Nardelli, E.: Scalable distributed data structures: A survey. In: WDAS, pp. 87–111 (2000)
7. Garcia-Molina, H., Ullman, J.D., Widom, J.: Database Systems: The Complete Book, 2nd edn. Prentice Hall Press, Upper Saddle River (2008)
8. Herlihy, M.P., Wing, J.M.: Linearizability: A correctness condition for concurrent objects. ACM Transactions on Programming Languages and Systems (TOPLAS) 12(3), 463–492 (1990)
9. Hidouci, W.K., Zegour, D.E.: Actor oriented databases. WSEAS Transaction on Computers 3(3), 653–660 (2004)
10. Hidouci, W.K., Zegour, D.E.: Act21: a parallel main memory database system. International Journal of Computing and Information Sciences (2006)
11. Hunt, P., Konar, M., Junqueira, F.P., Reed, B.: ZooKeeper: Wait-free coordination for internet-scale systems. In: USENIX Annual Technical Conference, vol. 8, pp. 1–14 (2010)
12. Łukawski, G., Sapiecha, K.: Fault tolerant record placement for decentralized sdds lh*. In: Wyrzykowski, R., Dongarra, J., Karczewski, K., Wasniewski, J. (eds.) PPAM 2007. LNCS, vol. 4967, pp. 312–320. Springer, Heidelberg (2008), http://dx.doi.org/10.1007/978-3-540-68111-3_33
13. Peng, D., Dabek, F.: Large-scale incremental processing using distributed transactions and notifications. OSDI 10, 1–15 (2010)
14. Sadalage, P.J., Fowler, M.: NoSQL distilled: a brief guide to the emerging world of polyglot persistence. Addison-Wesley, Upper Saddle River (2013)
15. Sapiecha, K., Lukawski, G.: Scalable distributed two-layer data structures (SD2DS). International Journal of Distributed Systems and Technologies (IJDST) 4(2), 15–30 (2013)
16. Thomson, A., Diamond, T., Weng, S.C., Ren, K., Shao, P., Abadi, D.J.: Calvin: fast distributed transactions for partitioned database systems. In: Proceedings of the 2012 ACM SIGMOD International Conference on Management of Data, pp. 1–12. ACM (2012)
17. Wei, Z., Pierre, G., Chi, C.H.: CloudTPS: Scalable transactions for web applications in the cloud. IEEE Transactions on Services Computing 5(4), 525–539 (2012)

Modern Temporal Data Models:
Strengths and Weaknesses

Dušan Petković[(✉)]

University of Applied Sciences,
Hochschulstr. 1, 83024 Rosenheim, Germany
petkovic@fh-rosenheim.de

Abstract. Time is generally a challenging task. All issues in relation to time can be better supported using a temporal data model than implementing them in user applications. More than two-dozen temporal data models have been introduced in time period between 1982 and 1998. After several years of stagnancy, the last couple of years brought the new revival of the topic and the emergence of new data models. Two temporal data models have been specified recently. The one is the SQL:2011 standard, published in 2011. The second one is from Teradata. In this article we present the temporal data model of the ANSI SQL standard on one side and the data model of an existing relational DBMS on the other. After that, we compare their support of several temporal concepts. Finally, we discuss strengths and weaknesses of both models and give suggestions for future extensions.

Keywords: Temporal data models · ANSI SQL standard · RDBMS

1 Introduction

Research of temporal data in relation to database systems has a very long history. Hence, there is a lot of work on this topic, and many temporal data models have been introduced. All these models can be evaluated in relation to general concepts. Hence, we will examine the following concepts in relation to the considered models: time dimensions, the PERIOD data type, coalescing, temporal joins, implicit vs. explicit timestamps and key constraints.

1.1 Time Dimensions

In relation to time dimensions, there are three different forms: user-defined time, valid time and transaction time. User-defined time is a time representation designed to meet the specific needs of users. Valid time concerns the time when an event is true in the real world. For this reason, an event is independent of the time when it is stored and can concern the past, present and future snapshots of it. Using timestamps, it is possible to form a history of an event, and this is the central aspect of the realization of temporal databases.

© Springer International Publishing Switzerland 2015
S. Kozielski et al. (Eds.): BDAS 2015, CCIS 521, pp. 136–146, 2015.
DOI: 10.1007/978-3-319-18422-7_12

Transaction time concerns the time when an event was present in the database as stored data. Therefore, transaction time of an event presents the correct database image of the modeled world. Timestamps of transaction time are defined according to the schedule adopted by the operating system. According to this, we can build the history of all such timestamps in relation to the past and current time, but not in relation to future. Furthermore, only the current values may be updated, and the updates cannot be retroactive, as in case of valid time.

A data model that supports only valid time is called valid-time model and one that supports only transaction time is called transaction-time model. If a data model supports both of them, it is called bitemporal. Therefore, a bitemporal table is a union of data with corresponding valid and transaction time, and each row in such a table is associated with both valid time as well as transaction time periods. There is also a special case of bitemporal model, when the valid and transaction times of an event are identical. The situation, where an event is recorded as soon as it becomes valid in reality is a simple example for this case.

1.2   PERIOD Data Type

The PERIOD data type is specified as a time interval, which comprises the set of subsequent time units. These units use a closed-open concept, meaning that the starting granule of the time period is included, while the end granule is excluded. The data type of the start and end of the period can be DATE, TIME or TIMESTAMP. The main advantage of the PERIOD data type is that it can be used in the natural way to represent time intervals.

The use of the PERIOD data type requires the specification and implementation of corresponding functions i.e. methods. First, a temporal constructor with the same name as the data type is implicitly defined, when the PERIOD type is specified. (A temporal constructor is an expression that returns a temporal value.) Second, unary and binary functions, which operate on the instances of the PERIOD data type are specified and implemented, too. The following are examples of such functions: CONTAINS, EQUALS, PRECEDES and OVER-LAPS. (The proposal, how the PERIOD data type should be specified, can be found in [18].)

1.3   Coalescing and Value Equivalence

Two tuples are value equivalent if they are identical. A tuple is coalesced, if overlapped or consecutive value-equivalent tuples are disallowed. When timestamps of tuples have temporal elements as values, the requirement of coalescing is identical to the requirement that there will be no value-equivalent tuples. The goal of coalescing is that the "merge" of the corresponding rows should be done by the DBMS system, and not by users in their applications.

The need for coalescing happens when a projection or union operation is performed during retrieval of data or INSERT i.e. UPDATE statements are executed. The general approach for implementation of coalescing is similar to the problem of computation of the transitive closure of a graph, with the subsequent deletion of non-maximal paths and can be implemented by iterating an INSERT statement that coalesces two paths and inserts a new tuple into the relation.

1.4 Temporal Joins

A temporal join is an important operation for applications that maintain time-evolving data. A typical temporal record has a key, one or more time-varying attributes and a time interval representing the record's time validity. The result set is fetched based on the time interval mapped using the temporal column. A record is only included in the result set if the considered time interval lies within the valid one. A time interval is assigned to each record in the result set. The records are valid for the duration of the interval to which they are assigned.

Several approaches have been proposed for implementing temporal joins. The most of them are based upon ordering of the involved tables. If the tables are ordered, the merging of value-equivalent rows can be handled in a way similar to the projection operation. The use of multi-way joins is generally not an issue, because the most temporal join in the practice can be evaluated using 2-way join algorithm. For instance, all queries defined in the temporal test suite are 2-way joins [8].

Like the convenient outer join operation, the temporal outer join retain tuples that do not participate in the join operation. However, in a temporal database, a tuple may dangle over a portion of its time interval. Hence, implementation of the outer join operations is performed using the convenient outer join operations and coalescing [6].

1.5 Implicit vs. Explicit Timestamps

The association of time with facts is different for existing temporal data models. Some of them specify that this association is explicit, meaning that temporal facts are displayed directly. Other models treat display of data so, that they are not displayed in the same way as other, temporal-invariant attributes. This issue has also consequences in relation to update languages in the following way: While transaction times of facts are supplied by the system itself, update operations in transaction-time models treat the temporal aspect of facts implicitly. On the other hand, the user is responsible to supply valid times of facts. Therefore, updating facts in valid time and bitemporal data models generally must treat time explicitly and are forced to represent a choice as to how the valid times of facts should be specified by the user [9].

1.6 Temporal Key Constraints

There are two temporal key constraints: one in relation to primary key (PK) and the other in relation to referential integrity. The requirements concerning primary key depends whether the table captures valid or transaction time. In the case of valid time, the convenient primary key of relational tables is not sufficient and the primary key has to include time-variant attributes, because there can be several rows with one value for the relational primary key and several time periods (see Example 2). On the other hand, tables capturing transaction time always include only one row concerning current time, and (possibly) several history rows, which cannot be modified. For this reason, in case of transaction time, relational primary key is also the temporal primary key of the corresponding table with time-variant attributes. In the case of temporal referential integrity, each value of the relational foreign key in the child table must correspond to some value of the relational primary key in the parent table at every point in time. Hence, it must be possible to forbid a row in a child table whose valid time is not contained in the time period of a corresponding row in the parent table.

Two temporal data models have been specified recently. The one is the SQL:2011 standard, published in December, 2011. The second one is from Teradata, implemented in Versions 13 and 14. The goal of this paper is to compare these models in relation to temporal concepts and to show strengths and weaknesses of both of them.

The rest of the paper is organized as follows. Section 2 introduces the Teradata's temporal data model. In the first subsection, basic definitions are given. The following five subsections discuss the support of the issues mentioned in the introductory part of the paper and explain how Teradata supports these concepts. Section 3 discusses the temporal model specified in the SQL:2011. The structure of this section is identical to the structure of the previous one. Section 4 deals with related work to the topics discussed in the paper. The last section shows our conclusions.

2   Teradata's Temporal Data Model

2.1   Teradata: Basic Definitions

To discuss Teradata's temporal model, we use two tables, shown in Example 1.

Example 1

```
CREATE MULTISET TABLE Dept( DNo INTEGER NOT NULL,
                            DName VARCHAR(30) NOT NULL,
                            Dept_period PERIOD(DATE) NOT NULL AS VALIDTIME
                          ) PRIMARY INDEX (DNo);
CREATE MULTISET TABLE Emp ( ENo INTEGER NOT NULL,
                            EDept INTEGER NOT NULL,
                            Emp_period PERIOD(DATE) NOT NULL AS VALIDTIME
                          ) PRIMARY INDEX (ENo);
```

The Dept table contains, among other attributes, the Dept_Period attribute, which is time-variant and as such is specified using the PERIOD data type. Also, the specification of this attribute contains the VALIDTIME option, meaning that this attribute comprises valid time. (The syntax for tables with transaction time as well as bitemporal tables will be given shortly.) Also, the second CREATE TABLE statement specifies that the Emp_period column is of the type PERIOD. Example 2 shows several INSERT statements that are used to load data in the Dept and Emp tables.

Example 2
```
INSERT INTO Emp (ENo, EDept) VALUES (22217, 3);
SEQUENCED VALIDTIME
INSERT INTO Emp VALUES(22218,4,PERIOD(DATE '2010-01-01',DATE'2010-11-12'));
SEQUENCED VALIDTIME
INSERT INTO Emp VALUES(22219, 3, PERIOD(DATE '2010-01-01',DATE'2011-11-12'));
SEQUENCED VALIDTIME
INSERT INTO Dept VALUES(3,'Test', PERIOD(DATE '2010-01-01',DATE'2011-12-31'));
INSERT INTO Dept (DNo, DName) VALUES (4, 'Development');
```

Example 2 contains two forms of the INSERT statement. The first one, which is equivalent to the convenient INSERT statement, inserts a row, which has a current time as its start time and "forever" as its end time. If the start and end time has to be explicitly defined in INSERT, the prefix "SEQUENCED TIME" has to be specified. (All other INSERT statements in Example 2 are specified in this manner.)

2.2 Teradata: Time Dimensions

Teradata supports all three time dimensions. As can be seen from Example 1, the VALIDTIME clause in the specification of the column with the PERIOD data type defines valid time. According to this, the TRANSACTIONTIME clause specifies the column which comprises transaction time periods. Bitemporal tables contain the TRANSACTIONTIME as well as the VALIDTIME clause (see Example 3).

Example 3
```
CREATE MULTISET TABLE Bi_Emp(
    ENo INTEGER NOT NULL, EDept INTEGER NOT NULL,
    Emp_VT_period PERIOD (DATE) NOT NULL AS VALIDTIME,
    Emp_TT_period PERIOD (TIMESTAMP(6)) NOT NULL AS TRANSACTIONTIME
  ) PRIMARY INDEX (ENo);
```

2.3 Teradata: PERIOD Data Type

One of main properties of Teradata's temporal data model is the support for the PERIOD data type. Teradata introduced this data type with Version 13.0, together with several methods, which can be applied to instances of this type. Teradata also supports the PERIOD constructor. The comparison of period values as well as the assignment of them is supported, too.

2.4 Teradata: Coalescing

Teradata does not support coalescing of the data. The only way to coalesce data is using the P_NORMALIZE() function, but it works only for the output and does not modify storage of the underlying rows. The function returns an instance of the PERIOD data type that is the combination of the two instances of the same data type, which overlap or are consecutive. Example 4 shows the use of the P_NORMALIZE() function. First, two rows, which are consecutive are inserted and after that, in the SELECT statement the P_NORMALIZE function coalesces the corresponding rows, before the result is displayed.

Example 4

```
SEQUENCED VALIDTIME
INSERT INTO Emp VALUES(22220, 3, DATE '2011-01-01', DATE '2011-12-31');
SEQUENCED VALIDTIME
INSERT INTO Emp VALUES(22220, 3, DATE '2012-01-01', DATE '2012-12-31');
SEQUENCED VALIDTIME
SELECT * FROM Emp AS E1 INNER JOIN Emp AS E2 ON E1.ENo = E2.ENo
       WHERE E1.Emp_period <= E2.Emp_period
       AND E1.Emp_period P_NORMALIZE E2.Emp_period IS NOT NULL;
```

The implementation of temporal coalescing in Teradata is under way and is based upon the paper of Al-Kateb et al [1]. The approach uses higher analytic functions together with runtime partitioning. That way, self-joins and nested queries are avoided.

2.5   Teradata: Temporal Joins

The temporal functionality of Teradata supports inner joins. Example 5 shows how tables from Example 1 can be joined when the join attributes are time-variant. (Teradata does not support temporal outer joins.)

Example 5

```
SEQUENCED VALIDTIME
SELECT ENo, EDept FROM Emp INNER JOIN Dept ON Emp.EDept = Dept.Dno
       WHERE DName = 'Test';
```

2.6   Teradata: Implicit vs. Explicit Timestamps

Teradata supports the implicit display of values of time-variant attributes. For temporal tables in general, only the values of non-temporal columns are returned, when an asterisk is used in a SELECT list. The first SELECT statement in Example 6 displays only values of time-invariant columns. To display all values, the syntax form with asterix is used to display values of non-temporal columns, followed by the list of all temporal columns, whose values should be returned. (The second query in Example 6 shows this issue.)

Example 6

```
SELECT * FROM Emp;
SELECT Emp.*, Emp_period FROM Emp;
```

2.7   Teradata: Temporal Key Constraints

Teradata supports the PRIMARY KEY option in the CREATE (ALTER) TA-BLE statement to specify that any given value in the constrained column will not exist in more than one row at any time granule. Temporal referential integrity is not enforced by Teradata. This means that these constraints have to be implemented by users using triggers. (The general discussion how temporal referential integrity can be implemented using trigger can be found in [16].)

3   Temporal Data Model of SQL:2011

3.1   SQL:2011: Basic Definitions

Before we start to discuss temporal concepts in relation to the SQL:2011 specification, let us see, how the necessary definitions of the Emp and Dept tables look like. Example 7 shows how both tables are created. (As in the case in section 2.1, we will first create tables with valid time.)

Example 7
```
CREATE TABLE Dept( DNo INT, DStart DATE, DEnd DATE,
                   DName VARCHAR(30), PERIOD FOR DPeriod (DStart, DEnd),
                   PRIMARY KEY (DNo, DPeriod WITHOUT OVERLAPS));
CREATE TABLE Emp ( ENo INTEGER, EStart DATE, EEnd DATE,
                   EDept INT, PERIOD FOR EPeriod (EStart, EEnd),
                   PRIMARY KEY (ENo, EPeriod WITHOUT OVERLAPS));
```

The definition of temporal tables in the SQL:2011 standard is significantly different than in the case of Teradata. The main difference is that begin and end of a time period has to be specified explicitly, using two columns. (In the case of the Dept table, these two columns are DStart and DEnd.) The syntax of the PERIOD clause is different, too, and this is the direct consequence of the explicit definition of the time period in the standard specification.

Both tables created in Example 7 show another property of the SQL:2011 specification: The PRIMARY KEY clause is extended with the WITHOUT OVERLAPS option. This option specifies an integrity constraint meaning that two tuples with the same value in the primary key cannot timely overlap.

3.2   SQL:2011: Time Dimensions

The SQL:2011 standard specifies tables for all three time dimensions. A table with valid time is called application-time period table. (The tables in Example 7 are application-time period tables.) Also, a table, which supports transaction time is called system-versioned table. The standard does not explicitly use the phrase bitemporal, but supports the union of valid and transaction time. The following example shows a bitemporal table according to the SQL standard.

Example 8
```
CREATE TABLE BI_Emp ( id VARCHAR(30) NOT NULL,
                      name VARCHAR (20) NOT NULL, dept_id VARCHAR(30),
                      g_start DATE NOT NULL, g_end DATE NOT NULL,
                      s_start DATE NOT NULL GENERATED ALWAYS AS ROW START,
                      s_end DATE NOT NULL GENERATED ALWAYS AS ROW END,
                      PERIOD FOR SYSTEM_TIME (s_start, s_end),
                      PERIOD FOR time_period (g_start, g_ende),
                      PRIMARY KEY (Id, time_period WITHOUT OVERLAPS)
                    ) WITH SYSTEM VERSIONING;
```

3.3   SQL:2011: PERIOD Data Type and Coalescing

The specification of the temporal data model in SQL:2011 does not include the PERIOD data type [7]. This feature is probably the most important one, and should be specified in one of the future versions of the standard, as proposed in [14].

The specification of the temporal data model in SQL:2011 does not include the specification of coalescing, too. (The standard uses the phrase "normalization of data" as a synonym for coalescing.)

3.4 SQL:2011: Temporal Joins

The SQL:2011 specification does not include the support for joins, where rows from one table are linked with rows from another one so, that their valid or transaction time periods satisfy a given condition. On the other hand, it is possible to apply inner join operations using the SQL:2011's OVERLAPS operator as Example 9 shows. (The description of this operator can be found in [7]. Outer joins require additional syntax built into the language, and are not supported, too.

Example 9
```
SELECT R.ENo, S.ENo FROM Emp R, Emp S WHERE R.Emp_period
    OVERLAPS S.Emp_period;
```

3.5 SQL:2011: Implicit vs. Explicit Timestamps

In 1994, during the standardization process of the temporal data model, Snodgrass et al proposed bitemporal data model called TSQL2 [17] to be the temporal extension of the SQL standard. This proposal has been criticized by Date and Darwen [3] primarily because of its treatment of timestamps. Date and Darwen state that access to implicit (i.e. hidden) columns cannot be done in regular relational fashion, and hence the whole table with time-variant attributes is not relational table any more.

The SQL:2011 standard, as a specification for a relational language, strongly supports explicit timestamps. In other words, the time-variant columns are associated with the rows of a temporal table using user-defined columns. Therefore, SELECT * FROM table displays time-invariant as well as time-variant columns.

It is interesting to note that temporal implementation in IBM DB2 supports explicit timestamps, but allows a user to hide them [12]. Example 10 shows the use of the IMPLICITLY HIDDEN option. When this option is used, the time-variant columns will not be displayed after executing the "SELECT * FROM .." statement.

Example 10
```
CREATE TABLE policy (
    id INT primary key not null, vin VARCHAR(10), annual_mileage INT,
    rental_car CHAR(1), coverage_amt INT,
    sys_start TIMESTAMP(12) GENERATED ALWAYS AS ROW BEGIN IMPLICILY HIDDEN,
    sys_end TIMESTAMP(12) GENERATED ALWAYS AS ROW END IMPLICITLY HIDDEN,
    trans_start  TIMESTAMP(12) GENERATED ALWAYS AS TRANSACTION
                              START ID IMPLICITLY HIDDEN,
    PERIOD SYSTEM_TIME (sys_start, sys_end));
```

3.6 SQL:2011: Temporal Key Integrity

The semantics of the primary key integrity is the same as in the case of Teradata, but the syntax is different. The requirement that each value of the relational

Table 1. Support of temporal concepts: Overview

x	Time dim	PERIOD Type	Coalescing	Temp joins	Impl/Expl	Key constraint
SQL:2011	+	-	-	-	Explicit	+
Teradata	+	+	-	+-	Implicit	+-

primary key is unique at any given time and hence forbids overlapping of time periods, is solved using the WITHOUT OVERLAPS clause in the CREATE or ALTER TABLE (see Example 7.) The temporal referential integrity can be prohibited using the statement:

```
ALTER TABLE Emp ADD FOREIGN KEY (Edept, PERIOD EPeriod)
    REFERENCES Dept (DNo, PERIOD DPeriod);
```

4 Related Work

The first two temporal data models have been founded in the doctoral thesis of Clifford [2]. The main contribution of Clifford was the introduction of the new data model called HRDM (historical relational database model). Besides that, he considered the traditional RDBMS as a special case of a historical database, when the time interval is reduced to (Now, Now).

After 1982, the research on temporal database models began to make significant progress. The most important result of this phase was the book written by Tansel, Clifford, Gadia, Jajodia, Segev and Snodgrass [20]. The first part of the book describes several temporal data models: the two models mentioned above as well as Gadia's TempSQL [5], Snodgrass TQuel [17] and Lorentzos IXRM model [11]. Also, the special temporal data models, such as Elmasri's temporal model called TEER have been discussed [4].

The era of modern temporal models started 2008, when the work on the time-variant extensions in the SQL standard has been resumed. The final specification of the standard has been published in December, 2011. After that, two vendors of RDBMSs - IBM DB2 [16] and Oracle [13] - implemented the temporal features published in the standard. In the years 2010-11, Teradata released its temporal extensions, which significantly differ from the published standard [15].

5 Conclusions

The main advantage of the Teradata system is support of the PERIOD data type. From our point of view, the cornerstone of temporal data support is the introduction of this data type with all necessary methods [19]. At the same time, probably the biggest weakness of the SQL standard is lack to support this data type. The reason, why the PERIOD data type is not specified in the standard can be found in [10], but this explanation is not helpful, knowing the importance of this type. On the other hand, Teradata's support for implicit timestamps is from our point of view not a good solution. There are several drawbacks in relation to implicit timestamps. First, the existence of implicit i.e. hidden columns is incompatible with the notion of the relational table. (Such a table requires all information associated with the table's rows to be caught explicitly as columns values.) Second, as can be seen in Example 6, all queries of the form: "SELECT * FROM temp_table" do not display time-variant information associated with the rows. (Teradata forces the user to explicitly name temporal column in projection, if it should be displayed.)

The both model lack the support for coalescing and temporal outer joins, but, according to our discussion in sections 2.4 and 2.5, it seems so that Teradata is on the way to support these concepts soon. (The publication of the next SQL standard is not yet known, and it is questionable whether it will contain these two concepts.) Table 1 gives an overview, how the discussed temporal concepts are supported with the SQL:2011 standard and Teradata DBMS.

References

1. Al-Kateb, M., Ghazal, A., Crolotte, A.: An efficient SQL rewrite approach for temporal coalescing in the teradata RDBMS. In: Liddle, S.W., Schewe, K.-D., Tjoa, A.M., Zhou, X. (eds.) DEXA 2012, Part II. LNCS, vol. 7447, pp. 375–383. Springer, Heidelberg (2012)
2. Clifford, J.: A Logical Framework for the Temporal Semantics and Natural-Language Querying of Historical Databases. Ph.D. thesis, Dept. of Computer Science, State University of New York (1982)
3. Darwen, H., Date, C.J.: An overview and analysis of proposals based on the tsql2 approach, http://www.dcs.warwick.ac.uk/~hugh/TTM/OnTSQL2.pdf (accessed: December 31, 2014)
4. Elmasri, M., Wu, G.: A temporal model and query language for eer dbs. In: Sixth International Conference on Data Engineering (1990)
5. Gadia, S.K.: A homogeneous relational model and query languages for temporal databases. ACM Trans. on Databases Systems 13, 418–448 (1998)
6. Gao, D., Jensen, C.S., Snodgrass, R.T., Soo, M.D.: Join operations in temporal databases. VLDB Journal 14 (2005)
7. ISO/IEC 9075-2:2011: Database languages: SQL, part 2 (2011)
8. Jensen, C.S.: A consensus test suite of temporal database queries (1993), http://people.cs.aau.dk/~csj/Papers/Files/1993_jensenCGTDC-TR93-2034.pdf (accessed: December 15, 2014)
9. Jensen, C.S., Snodgrass, R., Soo, M.D.: The tsql2 data model (1994), http://people.cs.aau.dk/~csj/Thesis/pdf/chapter12.pdf
10. Kulkarni, K., Michels, J.: Temporal features in sql:2011. SIGMOD Records 41 (2012)
11. Lorentzos, N.: Management of intervals and temporal data. Report 49 (1991)
12. Michels, J.E., Nicola, M.: Adopting temporal tables in db2 (October 2012), http://www.ibm.com/developerworks/data/library/techarticle/dm-1210temporaltablesdb2/dm-1210temporaltablesdb2-pdf.pdf (accessed: December 20, 2014)
13. Oracle: Temporal database functionality (2012), http://www.oracle.com/technetwork/database/application-development/total-recall-1667156.html (accessed: December 1, 2014)
14. Petkovic, D.: Temporal data in the sql standard (in german). Datenbank Spektrum 13, 131–138 (2013)

15. Sannik, G.: Enabling the temporal data warehouse (2010), `http://www.`
    `teradatamagazine.com/v11n02/Tech2Tech/Managing-Time/` (accessed: December
    15 2014)
16. Sarraco, S., Nicola, M.: A matter of time: Temporal data management in db2 10,
    http://www.ibm.com/developerworks/data/library/techarticle (accessed: Decem-
    ber 19, 2014)
17. Snodgrass, R.T.: The temporal query language tsql2. TODS 12 (1987)
18. Snodgrass, R.T.: Developing Time-Oriented Applications in SQL. Kluwer Aca-
    demic Publishers (1999)
19. Snodgrass, R.T.: A case study of temporal data (2012),
    http://www.teradatamagazine.com (accessed: December 10, 2014)
20. Tansel, A.U.: Temporal Databases. Benjamin/Cummings (1993)

Multiargument Relationships
in Possibilistic Databases

Krzysztof Myszkorowski[✉]

Institute of Information Technology,
Lodz University of Technology, Poland
kamysz@ics.p.lodz.pl
http://edu.ics.p.lodz.pl

Abstract. In the paper we consider the possible coexistence of associations between $k < n$ attributes of the n-ary relationship. The analysis is carried out using the theory of functional dependencies. We assume that attribute values are represented by means of possibility distributions. According to the representation of data the notion of fuzzy functional dependency has been appropriately extended. Its level is evaluated with the use of possibility and necessity measures. The dependencies between all n attributes describe the integrity constraints of the n-ary relationship and must not be infringed. They constitute a restriction for dependencies of fewer attributes. The paper formulates the rules to which fuzzy functional dependencies between $(n$-1$)$ attributes of the n-ary relationship must be subordinated.

Keywords: Possibilistic databases · Possibility distribution · Fuzzy implication operator · n-ary relationships · Fuzzy functional dependencies

1 Introduction

Conventional database systems are designed with the assumption of precision of information collected in them. The problem becomes more complex if our knowledge of the fragment of reality to be modeled is imperfect. In such cases one has to apply tools for describing uncertain or imprecise information [9]. One of them is the theory of possibility. In the possibilistic database framework attribute values are represented by means of possibility distributions [1, 2, 5, 12]. Each value x of an attribute X is assigned with a number $\pi_X(x)$ from the unit interval which expresses the possibility degree of its occurrence. The possibility distribution takes the form: $\{\pi_X(x_1)/x_1, \pi_X(x_2)/x_2, ..., \pi_X(x_n)/x_n\}$, where x_i is an element of the domain of X. At least one value must be completely possible i.e. its possibility degree equals 1. This requirement is referred to as the normalization condition. Different ways of determination of the possibility degree have been described in [7]. Possibilistic databases allow for a unified way of representing of precise and ill-known information [2]. A precise attribute value is represented by $\{1/x\}$. A fuzzy value expressed by a fuzzy set A with the membership function $\mu_A(x)$ is represented by the possibility distribution such that $\pi_X(x) = \mu_A(x)$.

S. Kozielski et al. (Eds.): BDAS 2015, CCIS 521, pp. 147–156, 2015.
DOI: 10.1007/978-3-319-18422-7_13

The main concepts of the possibility theory are possibility and necessity measures, denoted by Pos and Nec, respectively. A possibility measure is a function $P(U) \to [0,1]$, where $P(U)$ denotes the power set of a universe of discourse U, such that $Pos(\emptyset) = 0$, $Pos(U) = 1$ and $\forall A, B \in P(U)$ $Pos(A \cup B)$ $= \max(Pos(A), Pos(B))$. Based on the possibility distribution one can derive a possibility degree $Pos(A)$ for any event $A \in P(U)$: $Pos(A) = \sup_{x \in A} \pi_X(x)$. However, having $P(A)$ we cannot conclude to what extent A is certain. In order to improve the description of A the necessity measure has been introduced. Its value is equal to the impossibility of the opposite event $\bar{A}$: $Nec(A) = 1 -$ $Pos(\bar{A})$. Thus, the certainty degree of A can be equal 0 even if A is completely possible ($Pos(A) = 1$). The necessity measure cannot be greater than the possibility measure. Thus, $Nec(A) \le Pos(A)$. Moreover $Pos(A) < 1 \Rightarrow Nec(A) = 0$. By means of possibility and necessity measures one can compare values X and Y represented by possibility distributions. According to [12] $Pos(X = Y) =$ $\sup_u \min(\pi_X(u), \pi_Y(u))$ and $Nec(X = Y) = 1 - \sup_{u \ne u'} \min(\pi_X(u), \pi_Y(u'))$.

One of the most important notions in the database theory is the notion of functional dependency (FD) between attributes. Functional dependencies reflect integrity constraints and should be studied during the design process [15]. If attribute values are imprecise, one can say about a certain degree of data dependencies. The notion of functional dependency has to be extended according to the representation of fuzzy data [14]. Hence, different approaches have been described in professional literature. A number of different definitions concerning fuzzy functional dependencies (FFDs) in the possibilistic data model emerged (see for example [3, 11]). In the presented paper the level of FFD is evaluated with the use of possibility and necessity measures. In the definition there was applied the extended Gödel implication [11].

In database models, usually binary relationships occur between entity sets. When designing, it may be necessary to define relationships which involve three or more entity sets. They are referred to as n-ary or multiargument relationships. Within such connections there may exist relationships comprising fewer than n sets. However, there is no complete arbitrariness [4, 6, 8]. The relationships "embedded" in n-ary relationships are subjected to certain restrictions. This issue for ternary relationships in conventional databases was considered in [8]. The authors provided guidelines for database designers. For various types of ternary relationships there were formulated the rules to which cardinalities of binary relationships between pairs of sets are subjected. Pursuant to these, cardinalities of imposed binary relationships cannot be lower than the cardinality of the ternary relationship. The analysis may be also carried out by means of functional dependencies. In this paper multiargument relationships existing in possibilistic databases are analysed with the use of fuzzy functional dependencies.

Identifying relationships between attributes in databases plays an important role in the process of knowledge discovery. There are many possible ways of defining the approximation degree of data dependencies. In order to simplify a dependency one should reduce irrelevant attributes. In [13] Ślęzak introduced the notion of an association reduct as a pair of disjoint attribute sets A and B such that A determines B. The set A has to be minimal i.e. there is no a proper subset A' of A such that A' determines $(A \setminus A') \cup B$. Data dependencies are usually not exact. So the notion has been extended with the use of an approximation threshold to β-approximate association reduct, the definition of which is based on entropy. The concept of the approximate association reduct is to some extent similar to the concept of θ-key of relation scheme which was introduced for fuzzy databases [3].

The paper is an extension of the analysis presented in [10]. Section 2 discusses n-ary relationships in conventional databases. In section 3 we define a fuzzy functional dependency and fuzzy normal forms for possibilistic databases. Analysis of multiargument relationships is contained in section 4.

2 Multiargument Relationships

A multiargument relationship R between n entity sets E_1, E_2, ..., E_n may be formally presented using the relational notation: $R(X_1, X_2, \dots, X_n)$, where R is a relation scheme, and attributes X_i denote keys of entity sets E_i which participate in it. Cardinality of an n-ary relationship may be presented as $M_1 : M_2 : \dots : M_n$, where M_i denotes the number of entities e_i of E_i that can occur for each $(n\text{-}1)$-tuple $(e_1, e_2, \dots, e_{i-1}, e_{i+1}, \dots, e_n)$.

Let X, Y be subsets of attribute set $U = \{X_1, X_2, \dots, X_n\}$. The functional dependency (FD), denoted by $X \to Y$, holds in R if and only if in any relation r of R for any two tuples t_1, t_2 of r, if $t_1(X) = t_2(X)$ then $t_1(Y) = t_2(Y)$, where $t_1(X), t_2(X), t_1(Y), t_2(Y)$ denote X-values and Y-values for t_1 and t_2. The FDs describing a relationship $R(X_1, X_2, \dots, X_n)$ are of the following form:

$$U - \{X_i\} \to X_i, \quad i = 1, 2, \dots, n. \tag{1}$$

The maximal number of FDs (1) equals n. It occurs for relationships of cardinality 1:1: ... :1. If $M_i > 1$ for every i, the FDs (1) do not occur at all. The scheme R with $m \leq n$ FDs (1) has m candidate keys. The attributes occurring only on the left side of FDs (1) determine values of other attributes. They belong to every candidate key of R. The FDs (1) define the integrity constraints for R. They constitute a restriction for $(n\text{-}1)$-ary relationships.

The FDs describing relationships between $(n\text{-}1)$ attributes "embedded" in the n-ary relationship may be presented as follows:

$$U - \{X_i, X_j\} \to X_i, \quad i \neq j, \quad i, j = 1, 2, \dots, n. \tag{2}$$

The imposition of (2) disturbs integrity conditions if its consequence is the occurrence of a new FD (1). This takes place if X_i does not occur on the right side of any FD (1). Thus, the imposition of (2) is admissible if X_i does not belong to every candidate key of R [10].

3 Functional Dependencies in Possiblistic Databases

In conventional databases a relation r of the scheme $R(U)$ is a subset of Cartesian product of attribute domains: $r \subseteq D_1 \times D_2 \times \dots \times D_n$, where U denotes a set of attributes, $U = \{X_1, X_2,...,X_n\}$ and D_i denotes a domain of X_i. Let us assume that attribute values are given by means of normal possibility distributions:

$$t(X_i) = \{\pi_{t(X_i)}(x)/x : x \in D_i\} , \quad \sup_{x \in D_i} \pi_{t(X_i)}(x) = 1 ,$$

where t is a tuple of r and $\pi_{t(X_i)}(x)$ is a possibility degree of $t(X_i) = x$. Thus, a relation is a subset of Cartesian product $\Gamma_{D_1} \times \Gamma_{D_2} \times ... \times \Gamma_{D_n}$, where Γ_{D_i} is a set of possibility distributions of X_i on D_i.

The degree of closeness of $t_1(X_i)$ and $t_2(X_i)$, denoted by $\approx (t_1(X_i), t_2(X_i)) = (\approx (t_1(X_i), t_2(X_i))_N, \approx (t_1(X_i), t_2(X_i))_\Pi)$, is evaluated by means of possibility and necessity measures. For identical values $\approx (t_1(X_i), t_2(X_i)) = (1,1)$. Otherwise $\approx (t_1(X_i), t_2(X_i)) = (Nec(t_1(X_i) = t_2(X_i), Pos(t_1(X_i) = t_2(X_i)))$. Thus

$$\approx (t_1(X_i), t_2(X_i))_N = 1 - \sup_{x \neq y} \min(\pi_{t_1(X_i)}(x), \pi_{t_2(X_i)}(y)) ,$$
$$\approx (t_1(X_i), t_2(X_i))_\Pi = \sup_x \min(\pi_{t_1(X_i)}(x), \pi_{t_2(X_i)}(x)) .$$

For estimation of tuple closeness, denoted by $=_c(t_1(X), t_2(X)) = (=_c(t_1(X), t_2(X))_N, =_c(t_1(X), t_2(X))_\Pi)$, one must consider all the components X_i of X ($X_i \in X$) and apply the operation min:

$$=_c (t_1(X), t_2(X))_N = \min_i \approx ((t_1(X_i), t_2(X_i))_N ,$$
$$=_c (t_1(X), t_2(X))_\Pi = \min_i \approx ((t_1(X_i), t_2(X_i))_\Pi . \tag{3}$$

In [3] Chen defined a fuzzy functional dependency for the possibilistic data model with the use of Gödel implication operator. Its level was evaluated by means of the possibility measure of equality of two possibility distributions. If evaluation of closeness of imprecise values is made by means of both possibility and necessity measures the degree of FFD should be evaluated by two numbers from the unit interval. It requires the modification of the definition given by Chen. One can apply the following extension of Gödel implicator $I_{GE}(a,b)$ [11]:

$$I_{GE}(a,b) = (I_{GE}(a,b)_N, I_{GE}(a,b)_\Pi) , \quad a = (a_N, a_\Pi) , \quad b = (b_N, b_\Pi) , \tag{4}$$

where $a_N, a_\Pi, b_N, b_\Pi \in [0, 1]$ and

$$I_{GE}(a,b)_\Pi = \begin{cases} 1 & \text{if } a_\Pi \leq b_\Pi \\ b_\Pi & \text{otherwise} , \end{cases}$$

$$I_{GE}(a,b)_N = \begin{cases} 1 & \text{if } a_N \leq b_N \text{ and } a_\Pi \leq b_\Pi \\ b_\Pi & \text{if } a_N \leq b_N \text{ and } a_\Pi > b_\Pi \\ b_N & \text{otherwise} . \end{cases}$$

Definition 1. *Let $R(U)$ be a relation scheme where $U = \{X_1, X_2, ... , X_n\}$. Let X and Y be subsets of U: $X, Y \subseteq U$. Y is functionally dependent on X in $\theta = (\theta_N, \theta_\Pi)$ degree, denoted by $X \to_\theta Y$, if and only if for every relation r of R the following conditions are met:*

$$\min_{t_1, t_2 \in r} I(t_1(X) =_c t_2(X), t_1(Y) =_c t_2(Y))_N \geq \theta_N ,$$
$$\min_{t_1, t_2 \in r} I(t_1(X) =_c t_2(X), t_1(Y) =_c t_2(Y))_\Pi \geq \theta_\Pi , \tag{5}$$

where θ_N, $\theta_\Pi \in [0,1]$, $=_c$ *is the closeness measure* (3) *and I is the following implicator:*

$$I(a,b) = \begin{cases} I_c & \text{if } t_1(X) \text{ and } t_2(X) \text{ are identical} \\ I_{GE} & \text{otherwise ,} \end{cases} \qquad (6)$$

where I_c *is the classical implication operator and* I_{GE} *is defined by* (4).

Example 1. Relation *PES* (table 1) presents a relationship between the post held (P), education (E) and salary (S). The domains attached to attributes are categorical. Attributes of *PES* satisfy the following FFDs: $PE \rightarrow_{(0.5,0.8)} S$ and

Table 1. Relation *PES* for example 1

	P	E	S
t_1	{1/P1, 1/P2}	{0.4/E1, 1/E2}	{1/high, 0.8/very_high}
t_2	P1	{ 0.9/E2, 1/E3}	very_high
t_3	P1	E4	high
t_4	{1/P1, 0.3/P3}	{1/E4, 0.4/E5}	{1/high, 0.5/average}
t_5	{0.4/P4, 1/P5}	{0.3/E5, 1/E6}	{1/very_low, 0.2/low}
t_6	P5	E6	very_low
t_7	P4	E4	very_low

$SE \rightarrow_{(0.6,1)} P$. For we have:
1. for tuples t_1 and t_2: $=_c(t_1(PE), t_2(PE)) = (0,0.9)$, $=_c(t_1(S), t_2(S)) = (0,0.8)$
$=_c(t_1(SE), t_2(SE)) = (0,0.8)$, $=_c(t_1(P), t_2(P)) = (0,1)$
2. for tuples t_3 and t_4: $=_c(t_3(PE), t_4(PE)) = (0.6,1)$, $=_c(t_1(S), t_2(S)) = (0.5,1)$
$=_c(t_1(SE), t_2(SE)) = (0.5,1)$, $=_c(t_1(P), t_2(P))_L = (0.7,1)$
3. for tuples t_5 and t_6: $=_c(t_5(PE), t_6(PE)) = (0.6,1)$, $=_c(t_1(S), t_2(S)) = (0.8,1)$
$=_c(t_1(SE), t_2(SE)) = (0.7,1)$, $=_c(t_1(P), t_2(P))_L = (0.6,1)$
For other pairs t_i, t_j: $=_c(t_i(PE), t_j(PE)) = =_c(t_i(SE), t_j(SE)) = (0,0)$
Basing on formulas (4) and (5) we obtain the mentioned dependencies.

A FFD $X \rightarrow_\theta Y$ is partial if there exists a set $X' \subset X$ such that $X' \rightarrow_\theta Y$. In the opposite case $X \rightarrow_\theta Y$ fully i.e. Y is fully functionally dependent on X in θ degree. A FFD $X \rightarrow_\theta Y$, $\theta = (\theta_N, \theta_\Pi)$, is transitive if there exists attribute Z such that $X \rightarrow_\alpha Z$ and $Z \rightarrow_\beta Y$ with $\theta_N = \min(\alpha_N, \beta_N)$ and $\theta_\Pi = \min(\alpha_\Pi, \beta_\Pi)$. According to the definition 1 one can formulate the following set of axioms (extended Armstrong's axioms):

A1: $Y \subseteq X \Rightarrow X \rightarrow_\theta Y$ for all θ
A2: $X \rightarrow_\theta Y \Rightarrow XZ \rightarrow_\theta YZ$
A3: $X \rightarrow_\alpha Y \wedge Y \rightarrow_\beta Z \Rightarrow X \rightarrow_\lambda Y$, $\lambda = (\min(\alpha_N, \beta_N), \min(\alpha_\Pi, \beta_\Pi))$

From A1, A2 and A3 the following inference rules can be derived:

D1: $X \to_\alpha Y \wedge X \to_\beta Z \Rightarrow X \to_\gamma YZ$, $\gamma = (\min(\alpha_N, \beta_N), \min(\alpha_\Pi, \beta_\Pi))$;
D2: $X \to_\alpha Y \wedge WY \to_\beta Z \Rightarrow XW \to_\gamma Z$, $\gamma = (\min(\alpha_N, \beta_N), \min(\alpha_\Pi, \beta_\Pi))$;
D3: $X \to_\alpha Y \wedge Z \subseteq Y \Rightarrow X \to_\alpha Z$;
D4: $X \to_\alpha Y \Rightarrow X \to_\beta Y$ for $\beta_N \leq \alpha_N$ and $\beta_\Pi \leq \alpha_\Pi$.

According to the definition 1 one should extend the notion of relation key. Let F be a set of FFDs with respect to the relation scheme $R(U)$. Let us denote by F^+ a set of all FFDs logically implied by F.

Definition 2. *Let F be a set of FFDs (5) defined over a relation scheme $R(U)$. A set of attributes $K \subseteq U$ is called a θ-key of R, $\theta = (\theta_N, \theta_\Pi)$, if $K \to_\theta U \in F^+$ and $K \to_\theta U$ fully i.e. there is no proper subset K' of K such that $K' \to_\theta U \in F^+$.*

Attributes belonging to any θ-key of R are called θ-prime attributes, the others are called θ-nonprime attributes. The scheme PES from Example 1 has two candidate keys: PE - $(0.5, 0.8)$-key and SE - $(0.6, 1)$-key.

Based on the notions of FFD (5) and θ-key one can extend the definitions of classical normal forms. We assume that relations under consideration occur in the first fuzzy normal form (F1NF), which means that every attribute value is represented by an excluding possibility distribution.

Definition 3. *Scheme R (X_1, X_2, ... , X_n) is in θ-fuzzy second normal form (θ-F2NF), $\theta = (\theta_N, \theta_\Pi)$, if every θ-nonprime attribute is fully functionally dependent on a θ-key in α degree, where $\alpha = (\alpha_N, \alpha_\Pi)$.*

Example 2. Scheme PES from Example 1 is in $(0.5, 0.8)$-F2NF. (It is also in $(0.5, 0.8)$-F3NF.) Let us augment it by attribute A determining age and assume that its values are connected with values of attribute P. Let us assume that the relationship between P and A is expressed by $P \to_\phi A$, $\phi = (\phi_N, \phi_\Pi)$. In result of such modification PE is a θ-key with $\theta_N = \min(\phi_N, 0.5)$ and $\theta_U = \min(\phi_\Pi, 0.8)$. However, because of the introduced dependency the modified scheme is not in $(0.5, 0.8)$-F2NF. There is a nonprime attribute dependent on a part of the θ-key.

Definition 4. *Scheme R (U), $U = \{X_1, X_2, ... , X_n\}$, is in θ-fuzzy third normal form (θ-F3NF), $\theta = (\theta_N, \theta_\Pi)$, if for every FFD (5) $X \to_\phi Y \in F^+$, where X, $Y \subseteq U$, $Y \not\subseteq X$, X contains a θ-key of R or Y is a θ-prime attribute.*

Example 3. Let us consider scheme EES with attributes Em - employee, E - education, S - salary and FFDs: $Em \to_{(0.6, 0.8)} E$ and $E \to_{(0.7, 0.9)} S$. By A3 $Em \to_{(0.6, 0.8)} S$. The key of EES is Em. It is the $(0.6, 0.8)$-key. Conditions of Definition 4 are disturbed, because E is not a θ-key and S is not θ-prime. The θ-F3NF can be obtained by decomposition into schemes: $EE(Em, E)$ - $(0.6, 0.8)$-F3NF and $ES(E, S)$ - $(0.7, 0.9)$-F3NF. This decomposition maintains FFDs.

In relation occurring in θ-F3NF there are no θ-nonprime attributes that would be transitively dependent on any θ-key. The imposition of the requirement that the left side of any FFD must contain the θ-key leads to a stronger normal form. This is the θ-fuzzy Boyce-Codd normal form.

Definition 5. *Scheme R (U), $U = \{X_1, X_2, ... X_n\}$, is in θ-fuzzy Boyce-Codd normal form (θ-FBCNF), $\theta = (\theta_N, \theta_\Pi)$, if for every FFD (5) $X \to_\phi Y \in F^+$, where $X, Y \subseteq U$, $Y \not\subseteq X$, X contains a θ-key of R.*

4   Consistency Rules for Multiargument Relationships in Possibilistic Databases

Let us consider a ternary relationship $R(X,Y,Z)$ with the following FFDs:

$$XY \to_\alpha Z \ , \quad XZ \to_\beta Y \ , \quad YZ \to_\gamma X \ , \tag{7}$$

where $\alpha =(\alpha_N, \alpha_\Pi)$, $\beta =(\beta_N, \beta_\Pi)$, $\gamma =(\gamma_N, \gamma_\Pi)$. The scheme R has three candidate θ-keys and occurs in θ-FBCNF, where $\theta_N = \min(\alpha_N, \beta_N, \gamma_N)$ and $\theta_\Pi = \min(\alpha_\Pi, \beta_\Pi, \gamma_\Pi)$.

Let us define the following fuzzy sets of attributes:

$$\mathcal{L}_N = \{\gamma_N^c/X, \ \beta_N^c/Y, \ \alpha_N^c/Z\} \ , \quad \mathcal{B}_N = \{\gamma_N/X, \ \beta_N/Y, \ \alpha_N/Z\} \ ,$$
$$\mathcal{L}_\Pi = \{\gamma_\Pi^c/X, \ \beta_\Pi^c/Y, \ \alpha_\Pi^c/Z\} \ , \quad \mathcal{B}_\Pi = \{\gamma_\Pi/X, \ \beta_\Pi/Y, \ \alpha_\Pi/Z\} \ , \tag{8}$$

where $\alpha_N^c = 1 - \alpha_N$, $\beta_N^c = 1 - \beta_N$, $\gamma_N^c = 1 - \gamma_N$, $\alpha_\Pi^c = 1 - \alpha_\Pi$, $\beta_\Pi^c = 1 - \beta_\Pi$, $\gamma_\Pi^c = 1 - \gamma_\Pi$. Membership grades of attributes depend on levels of corresponding FFDs. If an attribute does not occur on the right side of any FFD (7) its degrees of membership to $\mathcal{L}_N$, $\mathcal{B}_N$, $\mathcal{L}_\Pi$ and $\mathcal{B}_\Pi$ are equal to 1, 0, 1, 0, respectively.

Theorem 1. *Within the fuzzy ternary relationship $R(X,Y,Z)$ with FFDs (7) there may exist binary relationships determined by FFDs of the form $V \to_\phi W$, $\phi =(\phi_N, \ \phi_\Pi)$, where $V, W \in \{X, Y, Z\}$, if $\phi_N \leq \alpha_N$ and $\phi_\Pi \leq \alpha_\Pi$ for $W = Z$, $\phi_N \leq \beta_N$ and $\phi_\Pi \leq \beta_\Pi$ for $W = Y$, $\phi_N \leq \gamma_N$ and $\phi_\Pi \leq \gamma_\Pi$ for $W = X$.*

Proof. Let us consider the imposition of $X \to_\phi Z$, where $\phi = (\phi_N, \ \phi_\Pi)$. Basing on A2 and D3 we obtain $XY \to_\phi Z$. If $\phi_N > \alpha_N$ or $\phi_\Pi > \alpha_\Pi$, the integrity constraints (7) will be disturbed. The membership degrees of Z in the fuzzy sets (8) will change. In the opposite case i.e. if $\phi_N \leq \alpha_N$ and $\phi_\Pi \leq \alpha_\Pi$ the imposition is allowed because then $XY \to_\alpha Z \Rightarrow XY \to_\phi Z$ (rule D4). Thus, the level of the imposed FFD is limited by values of α, β and γ in (7). The proof for other impositions is similar.   □

Let us notice that the imposition of $X \to_\phi Z$ introduces $X \to_\lambda Y$, where $\lambda_N = \min (\phi_N, \ \beta_N)$ and $\lambda_\Pi = \min (\phi_\Pi, \ \beta_\Pi)$. For we have: $X \to_\phi Z \wedge XZ \to_\beta Y \Rightarrow X \to_\lambda Y$. Thus, the new FFD results in additional limitations for the next impositions. One cannot impose $X \to_\psi Y$ with $\psi_N > \lambda_N$ or $\psi_\Pi > \lambda_\Pi$. The imposition of $Z \to_\psi Y$ is also not allowed because it would insert $Z \to_\eta X$ with $\eta_N = \min (\psi_N, \ \gamma_N)$ and $\eta_\Pi = \min (\psi_\Pi, \ \gamma_\Pi)$, which is in contradiction with the primary imposition. After imposition of $X \to_\phi Z$ X becomes a new θ-key with $\theta = (\min(\phi_N, \ \beta_N), \min(\phi_\Pi, \ \beta_\Pi))$. The scheme $R(X,Y,Z)$ remains in θ-FBCNF. However, if $\phi_N < \min(\alpha_N, \ \beta_N, \ \gamma_N)$ or $\phi_\Pi < \min(\alpha_\Pi, \ \beta_\Pi, \gamma_\Pi)$, the level of θ-FBCNF will decrease.

Let us consider the ternary relationship $R(X,Y,Z)$ with two FFDs:

$$XY \to_\alpha Z \ , \quad XZ \to_\beta Y \ . \tag{9}$$

The sets (8) are as follows: $\mathcal{L}_N = \{1/X, \beta_N^c/Y, \alpha_N^c/Z\}$, $\mathcal{B}_N = \{\beta_N/Y, \alpha_N/Z\}$, $\mathcal{L}_\Pi = \{1/X, \beta_\Pi^c/Y, \alpha_\Pi^c/Z\}$ and $\mathcal{B}_\Pi = \{\beta_\Pi/Y, \alpha_\Pi/Z\}$. Let us impose the binary relationship $Y \rightarrow_\phi Z$ with $\phi_N \leq \alpha_N$ and $\phi_\Pi \leq \alpha_\Pi$ (or $Z \rightarrow_\phi Y$ with $\phi_N \leq \beta_N$ and $\phi_\Pi \leq \beta_\Pi$) between attributes occurring on the right sides of dependencies (9). As a result of this imposition the scheme $R(X,Y,Z)$ will not occur in θ-FBCNF. The left side of the imposed FFD does not contain a key. It remains in θ-F3NF with the unchanged level.

Example 4. Relation scheme PES from Example 1 is in (0.5,0.8)-FBCNF. The sets (8) are as follows: $\mathcal{L}_N = \{0.6/P, 1/E, 0.5/S\}$, $\mathcal{B}_N = \{0.4/P, 0.5/S\}$, $\mathcal{L}_\Pi = \{1/E, 0.2/S\}$ and $\mathcal{B}_\Pi = \{1/P, 0.8/S\}$.

(1) Let us consider the imposition of $E \rightarrow_{(0.4,0.6)} S$. Since $E \rightarrow_{(0.4,0.6)} S$ implies $E \rightarrow_{(0.4,0.6)} P$, a new θ-key (attribute E) with $\theta = (0.4,0.6)$ is created. Scheme PES remains in θ-FBCNF with a smaller level $\theta = (0.4,0.6)$. Membership grades in sets (8) remain the same. The imposition is admissible. If the level ϕ of the imposed dependency did not satisfy the conditions $\phi_N \leq 0.5$ and $\phi_\Pi \leq 0.8$, the integrity constraints would be disturbed. The FFD $E \rightarrow_{(0.4,0.6)} S$ introduces limits for the secondary imposition of $E \rightarrow_\psi P$. Values of ψ_N and ψ_Π cannot be greater than 0.4 and 0.6, respectively. The initial limitation was (0.6,1).

(2) Let us consider the imposition of $P \rightarrow_{(0.4,0.7)} S$. No new key is created. The left side of $P \rightarrow_{(0.4,0.7)} S$ does not contain a key, which disturbs the definition of θ-FBCNF. Since S is a (0.6,1)-prime attribute, PES remains in θ-F3NF with the unchanged level ($\theta = (0.5,0.8)$).

Let us consider relationships comprising n attributes. The FFDs describing a relationship $R(X_1, X_2, \ldots, X_n)$ may be presented as follows:

$$U - \{X_i\} \rightarrow_{\alpha_i} X_i, \quad \text{where } \alpha_i = (\alpha_{i_N}, \alpha_{i_\Pi}), \quad i = 1, 2, \ldots, n \quad . \qquad (10)$$

The scheme R has n θ-keys in the form $U - \{X_i\}$ and occurs in θ-FBCNF, where $\theta_N = \min_i(\alpha_{i_N})$ and $\theta_\Pi = \min_i(\alpha_{i_\Pi})$.

Let us define the following fuzzy sets of attributes:

$$\mathcal{L}_N = \{\alpha_{1_N}^c/X_1, \ldots, \alpha_{n_N}^c/X_n\}, \quad \mathcal{B}_N = \{\alpha_{1_N}/X_1, \ldots, \alpha_{n_N}/X_n\},$$
$$\mathcal{L}_\Pi = \{\alpha_{1_\Pi}^c/X_1, \ldots, \alpha_{n_\Pi}^c/X_n\}, \quad \mathcal{B}_\Pi = \{\alpha_{1_\Pi}/X_1, \ldots, \alpha_{n_\Pi}/X_n\}, \quad (11)$$

where $\alpha_{i_N}^c = 1 - \alpha_{i_N}$ and $\alpha_{i_\Pi}^c = 1 - \alpha_{i_\Pi}$.

Theorem 2. *Within the fuzzy n-ary relationship $R(X_1, X_2, \ldots, X_n)$ with FFDs (10) there may exist (n-1)-ary relationships determined by the following FFDs:*

$$U - \{X_i, X_j\} \rightarrow_{\gamma_i} X_i, \quad \text{where } i \neq j, \quad i, j = 1, 2, \ldots, n, \quad \gamma_i = (\gamma_{i_N}, \gamma_{i_\Pi}), \quad (12)$$

in which $\gamma_{i_N} \leq \alpha_{i_N}$ and $\gamma_{i_\Pi} \leq \alpha_{i_\Pi}$.

Proof. The possibility to impose (n-1)-ary relationships is limited by values of α_{i_N} and α_{i_Π}. The membership degrees of attributes in the sets (11) cannot be changed. Due to the Armstrong's rules: $U - \{X_i, X_j\} \rightarrow_{\gamma_i} X_i \Rightarrow U - \{X_i\} \rightarrow_{\gamma_i} X_i$ and $U - \{X_i\} \rightarrow_{\alpha_i} X_i \Rightarrow U - \{X_i\} \rightarrow_{\gamma_i} X_i$ for $\gamma_{i_N} \leq \alpha_{i_N}$ and $\gamma_{i_\Pi} \leq \alpha_{i_\Pi}$. Thus, levels of FFDs (10) remain the same as well as the fuzzy sets (11). The imposed relationship does not disturb the integrity conditions. $\qquad \square$

The imposition of $U-\{X_i,X_j\}\rightarrow_{\gamma_i}X_i$ introduces $U-\{X_i,X_j\}\rightarrow_{\lambda_{i,j}}X_j$, with $\lambda_{i,j_N}=\min(\gamma_{i_N},\alpha_{j_N})$ and $\lambda_{i,j_\Pi}=\min(\gamma_{i_\Pi},\alpha_{j_\Pi})$. For we have:

$$U-\{X_i,X_j\}\rightarrow_{\gamma_i}X_i \wedge U-\{X_i\}\rightarrow_{\alpha_j}X_j \Rightarrow U-\{X_i,X_j\}\rightarrow_{\lambda_{i,j}}X_j$$
$$\text{with } \lambda_{i,j_N}=\min(\gamma_{i_N},\alpha_{j_N}) \text{ and } \lambda_{i,j_\Pi}=\min(\gamma_{i_\Pi},\alpha_{j_\Pi})\ . \quad (13)$$

As a result we cannot impose $U-\{X_i,X_j\}\rightarrow_{\phi_j}X_j$ with $\phi_{j_N}>\lambda_{i,j_N}$ or $\phi_{j_\Pi}>\lambda_{i,j_\Pi}$ although initially the admissible level was given by α_j.

If the number of FFDs (10) is lower than n there are attributes which do not occur on the right side of any dependency (10). Let us denote these attributes by $X_{m+1}, X_{m+2}, \dots, X_n$, where m denotes the number of FFDs (10) and assume that $m>1$. Attributes $X_{m+1}, X_{m+2}, \dots, X_n$ fully belong to $\mathcal{L}_N$ and $\mathcal{L}_\Pi$. The sets $\mathcal{L}_N$, $\mathcal{L}_\Pi$, $\mathcal{B}_N$, $\mathcal{B}_\Pi$ are as follows:

$$\mathcal{L}_N = \{\alpha_{1_N}^c/X_1, \dots, \alpha_{m_N}^c/X_m, 1/X_{m+1}, \dots, 1/X_n\}\ ,$$
$$\mathcal{B}_N = \{\alpha_{1_N}/X_1, \dots, \alpha_{m_N}/X_m\}\ ,$$
$$\mathcal{L}_\Pi = \{\alpha_{1_\Pi}^c/X_1, \dots, \alpha_{m_\Pi}^c/X_m, 1/X_{m+1}, \dots, 1/X_n\}\ ,$$
$$\mathcal{B}_\Pi = \{\alpha_{1_\Pi}/X_1, \dots, \alpha_{m_\Pi}/X_m\}\ . \quad (14)$$

The imposition of the relationship (12) may result in disturbance of conditions of θ-FBCNF (Definition 5).

Let us impose $U-\{X_i,X_j\}\rightarrow_{\gamma_i}X_i$, where $j\leq m$, $\gamma_{i_N}\leq\alpha_{i_N}$ and $\gamma_{i_\Pi}\leq\alpha_{i_\Pi}$. By (13) one can remove the attribute X_i from the left side of $U-\{X_j\}\rightarrow_{\alpha_j}X_j$. Thus, the left side of the imposed dependency forms the new θ-key with level $\lambda_{i,j}$. The scheme R is in θ'-FBCNF, where $\theta'_N=\min(\theta_N,\lambda_{i,j_N})$ and $\theta'_\Pi=\min(\theta_\Pi,\lambda_{i,j_\Pi})$. If $j>m$ the imposition of $U-\{X_i,X_j\}\rightarrow_{\gamma_i}X_i$ will disturb conditions of Definition 5. The left side of the imposed dependency does not contain the key. However, the scheme R will remain in θ-F3NF, because X_i is θ-prime. If $m=1$ there is only one candidate key, which is formed by the left side of the existing dependency (10). Imposition of (12) introduces a partial dependency of attribute X_i on the θ-key which means a disturbance in the conditions defining the θ-fuzzy second normal form.

5   Conclusions

In the paper we applied and extended the definition of fuzzy functional dependency which was formulated by Chen [3] within the possibilistic data model. Its level is evaluated by two numbers from the unit interval which correspond to the possibility and necessity measures. We used an extension of the Gödel implication operator. The concept of the fuzzy functional dependency has been applied in definitions of extended fuzzy normal forms which correspond to definitions in classical relational model. The subject of considerations was the possible occurrence - within the n-ary relationship with attributes represented by means of possibility distributions - of fuzzy functional dependencies (12) describing the relationships between $(n-1)$ attributes. The starting point are fuzzy functional dependencies (10) existing between all attributes of the relation scheme $R(X_1,$ $X_2, \dots, X_n)$. They determine the integrity constraints for R. According to Theorem 2 dependencies (12) are admissible if their levels do not exceed the levels of relevant dependencies (10). The imposition of them may disturb the normal form of scheme R.

References

1. Bosc, P., Pivert, O.: Querying possibilistic databases: Three interpretations. In: Yager, R.R., Abbasov, A.M., Reformat, M., Shahbazova, S.N. (eds.) Soft Computing: State of the Art Theory. STUDFUZZ, vol. 291, pp. 161–176. Springer, Heidelberg (2013)
2. Bosc, P., Pivert, O., Rocacher, D.: Tolerant division queries and possibilistic database querying. Fuzzy Sets and Systems 160(15), 2120–2140 (2009)
3. Chen, G.: Fuzzy Logic in Data Modeling - semantics, constraints and database design. Kluwer Academic Publishers, Boston (1998)
4. Cuadra, D., Martinez, P., Castro, E., Al-Jumaily, H.: Guidelines for representing complex cardinality constraints in binary and ternary relationships. Software and Systems Modeling 12(4), 871–889 (2013)
5. De Tré, G., Zadrożny, S., Bronselaer, A.: Handling bipolarity in elementary queries to possibilistic databases. IEEE Trans. on Fuzzy Systems 18(3), 599–612 (2010)
6. Dullea, J., Song, I., Lamprou, I.: An analysis of structural validity in entity-relationship modeling. Data and Knowledge Engineering 47(2), 167–205 (2003)
7. Galindo, J., Urrutia, A., Piattini, M.: Fuzzy Databases. Modeling, Design and Implementation. Idea Group Publishing, London (2005)
8. Jones, T., Song, I.: Analysis of binary/ternary relationships cardinality combinations in entity-relationship modeling. Data and Knowledge Engineering 19(1), 39–64 (1996)
9. Miłek, M., Małysiak-Mrozek, B., Mrozek, D.: A fuzzy data warehouse: Theoretical foundations and practical aspects of usage. Studia Informatica 31(2A-89), 489–504 (2010)
10. Myszkorowski, K.: Fuzzy functional dependencies in multiargument relationships. In: Rutkowski, L., Scherer, R., Tadeusiewicz, R., Zadeh, L.A., Zurada, J.M. (eds.) ICAISC 2010, Part I. LNCS, vol. 6113, pp. 152–159. Springer, Heidelberg (2010)
11. Nakata, M.: On inference rules of dependencies in fuzzy relational data models: Functional dependencies. In: Pons, O., Vila, M., Kacprzyk, J. (eds.) Knowledge Management in Fuzzy Databases. STUDFUZZ, vol. 39, pp. 36–66. Physica-Verlag, Heidelberg (2000)
12. Prade, H., Testemale, C.: Generalizing database relational algebra for the treatment of incomplete or uncertain information and vague queries. Information Science 34(2), 115–143 (1984)
13. Ślęzak, D.: Association reducts: A framework for mining multi-attribute dependencies. In: Hacid, M.-S., Murray, N.V., Raś, Z.W., Tsumoto, S. (eds.) ISMIS 2005. LNCS (LNAI), vol. 3488, pp. 354–363. Springer, Heidelberg (2005)
14. Tyagi, B.K., Sharfuddin, A., Dutta, R.N., Devendra, K.T.: A complete axiomatization of fuzzy functional dependencies using fuzzy function. Fuzzy Sets and Systems 151(2), 363–379 (2005)
15. Vucetic, M., Hudec, M., Vujosevic, M.: A new method for computing fuzzy functional dependencies in relational database systems. Expert Systems with Applications 40, 2738–2745 (2013)

Data Integration, Storage and Data Warehousing

Data Sharing and Exchange: General Data-Mapping Semantics

Rana Awada[✉] and Iluju Kiringa[✉]

University of Ottawa, EECS, Ottawa, Ontario, Canada
{rawad049,iluju.kiringa}@uottawa.ca

Abstract. Traditional data sharing and exchange solved the problem of exchanging data between applications that store same information using different vocabularies. We discuss in this paper the data sharing and exchange problem between applications that store *related* data which do not necessarily possess the same meaning. We first consider this problem in settings where source instances are *complete* – that is, do not contain unknown data. Then we address more collaborative scenarios where peers can store incomplete information. We define the semantics of these settings, and we provide the data complexity for generating solutions and the minimal among those. Also, we distinguish between *sound* and *complete* certain answers as semantics for conjunctive query answering.

1 Introduction

Integrating related data from independent sources which possess differences in both structure and vocabulary, is a problem that was first addressed in data coordination (DC) settings [6]. A DC setting [6] $\mathfrak{S}$ consists of two schemas $\mathbf{S}_1$ and $\mathbf{S}_2$, and a set of mapping tables $\{\mathcal{M}\}$. DC settings introduced so far considered mapping tables $\mathcal{M}$ with general interpretation of related data; that is, a pair (a, b) holds in $\mathcal{M}$ if a is *related* to b (where a belongs to the instance of $\mathbf{S}_1$ and b belongs to the instance of $\mathbf{S}_2$), and the *related* relationship does not possess any particular meaning. Although DC amalgamates data of different applications in a single result, it still uses in this set the different vocabularies of these applications. Such a property makes it harder in applications to analyse this data and make decisions about it. We elaborate this property in the following example.

Example 1. Let $\mathfrak{S} = (\mathbf{S}_1, \mathbf{S}_2, \mathcal{M})$ be a DC setting. Assume that schema $\mathbf{S}_1$ of university $Univ_a$ consists of the relation symbols Student(sname, sage) and Enroll(sname, cname, cgrade), and schema $\mathbf{S}_2$ of $Univ_b$ consists of the relation symbols St(sname, sage) and Take(sname, cname, cgrade). Relation *Student* (*St*) stores students' names and ages information. Relation *Enroll* (*Take*) stores the set of courses that each student completed with the final grades that he/she received in these courses. In addition, assume that $\mathbf{S}_1$ and $\mathbf{S}_2$ are connected by the mapping table $\mathcal{M}$ which contains the pairs: $\{(\text{Cid}_a,\text{Cid}_1), (\text{Cid}_a,\text{Cid}_2)\}$. Let $I_1 = \{\text{Student(Alex,18)}, \text{Enroll(Alex,Cid}_a,80)\}$ be an instance of $\mathbf{S}_1$, and I_2

© Springer International Publishing Switzerland 2015
S. Kozielski et al. (Eds.): BDAS 2015, CCIS 521, pp. 159–169, 2015.
DOI: 10.1007/978-3-319-18422-7_14

= {St(Ben,19), Take(Ben,Cid$_1$,B), Take(Ben,Cid$_2$,C)} be an instance of $\mathbf{S}_2$. To retrieve the list of students from both universities that completed course Cid$_a$ or those corresponding to it according to $\mathcal{M}$, authors in [6] rewrite a query Q posed to $Univ_a$ to a query Q' using $\mathcal{M}$ then pose it to $Univ_b$. Query Q' can be: Select $sname, cname$ From $Take$ Where $cname =$ Cid$_1$ Or $cname =$ Cid$_2$. The combined result of Q and Q' is {(Alex,Cid$_a$), (Ben,Cid$_1$), (Ben,Cid$_2$)}.

In Example 1, the combined result of Q and Q' has elements Cid$_a$, Cid$_1$, and Cid$_2$ which are different vocabularies. Such a property makes it hard to compare the performances of Alex and Ben in a certain course. In DC [6], one source element can be mapped in $\mathcal{M}$ to one or more target elements. Clearly, a higher number of mapped elements lead to bigger sizes of combined results, and more expensive query computations. For some queries, retrieving a portion of the result can be informative enough for users to escape the more expensive computation of the full result.Therefore, to control the size or related information returned, authers in [6] introduced the notions of complete and sound query translations that allow users compute the full set of correct answers of conjunctive queries (CQs), called *complete* answers, and a portion of those, named *sound* answers, respectively. From Example 1, query Q' is a complete query translation, while a sound query translation can be $Q'' =$ Select $sname, cname$ From $Take$ Where $cname =$ Cid$_1$ (since both Cid$_1$ and Cid$_2$ are mapped to Cid$_a$ in $\mathcal{M}$).

Our main interest in this work is to solve the heterogeneity problem in DC settings while allowing users to generate sound and complete answers of CQs. Intuitively, one possible solution is to exchange data from independent sources and store it in a separate target using a unified set of vocabularies, prior to applying queries. It turns out that data sharing and exchange (DSE) settings introduced recently in [3] provides such a functionality, and in particular, DSE solutions in this setting suits well our requirement to generate sound and complete answers. As given in [3], a DSE setting is a tuple $\mathfrak{S} = (\mathbf{S}, \mathbf{T}, \mathcal{M}, \Sigma_{st})$, where $\mathbf{S}$ and $\mathbf{T}$ are source and target schemas, respectively; $\mathcal{M}$ is an st-mapping table, and Σ_{st} consists of a set of mapping *source-to-target tuple-generating* dependencies (st-tgds). DSE [3] however supports applications that refer to same elements using different sets of vocabularies; that is, they use the equivalence semantics of related data in st-mapping tables.

We discuss in this paper a DSE setting, denoted by DSE$_G$, to exchange *related* information between applications that possess different schemas and different domains of constants. Our main contributions are the following: (1) we provide algorithms to generate DSE solutions in LOGSPACE, (2) we provide algorithms to generate more compact universal DSE solutions, minimal universal DSE solutions, in LOGSPACE, (3) we formally define sound and complete certain answers as semantics for CQ answering and we provide algorithms to generate those, (4) finally, we address the DSE$_G$ problem in more challenging settings where source instances contain unknown information in the form of *nulls*.

2   Preliminaries and Related Work

We start this section by defining the notations that we will use through our the paper. Then we give a brief summary of two settings defined previously in the literature to address the data exchange and data coordination problems. These are data exchange (DE) [5] and DSE [3] settings.

A *schema* $\mathbf{R}$ is a finite set $\{R_1, \ldots, R_k\}$ of relation symbols, with each R_i having a fixed arity $n_i > 0$. Let D be a countably infinite domain. An *instance* I of $\mathbf{R}$ assigns to each relation R_i of $\mathbf{R}$ a finite n_i-ary relation $R_i^I \subseteq \mathsf{D}^{n_i}$. Sometimes we write $R_i(\bar{t}) \in I$ instead of $\bar{t} \in R_i^I$, and call $R_i(\bar{t})$ a *fact* of I. The *domain* $dom(I)$ of instance I is the set of all elements that occur in any of the relations R_i^I. We often define instances by simply listing the facts that belong to them.

Data Exchange Settings. Data exchange [5] is the problem of extracting an instance over a source schema $\mathbf{S}$ and transform it to confirm to an independent target schema $\mathbf{T}$. More formally, a DE setting is a tuple $\mathfrak{S} = (\mathbf{S}, \mathbf{T}, \Sigma_{st})$, where $\mathbf{S}$ is a source schema, $\mathbf{T}$ is an independent target schema, Σ_{st} is a set of st-tgds which are FO sentences of the form $\forall \bar{x} \forall \bar{y} \forall \bar{z}\, (\phi(\bar{x}, \bar{y}) \rightarrow \exists \bar{w}\, \psi(\bar{z}, \bar{w}))$, where $\phi(\bar{x}, \bar{y})$ and $\psi(\bar{z}, \bar{w})$ are conjunctions of relational atoms over $\mathbf{S}$ and $\mathbf{T}$, respectively. It is customary in DE to exchange data based on st-tgds that consists of positive predicates and equality formulas, until authors in [4] introduced a solution that solved the problem of exchanging data using st-tgds which contain negated predicates and/or inequality formulas. In their solution, they extended the source schema $\mathbf{S}$ to $\hat{\mathbf{S}} := \mathbf{S} \cup \{\hat{R} : \neg R \in \Sigma_{st}\} \cup \{N\}$. They also re-constructed Σ_{st} to $\hat{\Sigma}_{st}$ by replacing each negated literal $\neg R(\bar{x})$ with $\hat{R}(\bar{x})$, and inequality $x \neq y$ over $\mathbf{S}$ with and $N(x, y)$ over $\hat{\mathbf{S}}$, respectively. Each table $\hat{R}$ is populated with the evaluation of $\neg R(\bar{x})$ on the instance I of $\mathbf{S}$, and $N(x, y)$ contains all the pairs of elements (a, b) in $dom(I)$ such that $a \neq b$.

An instance J of a target schema $\mathbf{T}$ is said to be a *solution* for a source instance I under $\mathfrak{S} = (\mathbf{S}, \mathbf{T}, \Sigma_{st})$, if the instance (I, J) of $\mathbf{S} \cup \mathbf{T}$ satisfies Σ_{st}. The DE literature identified a class of "good" solutions, called the *universal* solutions, that in a precise way represent all other solutions. To define those, authors in [5] used the notion of homomorphism between instances. Let K_1 and K_2 be instances of the same schema $\mathbf{R}$. A *homomorphism* h from K_1 to K_2 is a function $h : dom(K_1) \rightarrow dom(K_2)$ such that: (1) $h(c) = c$ for every $c \in \mathsf{Const} \cap dom(K_1)$, and (2) for every $R \in \mathbf{R}$ and tuple $\bar{a} = (a_1, \ldots, a_k) \in R^{K_1}$, it holds that $h(\bar{a}) = (h(a_1), \ldots, h(a_k)) \in R^{K_2}$. Let $\mathfrak{S}$ be a DE setting, I a source instance and J a solution under $\mathfrak{S}$. Then J is a *universal* solution under $\mathfrak{S}$, if for every solution J' for I under $\mathfrak{S}$, there exists a homomorphism from J to J'.

Data Sharing and Exchange Settings. DSE [3] exchanges data between applications that refer to same objects using different instance values. Formally, a DSE setting is a tuple $\mathfrak{S} = (\mathbf{S}, \mathbf{T}, \mathcal{M}, \Sigma_{st})$, where: (1) $\mathbf{S}$ and $\mathbf{T}$ are a source and a target schema, respectively; (2) $\mathcal{M}$ is a binary relation symbol that appears neither in $\mathbf{S}$ nor in $\mathbf{T}$, and that is called a *source-to-target (st-) mapping* and (3) Σ_{st} consists of a set of *mapping* st-tgds, which are FO sentences of the form $\forall \bar{x} \forall \bar{y} \forall \bar{z}\, (\phi(\bar{x}, \bar{y}) \wedge \mu(\bar{x}, \bar{z}) \rightarrow \exists \bar{w}\, \psi(\bar{z}, \bar{w}))$, where (i) $\phi(\bar{x}, \bar{y})$ and $\psi(\bar{z}, \bar{w})$ are conjunctions of relational atoms over $\mathbf{S}$ and $\mathbf{T}$, resp., (ii) $\mu(\bar{x}, \bar{z})$ is a conjunction of atomic formulas that only use the relation symbol $\mathcal{M}$.

While data mappings possessed no particular meaning in DC settings [6], authors in [3] adopted the *equivalence* semantics in st-mapping tables; that is two elements a and b are mapped in an st-mapping table $\mathcal{M}$ only if they possess the same meaning. Apparently, and as discussed in [3], such equivalence interpretation of related data can entail new equivalent elements in $\mathcal{M}$. For example, if an element a is related to elements a' and a'' in $\mathcal{M}$ and element b is related to elements a' and b' in $\mathcal{M}$, then as a consequence elements a'' and b' are considered equivalent according to the semantics of $\mathcal{M}$ and a can be mapped to b' in $\mathcal{M}$. This equivalence property in st-mapping tables made the knowledge exchange framework [2] be a natural representation for DSE [3]. A knowledge base (KB) over a schema $\mathbf{R}$ is a pair (K, Σ), where K is an instance of $\mathbf{R}$ (the explicit data) and Σ is a set of logical sentences over $\mathbf{R}$ (the implicit data).

In DSE [3], source and target instances store explicit data accompanied with implicit information in the form of full tgds[1] – denoted by Σ_s and Σ_t respectively– that complete the source KB, the st-mapping table, and the target KB with the entailed information. They also defined a class of "good" solutions – they called it universal DSE solutions – that could be generated in LOGSPACE.

Let $\mathfrak{S} = (\mathbf{S}, \mathbf{T}, \mathcal{M}, \Sigma_{st})$ be a DSE setting [3], I a source instance, $\mathcal{M}$ an st-mapping table, J a target instance. Denote by $K_{\mathbf{R}'}$ the restriction of instance K to a subset $\mathbf{R}'$ of its schema $\mathbf{R}$. Recall that Σ_s, Σ_t are the source and target completions of $\mathfrak{S}$, respectively. Then, J is a DSE solution for I and $\mathcal{M}$ under $\mathfrak{S}$, if for every $K \in \mathsf{Mod}((J \cup \{\mathcal{M}\}), \Sigma_t)^2$ there is $K' \in \mathsf{Mod}((I \cup \{\mathcal{M}\}), \Sigma_s)$ such that the following hold: (a) $K'_{\mathcal{M}} \subseteq K_{\mathcal{M}}$, and (b) $K_{\mathbf{T}} \vDash ((K'_{\mathbf{S}} \cup \{K'_{\mathcal{M}}\}), \Sigma_{st})$ under $\mathfrak{S}$. In addition, J is a *universal* DSE solution for I and $\mathcal{M}$ under $\mathfrak{S}$, if J is a DSE solution, and for every $K' \in \mathsf{Mod}((I \cup \{\mathcal{M}\}, \Sigma_s)$ there is $K \in \mathsf{Mod}((J \cup \{\mathcal{M}\}, \Sigma_t)$ such that (a) $K_{\mathcal{M}} \subseteq K'_{\mathcal{M}}$, and (b) $K_{\mathbf{T}} \vDash ((K'_{\mathbf{S}} \cup \{K'_{\mathcal{M}}\}), \Sigma_{st})$ under $\mathfrak{S}$.

3   General Data Sharing and Exchange

As before, instances of $\mathbf{S}$ (resp. $\mathbf{T}$) are called source (resp. target) instances[3]. Instances of $\mathcal{M}$ are called st-mapping tables[4]. Also, as the case in [3], we assume the existence of two (not necessarily disjoint) countably infinite sets of constants $\mathsf{Const}^{\mathbf{S}}$ and $\mathsf{Const}^{\mathbf{T}}$, that denote the set of source and target constants, respectively. We also assume the existence of a countably infinite set Var of labeled nulls (that is disjoint from both $\mathsf{Const}^{\mathbf{S}}$ and $\mathsf{Const}^{\mathbf{T}}$). For the time being we assume that $dom(I) \in \mathsf{Const}^{\mathbf{S}}$. We will drop this assumption later in section 4. Also, as the case in [3], st-mapping tables are over $(\mathsf{Const}^{\mathbf{S}}, \mathsf{Const}^{\mathbf{T}})$.

---

[1] A full tgd is a tgd that does not contain existentially quantified variables.

[2] $\mathsf{Mod}((J \cup \{\mathcal{M}\}), \Sigma_t)$ is defined as the set of instances of $(\mathbf{T} \cup \{\mathcal{M}\})$ that contain the explicit data in $(J \cup \{\mathcal{M}\})$ and satisfy the implicit data in Σ_t.

[3] We denote source instances by $I, I', I_1, \ldots$ and target instances by $J, J', J_1, \ldots$.

[4] We slightly abuse notations and denote both st-mapping relations and st-mapping tables by $\mathcal{M}$.

Formally, a DSE_G setting $\mathfrak{S}$ is a setting where the definition of a DSE setting introduced in [3] applies to $\mathfrak{S}$. We consider in DSE_G source KBs of the form $((I \cup \{\mathcal{M}\}), \emptyset)$, which correspond to data in the source instance I and the st-mapping table $\mathcal{M}$. On the other hand, the target KBs are of the form $((J \cup \{\mathcal{M}\}), \Sigma_t)$, where J contains a portion of the exchanged data, and Σ_t contains a set of FO sentences, of type full tgds, over a schema that includes $\mathbf{T}$ and $\mathcal{M}$.

We assume in a DSE_G setting that the st-mapping table $\mathcal{M}$ possesses the following particular characteristic: for each value $a \in (dom(\mathcal{M}) \cap \mathsf{Const}^{\mathbf{S}})$, there exists an element $a' \in (dom(\mathcal{M}) \cap \mathsf{Const}^{\mathbf{T}})$ such that $\mathcal{M}(a, a')$ holds, and for no $b \in (dom(\mathcal{M}) \cap \mathsf{Const}^{\mathbf{S}})$ – different than a – $\mathcal{M}(b, a')$ holds. We say a' uniquely identifies a in $\mathcal{M}$. More formally, we assume $\mathcal{M}$ satisfies the following constraint: $\forall x \exists y \forall z (\mathcal{M}(x, y) \wedge \mathcal{M}(z, y) \rightarrow x = z)$. To be used later, we store in a fresh table C, defined in both schemas $\mathbf{S}$ and $\mathbf{T}$, the set of values in $(dom(\mathcal{M}) \cap \mathsf{Const}^{\mathbf{T}})$ that uniquely identify source elements mapped in $\mathcal{M}$. So, following Example 1, and assuming that $\mathcal{M}$ includes the pair $\{(\mathrm{Cid}_b, \mathrm{Cid}_2), (\mathrm{Cid}_b, \mathrm{Cid}_3)\}$, then Cid_1 uniquely identifies Cid_a and Cid_3 uniquely identifies Cid_b, and $C = \{\mathrm{Cid}_1, \mathrm{Cid}_3\}$.

We adopt (universal) DSE solutions introduced in [3] with the restriction of $\Sigma_s = \emptyset$ to define the semantics of DSE_G. We illustrate DSE solutions in DSE_G settings in the following example.

Example 2. (Example 1 cont.) In reference to the DC setting $\mathfrak{S}$ given in Example 1, let I_1 be the source instance and $\mathcal{M}$ the st-mapping table in $\mathfrak{S}$. Also, assume that the set of mapping st-tgds Σ_{st} is the following:
(a) $Student(x, y) \wedge \mathcal{M}(x, x') \wedge \mathcal{M}(y, y') \rightarrow St(x', y')$.
(b) $Enroll(x, w, u) \wedge \mathcal{M}(x, x') \wedge \mathcal{M}(w, w') \wedge \mathcal{M}(u, u') \rightarrow Take(x', w', u')$.
Then a possible DSE solution J for I and $\mathcal{M}$ under $\mathfrak{S}$ would be $J = \{St(Alex, 18), Take(Alex, Cid_1, B), Take(Alex, Cid_2, B)\}$.

Clearly, C is the sole set of target values in a DSE_G instance that correctly captures the set of source values exchanged to a target instance. Therefore, we use C as a fundamental part of the FO sentences – the implicit data – in Σ_t. The set of FO sentences d over a schema that includes $\mathcal{M}$, C, and a fresh table C', which specify the target elements in $(dom(\mathcal{M}) \cap \mathsf{Const}^{\mathbf{T}})$ that are in C, are the following: (1) $\forall x \forall y \forall z (\mathcal{M}(x, y) \wedge \mathcal{M}(z, y) \wedge x \neq z \rightarrow C'(y))$, (2) $\forall x \forall y (\mathcal{M}(x, y) \wedge \neg C'(y) \rightarrow C(y))$.

Authors in [4] addressed the problem of chasing dependencies which contain negated predicates and inequality formulas. We adopt this solution and apply it to the set of rules d to populate the table C. We prove the following result:

Theorem 1. *Let $\mathfrak{S}$ be a fixed DSE_G setting and $\mathcal{M}$ be an st-mapping table. Then generating C is in* LOGSPACE.

We define below a target completion program as a set of full tgds, denoted Σ_t, over a schema that includes $\mathbf{T}$ and $\mathcal{M}$, such that applying Σ_t to a universal DSE solution J generates a universal DSE solution J' for I and $\mathcal{M}$ under $\mathfrak{S}$ which contains the complete set of exchanged data; that is $((I \cup \{\mathcal{M}\}), J') \vDash \Sigma_{st}$.

Let $\mathfrak{S} = (\mathbf{S}, \mathbf{T}, \mathcal{M}, \Sigma_{st})$ be a DSE_G setting. Σ_t, is the following set of FO sentences over $\mathbf{T} \cup \{\mathcal{M}, \mathrm{REL}\}$, and REL is a fresh binary table:

1. For each $T \in \mathbf{T} \cup \{\mathcal{M}\}$ of arity n and $1 \leq i \leq n$:
   $\forall x_1 \cdots \forall x_n (T(x_1, \ldots, x_i, \ldots, x_n) \to \text{REL}(x_i, x_i))$.
2. $\forall x \forall y \forall z (\mathcal{M}(z, x) \wedge \mathcal{M}(z, y) \wedge C(x) \to \text{REL}(x, y))$.
3. For each $T \in \mathbf{T}$ of arity n:
   $\forall x_1, y_1 \cdots \forall x_n, y_n \, (T(x_1, \ldots, x_n) \wedge \bigwedge_{i=1}^{n} \text{REL}(x_i, y_i) \to T(y_1, \ldots, y_n))$.

The first rule defines the reflexive relation REL on the domain of the target instance. The second rule captures the target elements that are related to the same source value in the st-mapping table $\mathcal{M}$. The last rule allows to complete the symbols of $\mathbf{T}$, by adding elements declared to be related in REL.

Intuitively, applying a procedure (based on the *chase* [4]) to the instance $(I \cup \{\mathcal{M}\})$, generates a universal DSE solution in DSE_G. Thus, since the chase runs in LOGSPACE [1] and following Theorem 1, we can conclude that there exists a LOGSPACE algorithm that generates DSE solutions in fixed DSE_G settings.

Most compact solutions in a DSE_G setting can be identified by the class of *minimal* universal DSE (MUDSE) solutions introduced in [3]. A MUDSE solution J for a source instance I and an st-mapping table $\mathcal{M}$ in DSE_G is such that (1) there exists no proper subset J' of J where J' is a universal DSE solution for I and $\mathcal{M}$ under $\mathfrak{S}$, and (2) there exists no universal DSE solution J' such that $(dom(J') \cap \mathsf{Const}^{\mathbf{T}})$ is properly contained in $(dom(J) \cap \mathsf{Const}^T)$.

Example 3. (Example 2 cont.) In reference to Example 2, a possible MUDSE solution J for I and $\mathcal{M}$ under $\mathfrak{S}$ would be $J = \{St(Alex, 18), Take(Alex, Cid_1, B)\}$.

We provide a procedure $\mathtt{CompMUDSE_Gsol_{\mathfrak{S}}}$, a variant of $\mathtt{CompMUDSEsol_{\mathfrak{S}}}$ [3], that given an instance I and an st-mapping table $\mathcal{M}$ in a DSE_G setting $\mathfrak{S}$, generates a MUDSE solution J for I and $\mathcal{M}$ under $\mathfrak{S}$.
$\mathtt{CompMUDS_Gsol_{\mathfrak{S}}}$:
Input: A source instance I, an st-mapping table $\mathcal{M}$, and a set Σ_{st} of st-tgds.
Output: A Canonical MUDSE solution J for I and $\mathcal{M}$ under $\mathfrak{S}$.

1. Populate in the table C elements from $\mathcal{M}$ using the set of FO sentences d.
2. Apply a procedure (based on the *chase* [4]) to the instance $(I \cup \{\mathcal{M}\})$, and generate a target instance J.
3. Compute a set of classes $\{C_1, \ldots, C_m\}$ over $dom(C)$ such that c_1 and c_2 exist in C_i if there exists a constant a such that $\mathcal{M}(a, c_1)$ and $\mathcal{M}(a, c_2)$ hold.
4. Choose a set of witnesses $\{w_1, \ldots, w_m\}$ such that $w_i \in C_i$, for $1 \leq i \leq m$.
5. Compute from J the instance $J' := \mathtt{replace}(J, w_1, \ldots, w_m)$ by replacing each occurrence of target constant $c \in C_i \cap dom(J)$ $(1 \leq i \leq m)$ with $w_i \in C_i$.
6. Apply a procedure (based on the core [5]) to the target instance J' and generate the target instance J_1 that is the core of J'.

We prove the correctness of $\mathtt{CompMUDSE_Gsol_{\mathfrak{S}}}$ and that it runs in LOGSPACE in the following result:

Theorem 2. *Let $\mathfrak{S}$ be a fixed DSE_G setting, I a source instance, and $\mathcal{M}$ an st-mapping table. Let J^* be an arbitrary result for $\mathtt{CompMUDSE_Gsol_{\mathfrak{S}}}$. Then, J^* is a MUDSE for I and $\mathcal{M}$ under $\mathfrak{S}$. Also, $\mathtt{CompMUDSE_Gsol_{\mathfrak{S}}}$ runs in LOGSPACE.*

4 DSE_G and Incomplete Source Data

We discuss in this section DSE_G in collaborative settings in which several sources interact, we denote it by DSE_I to avoid any confusion. In such scenarios, it is most likely that data in a source instance is obtained by an earlier exchange from a different source, and hence may contain null values. The most intuitive tables with null values are called naïve tables. Arenas et al. addressed in [2] the DE problem in such collaborative settings. They proved that target instances with positive conditional tables, named also as *pc-tables* [5], are the best tool to represent source instances that contain naïve tables and/or pc-tables.

Clearly, DSE_I combines both the KB exchange and the DE with incomplete information [2] concepts, and hence require applying the techniques from both settings. Authors in [2] defined a variant *chase* algorithm which generates in fixed DE settings universal solutions with pc-tables in PTIME. We combine this fact with the result given in Theorem 1 to deduce that there exists a PTIME algorithm that generates a universal DSE solution (J, Σ_t) for a source knowledge base $(I, \emptyset)$ in a fixed DSE_I setting. On the other hand, we show in the following Theorem that checking whether a given pc-table target instance is a universal DSE solution in a fixed DSE_I setting is of a higher complexity.

Theorem 3. *Let $\mathfrak{S} = (\mathbf{S}, \mathbf{T}, \mathcal{M}, \Sigma_{st})$ be a fixed DSE_I setting where Σ_{st} is a set of fixed st-tgds, I a pc-table over $\mathbf{S}$, and J a pc-table over $\mathbf{T}$. Also let the set of implicit data Σ_t be a fixed set of full tgds. Then, checking if J is a universal DSE solution for I is DP_2^P-complete.*

The class of MUDSE solutions discussed so far in this work and in DSE settings [3], contained solely of naïve tables. We prove in the following result that the best tool to represent the most compact DSE solutions for target instances with naïve tables in DSE_I is indeed a naïve table. That is, there does not exist a DSE solution with pc-tables that is more compact under $\mathfrak{S}$.

Theorem 4. *Let $\mathfrak{S} = (\mathbf{S}, \mathbf{T}, \mathcal{M}, \Sigma_{st})$ be a DSE_I setting, I a source instance, $\mathcal{M}$ an st-mapping table, and J a MUDSE solution (with naïve tables) for I and $\mathcal{M}$ under $\mathfrak{S}$. Then there does not exist a solution J' under $\mathfrak{S}$ such that J' contains pc-tables, $\mid J' \mid$ [6] $< \mid J \mid$, and $Rep(J') = Rep(J)$ [7].*

Given a DSE_I setting $\mathfrak{S}$ and a DSE solution J with pc-tables. At first, one can think that generating the smallest subset instance of J, in the favor of generating a MUDSE solution J' of J under $\mathfrak{S}$, is possible by applying the Greedy algorithm introduced in ordinary DE settings [5]. Briefly, a Greedy algorithm

---

[5] A pc-table is a naïve table extended with a *local* condition for each tuple. This local condition is a positive boolean combination of formulas of the form $\bot = \bot'$ and $\bot = a$, where $\bot, \bot' \in \mathsf{Var}$ and $a \in \mathsf{Const}$.

[6] $\mid T \mid$ define the number of tuples in a table T.

[7] $Rep(R) = \{\rho(R) \mid \rho : \mathsf{Var}(R) \to \mathsf{Const}$ is a valuation for the variables in table R (in case R is a pc-table, then a fact $\rho(t) \in \rho(R)$ only if the condition ψ in t is such that $\rho(t) = \mathrm{true}$).

R

A	B	C	Condition
x	y	z	$true$
1	1	1	$x = 1$
2	2	2	$x = 1$
3	3	3	$x = 1$
1	1	1	$x = 2 \wedge y = 2 \wedge z = 2$

Fig. 1. DSE solution J

R

A	B	C	Condition
x	y	y	$y = z$
x	x	z	$x = y$
x	a	b	$true$
c	y	d	$true$
e	f	z	$true$

Fig. 2. DSE solution J_1

R'

A	B	C	Condition
x	y	z	$x = y \vee y = z$
x	a	b	$true$
c	y	d	$true$
e	f	z	$true$

Fig. 3. DSE solution J'_1

initially initializes a fresh instance J' to J, then checks for each tuple $t \in J'$ whether $(J' - t) \equiv J$ holds (or in other words $Rep(J' - t) = Rep(J)$). If so, it updates J' to be $J' - t$. However, we show in Example 4 that this is not the case.

Example 4. Let $\mathfrak{S} = (\mathbf{S}, \mathbf{T}, \mathcal{M}, \Sigma_{st})$ be a DSE_I setting and J be a DSE solution under $\mathfrak{S}$. Assume that J consists of the pc-table R given in Fig. 1. Notice that $[R - \{\langle 1, 1, 1 : x = 1 \rangle, \langle 2, 2, 2 : x = 1 \rangle\}] \equiv R$ and $[R - \{\langle 3, 3, 3 : x = 1 \rangle\}] \equiv R$. However, $[R - \{\langle 1, 1, 1 : x = 1 \rangle, \langle 2, 2, 2 : x = 1 \rangle, \langle 3, 3, 3 : x = 1 \rangle\}] \not\equiv R$. Therefore, to check whether pc-table R' is the smallest subset pc-table of R, we need to non-deterministically determine whether there exists an instance $R'' \subset R$ such that $|R''| < |R'|$ and $R'' \equiv R$.

Following the intuition in Example 4 we provide the result in Theorem 5

Theorem 5. *Let $\mathfrak{S} = (\mathbf{S}, \mathbf{T}, \mathcal{M}, \Sigma_{st})$ be a DSE_I setting where Σ_{st} is a fixed set of st-tgds, I a source instance with pc-tables over $\mathbf{S}$ and J a DSE solution with pc-tables over $\mathbf{T}$. Also, let Σ_t be a fixed set of full tgds. To check if there exists a DSE solution J' for I and $\mathcal{M}$ under $\mathfrak{S}$ such that $J' \subset J$ is $\Pi_2^{P^{\parallel}}$-complete.*

In some DSE_I instances $\mathfrak{S}$, MUDSE solutions are not the most compact solutions under $\mathfrak{S}$. Consider the DSE solution J_1 given in Fig. 2. We can see that the DSE solution J'_1 given in Fig. 3 is such that $J_1 \equiv J'_1$, $J'_1 \not\subset J_1$, and $|J_1| > |J'_1|$. Although pc-tables are proved in [2] to be the best tool to represent source instances with incomplete information, we show in the following proposition that for some DSE solutions with pc-tables there exist target instances with *conditional tables* (c-table) [8] that are equivalent and more compact than those.

Proposition 1. *Let $\mathfrak{S} = (\mathbf{S}, \mathbf{T}, \mathcal{M}, \Sigma_{st})$ be a DSE_I setting, I a source instance, $\mathcal{M}$ an st-mapping table, and Σ_{st} be a set of st-tgds. Let J be a DSE solution for I and $\mathcal{M}$ under $\mathfrak{S}$. Then, the right tool to represent the most compact DSE solution J' of J is a c-table.*

5   Query Answering

We distinguish in DSE_G between *sound* and *complete* certain answers as semantics for CQs answering. Let $\mathfrak{S}$ be a DSE_G setting, I a source instance, and $\mathcal{M}$

---

[8] c-tables are more general pc-tables that can include inequality formulas, along with equalities, in their conditions.

an st-mapping table. A complete certain answer for a CQ Q over I and $\mathcal{M}$ under $\mathfrak{S}$, denoted by complete-certain$_{\mathfrak{S}}((I \cup \{\mathcal{M}\}), Q)$, corresponds to the set of tuples that belongs to the evaluation of Q over $K_{\mathbf{T}}$, for each DSE solution J for I and $\mathcal{M}$ under $\mathfrak{S}$ and $K \in \mathsf{Mod}((J \cup \{\mathcal{M}\}), \Sigma_t)$. A sound certain answer, on the other hand, denoted by sound-certain$_{\mathfrak{S}}((I \cup \{\mathcal{M}\}), Q)$, would be a set P of tuples that belongs to the evaluation of Q over a DSE solution J such that applying a query completion program, denoted by Σ_q, in the style of the target completion program Σ_t, to P would regenerate the set complete-certain$_{\mathfrak{S}}((I \cup \{\mathcal{M}\}), Q)$.

Computing the set complete-certain$_{\mathfrak{S}}$ answers using MUDSE solutions proved in [3] to be less expensive in run times than when using universal DSE solutions completed with Σ_t. Therefore, we introduce in what follow the method to compute complete-certain$_{\mathfrak{S}}$ answers of CQs using MUDSE solutions in DSE$_G$.

Let $\mathfrak{S}$ be a DSE$_G$ setting, I a source instance, $\mathcal{M}$ an st-mapping table, and $Q(\bar{x}) = (\bar{x})\exists \bar{y} \; \phi(\bar{x}, \bar{y}) \wedge \psi(\bar{y})$ be a CQ over $\mathbf{T}$ where: $\bar{x}$ is a set of distinguished variables, $\phi(\bar{x}, \bar{y})$ is a conjunction of predicate formulas with distinct variables, and $\psi(\bar{y})$ is a conjunction of formulas of the form $y_1 = y_2$ where $y_1, y_2 \in \bar{y}$.

To compute sound-certain$_{\mathfrak{S}}((I \cup \{\mathcal{M}\}), Q)$, we adopt a method similar to the combined approach given in description logic (DL) KB-exchange settings [7]. To do so, it first applies rules 1 and 2 in the target completion process Σ_t to populate the table REL with elements entailed to be related by $\mathcal{M}$. Then, re-writes query $Q(x_1, \ldots, x_n)$ to query: $Q'(x_1, \ldots, x_n) = (\bar{x})\exists \bar{y} \exists \bar{w} \; \phi(\bar{x}, \bar{y}) \wedge \psi'(\bar{y})$ where $\psi'(\bar{y})$ constitutes the formula REL $(y_1, y') \wedge$ REL(y_2, y') for each formula $y_1 = y_2$ in $\psi(\bar{y})$, to generate the set of sound certain answers.

Intuitively, the set complete-certain$_{\mathfrak{S}}((I \cup \{\mathcal{M}\}), Q)$ cannot be simply obtained by posing Q to the MUDSE solution J, since J might be incomplete with respect to $\mathcal{M}$. Therefore, we present below two possible methods for computing those complete certain answers using J.

The first method would be to complete J with the information entailed by the target completion program Σ_t as a first step, and this is done by applying Σ_t to J (denoted as $\Sigma_t(J)$) and generate a complete target instance $\hat{J}$, then apply Q to $\hat{J}$ as a second step and discard tuples with null values.

A second method, on the other hand, leverages the approach we used to compute sound-certain$_{\mathfrak{S}}((I \cup \{\mathcal{M}\}), Q)$. It first populates the table REL and re-writes Q to a query Q' the same way we did to compute sound-certain$_{\mathfrak{S}}((I \cup \{\mathcal{M}\}), Q)$. Then it completes the evaluation of Q' on J by returning the answer of $\hat{Q}'(z_1, \ldots, z_n) = Q'(x_1, \ldots, x_n) \wedge \bigwedge_{i=1}^{n} \text{REL}(x_i, z_i)$. We prove the correctness of the above query re-writing methods in the below proposition.

Proposition 2. *Let $\mathfrak{S} = (\mathbf{S}, \mathbf{T}, \mathcal{M}, \Sigma_{st})$ be a DSE$_G$ setting, I a source instance, $\mathcal{M}$ an st-mapping table, J a MUDSE solution, and Q a fixed conjunctive query over $\mathbf{T}$. Then, complete-certain$_{\mathfrak{S}}((I \cup \{\mathcal{M}\}), Q) = \hat{Q}(J)$ where $\hat{Q}(z_1, \ldots, z_n) = (\bar{x})\exists \bar{y} \exists \bar{w} \; \phi(\bar{x}, \bar{y}) \wedge \text{REL}(y_1, w_1) \wedge \text{REL}(\; y_2, w_1) \wedge \cdots \wedge \bigwedge_{i=1}^{n} \text{REL}(x_i, z_i), w_i \in \bar{w}$.*

Following Proposition 2, we deduce that the query re-writing method to compute sound-certain$_{\mathfrak{S}}$ is correct. Also, based on the result in Theorem 2 that MUDSE solutions can be generated in LOGSPACE in a fixed DSE$_G$ setting, and

Table 1. List of Queries

Q1	Fetch the name and age of each student enrolled in a course
Q2	Fetch the name and age of each student with the names of courses he completed
Q3	Fetch the name of each student with the name and grade of each course he finished
Q4	Fetch the list of teachers that already taught a course
Q5	Fetch the list of teachers' names and the list of courses they taught
Q6	Fetch the names of students with the names of courses they completed and the name of the teacher that taught each course
Q7	Fetch the list of pairs of students' names and ages that took the same course

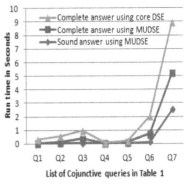

Fig. 4. Queries of table 1 run times

since checking if a fixed CQ is satisfied in a database is in LOGSPACE [2], we can deduce that computing complete-certain$_\mathfrak{S}$ and sound-certain$_\mathfrak{S}$ is in LOGSPACE.

6   Experiments

We conducted our experiments on a Lenovo workstation with a Dual-Core Intel(R) 1.80GHz processor, 4GB of RAM, and a 297 GB hard disk. We used PostgreSQL(v9.2) database system. We considered in our experiments the DSE$_G$ setting of Example 1 extended with relations $Teacher\ (tname, tage)$ and $Teach\ (tname, cname)$ that specify teachers' information and the list of courses they teach, respectively. We used a DSE solution (that is a core [5]) with 55,000 tuples, where each course in the source is related to 5 courses in the target. It is clear in Fig. 1 how MUDSE solutions compute efficiently both sound and complete certain answers for CQs.

7   Concluding Remarks

We defined in this paper a DSE$_G$ setting with general semantics of related data in an st-mapping table $\mathcal{M}$. We defined algorithms to generate DSE and MUDSE solutions. Also, we distinguished between sound and complete certain answers of CQs and we showed how to compute those. Finally, we addressed DSE$_G$ in a setting where the source instance contain unknown information.

References

1. Arenas, M., Barceló, P., Libkin, L., Murlak, F.: Relational and xml data exchange (2010)
2. Arenas, M., Perez, J., Reutter, J.: Data exchange beyond complete data (2011)

3. Awada, R., Barceló, P., Kiringa, I.: Sharing and exchanging data (2013)
4. Deutsch, A., Nash, A., Remmel, J.: The chase revisited (2008)
5. Fagin, R., Kolaitis, P.G., Popa, L.: Data exchange: getting to the core (2005)
6. Kementsietsidis, A., Arenas, M., Miller, R.J.: Mapping data in peer-to-peer systems: Semantics and algorithmic issues (2003)
7. Lutz, C., Toman, D., Wolter, F.: Conjunctive query answering in el using a database system (2008)

A Universal Cuboid-Based Integration Architecture for Polyglotic Querying of Heterogeneous Datasources

Michał Chromiak[1]([✉]), Piotr Wiśniewski[2], and Krzysztof Stencel[3]

[1] Institute of Informatics
Maria Curie Skłodowska University, Lublin, Poland
mchromiak@umcs.pl
[2] Faculty of Mathematics and Computer Science
Nicolaus Copernicus University, Toruń, Poland
pikonrad@mat.uni.torun.pl
[3] Institute of Informatics
University of Warsaw, Warsaw, Poland
stencel@mimuw.edu.pl

Abstract. Fortunately, the industry has eventually abandoned the old "one-size fits all" relational dream and started to develop task-oriented storage solutions. Nowadays, in a big project a devotion to a single persistence mechanism usually leads to suboptimal architectures. A combination of appropriate storage engines is often the best solution. However, such a combination implies a significant growth of data integrity maintenance. In this paper we describe a solution to this problem, i.e. a cuboid-based universal integration architecture. It allows hiding the peculiarities of integration so that it is transparent to the application programmer. We use graphs as an example of data that needs a task-oriented database in order to be efficiently processed. We show how graph queries can be effectively executed with the help of a graph database assisting a relational database. The proposed solution does not impose any additional complexity for programmers.

1 Introduction

Architects of software systems are reluctant to use heterogeneous data sources. They are usually afraid of high cost of integrity maintenance. On the other hand, relational database systems are still perceived as universal storage solutions. However, the relational database model is devoted to process flat collections of business objects.

In recent years, there have emerged database systems designed for particular applications. For instance, Cassandra [6] realizes fast column family oriented reads. MongoDB [12] effectively manages documents. Neo4j [8] efficiently stores and processes graph data.

Due to the abundance of task-oriented database systems architects face severe dilemmata. The universality of relational databases allows modelling any application domain. However, a decision to use such a database as the only storage

© Springer International Publishing Switzerland 2015
S. Kozielski et al. (Eds.): BDAS 2015, CCIS 521, pp. 170–179, 2015.
DOI: 10.1007/978-3-319-18422-7_15

can negatively impact the performance. An interesting example of data causing such impact for relational databases is *graph data*. We will use it as a running example in this article. At the end of the last century, the SQL:1999 standard introduced SQL means to query graph data (called also *recursive queries*) stored in relational tables. An interesting account on efficiency of SQL:1999 recursive queries can be found in [13].

In this article we use the case of graph data to show how useful are intelligent integrators of heterogeneous databases. Such tools allow exploiting the advantages of task-oriented database systems as they combine results returned by a number of databases into the form needed by an application. In our proposal, it is the integrator that abstracts the graph structure from a relational database and transfers it to a dedicated graph database (in this article Neo4j). Then relational queries to graph data are mapped to graph queries for appropriate identifiers of nodes. The result is the augmented with heavy relational data (attributes of tables). As we show in section 5 this execution method is faster than just querying a relational database. This hybrid (graph-relational) method was prototyped in the Cuboid framework [3–5].

The contributions of this article are as follows:

- a development of an automatic method to integrate Neo4j with a relational database that stores graph data,
- a mapping of recursive SQL:1999 queries onto a combination of a graph query and a simple relational query, and
- a proof-of-concept implementation of this mapping in the Cuboid that offers full transparency for an application programmer.

The article is organized as follows. Section 2 rolls out the motivation for our proposal. Section 3 presents the Cuboid. Section 4 describes the testing scenarios used to evaluate our proof-of-concept implementation. Section 5 reports the results of experimental experiments on the performance of the proposed solution. Section 6 concludes.

2   Motivating Example

We will use a table on employees of a network marketing company. The schema of this table is shown on Fig. 1. The employees of such a company form a natural hierarchy (a tree). Each salesman has a twofold income. The first part is the premium of its own sales, while the second part is a fraction of the sales of this employee's team.

The table **Employee** holds graph data that can be reconstructed by a self-join on columns **b_id** and **e_id**. The administration of our sample company often issues a query for all employees in the tree spanned by a particular employee. The first database management system to offer such query facilities was Oracle (from 1985). Listing 1.1 shows the appropriate query formulated in Oracle's SQL dialect.

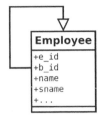

Fig. 1. An example database schema with graph data

Listing 1.1. A query for all employees under *Smith* in Oracle's early SQL dialect

```
SELECT * FROM Employee                                1
  WHERE level > 1                                     2
  START WITH sname = 'Smith'                          3
  CONNECT BY b_id = PRIOR e_id ;                      4
```

In 1999 queries for graph data were included into the SQL standard. The standardization committee decided to use a more flexible solution than Oracle's CONNECT BY. This solution was based on *recursive common table expressions*, i.e. a form of functional *let rec* for relational data. Listing 1.2 shows an SQL:1999 query for all employees reporting to a given employee.

Listing 1.2. A query for all employees under *Smith* using SQL:1999 recursive common table expression

```
WITH RECURSIVE emprec AS (                            1
    SELECT e_id , name, sname                         2
      FROM Employee                                   3
      WHERE sname = 'Smith'                           4
  UNION                                               5
    SELECT e.e_id , e.name, e.sname                   6
      FROM Employee e                                 7
        JOIN emprec r                                 8
        ON (r.e_id = e.b_id)                          9
)                                                     10
SELECT *                                              11
  FROM emprec;                                        12
```

The facilities to search graphs introduced into SQL:1999 significantly enriched graph processing tools on the level of relational databases. An application programmer could formulate a single query where formerly a series of queries had been unavoidable. The authors of [13] thoroughly examine the functionality and performance of recursive queries in major relational database systems. In modern application architectures that use object-relational mapping libraries the problem of querying graph data is even more complex [1, 17, 14, 15].

Although relational databases implement recursive queries and there is ongoing research on their optimization [7, 2, 11], relational databases only *support* graph queries. They do not implement them *natively*. A real efficiency improvement of big graph search is possible provided dedicated task-oriented database systems are used. Neo4j [16] is an example of such a database.

The efficiency of graph tasks in dedicated graph databases is noteworthy higher than in relational databases. Thus, an architect is tempted to consider an implementation in which the graph structure is also stored in a graph database. A graph search is then partitioned into two subtasks. The first of them is the pure traversal among nodes. The second one augments the result of traversal with mass attributes retrieved from a relational storage. Note, however, that such a solution is notably more costly from a maintainer's point of view.

In this article, we show that such an architecture can be totally transparent to an application programmer. Moreover, a middleware can take care of the whole logic that synchronizes a graph database, splits queries and merges their results. In following sections we present our proposal. In particular, section 5 attests satisfactory performance of the proof-of-concept implementation.

3   Cuboid - Explained

The middleware that we focus around in this paper is the cuboid based architecture. The Cuboid idea has been developed and expanded for a couple of years now and has become a multi-aspect tool for data integration. An abstract layer for data access and presentation is not a new idea. However, all solutions we know do not have the level of agility, flexibility and capabilities provided by the Cuboid. The idea of Cuboid has emerged from an idea of a universal, abstract platform for data processing. The uniqueness of this approach is about its independence by design. This implies that the discussed solution provides unified access for data, regardless of its storage particularities. The diversity of issues that has to be overcome due to this goal is extensive. Here is a list of a few most important and profound aspects of the data storage particularities:

- spatial distribution of data,
- differed data sources,
- different data models,
- open or proprietary data storage engine,
- data security policies, etc...

The intuition of data representation in our approach has adopted a form of a cuboid. This is due to it straightforward association with horizontal and vertical fragmentation of a distributed and replicated data. Those three data "dimensions" have been visualized as a cuboid shape in Fig. 2 as a Distributed Resource Universal Map (DRUM). Such representation can be easily tied to simple, record based model known from relational paradigm (see Fig. 3).

Following this model we can simply represent arbitrary data fragmentation and replication pattern. The DRUM however, can only cover the structure of a

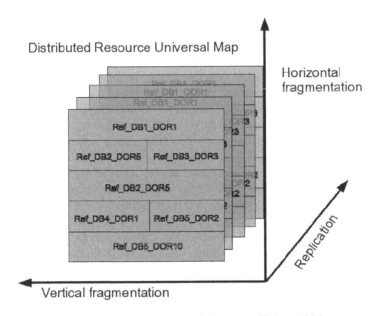

Fig. 2. DRUM - Distributed Resource Universal Map

data granular representation. Thus, the Cuboid must additionally define access methods for such data representation. We have proposed and presented an agile structure for representing data DRUM including dedicated access methods [3–5]. This structure - in form of interoperable Data Access Object (iDAO) - has concealed from the Cuboids' client all of the data that is required to access the actual data present in the grid.

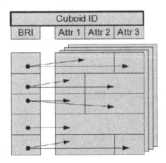

Fig. 3. Cuboid as the *Best Record ID* - based metadata model for data representation

Each structural element (e.g. element responsible for accessing a table, set of rows, a single record, replication of a table part etc.) of the DRUM contains one iDAO responsible for storing all of the information about actual data access method. Thus, each client call can be considered as a call for multiple iDAOs. Obviously the actual data represented in the Cuboid in the form of metadata is stored in legacy data sources that are being integrated. To be precise, thanks to Cuboid, we are able to integrate the data, not the data sources. Therefore, we can conform to the data-oriented paradigm of the Cuboid based architecture. The target data stored at the data source site has to be registered in the cuboid architecture in the process of data source registration. See Fig. 4 for integration procedure details.

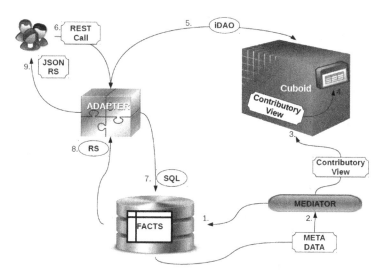

Fig. 4. Resource registration at the Cuboid integration facility

Thanks to registration procedure the Cuboid gets the notion of the actual integrated data and the meta information about its legacy schema and can place this information in form of iDAOs at the Cuboid's own virtual schema. In other words Cuboid can be considered as a virtual integration view.

This brings us to the point where the Cuboid can synthetically be called a polyglot master metadata repository interface for accessing integrated data. However, integration is just part of Cuboids' potential applications. As a general integrating instance Cuboid is also a suitable place to apply optimization methods based on native queries stored at iDAOs.

4   Testing Scenarios

The Cuboid-based architecture can aid a significant improvement of big graph-based searches. Furthermore, it does it transparently, since an application programmer is not aware of the actual data structure used. As mentioned in section 2 querying the data of a graph nature stored in graph database leads to a simple traversals among nodes. In order to verify it, we have created three testing scenarios. They are based on the example schema presented in Fig. 1. In every scenario a fixed number of records is retrieved from the result of a recursive query. These results are consumed by client calls to dedicated Cuboid REST API.

For the first test scenario we have generated 18 GB `Employee` table (Fig. 1) and loaded it into PostgreSQL. Then, using the `JDBCTemplate` as a lightweight wrapper for pure JDBC driver, we have commenced multiple recursive queries (Listing 1.2) that resulted in about 8 MB per 10 returned records. The results are summarized in table 3.

In the second test scenario the recursive part of the `Employee` schema and its data has been replicated into a separate table in PostgreSQL. This table is called `EmpBase` and has the minimal useful set of attributes including the primary and foreign keys. This set of attributes is enough to construct the tree structure of the self referencing recursive hierarchy for the entire `Employee` table. Remaining attributes have no influence on this hierarchy. Therefore, calling the recursive query based only on the `EmpBase` table produces the same hierarchy as if it was run on the `Employee` table. Then, it remains to join the hierarchy with the `Employee` table in order to collect the rest of attributes. This test shows that such a way to execute this query is notably faster than just running the original query on the `Employee` table (see section 5).

The third test case, similar to the second scenario, involved moving the recursive part of the structure and its data from the relational testing schema. However, this time, instead of creating a separate table in PostgreSQL, we have copied the data into an instance of Neo4j, i.e. a graph database. Likewise the second scenario, the non-recursive and heavyweight attributes of the `Employee` schema stayed in PostgreSQL in the table `Employee`. In other words, the parts of the schema that define the relation like the primary key and the foreign keys has been copied over to Neo4j. Thus, we have the entire schema and data in PostgreSQL and parts of the schema that describe the recursive relation moved to the Neo4j. The migration process has been commenced with regard to official recommendations [9, 10]. We hoped that the efficiency of such a database with respect to our sample recursive query would be significantly better, because of the Neo4j graph model supposed to handle graph-based data natively. In contrast to the second test scenario, this time the parts of the `Employee` schema defining recursive hierarchy has been presented as graph nodes' properties. As a result of this scenario, the remaining heavy attributes left in relational database are being augmented to data retrieved from Neo4j by means of the Cuboid mapping. Thus, the hierarchy structure was built efficiently by Neo4j and then combined in Cuboid with the attributes from PostgreSQL. That resulted in a significant acceleration of the query execution.

In this scenario the role of Cuboid is extremely important. Because of its agility the client calls for Cuboid's REST API were transparent. Therefore, they enabled the client to focus on the delivered data and not the data delivering

source. The client REST call tends to call for the `Employee` data regardless of
the actual internal implementation of data structures.

5   Experimental Evaluation

The tests has been performed using the hardware characterised in table 1 and
the software listed in table 2.

Table 1. The hardware configuration used in the experimental evaluation

CPU	Intel Core i7-3612QM CPU @ 2.10 GHz x 8
RAM	15,6 GiB
Disk	SAMSUNG SSD PM830 2.5" 7mm 512GB
OS	Ubuntu14.04 LTS
Kernel	3.13.0-30-generic
Arch.	x86_64 GNU/Linux

Table 2. The software used in the experimental evaluation

Java	Java version 1.7_60
	Java(TM) SE Runtime Environment (build 1.7.0_60-b19)
	Java HotSpot(TM) 64-Bit Server VM (build 24.60-b09, mixed mode)
REST Testing Client	ApacheBench, Version 2.3
Http Server	Apache Tomcat/6.0.29

Table 3. The execution times of test queries related to the used data model

Request No	Result Records No	Total (Time per request- mean) [ms]
EmpFull (17GB-pgsql) (recursive)		
1	10	38 850,33
	100	188 414,76
	1 000	274 333,25
EmpBase (1GB-pgsql) + Widedata (17GB- pgsql)		
1000	10	1 783,06
	100	9 367,09
	1 000	17 069,04
EmpBase (1GB-neo4j)(recursive) + Widedata (17GB pgsql)		
1000	10	73,37
	100	774,33
	1 000	8 284,40

We measured response times for the REST client calls to three test queries. A
testing REST client called REST API that used different ways of retrieving the
data from the table `Employee` . Each REST call conformed the following schema:
(...){strategy}/dbSchema/{dbEntityName}.json/limit={value1}&idoffset={value2}

The `strategy` variable has three possible values $1, 2, 3$ depending on the test scenario. The `limit` and `offset` variables represented the number of resulted records and the offset just as SQL syntax.

We have commenced 1000 requests for each limited amount of result records. In first test scenario we had limited the number of requests to one, because of the long times needed to retrieve data. In two other scenarios we did not impose such a limit. The size of data for every response of 10 records was about 8 MB.

6   Conclusions

Cuboid has already proven its utility for query integration [3] and optimization [5]. However, in this article we have focused on combining both these aspects. We have used the possibility to integrate data from heterogeneous data sources. Furthermore, at the same time we facilitated optimization of the query execution using the Cuboid based mapping.

The presented results are promising. However, the Cuboid has provided full transparency for a client to get arbitrary data regardless of its the paradigm of its original data source. The Cuboid has enabled fitting the data representation to the appropriate paradigm of data storage and processing. The method to achieve this is based on both SQL and NOSQL engines. This way the Cuboid tends to face the NoSQL as the *"Not Only SQL"* trend in database development.

References

1. Burzańska, M., Stencel, K., Suchomska, P., Szumowska, A., Wiśniewski, P.: Recursive queries using object relational mapping. In: Kim, T.-H., Lee, Y.-H., Kang, B.-H., Ślęzak, D. (eds.) FGIT 2010. LNCS, vol. 6485, pp. 42–50. Springer, Heidelberg (2010)
2. Burzańska, M., Stencel, K., Wiśniewski, P.: Pushing predicates into recursive SQL common table expressions. In: Grundspenkis, J., Morzy, T., Vossen, G. (eds.) ADBIS 2009. LNCS, vol. 5739, pp. 194–205. Springer, Heidelberg (2009)
3. Chromiak, M., Stencel, K.: The linkup data structure for heterogeneous data integration platform. In: Kim, T.-H., Lee, Y.-h., Fang, W.-C. (eds.) FGIT 2012. LNCS, vol. 7709, pp. 263–274. Springer, Heidelberg (2012), http://dx.doi.org/10.1007/978-3-642-35585-1`36
4. Chromiak, M., Stencel, K.: A data model for heterogeneous data integration architecture. In: Kozielski, S., Mrozek, D., Kasprowski, P., Małysiak-Mrozek, B. (eds.) BDAS 2014. CCIS, vol. 424, pp. 547–556. Springer, Heidelberg (2014), http://dx.doi.org/10.1007/978-3-319-06932-6`53
5. Chromiak, M., Wisniewski, P., Stencel, K.: Exploiting order dependencies on primary keys for optimization. In: Proceedings of the 23rd International Workshop on Concurrency, Specification and Programming, Chemnitz, Germany, September 29 - October 1, pp. 58–68 (2014), http://ceur-ws.org/Vol-1269/paper58.pdf (accessed: February 06, 2015)
6. Cloudkick: 4 months with Cassandra, a love story (March 2010), https://www.cloudkick.com/blog/2010/mar/02/4_months_with_cassandra/ (accessed: November 12, 2013)

7. Ghazal, A., Crolotte, A., Seid, D.Y.: Recursive SQL query optimization with k-iteration lookahead. In: Bressan, S., Küng, J., Wagner, R. (eds.) DEXA 2006. LNCS, vol. 4080, pp. 348–357. Springer, Heidelberg (2006)

8. Holzschuher, F., Peinl, R.: Performance of graph query languages: Comparison of cypher, gremlin and native access in neo4j. In: EDBT/ICDT 2013, pp. 195–204. ACM, New York (2013), http://doi.acm.org/10.1145/2457317.2457351

9. Hunger, M.: Load csv with success (2014), http://jexp.de/blog/2014/10/load-cvs-with-success/ (accessed: February 06, 2015)

10. Neo4j: Load csv into neo4j quickly and successfully (2014), http://jexp.de/blog/2014/06/load-csv-into-neo4j-quickly-and-successfully/ (accessed: February 06, 2015)

11. Ordonez, C.: Optimization of linear recursive queries in sql. IEEE Trans. Knowl. Data Eng. 22(2), 264–277 (2010)

12. Plugge, E., Hawkins, T., Membrey, P.: The Definitive Guide to MongoDB: The NoSQL Database for Cloud and Desktop Computing, 1st edn. Apress, Berkely (2010)

13. Przymus, P., Boniewicz, A., Burzańska, M., Stencel, K.: Recursive query facilities in relational databases: A survey. In: FGIT-DTA/BSBT, pp. 89–99 (2010)

14. Szumowska, A., Burzańska, M., Wiśniewski, P., Stencel, K.: Efficient implementation of recursive queries in major object relational mapping systems. In: Kim, T.-h., Adeli, H., Slezak, D., Sandnes, F.E., Song, X., Chung, K.-i., Arnett, K.P. (eds.) FGIT 2011. LNCS, vol. 7105, pp. 78–89. Springer, Heidelberg (2011)

15. Szumowska, A., Burzańska, M., Wiśniewski, P., Stencel, K.: Extending HQL with plain recursive facilities. In: Morzy, T., Härder, T., Wrembel, R. (eds.) Advances in Databases and Information Systems. AISC, vol. 186, pp. 265–272. Springer, Heidelberg (2013)

16. Van Bruggen, R.: Learning Neo4j. Packt, Birmingham (2014)

17. Wiśniewski, P., Szumowska, A., Burzańska, M., Boniewicz, A.: Hibernate the recursive queries - defining the recursive queries using Hibernate ORM. In: ADBIS (2), pp. 190–199 (2011)

Efficient Multidisk Database Storage Configuration

Mateusz Smolinski[✉]

Institute of Information Technology,
Technical University of Lodz,
Wolczaska 215, 90-924 Lodz, Poland
mateusz.smolinski@p.lodz.pl
http://it.p.lodz.pl

Abstract. Simple database storage configuration includes block device and filesystem. In advanced multi-disk database storage configuration also disk manager is required. Article presents multi-disk storage configuration impact on relational database performance. Research results include various local multi-disk storage space configuration scenarios for database cluster in modern database management systems with popular disk managers like software RAID and standard or thin provisioned logical volume equipped with disk allocation policy. The research conclusions facilitate the local storage space configuration for efficient transaction processing in relational databases.

Keywords: Database storage configuration · Transaction processing performance in MariaDB and PostgreSQL · Oltp workload efficiency

1 Introduction

Relational databases are popular data sources for various applications and information systems. Data access efficiency has a substantial impact on performance of software, which bases on relational database. Relational databases characteristic includes data structuring and typing, integrity and security. Database handling is performed by database management system implementation, which provides various data types, structure, integrity constraint and storage space management.

1.1 Transaction Processing in Relational Databases

Selection and configuration of DBMS affects relational database performance. Data security aspect is assured by authorization, authentication and transaction, which are provided by DBMS. However, not all DBMS have to support of all security aspects. Not all DBMS engines supports transaction processing, which allows database operation execution according to ACID properties. Transactional databases allow secure and parallel execution of operations on stored data, that is base of multiple database software design in particular popular

© Springer International Publishing Switzerland 2015
S. Kozielski et al. (Eds.): BDAS 2015, CCIS 521, pp. 180–189, 2015.
DOI: 10.1007/978-3-319-18422-7_16

on-line transaction processing systems. The base performance measure of OLTP systems is a number of committed transactions in time [9,6].

Efficient transaction processing in OLTP systems requires use high performance transactional data sources. For a fixed database structures and data choice of DBMS and its storage space configuration is crucial for performance of local transaction processing in database. Base elements of storage space are a block device and filesystem. Its configuration has impact on databases performance supported by DBMS. In advanced storage space configuration many physical block devices can be used to store DBMS tablespaces. In this type of storage configuration additional disk manager element is required. Its role is to manage disk block addressing and provide logical block device with fixed addressing policy. Disk manager can be hardware or software implemented, its implementation may offer various addressing policies for created logical block devices. The disk manager, number of disks and allocation policy are base parameters in advanced multi-disk database storage space configuration.

The research of transaction processing efficiency in relational database includes modern open source DBMS like PostgreSQL and MariaDB with local multi-disk storage space configurations scenarios. Analysis of research results reveals storage space configuration parameters and its impact on performance of transaction processing in relational database. Knowledge of the impact of multi-disk storage configuration parameters on the transaction processing processing allows to establish an efficient storage configuration for relational database.

2 Unified Environment for Transaction Performance Testing

2.1 Environment Configuration

Computer used in database transaction performance testing was equipped with i5-2400 CPU, 8GB memory and five identical, connected SATA-3 Western Digital disks model WD5000AZRX with 466GB capacity and 64MB cache. The 64 bit computer architecture is managed by Linux operating system from Fedora 20 distribution (kernel version 3.17.2-200.fc20.x86_64). Operating system was installed on additional disk connected via USB, therefore all connected SATA disks were used only for database storage configurations (known as scenarios). All disks have default I/O scheduler the Completely Fair Queuing [2].

In local multi-disk storage configuration two software disk managers were used: software Redundant Array Of Independent Disks configured by MDADM software in version 3.3-7 and Logical Volume Manager in version 2.02.106-1. LVM bases on another Device Mapper software, which version 1.02.85-1 was used. Device Mapper offers targets with various allocation policy implemented as Linux kernel dynamic modules. In all multi-disk storage space configuration managed by LVM was used the default size 4MB extent.

Each storage space configuration scenario includes creation of local multi-disk logical block device configuration with fixed disks manager and filesystem.

In created files storage space was prepared a database cluster with all DBMS tablespaces for testing database transaction performance.

In operating system were installed two popular open source database management systems: MariaDB version 5.5.39-1 and PostgreSQL version 9.3.5-2. In any performed transaction processing test a single relational database was created in each DBMS. Database cluster was localized on multi-disk block device configured according to storage configuration scenario.

2.2 Storage Configuration Scenarios

Testing efficiency of database transaction processing in multi-disk storage space configurations scenario includes fixing filesystem type (from set: FAT32, EXT3 and EXT4, XFS and BTRFS), number of disks and one of software disk management mechanism implemented in Linux kernel. All used filesystem have identical 4KB allocation unit size. BTRFS filesystem was used without its internal disk manager [7]. Additional parameters for scenario according to software RAID (ident. MDADM) and Logical Volume Manager (ident. LVM) is allocation policy and allocation unit size (chunk size). In each scenario with software RAID allocation policy was determined by RAID level, which fixing block addressing method. The tests include only RAID level 0, 5 and 6 allocation policies [1].

Storage scenario with LVM includes creation of logical volume with allocation policy, which controls location of data blocks between physical volumes, each one placed on a separate disk. Allocation policies for logical volumes used in storage scenarios include striping in raid0, raid5 and raid6. Some scenarios was created using thin provisioned logical volume with data striping across fixed number of physical volumes.

2.3 Unified Transaction Performance Testing

Performance tests of database transaction processing in different multi-disk storage configurations were realized in identical environment with the same method of measurement. For this purpose was used a sysbench software tool. This tool is indicated on the MariaDB project pages and works with both open source DBMS used in transaction performance testing. From wide range of sysbench tests a OLTP test were selected. This type of sysbench test unifies database operations performed within the transaction. Single transaction operations were performed on non-sequential database records. All OLTP tests were performed by single sysbench thread, which excludes deadlock problem. Limitations of used in tests database engines in multicores environment were analized in other research [8]. In each test single sysbench thread performs workload of database transactions in fixed period of time. When transaction processing test finishes the sysbench tool provides statistics which include averaged number of commited transaction per second. Sysbench tool does not provide advanced models for various transaction processing characteristic like TPC standards, but is sufficient to demonstrate the database storage configuration impact on transaction processing performance [9,4].

The selected software tool in addition to standardizing the method of measurement allows also preparation of identical relational database, that was used in every transaction processing test. Prepared database includes single multi-column table with 10 million rows. In both used DBMS database size was over 2GB. Unification of used relational database and transaction processing measurement method allows to compare the results for various database storage configurations. A reference point for fixed DBMS and filesystem was database storage configuration equipped with only single disk (Fig. 1). In presented scenarios XFS and EXT, in which database cluster was localised, provide better performance for database transaction processing in PostgreSQL and MariaDB.

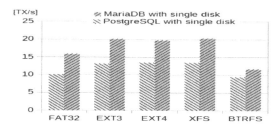

Fig. 1. Database transaction processing performance for single disk storage space configuration scenario depending on filesystem type

3   Performance Analysis of Transaction Processing

The results obtained in research of emerge the filesystem impact on performance of database transaction processing. The relationship between number of committed database transactions and type of filesystem storing database cluster presents in reference to single disk and also multi-disk storage configuration and is dependent of disk manager, number of disks, allocation unit size and policy.

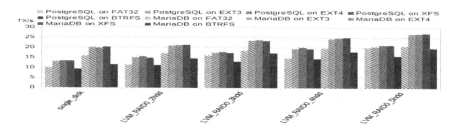

Fig. 2. Performance of database transaction processing for logical block device managed by LVM with striping policy with 64 KB chunk size depending on number of disk

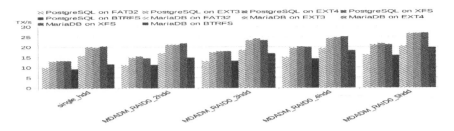

Fig. 3. Performance of database transaction processing for logical block device managed by MDADM with striping policy with 64 KB chunk size depending on number of disk

The choose of filesystem type fixes data structures and block allocation, which affect data access time [3]. Therefore increase of disk number or disk manager allocation unit size has different impact on filesystem operation performance depending on its type. Figures 2 and 3 show performance of database transaction processing for various filesystem types, which was formatted on logical volume managed by LVM or MDADM with striping policy and default 64 KB chunk size.

3.1   Multi-disk Storage Space Configuration

Each multi-disk storage space configuration requires dedicated software as disk manager and selection its base parameters like number of used active disks, block allocation policy and size. For tests of database transaction performance in multi-disk storage configurations only two software disk managers were used: LVM and software RAID. Each one offers various allocation policy for managed logical block devices. In storage configuration scenarios for both disk managers was used: data striping policy (ident. RAID0), also RAID level 5 policy (ident. RAID5) and RAID level 6 policy (ident. RAID6). In the case of RAID6 scenarios managed by LVM five disks were used, as a result of the limitation of this disk manager. In all scenarios with RAID5 and RAID6 policy before the filesystem formatting disk synchronization phase was performed. The RAID5 and RAID6 policy uses additional metadata to increase data security and reliability, which retains data even disk fails.

In addition to the applied block allocation policy also the number of used disks affect performance of database transaction processing. Figures 4 and 5 present database performance growth for various filesystems localized on logical device managed by LVM or MDADM with RAID5 and RAID6 allocation policy with the increasing number of disks (scenarios with default 64KB chunk size).

For each filesystem the highest performance growth offers striping policy when five disks were used. This is due to the use of additional metadata in RAID5 and RAID6 allocation policy. But not every multi-disk configuration offers better performance of database transaction processing than that in single disk scenario. Drop in transaction processing performance was observed generally in scenarios with minimal disk numbers for allocation policy (i.e. RAID5 with 5 disks).

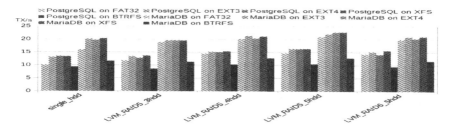

Fig. 4. Performance of database transaction processing for logical block device managed by LVM with RAID5 and RAID6 policy with 64 KB chunk size depending on number of disk

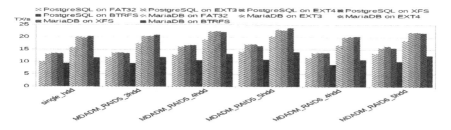

Fig. 5. Performance of database transaction processing for logical block device managed by MDADM with RAID5 and RAID6 policy with 64 KB chunk size depending on number of disk

3.2 Allocation Unit Used by Disk Manager

Another parameter of local multi-disk database storage space configuration is allocation unit size used by disk manager. Both LVM and MDADM software disk managers offer fixing chunk size when logical block device is created. Chunk size is always a multiple of two, and should be greater then filesystem blok size. In performance testing for database transaction processing a chunk size from range 8KB to 512KB was used. The upper limit of chunk size in multi-disk storage configuration results from LVM constraint for used allocation policies.

Figures 6 and 7 present performance in processed transaction for filesystems and striping allocation policy for LVM or MDADM disk manager with the various chunk size in five disk storage space configurations. Research results also show that storage space configurations scenarios with less disks also have the same trend in transaction processing performance according to chunk size.

For fixed disk number in all filesystem increasing the size of allocation unit used by disk manager causes higher performance of database transaction processing. This trend is observable regardless of used disk manager and allocation policy. If chunk size is less then 64KB a performance of database transaction in scenarios with RAID5 and RAID6 allocation policy can be lower then in scenario with single disk (For MDADM figures 8 and 9).

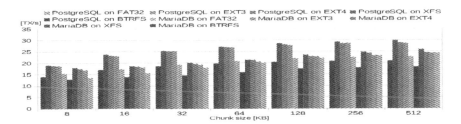

Fig. 6. Performance of database transaction processing for logical block device managed by LVM with striping allocation policy for five disk storage configuration depending on disk allocation unit size

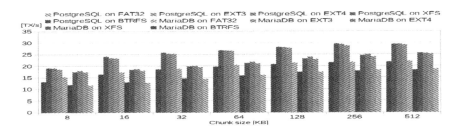

Fig. 7. Performance of database transaction processing for logical block device managed by MDADM with striping allocation policy for five disk storage configuration depending on disk allocation unit size

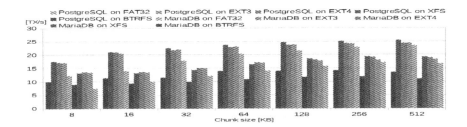

Fig. 8. Performance of database transaction processing for logical block device managed by MDADM with RAID5 allocation policy for five disk storage configuration depending on disk allocation unit size.

To avoid performance degradation of database transaction processing in multi-disk storage with striping policy recommended chunk size for LVM and MDADM disk manager should not be less than 32KB.

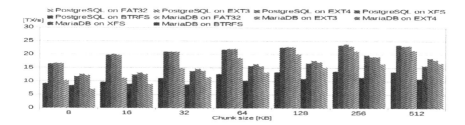

Fig. 9. Performance of database transaction processing for logical block device managed by MDADM with RAID6 allocation policy for five disk storage configuration depending on disk allocation unit size

3.3 Disk Manager

Software disk manager implementation in Linux kernel also has impact on database transaction processing performance. Database storage space configuration is created with FAT32 filesystem and disk striping a LVM disk manager provides even 23% better performance in database transaction processing. This trend for FAT32 is also observed for RAID5 and RAID6 allocation policy. Other filesysstem show up to 5% performance grow in database transaction processing when MDADM disk manager is used insted of LVM (Fig. 10).

Differences in database transaction performance were also observed between standard and thin provisioned logical volume with the striping policy for fixed disk number and chunk size, in particular 37% decrease in performance if BTRFS was used on thin provisioned logical volume with 512KB chunk size (Fig. 11). For FAT32 localized in thin provisioned volume this performance loss was 26% in relation to standard logical volume. For database storage space configuration with EXT and XFS filesystem localized in thin provisioned logical volume observed database transaction performance degradation was up to 10%.

3.4 Database Managment System

Performance of database transaction precessing depends on the DBMS implementation that supports relational database. Both PostgreSQL and MariaDB support transaction processing and use multi-version concurency control model. Used configurations of each of them based on the default covention. In MariaDB configuration local socket localization was changed, which allows to change filesystem for database cluster. If MariaDB control the relational database, then only InnoDB engine with Percona XtraDB enhancement was used. Engine enhancements increase its performance in relation to the orginal InnoDB implementation in MySQL project [5].

The DBMS storage space was configured by mounting in default database cluster localization a filesystem created each time on multi-disk logical block device. Before each test of transaction performance, DBMS cluster was restored. The results of the measurement show that depending on the database cluster storage

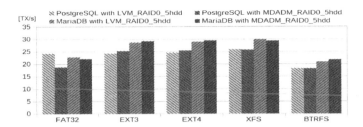

Fig. 10. Performance of database transaction processing for logical block device managed by LVM and MDADM with striping policy with 512 KB chunk size

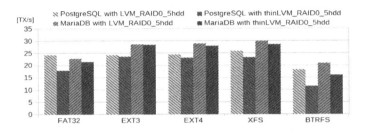

Fig. 11. Performance of database transaction processing for logical block device configured as normal and thin provisioned LVM volume with striping policy with 512 KB chunk size

space scenario the number of performed transactions in MariaDB database was up to 34% higher than in PostgreSQL database. In most storage space configuration scenarios the number of transactions was 19% and higher for database in MariaDB, even in single disk reference storage configuration.

4   Conclusions

Relational database storage space configuration has significant impact on transaction processing performance, in particular when database storage space is created with many disks. Selection of parameters like number of used disks, disk manager and its allocation unit size and policy affect the input and output operations on block device. Choice of database storage space parameters is not obvious, not all increase of disk number provide efficiency growth in transaction processing. Also selection of filesystem for database storage is crucial to achieve high performance level in database transaction processing.

Among filesystems used in database transaction performance testing in unified environment, obtained results showed that for fixed disk number, allocation unit and policy only XFS and EXT4 offers highest number of processed transaction in relational database. Also in single disk database storage configurations those filesystems provide better results as BTRFS or even not journaled FAT32.

Additionally for EXT and XFS filesystem type both disk managers for fixed allocation unit and policy software RAID and LVM offers similar performance in database transaction processing. However using thin provisioned logical volume can only decrease number of transaction processed in database, which is localized in filesystem stored on it. Observed performance loss in database transaction processing, where database was localized in thin provisioned logical volume, was lowermost for EXT and XFS filesystem.

Multi-disk storage configuration include also choice of allocation unit, which is important parameter influencing on performance in database transaction processing. Results of performance test shows that increasing allocation unit size causes performance growth in database transaction processing. It was also showed that smaller than 64KB allocation unit for RAID5 and RAID6 policy and 32KB for striping policy can decrease performance of database transaction processing in multi-disk storage space configuration to level less than in single disk storage space configuration.

The difference in performance of database transaction processing is noticeable according to database management system choice and its engine. Research results highlight that for the same storage space configuration a relational database under MariaDB with modified InnoDB engine is more efficient in transaction processing than under PostgreSQL. High performance in database transaction processing in MariaDB was observed in single and multi-disk storage configurations.

References

1. Anderson, E., Swaminathan, R., Veitch, A., Alvarez, G., Wilkes, J.: Selecting raid levels for disk arrays. In: Proceedings of the 1st USENIX Conference on File and Storage Technologies (2002)
2. Axboe, J.: Linux block io present and future. In: Proceedings of Ottawa Linux Symposium, pp. 51–61 (2004)
3. Bryant, R., Forester, R., Hawkes, J.: Filesystem performance and scalability in linux 2.4.17. In: Proceeding of the FREENIX TRACK, pp. 259–274 (2002)
4. Hsu, W.W., Smith, A.J., Young, H.C.: I/o reference behavior of production database workloads and the tpc benchmarks an analysis at the logical level. ACM Transactions on Database Systems 26, 96–143 (2001)
5. Percona Benchmark: Benchmanrks For Percona Server, 2015, http://www.percona.com/software/percona-server/benchmarks
6. Llanos, D.R.: Tpcc-uva: An open-source tpc-c implementation for global performance measurement of computer systems. ACM SIGMOD RECORD 35, 6–15 (2006)
7. Rodeh, O., Bacik, J., Mason, C.: Btrfs: The linux b-tree filesystem. ACM Transactions on Storage 9, 1–32 (2013)
8. Salomie, T.I., Subasu, I.E., Giceva, J.: Database engines on multicores, why parallelize when you can distribute? In: Proceedings of the EUROSYS Conference, pp. 17–30 (2011)
9. TPC: Transaction Performance Council (2015), http://www.tpc.org

E–LT Concept in a Light of New Features of Oracle Data Integrator 12c Based on Data Migration within a Hospital Information System

Lukasz Wycislik[✉], Dariusz Rafał Augustyn, Dariusz Mrozek, Ewa Pluciennik, Hafed Zghidi, and Robert Brzeski

Institute of Informatics, Silesian University of Technology
16 Akademicka St., 44-100 Gliwice, Poland
{lukasz.wycislik,dariusz.augustyn,dariusz.mrozek,ewa.pluciennik,
zghidi.hafed,robert.brzeski}@polsl.pl

Abstract. The paper, presents an approach to developing and running ETL processes implemented in Oracle Data Integrator (ODI), on the example of a hospital information system. Thanks to inversion of the classical order of ETL stages to Extract-Load-Transform (ELT) sequence, ODI simplifies and improves efficiency of most common ETL tasks. Several new features introduced in 12c version positively affects productivity, efficiency and functionality as well.

Keywords: Oracle · ETL · ELT · ODI · Data integration · Integration domain model

1 Introduction

Information plays a crucial role in all areas of the world today. Almost everyone wants often to check bus schedule or fuel prices in one's city. Amount of information stored and processed is growing and it does not seem that this trend would have changed for the next few years.

Today's information systems are rarely separated islands cut off the outside world. The data often are collected from multiple sources and only merged make usable logical unit. The whole basic concept of ETL process is presented for example in [6,4]. These articles cover such aspects as data warehouse architectures, data mapping, ETL modeling, data quality issues including consistency, validity, conformity, accuracy, integrity etc.

Widely used for data migration XML format, proved to have too much overhead on the transmitted data volume, that is discussed in [10]. This resulted in situation where tuning of data extraction and migration is an integral part of the implementation of ETL solutions. This applies in particular for systems such as real-time data warehousing [4] or continuous monitoring systems [3].

In this article the ETL Oracle ODI solution was introduced. It uses opposite to XML-based as transport layer technologies and strategies for data migration. Information about other Oracle tools supporting ETL processes can be found in [8].

© Springer International Publishing Switzerland 2015
S. Kozielski et al. (Eds.): BDAS 2015, CCIS 521, pp. 190–199, 2015.
DOI: 10.1007/978-3-319-18422-7_17

2   ETL Processes

ETL abbreviation stands for the extract-transform-load and applies to processes of data migration from a source or sources to a destination, where most often both source and destination are relational databases. Sometimes different technologies of data storing may by involved including LDAP, web services, flat files, XML, XLS, JMS, nonrelational databeses etc. The extract phase covers the data extraction from the source system to make it accessible for further processing. Here the challenge is to retrieve all the required data from the source system with as little resources as possible. The extracted data is usually stored as a structured file (e.g. XML) on the ETL server side. After that the data needs to be cleaned to be sure it meets all requrements and contraints defined for a given ETL process. Clean data is ready for transformations required by target systems. Both cleaning and transforming processes are usually run on a dedicated ETL host. Loading data to a destination should consider a target system availability and how to update an existing data.

3   Oracle Data Integrator

Oracle Data Integrator is an extract, load and transform tool produced by Oracle company that is equiped with a graphical environment that allows to define, manage nad maintain data integration processes. Formerly it was developed by Sunopsis software company that was acuired by Oracle in October 2006. It supports many database systems vendors and is equipped with a powerful graphical user interface that enables defining graphical models of data flow and transformation in a declarative way. The software is developed in Java technology and ETL projects are stored in relational databases (e.g. Oracle, Postgresql). ODI is in fact a part of a much larger software package delivered under the name of Oracle Fusion Middleware. It may raise some doubt the existence in Oracle product list similar tool called Oracle Warehause ETL Option Builder (OWB), but the latter is only dedicated Oracle database platform and the company gives strong recommendations to use ODI when the main goal is data migration, in particular, when this migration involves sources or destination based on other than Oracle's databases [7].

3.1   ELT Concept

Thanks to inversion of the classical order of ETL stages to Extract-Load-Transform (ELT) sequence, ODI simplifies and improves efficiency of most common ETL tasks. Traditional ETL tools perform complex data transformations using proprietary, middle-tier ETL engines. Instead, ODI uses the E-LT (Extract - Load and Transform) approach, wherein all data transformations are executed by the existing RDBMS engine(s). Unlike traditional ETL tools that often closely mix data transformation rules with the procedural steps of the integration process and require the development of both data transformations and data flow, ODI clearly separates the business rules from the actual implementation. Business rules describing mappings and transformations are defined graphically and stored independently from the implementation. ODI generates the data flow automatically using code templates, which can be fine tuned if required. This architecture often decreases the learning curve for database developers that could immediately be able to create transformation using familiar SQL syntax.

3.2 Domain Model

The ODI consistently seperates the logic of data migration and transformation from the physical layer of data sources. However, this abstraction is paid by fairly complicated approach to a data flow modeling. The complexities of this process will be easiest to introduce by means of conceptual and simplified class diagram in Fig. 1.

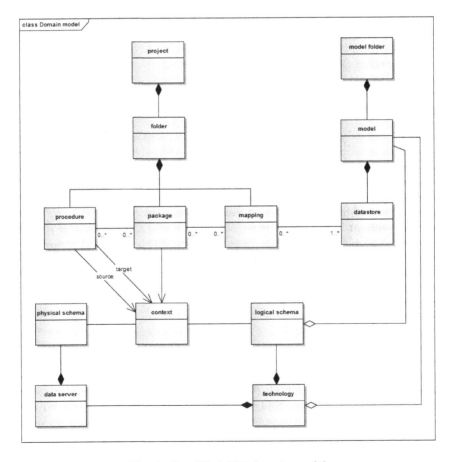

Fig. 1. Simplified ODI domain model

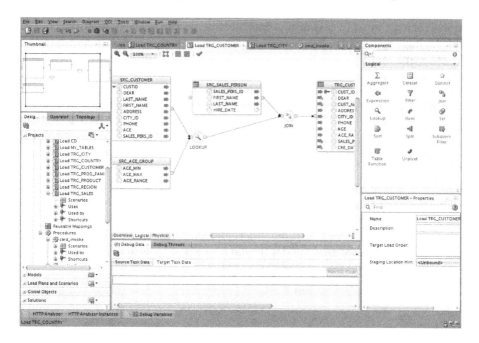

Fig. 2. ODI-an example screen of data mappings

The ODI supports many different *technologies* (e.g. Attunity, Axis2, BTrieve, Complex File, DBase, Oracle ...). In order for a technology to be the source or target of data flow is the existence of jdbc driver, which provides the data in a relational way. For a given technology one defines *data servers* i.e. hosts that share data services usually grouped into *physical schemas* that have dedicated connection parameters (e.g. username, password, url, network port). On the other hand, for a given technology one defines logical schemas, that allows the developer to define 'data metaflows' that are independent of the physical parameters of the existing technical infrastructure. With this approach, the designer can define a flow for the given domain issue only once and the run in a particular environment only requires the operator to configure the physical parameters. *Models* that provide a set of *datastore* are organized into *model directories*, must be explicitly assigned to a *logical schema*. The *datastores* being made available throught jdbc technology (e.g. tables or views in relational database) could be both *datasources* and *datatargets* and could be defined by developer from scratch or reverse engineered from given *physical schema*. Data flows and transformations are defined within the *projects* and are grouped by *folders*. *Mapping* is a fundamental technique for defining a data transformation and has replaced the former concept named in ODI 11 as interfaces. Here the developer has the possibility to determine in an intuitive and graphical way how data is mapped from one *datastore* is mapped to another one (e.g. screen of ODI Studio shown in Fig. 2. When defining the mapping developer has range of specialized functions operating on individual attributes (e.g. substring, upper), on several attributes (e.g. concat, sum) but also on the whole data streams (e.g. join, lookup, split, pivot, aggregate). All special cases, which can not describe the standard *mapping* can be implemented as a *procedures* that use specialized functions, available from the library grouped into subject specific categories. Both *mappings* and *procedures* in order to be run in a physical environment must be instanced as *package* with defined *context* which binds logical schema with physical one.

3.3   Architecture

ODI 12c is developed around a modular repository and accessed in client/server (ODI Studio) or thin client mode by applications that are build entirely in Java. The physical architecture is shown as deployment diagram in Fig. 3. There are two or more seperate environments – development and one or more production. The basic development tool is ODI Studio (see Fig. 2) which operates on devel-

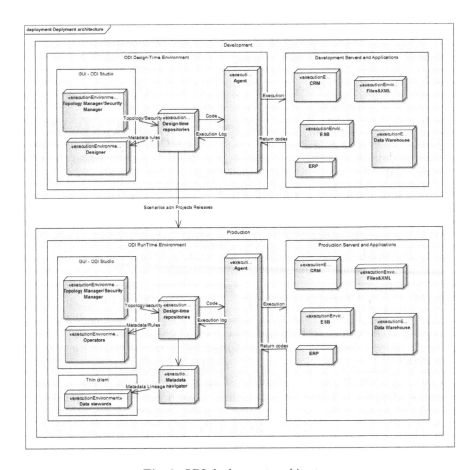

Fig. 3. ODI-deployment architecture

opment repositories. The ODI Studio is used by designers (for project developing, reverse-engineering, scenarios realising), operators (for sessions monitoring, production operating), topology managers (information systems infrastructure defining) and security managers (users privileges managing). Agents are the executive part of ODI system and are responsible for task flow, processes monitoring and event logging. When the data flow is defined not only by mappings but by procedures as well, the agents take also responsibility for some additional data processing. When the data flow project is developed it may be deployed to production environment. Thanks to separation physical and logical aspects of projects, once developed project can be reused in many production environments as soon as operator customizes parameters describing physical environment. Additionally, there is a possibility of access to the production repository by a thin client application. This is useful in case one needs a small interference in the production environment or wants to monitor processes remotely with standard web browser. It is worth emphasizing the undisputable advantage of the ELT concept implemented in the ODI application – when mappings and flows are built using generic functions than the role of agents is limited only to controlling and monitoring of data migration processes. The proof of this is shown on trivial listing 1.1 generated by ODI as script to be executed by an agent.

3.4   Hospital Information System

ETL processes are frequently performed in medical systems that require integration of information coming from many heterogeneous systems and data sources. Medical centre is a complex organization that requires management is many aspects of its operation. Component diagram (Fig. 4) shows the arbitrary way of grouping related aspects together in a domain way, or by the actors responsible for them.

The core of the system is an admission, discharge, and transfer module (ADT) that includes a monitoring of patient treatment during a hospitalization. Dictionaries and registers are a service components that are open to all other modules. Complementary processes – drugs providing, laboratory diagnosing – support pharmacy and respectively laboratory information system (LIS) modules. This functionality group may include other modules that implement the functionality specific to the type of treatment and support core processes. These are typical transactional systems.Picture archiving and communication system (PACS) and radiology information system (RIS) are specialized products that implement the functionality of diagnostic image processing and radiological diagnosing equipment handling. They are usually supplied by the manufacturers of these devices, of which the latter can be classified as embedded in the device. This group would include other embedded systems, e.g., a devices for maintaining vital processes. CDMS modules (clinical document management system), EHR (electronic healthcare record) and clinical statistics (reporting and data consistency controlling) support functions that are derived from data collected by the ADT module. Settlements, CMS (content management system) and e-patient modules support functions dedicated to the contact with the external actors and pursue

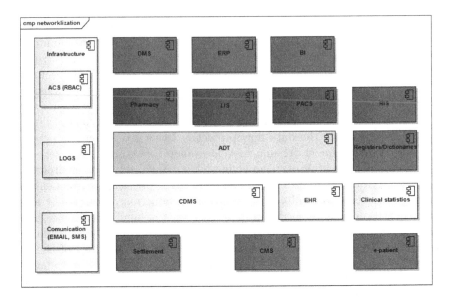

Fig. 4. Business component model diagram of a Hospital Information System

a settlement with the payer, presenting structured web content and provide patients with the possibility of registering for medical services or viewing treatment history and billings. DMS (document management system), ERP (enterprise resource planning) and BI (business intelligence) functionalities are loosely related to the core of the subject areas. These are mostly major player, prepacked products adapted to the specifics of the medical domain by customization. All of these functions must be available within certain infrastructural framework borders of authentication, access control, technical processes, monitoring and logging, etc.

Computer systems are used in hospitals from a long time what leads to aging these products both in a functional and a technical dimension. Replacing old system with new ones should ensure proper data transfer and transformation from the old system with a new one, which is usually not a trivial process. Thanks to the flexibility and very good performance the ODI system seems to be successful proposal of ETL solution.

An example of the use of the tools will be shown below. For simplicity, it is based on a small subset of entire medical system – ACS (access control system) that was implemented in accordance with the concept of RBAC (role based access controll). This particular implementation of RBAC was widely descibed in [9].

The considered task was designated to migrate information about users from legacy DBAP access control system to ACS mensioned above. The script generated by ODI tool is shown in listing 1.1.

Listing 1.1. Simple data flow script generated by ODI tool

drop database link AMMS_INSTANCE

create database link AMMS_INSTANCE **connect** to nthsadm
identified **by** <@=odiRef.getInfo("SRC_PASS")@> **using** 'orcl'

INSERT
    /*+ APPEND   PARALLEL */
    INTO ACSADM.USERS
    (
        UNAME ,
        FNAME ,
        LNAME ,
        PASSWD,
        FLAGS ,
        PASSWD_TIME
)
SELECT
    DBAP_UZYTKOWNICY.KOD ,
    DBAP_UZYTKOWNICY.IMIONA ,
    DBAP_UZYTKOWICY.NAZWISKO,
    'ETL' ,
    1,
    sysdate
FROM
    SYSADM.DBAP_UZYTKOWNICY@AMMS_INSTANCE DBAP_UZYTKOWNICY

drop database link AMMS_INSTANCE

In this case both source and destination are Oracle databases, but source data is located in different database instance and even different host than a transformed data should be loaded. As it can be seen, a destination database establishes direct communication path to the source (using database link technique) that bypass agent and allows direct data migration between this two databases. The necessary data transformations are done by native SQL functions specific to a given database vendor. In addition, what is new to ODI 12c, it supports the acceleration of data loading by enabling it in parallel thanks to using /*+ APPEND PARALLEL */ hint. How parallelism can positively affect an efficiency of SQL query execution can be find in [1].

3.5   Other New Features

Besides new excellent declarative flow-based user interface and accelerating data migration thanks to use *parallel* option, Oracle in ODI 12c introduced also other new features among them the most important are:

- multiple target support that enables in one mapping to define multiple targets as a part of single flow,
- reusable mappings that encapsulate partial flow, which can be reused in multiple mappings,
- step-by-step debugger that enables manually traverse task execution and set breakpoints to interrupt execution at pre-defined locations.

4   Summary

This paper presents new features of Oracle Data Integrator 12c. Since the acquisition of the tool in 2006, it was gradually improved by Oracle Company. However, only changes in current version made this tool, thanks to declarative flow-based user interface, as user friendly as other leading products. Consistently developed ELT approach makes it one of the most powerful (in terms of productivity) tools of this type is available on the market today. Taking into account that migration projects are reusable on diffrent physical environments it can be concluded that the productivity of this solution is high too. However, if there are infrequent needs of data migration and are limited to running on one dedicated physical infrastrucure one may consider to use some other tool, especially that there exist several ETL solutions available for free, for example Pentaho Data Integration (Kettle) or Talend Open Studio for Data Integration. Both of these tools can be an alternative to ODI [5,2].

Acknowledgements. This work was supported by NCBiR of Poland (No INNOTECH-K3/IN3/46/229379/NCBR/14).

References

1. Burns, D.: Tuning parallel execution, http://oracledoug.com/px.pdf (visited on November 2014)
2. Espinosa, R.: ETLs: Talend open studio vs pentaho data integration (kettle). comparative, https://churriwifi.wordpress.com/2010/06/01/comparing-talend-open-studio-and-pentaho-data-integration-kettle/ (visited on November 2014)
3. Jestratjew, A., Kwiecień, A.: Using cloud storage in production monitoring systems. In: Kwiecień, A., Gaj, P., Stera, P. (eds.) CN 2010. CCIS, vol. 79, pp. 226–235. Springer, Heidelberg (2010)
4. Kakish, K., Kraft, T.A.: ETL evolution for real-time data warehousing. In: Proceedings of the Conference on Information Systems Applied Research (2012)
5. Levin, J.: Open source ETL tools vs commercial ETL tools http://www.jonathanlevin.co.uk/2008/03/open-source-etl-tools-vs-commerical-etl.html (visited on November 2014)
6. Shaker, H., El-Sappagha, A., Abdeltawab, M., Hendawib, A., El Bastawissyb, A.H.: A proposed model for data warehouse ETL processes. Journal of King Saud University - Computer and Information Sciences 23, 91–104 (2011)

7. Testut, J.: Oracle data integrator 12c architecture overview (February 2014), http://www.oracle.com/technetwork/middleware/data-integrator/overview/oracledi-architecture-1-129425.pdf (visited on November 2014)
8. Wiak, S., Drzymała, P., Welfle, H.: Using ORACLE tools to generate multidimensional model in warehouse. Przeglad Elektrotechniczny, 0033–2097 (2012)
9. Wycislik, L.: Access control system based on generalization of RBAC model-concept, architecture and sample implementation. Technologie przetwarzania danych. TPD 2010. III Konferencja naukowa, Poznan, 21-23 czerwca 2010. Materialy konferencyjne (2010)
10. Wycislik, L.: Performance issues in data extraction methods of ETL process for XML format in oracle 11g. In: Kozielski, S., Mrozek, D., Kasprowski, P., Małysiak-Mrozek, B. (eds.) BDAS 2014. CCIS, vol. 424, pp. 581–589. Springer, Heidelberg (2014)

Automating Schema Integration Technique Case Study: Generating Data Warehouse Schema from Data Mart Schemas

Nouha Arfaoui[✉] and Jalel Akaichi

BESTMOD - Institut Suprieur de Gestion,
41, Avenue de la libert, Cit Bouchoucha, Le Bardo 2000, Tunisie
arfaoui.nouha@yahoo.fr, jalel.akaichi@isg.rnu.tn

Abstract. The schema integration technique offers the possibility to unify the representation of several schemas into one global schema. In this work, we present two contributions. The first one is about automating this technique to reduce human intervention. The second one is about applying this technique to generate data warehouse schema from data mart schemas. To response to our goals, we propose a new methodology that is composed by schema matching and schema mapping. The first technique compares the elements of the two schemas using a new semantic measure to generate the mapping rules. The second one transforms the mapping rules into queries and applies them to ensure the automatic merging of the schemas.

Keywords: Schema Integration · Schema matching · Schema Mapping · Data Warehouse · Data Mart

1 Introduction

Schema Integration (SI) is defined as the activity of integrating the schemas of existing or proposed databases into a global unified schema to find a unified representation, called the integrated schema [3]. The unification is done through extracting all relationships between the different schemas to be merged. It is used also to resolve the problems related to the inconsistencies and redundancies of the data when it passes from the application oriented operational environment to DW [11].

In this paper, we propose automating the schema integration technique, and applying it in the Data Warehouse (DW) to generate its schema from a set of Data Mart (DM) schemas. We opt, then, for the bottom-up approach that is about building the DW incrementally by designing and implementing one DM at a time [4]. DM "is defined as a flexible set of data, ideally based on the most atomic (granular) data possible to extract from an operational source, and presented in a symmetric (dimensional) model that is most resilient with unexpected user queries" [12].

Applying the SI to generate one final schema is not an easy task. It requires choosing the appropriate strategy and methodology. Concerning the strategies,

© Springer International Publishing Switzerland 2015
S. Kozielski et al. (Eds.): BDAS 2015, CCIS 521, pp. 200–209, 2015.
DOI: 10.1007/978-3-319-18422-7_18

there are two types which are "bottom-up" and "top-down". The use of one of them depends on the existence or not of the global schema. So, in the first strategy, the global schema does not exist, and the integration process involves both the definition of a global schema, as well as the definition of the mappings between the data source schemas and the global schema [3]. In the top-down integration setting the global schema exists, and the mapping rules need to be defined between the data source schemas and the global one [16]. The bottom up strategy is appropriated in the case of schema integration process while the top-down is more suited from the perspective of domain engineering [9]. So, and according to our goal we will adapt bottom up strategy to create our global schema. There are different ways to apply this strategy. In fact, it depends on how we will merge the local schemas i.e. using as input two schemas (binary) or all-at-once (n-ary). The binary can be divided into "ladder" [1] and "balanced" [7]. The n-ary is composed by "one-shot" [23] and "iterative" [19].

Our proposed solution uses the Binary Ladder. It is iterative. It takes as input two schemas to give one as output. The latter is used next as input, and so on until having one final schema.

Concerning the methodologies, there are several propositions such as: [8,20], etc. In our work, we propose a new one that is used mainly with DW and DM schemas. It is composed by two steps: the schema matching and the schema mapping. It deals with star and/or snowflake schemas because they are the most used models [17]. They are unified from the beginning since we propose the use of unified interface to specify the elements of the schemas (fact table, dimension tables, measures, attributes and/or parameters).

The remainder of this article is organized as follows:

– In the next section, we present the state of the art. We summarize some work using the schema integration and especially those which deal with the semantic aspect related to the application of this technique.
– In the third section, we define our methodology that is composed by schema matching and schema mapping. We detail the schema matching technique and we present the possible existing conflicts as well as the necessary steps as we apply to our case. We detail also the schema mapping technique. We clarify each point using an example.
– We finish our work with the conclusion and future work.

2   State of the Art

In the literature several studies focus on schema integration technique. In the following, we summarize some of them.

In [13], the authors focus on capturing the semantic of data stored in databases to integrate the different data sources. They define, then, the global dictionary that provides standardized terms for referencing and categorizing data. These terms are stored in record-based semantic specifications that store metadata and semantic descriptions of the data.

According to the authors, in [2], it is very important to deal with the semantic relationships of the elements when integrating the data from a set of databases. In this context, they propose Sphinx which is a new prototype system used to extract the knowledge without requiring any skills from the user. It develops a linguistic meta-model of integrating view definitions enabling an active learning algorithm.

In [15], the authors address the problem of constraint conflicts while integrating the conceptual schemas of multiple autonomous Entity-Relationship databases. They propose a new framework used to resolve three types of constraint conflicts: domain constraint conflicts; attribute constraint conflicts and relationship constraint conflicts. Concerning the convertible domain constraints conflict, there are two types which are reversible and irreversible.

The authors, in [22], discuss the applicability of schema integration techniques developed for tightly-coupled database interoperation to interoperate the databases stemming from different modeling contexts. Indeed, in absence of full knowledge on the semantics of remotely defined classes, instance level semantic relationships form an appropriate basis for database interoperation.

In [10], the authors focus on the part of semantics related to the meanings of the terms used as identifiers in schema definitions. They propose, in this context, an approach to integrate schemas from different communities based on merging the ontologies taking into account the similarity of relations among concepts of different ontologies. Each such community is using its own ontology.

The authors, in [8], propose a novel schema integration method "Clio" that enumerates multiple interesting integrated schemas. It provides easy-to-use capabilities for searching and refining the enumerated schemas via user interaction. It offers a visual interface that facilitates the users' understanding of the schema integration task at hand.

3   Our Proposed Methodology

Several methodologies are proposed to ensure the application of schema integration technique. To deal with DW/DM schemas, we suggest a new one that is composed by two steps which are the schema matching and the schema mapping. The first one is about comparing the elements of two schemas using a new semantic measure to generate a set of mapping rules that will be used by the schema mapping to merge the schemas. In fact, the second technique transforms the mapping rules into queries to automate this step.

3.1   Measure of Semantic Similarity

In order to compare the elements, we need a measure of semantic similarity.

Let Sch1 and Sch2 be two schemas belonging to the same cluster.

Let Ci be the categories of elements existing in the schema.

$Ci = \{fact, dimension, measure, attribute, parameter\}$.

$\forall ei \in Sch1, \exists ej \in Sch2$, so that ei and ej belong to the same category Ci.

When we calculate the similarity between the elements of the two schemas, we should take into consideration the following points:

- Identical: We use the same elements name in the two schemas.
  DeId (ei, ej) = 1 if ei and ej are identical and 0 if not.
- Synonymous: We use two different names that have the same meaning.
  DeSy (ei, ej) = 1 if ei and ej are synonymous, and 0 if not.
- Typos: The user makes mistakes when writing the name of the element. In this case, we calculate the degree of error. If it is low, we are in the case of typing error. If it is high we are in the case of two different words. In the following we only take into consideration the first case.
  DeTy (ei, ej) =1 if ei and ej are the same with the existence of typing error.
- Postfix: We use the postfix.
  DePost (ei, ej) = 1 if ei is the postfix of ej, and 0 if not.
- Prefix: We use the prefix.
  DePre (ei, ej) = 1 if ei is the prefix of ej, and 0 if not.
- Abbreviation: We use the abbreviations when writing the names.
  DeAbb (ei, ej) = 1 if ei is the abbreviation of ej, 0 if not.

The degree of similarity of ei and ej (DeSim (ei, ej)) is measured by the numeric value $\{0\}$, or $\{1\}$, and it is calculated using the formula (1):

$$DeSim\ (ei,\ ej) = [\ DeId\ (ei,\ ej) + DeSy\ (ei,\ ej) + DeTy\ (ei,\ ej) + DePost\ (ei,\ ej) + DePre\ (ei,\ ej) + DeAbb\ (ei,\ ej)\ ]\ (1)$$

3.2 Schema Matching

The schema matching is considered as one of the basic operations required by the process of data integration [6]. It is used to solve the problem related to the heterogeneity of the data sources by finding semantic correspondence between the elements of two schemas. This phase takes as input two or many schemas to get as output a set of mapping rules facilitating the merging of the schemas later.

The Conflicts Detection. The schema integration requires dealing with possible existing conflicts. According to [9] there are three types which are: extensional, structural and naming conflicts.

- Extensional conflict: It refers to the redundancies among different classes [21].

- Structural conflict: In the context of the schemas of databases, the structure conflict "occurs when related real world concepts are modeled using different constructs in the different schemas" [14]. In the context of our work, we do not need to focus on this kind of conflict since we will keep each element as it is defined in the global schema.

– Naming conflict: It "refers to the relationship between the object attribute or instance names" [18]. The relationship between the names is commutative i.e. term1 is homonyms of term2 implies also term2 is homonyms of term1. In this part, we treat homonyms and synonyms. The homonyms occur if one name is used for two or more concepts [5], and the synonyms occur if two or more names are used for the same concept [5], it can exist in any category. It is solved using the generalization [18]. This conflict is determined using different tools such as wordnet, thesaurus, etc. Their specification depends on their context.

The different conflicts are resolved using the semantic measure as described previously.

Schema Matching Steps. The schema matching serves to generate the mapping rules by extracting the closest elements belonging to the same category and the possible existing conflicts. To achieve this task, we go through the following steps:

– Categorization: It is to specify the category of each element. This can reduce the risk of error which provides a gain of time. As it was already mentioned, we have as categories: fact, measure, dimension, attribute and parameter.
– Construction of the similarity matrix: It is to assign a coefficient to two elements taking into consideration the existence of conflict, already cited. This coefficient is calculated using the formula (1) and it takes also into consideration the similarity of types when dealing with the measures and the attributes. In case of type difference, we need a human intervention. The similarity matrix is a way to find the closest elements. The cells contain the coefficient of similarity of the different elements belonging to the same category.
– Generation of the mapping rules: The rules visualize the possible instantiations of the elements belonging to the same category. They are expressed as: "If **Similar** (X, Y) then **Action** (X or Y) and **Save** (X, Y)", with:
    • X and Y: two elements belong to the same category (fact, measure, dimension, attribute or parameter).
    • Similar ( ): It is a function that specifies if the two inputs are similar or not. It uses the similarity matrix determined in the previous step.
    • Action ( ): It specifies the actions to perform. They can be union, or intersection. They are specified during the mapping.
    • Save ( ): It saves the two elements.
Example: Let us take the following example (Fig. 1) where it presents two DM schemas.
– Categorization:
    • Sch1. Fact = {SalesFact}; Sch1.PrimaryKey = {ProductId, DateId}; Sch1. Measures = {TotalSale {Double} }; Sch1.Dimension&Attribute = {Date {DateId{Integer}, Day{Integer}, Month{Integer}, Year{Integer}}; Product {ProductId {Integer}, ProductName {String}, Quantity {Double}, Price {Double}}} ; Sch1.Parameter = {Category {CategoryId {Integer}, CategoryName {String}}}.

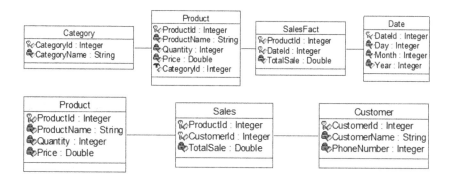

Fig. 1. Example of two DM schemas

- Sch2. Fact = {Sales}; Sch2. PrimaryKey = {ProductId, CustomerId};
  Sch2. Measures = {TotalSale {Double}}; Sch2.Dimension&Attribute =
  {Product {ProductId {Integer}, ProductName {String}, Quantity
  {Double} Price{Double}}; Customer {CustomerId {Integer}, Customer-
  Name {String}, PhoneNumber{Integer}}}.
- Construction of the similarity matrix:
  In table 1, the two fact tables are similar. They have the value "1".

Table 1. Fact similar matrix

Fact	Sales	Max
SalesFact	1	1

In table 2, the measures of the two fact tables are similar and they have the
same type "Double". If their types are different, the user intervenes.

Table 2. Measures similar matrix

Measure	TotalSale (Double)	Max
TotalSales (Double)	1	1

In table 3, the two schemas have as one similar dimensions: {Product,
Product}.
In table 4 the dimensions "Product and Product" have as similar attributes the
following pairs :{{ProductId, ProductId}, {ProductName, ProductName},
{Quantity, Quantity}, {Price, Price}} with the same types. Concerning "Cate-
groyId", it is a foreign key which implies the existing of a hierarchy.
- Generation of the mapping rules:
  In the following, we give some examples of mapping rules.

Table 3. Dimension similar matrix

Dimension	Product	Customer	Max
Product	1	0	1
Date	0	0	0

Table 4. Attributes, primary keys and types similarity matrix of the dimensions Product/Product

Attributes of the dimensions /ProductProduct	ProductId (Integer)	ProductName (String)	Quantity (Double)	Price (Double)	Max
ProductId (Integer)	1	0	0	0	1
ProductName (String)	0	1	0	0	1
Quantity (Double)	0	0	1	0	1
Price (Double)	0	0	0	1	1
CategoryId (Integer)	0	0	0	0	0

- If Similar (Sale, SalesFact), then Action (Sale or SalesFact) and Save (Sale, SalesFact).
- If Similar (TotalSales, TotalSales), then Action (TotalSales or Total-Sales) and Save (TotalSales, TotalSales).
- If Similar (Product, Product), then Action (Product or Product) and Save (Product, Product).
- If Similar (ProductId, ProductId), then Action (ProductId or ProductId) and Save (ProductId, ProductId).
- If Similar (ProductName, ProductName), then Action (ProductName or ProductName) and Save (ProductName, ProductName).

3.3 Schema Mapping

Now, we move to the next step that is about transforming each mapping rule into query and executing it to automatically merge the schemas.

The schema mapping is a qua-triple M = (sch1; sch2; T; δi). "sch1" is the first schema, "sch2" is the second schema, "T" is the target schema, and "δi" is a set of formulas over <sch1, sch2; T>.

An instance of M is an instance of <s1, s2; t; δi > over <sch1, sch2; T; δ > that has a specific formula in the set δi.

The formulas existing in δi correspond to one of the following functions:

- Union: R = union (ei, ej) implies that R contains all the components of "ei" and all components of "ej". It is applied when the two elements are identical.
- Intersection: R= intersection (ei, ej) implies that R contains the components that exist in "ei" and "ej". It is applied when the two elements are similar but not identical.

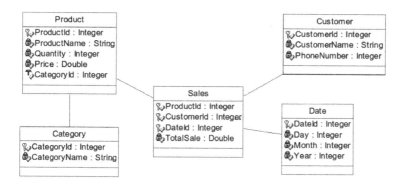

Fig. 2. The schema of the final Data Warehouse schema

Example: In the following, we present some examples of queries.

- Query ="Insert into Schema (idSchema) values ("+schemaId +")" ;
- Query1 = "Insert into Fact (FactName, idSchema) values ('Sales', "+ schemaId +")";
- Query2 = "Insert into Measure (MeasureName, idFact) values ('TotalSales', "+ factId +")" ;
- Query3 = "Insert into Dimension (DimensionName, idHierarchy, idSchema) values ('Date',"+ null+","+schemaId+")";

The different queries are used by the following algorithms, where "Extract_Rule" algorithm verifies the existence of the rule into the corresponding database, "Generate_Rule" algorithm generates the corresponding rule. Here, we present only the rule of the "Measure" element. The final algorithm "Apply_Rule" applies the rules through the execution of the corresponding query. The same algorithms are used with the rest of elements to ensure the automatic merging of the schemas.

```
public String Extract_Rule (String elem)
{ String rule="";
try { //Connect to the DataBase
String requete2 = "Select * from integrationrule where Elements
='"+elem+"';";
Statement requete = conn.createStatement();
ResultSet resultat = requete.executeQuery(requete2);
while (resultat.next()) { rule = resultat.getString("Rule");}
    }catch(Exception e){e.getMessage();}
        return rule; }
    public String Generate_Rule(String Type, String elem, int IdParent)
{String rule="";
if (Type.equals("Measure"))
rule="Insert into Measure (MeasureName, MeasureType, idFact) values
('"+ elem + "', '"+ type + "',"+IdParent+")";
```

```
return rule;}
   public void Apply_Rule}(String rule)
{try{ //Connect to the DataBase
java.sql.Statement stm = conn.createStatement();
int resultats = stm.executeUpdate(rule);
}catch (Exception e){e.getMessage();}}
```

Once the previous queries are executed, we get the following schema (Fig. 2) that is composed by one fact table "Sales" surrounded by three dimension tables "Product, Date and Customer". The schema has one hierarchy connected to the dimension "Product" and contains one parameter "Category".

4   Conclusion and Future Work

The schema integration technique offers the possibility to unify the representation of several schemas into one global schema. This technique is used, in the database field, to capture the semantic of data to facilitate their integration later. It is used also with the DW to resolve the problems related to the inconsistencies and redundancies of the data related when it is loaded.

In this paper, we proposed a new methodology to automate this technique and we proposed a new application. In fact, the SI is used to ensure the construction of the DW schema from the set of DM when using the bottom-up approach. Our methodology is composed by two steps. The first one is the matching. It is about comparing the schemas and generating the mapping rules. The second step is the mapping. It is about merging the schemas by converting the mapping rules into queries and executing them.

As perspective, we propose applying some algorithms in order to improve the extraction of rules form the database to accelerate the response time.

Acknowledgments. I would like to thank everyone!

References

1. Batini, C., Lenzerini, M.: A methodology for data schema integration in the entity relationship model. IEEE Transactions on Software Engineering 6, 650–664 (1984)
2. Barbançon, F., Miranker, D.P.: Interactive schema integration with sphinx. In: Christiansen, H., Hacid, M.-S., Andreasen, T., Larsen, H.L. (eds.) FQAS 2004. LNCS (LNAI), vol. 3055, pp. 175–190. Springer, Heidelberg (2004)
3. Batini, C., Lenzerini, M., Navathe, S.B.: A comparative analysis of methodologies for database schema integration. ACM Computing Surveys (CSUR) 18(4), 323–364 (1986)
4. Battaglia, A., Golfarelli, M., Rizzi, S.: Qbx: A case tool for data mart design. In: De Troyer, O., Bauzer Medeiros, C., Billen, R., Hallot, P., Simitsis, A., Van Mingroot, H. (eds.) ER Workshops 2011. LNCS, vol. 6999, pp. 358–363. Springer, Heidelberg (2011)
5. Bellström, P.: Bridging the gap between comparison and conforming the views in view integration. In: 10th East-European Conference on Advances in Databases and Information Systems, pp. 184–199 (2006)

6. Bernstein, P.A., Melnik, S.: Meta data management. In: 2013 IEEE 29th International Conference on Data Engineering (ICDE), pp. 875–875. IEEE Computer Society (2004)
7. Casanova, M.A., Vidal, V.M.P.: Towards a sound view integration methodology. In: Proceedings of the 2nd ACM SIGACT-SIGMOD Symposium on Principles of Database Systems, pp. 36–47. ACM (1983)
8. Chiticariu, L., Hernández, M.A., Kolaitis, P.G., Popa, L.: Semi-automatic schema integration in clio. In: Proceedings of the 33rd International Conference on Very Large Data Bases, pp. 1326–1329. VLDB Endowment (2007)
9. Dang, L.H., Feldmann, D.I.M.: A Guideline for the Conduction of Data Integration for Heterogeneous Information Systems. Ph.D. thesis, Diploma Thesis (2010)
10. Hakimpour, F., Geppert, A.: Resolving semantic heterogeneity in schema integration. In: Proceedings of the International Conference on Formal Ontology in Information Systems, vol. 2001, pp. 297–308. ACM (2001)
11. Inmon, W.H.: Building the data warehouse. John Wiley & Sons (2005)
12. Kimball, R., Ross, M.: The data warehouse toolkit: the complete guide to dimensional modeling. John Wiley & Sons (2011)
13. Lawrence, R., Barker, K.: Integrating data sources using a standardized global dictionary. In: Knowledge Discovery for Business Information Systems, pp. 153–172. Springer (2002)
14. Lee, M.L., Ling, T.W.: Resolving structural conflicts in the integration of entity-relationship schemas. In: Papazoglou, M.P. (ed.) ER 1995 and OOER 1995. LNCS, vol. 1021, pp. 424–433. Springer, Heidelberg (1995)
15. Lee, M.L., Ling, T.W.: Resolving constraint conflicts in the integration of entity-relationship schemas. In: Embley, D.W., Goldstein, R.C. (eds.) ER 1997. LNCS, vol. 1331, pp. 394–407. Springer, Heidelberg (1997)
16. Lenzerini, M.: Data integration: A theoretical perspective. In: Proceedings of the Twenty-First ACM SIGMOD-SIGACT-SIGART Symposium on Principles of Database Systems, pp. 233–246. ACM (2002)
17. Levene, M., Loizou, G.: Why is the snowflake schema a good data warehouse design? Information Systems 28(3), 225–240 (2003)
18. Naiman, C.F., Ouksel, A.M.: A classification of semantic conflicts in heterogeneous database systems. Journal of Organizational Computing and Electronic Commerce 5(2), 167–193 (1995)
19. Navathe, S.B., Gadgil, S.G.: A methodology for view inegration in logical database design. In: Proceedings of the 8th International Conference on Very Large Data Bases, pp. 142–164. Morgan Kaufmann Publishers Inc. (1982)
20. Parent, C., Spaccapietra, S.: Integration de bases de donnees: Panorama desproblemes et des approches. In: Ingenierie des Systemes dInformation 4(LBD-ARTICLE-1996-002), pp. 333–359 (1996)
21. Schwarz, K., Schmitt, I., Türker, C., Höding, M., Hildebrandt, E., Balko, S., Conrad, S.: Design support for database federations. In: Akoka, J., Bouzeghoub, M., Comyn-Wattiau, I., Métais, E. (eds.) ER 1999. LNCS, vol. 1728, pp. 445–460. Springer, Heidelberg (1999)
22. Vermeer, M.W.W., Apers, P.M.G.: On the applicability of schema integration techniques to database interoperation. In: Thalheim, B. (ed.) ER 1996. LNCS, vol. 1157, pp. 179–194. Springer, Heidelberg (1996)
23. Yao, S.B., Waddle, V.E., Housel, B.C.: View modeling and integration using the functional data model. IEEE Transactions on Software Engineering 8(6), 544–553 (1982)

Proposal of a New Data Warehouse Architecture Reference Model

Dariusz Dymek, Wojciech Komnata, and Piotr Szwed[✉]

AGH University of Science and Technology, University in Kraków, Poland
eidymek@kinga.cyf-kr.edu.pl, {wkomnata,pszwed}@agh.edu.pl

Abstract. A common taxonomy of data warehouse architectures comprises five basic approaches: Centralized, Independent Data Mart, Federated, Hub-and-Spoke and Data Mart Bus. However, for many real world cases, an applied data warehouse architecture can be their combination. In this paper we propose a Data Warehouse Architecture Reference Model (DWARM), which unifies known architectural styles and provides options for adaptation to fit particular purposes of a developed data warehouse system. The model comprises 11 layers grouping containers (data stores, sources and consumers), as well as processes, covering typical functional groups: ETL, data storage, data integration and delivery. Actual data warehouse architecture can be obtained by tailoring (removing unnecessary components) and instantiating (creating required layers and components of a given type).

Keywords: Data warehousing · Data warehouse architecture · Reference model

1 Introduction

For many companies data warehouses are basic sources of clean, accurate, timely and integrated data used for reporting, analyses and decision making [3]. They integrate data from different sources and ensure effective access to information stored in one place and sharing a single data model. Data warehouses include systems with very different characteristics, with respect to their architectures (from centralized systems to distributed ones), in terms of the data model and database tools used. A common feature of data warehouses is a strong focus on provision of information to users, while assuring full control over data processing.

The motivation for our work were two projects extensively using data warehousing technology conducted at the Department of Applied Computer Science at AGH University of Science and Technology . The first aimed at designing a platform for exchange of confidential data dedicated for security agencies [5,14]. The platform was based on a federated data warehouse architecture assuring safety of communication, as well as supervisory and analytical functions provided by a data warehouse component storing anonymized information describing communications.

© Springer International Publishing Switzerland 2015
S. Kozielski et al. (Eds.): BDAS 2015, CCIS 521, pp. 210–221, 2015.
DOI: 10.1007/978-3-319-18422-7_19

The second project, Green AGH Campus [13], falls into the domain of widely understood energy industry and solutions for a Smart City initiative. It is developed as a result of cooperation between AGH Universtity, the Office of Malopolska Region and leading representatives of the industry. Within this project a data warehouse is to be used to store aggregated information related to energy consumption, environment conditions and decisions made by deployed energy distribution management systems.

One of challenges encountered during the work on both projects was a problem of choosing data warehouse architecture. The complexity of analyzed systems, legacy software to be integrated, security and confidentiality issues and various business needs prompted us to envisage complex architectural patterns, which cannot be directly mapped onto typical data warehouse architectures. In consequence, we decided to propose a flexible Data Warehouse Architecture Reference Model (DWARM), which can be mapped on known architectural styles and provide options for adaptation to fit particular purposes of a developed data warehouse system. The model comprises 11 layers grouping containers (data stores, sources an consumers) and processes, covering typical functional groups: ETL, data storage, data integration and delivery. Actual data warehouse architecture can be obtained by tailoring (removing unnecessary components) and instantiating (creating required layers and components of a given type).

The paper is structured as follows: in section 2 we discuss basic topis related to data warehouse architectures. Section 3 describes the DWARM model. In section 4 a mapping between DWARM and common architectural styles is provided. Section 5 gives concluding remarks.

2   Related Works

A practical reason for the construction of complex information systems collecting massive amounts of operational data is to support a decision-making processes. For nearly three decades data warehouses have been an integral component of modern decision support systems [6,15,11]. Data warehouses allow for efficient execution of complex analytic queries, including various cross sectional data and at various degrees of aggregation. This makes it possible to provide information on the desired level of detail.

The concept of the data warehouse is directly related to the work of William Inmona and Ralph Kimball. Data warehouses, according to Inmon, are hearts of all decision support systems [8] and they become major tools for an analytic work within many organization. Ralph Kimball, in contrast to Bill Inmon, is focused on a functional aspect of data warehousing, virtually combined with the concept of business intelligence [9].

Data warehouses are much more than simple reporting systems, they are usually characterized by two features: (1) they store only one version of data items and (2) they provide ability to access all the data, whenever it is required [8,9]. According to Laura Reeves, these two features are not only the sufficient conditions related to data warehouses, they are far-reaching requirements on the discipline and organization-wide commitment to well-defined data management [12].

A typical, very simple architecture of a data warehouse [4,7,3,1,5] is presented in Fig. 1. It comprises: source systems, components implementing data collection process, stores (a central repository and data marts), as well as consumers: reporting and analytic applications.

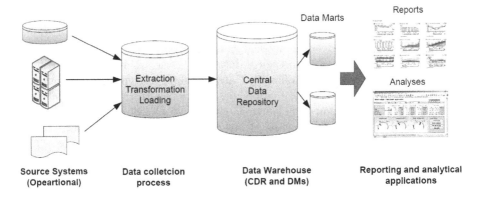

Fig. 1. A simple architecture of data warehouse

The literature discusses two classifications of data warehouse architectures. The first [7] describes a data warehouse structure as a set of linked layers i.e. it distinguishes single-layer, two-layer and three-layer architectures. The second classification takes into account such factors as roles of individual components (in particular those storing data: central repository and data marts), data flows allowing to access data for the entire enterprise or individual departments. The commonly recognized arhcitectural styles [3,4,10,1] are the following: Centralized, Independent Data Marts, Federated Hub-and-Spoke and Data Bus-Mart. They are shown in Fig. 2.

Selection of data warehouse architecture may have a critical impact on the execution time, organization and efficiency of queries generated at the level of a given department or entire organization [2]. Therefore, before starting a data warehouse project, a research should be conducted to learn core areas of business activity, challenges that an organization faces and its strategic goals, but also existing assets and resources [7,12,3,9]. Golfarelli and Rizzi [7] indicated the following key quality attributes relevant for a data warehouse system architecture: separation, scalability, flexibility, security, readability and easy management.

3   Data Warehouse Architecture Reference Model

In this section we present our proposal of a Data Warehouse Architecture Reference Model (DWARM), which is intended to unify the dominant approaches used

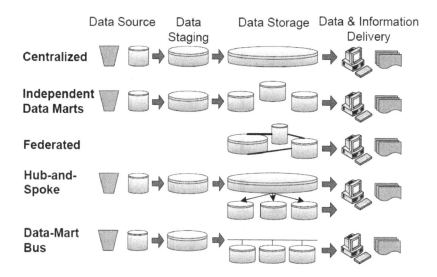

Fig. 2. Basic five data warehouse architectures identified

to describe data warehouse architectures discussed in the previous section, i.e. the layer-oriented [7] and based on 5 commonly recognized architectural styles.

Selecting a language that will be used to describe architecture variants we intentionally focused on a limited set of concepts. Its basic elements are: *processes* (abstractions of behavior) and *containers* representing components capable of storing, delivering or consuming data, e.g. repositories, data stores, sources. Such choice reflects well established approach to modeling complex systems defined in various standards or methodologies, e.g IDEF0, DFD and Structured Analysis [16].

The DWRAM is organized into eleven layers (see Fig. 3) grouping processes, containers (and implicitly data) sharing a common characteristics or functions. The distinctions made were based on multivariate logical organization of various analyzed data warehouse architectures. The criteria used to establish the set of proposed layers aimed at:

1. Distinguishing processes realizing similar functions, while preserving succession relations and optional character in a particular data warehouse architecture.
2. Identifying components responsible for data collection and storage, based on their role in a particular architectural style or pattern.

Based on those criteria we have identified six types of containers and thirteen types of processes, which are depicted in Fig. 3. For a particular architecture model, they should be instantiated, e.g. a number of data marts and in consequence processes feeding and delivering data to them should be created.

The obtained multilayer model comprises all salient types of processes and data stores appearing in various architectural models discussed in section 2. It is

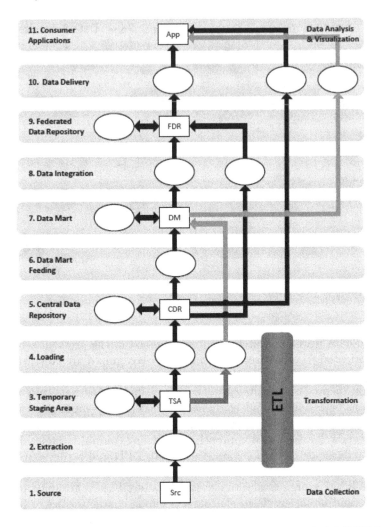

Fig. 3. Layers in the Data Warehouse Architecture Reference Model (DWARM)

fine-grained enough to capture architecture variants. An example can be Extraction Transformation Loading (ETL) process (see Fig. 3), which is represented by three layers (including the data store within the Temporary Staging Area). The selected granularity level allows to adapt the model by *tailoring*: removing particular containers and processes.

An important feature of the presented reference model is the discretionary character of its components, with a stipulation that this optionality is restricted by the expected data warehouse functionality. For example layers 1–4 must appear in every data warehouse architecture instance, although the multiplicity of their elements (e.g. the number of source systems) is not specified by the reference model. Further, layers 5 (Central Data Repository) and 7 (Data Marts) are optional, but at least one of them must appear in a valid architecture instance. Inclusion of one of these layers into an architecture model implies that it should embrace also processes of level 4 (Loading) and/or 6 (Data Mart Feeding).

This multivariate character of DWARM model allows not only to reflect commonly identified data warehouse architectures, but also to describe new emerging solutions, especially in case of complex data integration systems, where multiple stacks of layers can be used.

It should be mentioned, that the proposed DWARM model was formally described as an ontology [5] defining its main concepts and relations and also extending it to cover basic architectural decisions and dependencies between data.

The rest of the sections provides details of particular layers belonging to the reference model.

(1) Source. Source layer comprises entities that can deliver data to the data warehouse, including various enterprise information systems and registering devices. Typically, processes of data collection can be automatic or manual, synchronous or asynchronous, organized according to the specificity related to the domain of usage. Information can be represented in both forms: structured and unstructured, stored in various kinds of data bases or other forms of digital data storage. Data models used are related to singular computer systems area of usage.

(2) Extraction. Extraction layer comprises processes of data acquisition, transformation to transport formats and sending them to a data warehouse (DW). Typically, extraction processes are specific for particular software applications (sources). They can be started manually or automatically based on metadata. The extracted information can be structured according to various models and formats. The data can be either complete or incremental (in relation to a source). Transport formats include some additional information that can be used to identify source, time, completeness of data, etc.

(3) Temporary Staging Area (TSA). TSA is responsible for storing data extracted from sources and their transformation (including model and structure transformation, quality checking, etc.) to the form suitable for DW Loading. Processes, fully controlled by metadata, cover data transformation (model, structure, values) and data quality assurance (completeness, internal and external consistency, compliance with control rules: technical, law, business, etc.). From this standpoint, TSA is probably one of the most complex DW layers. Aside from various models and formats of source data sets, some temporary and technical models and formats can appears. All data have the time and source stamps, which identify their sources and applied extraction processes.

(4) Loading. Processes belonging to the loading layer are responsible for transferring data from TSA to a Central Data Repository (CDR) or a Data Mart (DM), depending on the architecture type. They are usually periodical, fully controlled by metadata (though they can be started manually). The processes include data transfer (copying) and automatic processing (integration, aggregation, etc.). Data models, formats and structures mainly depend on the tools, which are used within the TSA and Central Data Repository (CDR) layer. Primary data after loading are treated as non-modifiable.

(5) Central Data Repository (CDR). CDR is responsible for data management (in *centralized* and *hub-and-spoke* architectures). Inner processes of CDR are fully controlled by metadata. They are started automatically or on demand. We can distinguish processes responsible for internal data processing and providing data to processes responsible for loading other layers: Data Marts (7) DM, Federated Data Repository (9) FDR or Consumer Applications (11) CA. All data are structured according to a common schema. Modeling approaches are based on concepts of facts and dimensions (star, snow flake or constellation), rarely flattened model. Implementation depends on database tools; a relational implementation is the most often used.

(6) Data Mart Feeding. In the *hub-and-spoke* architecture the DMs are supplied with the data from CDR. The layer contains processes performing selection of data from CDR, then applying required transformations and loading to separate Data Marts. Processes, controlled by metadata, are started periodically, on-action or on-demand.

(7) Data Mart (DM). In *independed DM* and *DM bus* architectures the DMs are the primary data storage for all delivery processes (can be treated as local CRDs). In case of *hub-and-spoke* architecture a DM plays the role of a local data storage (a materialized view of CDR) dedicated for a particular purpose or a group of users. Characteristic similar to CDR. We can distinguish the processes of data processing (e.g. some OLAP processes) and data providing for other layers. Each DM can have its own data model or can share the main model of the data warehouse. Smaller volumes and domain-orientation of data can be an indication for using more redundant data models and structures. Often multidimensional data bases (data cubes) are used for storing data.

(8) Data Integration. This layer is responsible for integration of data originating from different data marts or even different data warehouses, depending on business needs. Applying data integration can be an efficient solution in case of organizational changes within the business or company mergers and acquisitions. If several data marts share the common data model, the process of data integration is similar to joining databases with additional selection and transformation operations (optional). In the case of different data models the process of integration is similar to supply process of CDR (the ETL process) and the data characteristic is similar to layers 2,3 and 4 of the presented model.

(9) Federated Data Repository (FDR). FDR can be implemented as a virtual or materialized data storage. It plays the role of additional source of data satisfying various need of Consumer Applications, in opposition to Central Data Repository, which is the primary source of data. In the case of materialized implementation, the processes of FDR are similar to CDR. In the case of virtual implementation, they are closely connected to processes of Data Integration (8) and are based on concepts and tools of middleware and multi-databases. Virtual implementation implies that any access to a data item at the FDR level must be transformed to an access to the primary data source (with all the consequences: model, structure and value transformations) In the case of materialized implementation, the data characteristic of FDR is similar to this of CDR. In the case of virtual implementation, the physical data have the characteristic of their sources (DM or CDR). The logical characteristic of data is similar to data characteristic of CDR.

(10) Data Delivery (DD). The layer is a link between Consumer Applications (11) and all needed sources of data. Depending on architecture, it can pull data from a CDR (5), DMs (7) or FDRs (9). The main goal of this layer is bringing the demanded data to any application on CA layer. Data Delivery (DD) processes can be assigned to CDR (5), DM (7) or FDR (9), however, their common characteristics allows to group them as a separate layer. Basic DD processes are related to data selection and provision. However, this group includes also processes providing security functions (granting data access, logging, etc). As DD processes works directly on data storages, they must be fully synchronized with processes of others layers and controlled by metadata. DD layer does not impose a special data characteristic. Delivered data have the characteristics of their sources either CDR (5) or DM (7) or FDR (9).

(11) Consumer Applications. This layer includes all end-user applications, which use the data from data warehouse for any purpose. They can be divided into two groups. The first includes applications, which are controlled by metadata, e.g. reporting systems with specified data formats, periodicity and recipients. Applications comprised in the second group are not described by metadata (e.g. Excel which use the data exported from a data warehouse). There is no common process characteristic for all consumer applications, each of them can have its own. There is no special data characteristic either.

4   Mapping – Example Models

The described reference model architecture of a data warehouse DWARM allows for representation of the various types of architectures commonly used [3]. They can be obtained by removing or multiplying particular layers, as well as their components (processes and data stores), while preserving the integrity of the whole architecture model.

Table 1 provides a mapping between five commonly recognized architecture models and the layers of the DWARM. The multiplicity of layers (0, 1, many) for a given architectural style should be interpreted as a number of layer instances of a given kind, and not as a number of its components. For instance, for the Independent Data Mart architectural style it is possible that each data mart is fed from separate sets of data sources (which can comprise several operational software systems). This means that the several source layers can be distinguished for this architectural style. Multiplicity "0" means that a layer does not appear in a given architectural style, the symbol "1" means that a layer appears exactly once (have to appear), the symbol "many" means that a layer can appear many times (have to appear at least once).

Table 1. Mapping between DWARM layers and five architectural styles

Layer	Centralized	Independent Data Mart	Federated	Hub and Spoke	Data Mart Bus
1. Source	1	many	many	1	1
2. Extraction	1	many	many	1	1
3. Temporary Staging Area	1	many	many	1	1
4. Loading	1	many	many	1	many
5. Central Data Repository	1	0	0 or many	1	0
6. Data Mart Feeding	0	0	0 or many	1	0
7. Data Mart	0	many	0 or many	many	many
8. Data Integration	0	0	1	0	0
9. Federated Data Repository	0	0	1	0	0
10. Data Delivery	1	many	many	many	many
11. Consumer Application	1	many	many	many	many

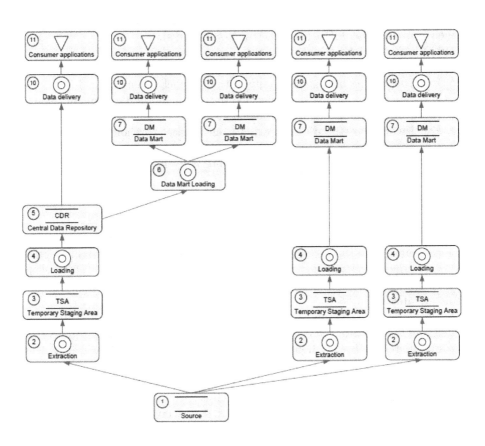

Fig. 4. A combination of *Data Mart Bus* and *Hub and Spoke* architectures

Analyzing organization of various data warehouses, two basic functional groups of components can be identified: the first is responsible for loading data and the second is responsible for storing and sharing data with the analytic part. Based on the these functions various architectural models can be mapped to the defined DWARM layers. Described types of data warehouse architectures are kinds of mock-ups, which are based on the basic variants of loading area and the area of data storage and sharing.

Architectures of real-world data warehouses can differ significantly from basic five models referred in table 1. Indeed, we prefer to call them styles, as actual architecture can be a combination of two or more of them. Very often data warehouses are developed during an evolutionary process, following organizational changes and business needs. Thus, the resulting architecture is more influenced by the functionality required, than technological issues. The proposed DWARM model is expressive enough to describe hybrid architectures that combine basic architectural styles. An example given in Fig. 4, shows a combination of two types of data warehouse architecture: Hub-and-Spoke and Data-Mart Bus. Such architecture can be developed due to historical circumstances. For example, a system was initially designed as a Data-Mart Bus and several data marts were created. Then an architect decided to modify the design and develop further system components following a centralized architecture.

The DWARM model was used in the project aiming at designing a platform dedicated for security agencies supporting exchange of confidential data [5]. The model was used as a tool for analysis of architectures of existing systems and for designing the architecture of the final system. Obtained results show that the DWARM model can constitute a basis for developing a formal method allowing evaluation of architectures of existing DW systems (in terms of architecture consistency, effectiveness of data processing and storage, etc.) and supporting selection of data warehouse architectural style during the development of a new DW system.

5   Conclusions

Data warehouses are complex information systems that aims at providing analytical data from various areas within an organization to such end users as: analysts and decision makers. Challenges for data warehousing involve many aspects: data modeling, designing internal warehouse processes, providing storage for growing volumes of data, optimizing access by creating dedicated repositories containing selected subsets of data (data marts), integration of data from multiple sources, secure access and finally management of metadata describing all of these elements.

The goal of our work was to define a consistent model that can be used to express various data warehouse architectures, which are developed to meet very diverse technical and business needs. In particular, the latter concerns more and more frequent problems of sharing and integrating data from many sources, both transactional systems and distributed repositories. Typical examples are integration solutions providing access to distributed data repositories e.g. medical or pharmaceutical resources, secure information sharing within security agencies and legacy systems integration in the case of company mergers and acquisitions.

The proposed reference model of data warehouse architectures DWARM is divided into 11 layers grouping processes and containers (data stores) sharing common characteristics and roles. An instance of data warehouse architecture model can be obtained by tailoring (removing selected elements or layers) and creating instances of appropriately connected components. Starting from DWARM and applying these operations may lead to well known architectural models, but also to hybrid architectures (c.f. Fig. 4). It should be mentioned that the DWARM is formalized as an ontology [5] defining components, their relations and expected connections.

References

1. Alsqour, M., Matouk, K., Owoc, M.L.: A survey of data warehouse architectures - preliminary results. In: 2012 Federated Conference on Computer Science and Information Systems (FedCSIS), pp. 1121–1126. IEEE (2012)
2. Ariyachandra, T., Watson, H.J.: Which data warehouse architecture is most successful? Business Intelligence Journal 11(1), 4 (2006)
3. Ariyachandra, T., Watson, H.J.: Key organizational factors in data warehouse architecture selection. Decision Support Systems 49(2), 200–212 (2010)
4. Connolly, T.M., Begg, C.E.: Database Systems: A Practical Approach to Design, Implementation, and Management. International computer science series. Addison-Wesley (2005), http://books.google.pl/books?id=jJbnDxiJ4joC
5. Dymek, D., Komnata, W., Kotulski, L., Szwed, P.: Data Warehouse Architectures. Reference model and formal architecture description. In: Polish: Architektury Hurtowni Danych. Model referencyjny i formalny opis architektury. AGH University of Science and Technology Press (2015)
6. Finlay, P.N.: Introducing decision support systems. NCC Blackwell (1994), http://books.google.pl/books?id=RhBPAAAAMAAJ
7. Golfarelli, M., Rizzi, S.: Data Warehouse Design: Modern Principles and Methodologies. Mcgraw-hill (2009), http://books.google.pl/books?id=VVRLa0dYmxoC
8. Inmon, W.H.: Building the Data Warehouse. Wiley (2005), http://books.google.pl/books?id=QFKTmh5IFS4C
9. Kimball, R., Margy, R.: The Data Warehouse Toolkit: The Definitive Guide to Dimensional Modeling. Wiley (2013), http://books.google.pl/books?id=WMEqTf21K84C
10. Ponniah, P.: Data Warehousing Fundamentals. Wiley India Pvt. Limited (2006), http://books.google.pl/books?id=z4QjAw1YmxgC
11. Power, D.J.: A brief history of decision support systems (2007), http://dssresources.com/history/dsshistoryv28.html
12. Reeves, L.: A Manager's Guide to Data Warehousing. Wiley (2009), http://books.google.pl/books?id=aCNxzTV0U3UC
13. Szmuc, T., Kotulski, L., Wojszczyk, B., Sedziwy, A.: Green AGH campus. In: SMARTGREENS 2012 - Proceedings of the 1st International Conference on Smart Grids and Green IT Systems, Porto, Portugal, April 19 - 20, pp. 159–162 (2012)

14. Szwed, P.: Belief propagation during data integration in a P2P network. In: Rutkowski, L., Korytkowski, M., Scherer, R., Tadeusiewicz, R., Zadeh, L.A., Zurada, J.M. (eds.) ICAISC 2014, Part I. LNCS (LNAI), vol. 8467, pp. 805–816. Springer, Heidelberg (2014), http://dx.doi.org/10.1007/978-3-319-07173-2_69

15. Turban, E.: Decision Support and Expert Systems: Management Support Systems. MacMillan Series in Information Systems. Macmillan (1990), http://books.google.pl/books?id=k2ZaAAAAYAAJ

16. Yourdon, E.: Modern Structured Analysis, 2nd edn. Prentice Hall PTR, Upper Saddle River (2000)

DWARM: An Ontology of Data Warehouse Architecture Reference Model

Piotr Szwed[✉], Wojciech Komnata, and Dariusz Dymek

AGH University of Science and Technology, University in Kraków, Poland
{wkomnata,pszwed}@agh.edu.pl, eidymek@kinga.cyf-kr.edu.pl

Abstract. This paper describes DWARM, an ontology formalizing a new data warehouse architecture reference model intended do capture common five architectural approaches, as well as to provide means for describing complex hybrid architectures that emerge due to observed evolution of business and technology. The ontology defines concepts, e.g. layers, processes, containers and property classes, as well as relations that can be used to construct precise architectural models. Formalization of architecture description as an ontology gives an opportunity to perform automatic or semiautomatic validation and assessment.

Keywords: Data warehousing · Data warehouse architecture · Ontology · Architecture evaluation

1 Introduction

Mature data warehousing technologies and solutions have been present on the market for over 20 years. They are developed within companies to support various analytical functions, reporting and provide accurate data for decision systems. A variety of applications, business needs, technical reasons or requirements related to performance, scalability, modifiability causes that the problem of selection of data warehouse architecture attracts many discussions and debates.

The literature reports five dominant architectural approaches: *Independent Data Marts, Data Mart Bus* architecture, *Hub-and-spoke, Centralized* and *Federated* [3,2]. Empirical studies of the field, e.g. [1] show that the most popular are Centralized and Hub-and-spoke recommended by Bill Inmon [9] and Data Mart Bus proposed by Ralph Kimball [12].

In many practical situations a particular architecture of a data warehouse is developed as a result of evolutionary process stimulated by business needs emerging during the system lifetime. This often leads to hybrid solutions or promotes federated architectures. In particular, the latter can be applied in such situations as company mergers or acquisitions, when it is usually more efficient to integrate legacy systems with an additional middleware layer providing schema mapping and reconciliation [4], than to build a completely new infrastructure.

In [5,6] we have proposed DWARM, a data warehosue architecture reference model that was intended to unify known architectural styles. The model distinguishes eleven layers comprising processes loading and transforming data and

© Springer International Publishing Switzerland 2015
S. Kozielski et al. (Eds.): BDAS 2015, CCIS 521, pp. 222–232, 2015.
DOI: 10.1007/978-3-319-18422-7_20

containers (data stores, source systems and consumer applications). Its layers are the following: (1) Source, (2) Extraction, (3) Temporary Staging Area (TSA), (4) Loading, (5) Central Data Repository, (6) Data Mart Feeding, (7) Data Mart, (8) Data Integration, (9) Federated Data Repository, (10) Data Delivery, and (11) Consumer Applications.

The reference model can be used to express known architectures by tailoring (removing whole layers or individual components) and multiplying instances of selected layers. Also hybrid architectures combining basic approaches can be expressed with DWARM, provided they respect some consistency rules. A number of examples can be found in [5].

The language used to specify the reference model follows the concepts of Structured Analysis [21]. It distinguishes processes, data stores, etc. It is expressive enough to give an overall view of components structure. However, we found that for more complex tasks, as automatic architecture validation or assessment, a more formal and rigorous description is required. This was the motivation for developing the DWARM ontology, which provides a taxonomy of processes and containers, assigns them to appropriate layers, defines feasible connections between components, as well as additional properties that can be used for the architecture evaluation purposes.

The paper is organized as follows: section 2 discusses related works, it is followed by section 3 specifying ontology goals. Next section 4 presents the ontology content. A few partial examples of models are given in section 5, finally section 6 provides concluding remarks.

2   Related Works

An application of ontologies to provide a systematic and formal description of software architectures was first proposed by Kruchten in [13]. The ontology distinguished several types of decisions that can be applied to software architecture and its development process. Main categories included: Existence, Ban, Property and Executive decisions. The ontology defined also attributes, which were used to describe decisions, including states (Idea, Tentative, Decided, Rejected, etc.). In [7] an ontology supporting Architecture Tradeoff Analysis Method (ATAM) based evaluation was proposed. The ontology specified concepts covering the ATAM [10] model of architecture, quality attributes, architectural styles and decisions, as well as influence relations between elements of architectural style and quality attributes. The effort to structure the knowledge about architectural decisions, was accompanied by works aimed at a development of tools enabling the edition and graphical visualization of design decisions, often in a collaborative mode, e.g. [14].

Such tasks as software architecture assessment face the problem of gathering information related both to overall design, and the detailed architectural decisions made [20]. An ontology supporting ATAM based assessment of systems following SOA paradigm was proposed in [18,19]. Its main goal was to define types of components and their attributes, with an intent of using them while representing architecture instances as ontological models. The obtained rich descriptions that can be then used for reasoning about quality attributes. An example of security risk evaluation, that reuses parts of formal architecture description can be found in [17].

The problem of data warehouse assessment was widely discussed [3,2] and even several guidelines of dedicated questionnaires can be found [8]. Surprisingly, up to our best knowledge there are no references to direct application of ontologies to describe architectural models and design decisions. A number of papers related to application of ontologies in a data warehouse design process can be found, e.g. [15,11]. They are focused, however, on important, yet quite different problem of building data models based on ontologies and semantic integration within data warehouses.

3   Ontology Goals

The main goal of the DWARM (*Data Warehouse Architecture Reference Model*) ontology is to provide a formal description of data warehouse architectures comprising concepts and relations used to define them, as well as constraints that should be satisfied by the components. It was also assumed that a concrete architecture instance could be defined using the ontology elements. This allows to apply common Semantic Web tools and methods, e.g. reasoners, to evaluate consistency of models and compliance with restrictions and data warehouse design guidelines.

Ontologies are often described as unions of two layers: terminological (*TBox*) and assertional (*ABox*). The *TBox* defines concepts and types of relation including: taxonomic relations between concepts, *object properties* and *datatype properties*. The *ABox*, in turn, gathers facts about individuals and existent relations.

While designing the DWARM ontology we assumed that the reference model of data warehouse will be reflected as the TBox layer. It comprises two groups of entities: *core components*, which correspond to elements defined in architectural views and classes defining *component properties*. Their individuals, which can be treated as constants, are also defined in the ontology. A concrete architecture instance, including processes, data stores, source systems and analytic tools, together with assigned properties, is to be defined as an ABox layer (c.f. Fig. 1).

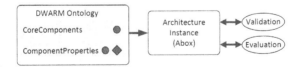

Fig. 1. A concept of DWARM ontology application

An intended application of DWARM ontology, apart from providing precise description language, is to use it within a software tool allowing to document,

validate and assess data warehouse architecture. The ontology was populated
with classes, relations and individuals identified based on literature review summarized in [5]. The ontology is formalized in OWL language. At present, it
defines about 100 classes and 40 relations.

4 Ontology Content

The ontology is built around a foundational model that specifies core classes and
relations. It can be treated as a metamodel of the language used to describe the
reference architecture. The class diagram in Fig. 2 depicts an ontology skeleton that was further extended by subclassing main classes: *Containers*, *Process*,
Data, etc. They are described in the rest of this section.

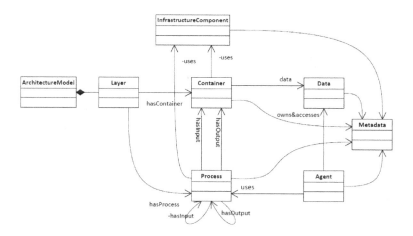

Fig. 2. Top-level ontology classes (CoreCompnents)

4.1 Containers

Class *Container* is an abstraction of architecture component capable of storing,
sending or receiving data. The ontology distinguishes 6 classes of containers:
SourceSystem, *TemporaryStagingArea*: temporary data for ETL processes, *CentralDataRepository*, *DataMart*, *FederatedDataRepository* and *ConsumerApplication*. Containers can be assigned with such properties as stored data, subject,
type (*RelationalDataBase*, *NOSQLDataBase*, *PlainFile*, *RecordOrientedStorage*,
XML, *VirtualContainer*), database engine, interface type (*SQL*, *WebService*,
Propriety), implementation technology and metadata.

4.2   Processes

Process represents an activity consisting in data transformation and transfer between containers. The DWARM ontology defines 15 classes of process. Basic relations that can be used to connect them with containers are: *hasInput* and *hasOutput*. Restrictions for valid architecture models are similar to those, defined for Structured Analysis [21]: processes without inputs (data generators) or outputs (black holes) are not allowed.

Processes can be described using additional properties defined in the ontology: metadata, language, execution engine, trigger and data read policy. For the later, appropriate dictionaries (classes and individuals) are defined: *ProcessTrigger* (*Time, DataArrival, Manual, Process*) and *DataReadPolicy* (*Incremental, FullRefresh*). Information on defined processes is summarized in table 1.

Table 1. Processes

Process	Description
Extraction	A component of *L02-ExtractionLayer*. Reads data from a source system and stores in the *TemporaryStagingArea* container.
Transformation	Belongs to the layer *L03-TemporaryStagingAreaLayer*. Responsible for data transformation within temporary staging area.
LoadingFromTSA	Part of *L04-LoadingLayer*. Used to model data loading. In an architecture model its subclasses should be used: *LoadingFromTSAToCDR* (loading to a central repository) or LoadingFromTSAToDM (loading to a data mart).
ETL.	The process does not belong to any of layers. It was introduced to preserve the terminological convention. The process is a composition of extraction, transformation and loading processes.
CDRTransformation.	Data transformation within a central repository, e.g. calculation of aggregated values. Constitutes an element of the layer *L05-CentralDataRepositoryLayer*.
DMFeeding.	Part of *L06-DataMartFeedingLayer*. The process feeds data marts from a central repository.
DMTransformation.	Data transformation for data marts, e.g. a process calculating totals for various data dimensions. Belongs to *L07-DataMartLayer*.
Integration.	Process belonging to the layer *L08-DataIntegrationLayer*. It is allowed to read data either from a central repository or a data mart, then it delivers it to a container *FederatedDataRepository*.
FDRTransformation	A process of data transformation undergoing in federated data repository (the layer *L09-FederatedDataRepositoryLayer*).
DataDelivery.	An abstract superclass of three classes of processes intended to be used: *DMDataDelivery, CDRDataDelivery* and *FDRDataDelivery*. They provide data for applications of the top layer reading them from correspondingly: data marts, a central repository of a federated repository. All processes belong to *L10-DataDeliveryLayer*.

4.3   Layers

Following the data warehouse reference model [5], the ontology defines 11 classes of layers. To increase readability their numbers were included into names. Layers are abstract elements of the model, not having tangible counterpart in an implemented software. They constitute an integrating and ordering element. An individual representing an architecture model is linked with layer instances. They, in turn, are linked by *containsContainer* and *containsProcess* with container and process instances (Fig. 3).

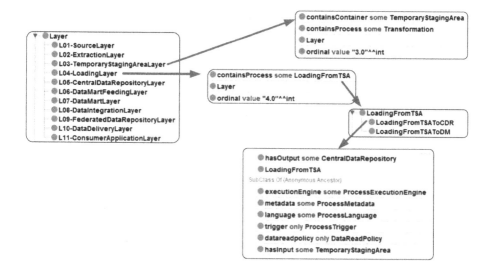

Fig. 3. Layers and their definitions

The ontology content can be browsed staring from layer classes using a software for ontology modeling and visualization, e.g. Protégé tool. Fig. 3 shows a typical structure of links within the DWARM ontology. The layer *L04-LoadingLayer* contains the abstract process *LoadingFromTSA*. One of its subclasses *LoadingFromT-SAToCDR* describes a concrete process, whose individual may appear in a concrete data warehouse architecture instance. The figure shows properties defined at the class level (output to *CentralDataRepository*) and properties inherited from super-classes.

4.4   Data

The class *Data* is intended to model data types, which can be assigned to containers. Data cane be characterized by such properties, as: *model* (relational, dimensional, object-oriented), *owner, materialization, type* (*primary* or *secondary*) and *representation* (record, file, XML, etc.)

The ontology defines three object properties for the *Data* class. The relation *subclass* corresponds to inheritance, whereas *references* to association. The third introduced relation: *originates* is intended to model dependencies between source data and data resulting from their transformation (e.g. filtering, joining). Hence, it allows to trace, from where the data originate, as well as to check, whether the process topology corresponds to data dependencies.

4.5   Metadata

The ontology specifies 25 classes of metadata arranging them into taxonomies corresponding to three commonly used classifications [16]: business and technical, front-end and back-end, as well as functional: *InfrastructureMetadata, ModelMetadata, ProcessMetadata, QualityMetadata, ProcessMetadata* and *Administration-Metadata*. To support classification the ontology defines three classes of metadata properties: *MetadataUsage, MetadataApplicationArea, MetadataFunction*. They come along with predefined individuals that can be used as constants in `owl:hasValue` restrictions, e.g. *QualityMetadata = Metadata ⊓ (hasFunction DataAssessmentAndMetrics)*.

4.6 Component Properties

Component properties are, beside core components, the second group of classes in DWARM ontology. Their taxonomy corresponds to basic types of components, hence, the top level properties are: *ContainerProperty, ProcessProperty, DataProperty, MetadataProperty* and *AgentProperty*.

In many cases property classes have predefined individuals that enumerate possible architectural decisions. A designer documenting the system architecture for a particular component may select one or a few dictionary values, e.g. for a *Data* object they can choose from the following *DataRepresentation* individuals: {*File, Record, Structured, Unstructured, XML*}.

The classes of properties are treated as a part of ontology that is intended for future extensions. Documentation of architecture of a particular system may require using new classes of design decisions, which are difficult to establish at the stage of ontology formalization. Based on an experience related to assessment of software architectures with ATAM method [20] it can be concluded that a certain freedom in describing design decisions and a capability to introduce new traits of components, e.g. referencing particular technologies, may be a key point while collecting information required to perform architecture evaluation.

5   Examples of Models

Fig. 4 depicts an excerpt from an ontology describing an example architecture of a data warehouse for a fictitious insurance company. The company activities concentrate within two areas. The first is selling insurances in branches and by mobile agents. The second is claim handling with a central system allowing to deposit notifications by clients, as well as agents.

The diagram was prepared following the convention of OntoGraph plugin for Protégé ontology editor. Elements symbolizing classes are marked with circles, whereas individuals (actual architecture elements) with diamonds. Dashed line arrows are used to show *has individual* relations (target elements are individuals belonging to the classes placed at arrows starts), whereas continuous line is used for *hasInput* and *hasOutput* object properties. It should be noted, that arrows reflect directions of relations and not data flows.

The class *SourceSystem* has three individuals: *Branches, Claims* and *MobileAgents*. The system comprises a single instance of temporary data container *TSA*, belonging to the class *TemporaryStagingArea*. It is fed by separate processes of *Extraction* type: *BranchesExtr*, *ClaimsExtr* and *MobiAgExtr*. For greater clarity, data transformation processes were omitted. The next architecture element is *Loading*, an instance of loading process belonging to the class *LoadingFromTSAToCDR*. It is linked by *hasInput* relation with *TSA* and by *hasOutput* with the central data repository *CDR*. The system architecture includes two data marts: *SalesDM* and *ClaimsDM*. They are fed by two processes of *DMFeeding* type: correspondingly *SalesFeed* and *ClaimsFeed*. The consumer application layer comprises three elements: *SalesReporting*, *ClaimsReporting* and *NewProductsBenchmarking*. It is fed by processes belonging to classes *DMDataDelivery* or *CDRDataDelivery*. Processes *SalesOLAP* and *ClaimsOLAP* are instances of *DMDataDelivery*. Their inputs are data marts *SalesDM* and *ClaimsDM*, whereas their outputs are analytic applications for sales and claim handling. The tool *NewProductsBenchmarking* requires aggregated information related both to sales and claims, hence it is fed directly from the central repository *CDR* by the process *CrossChk* (member of *CDRDataDelivery*).

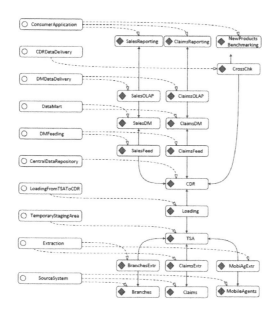

Fig. 4. An excerpt of data warehouse architecture model in a form of ontology

Fig. 5 shows another architectural view comprising containers and the data, which they store. The relation *originates* is used to capture data dependencies, e.g. data related to *AgentActivity* stored in the *CDR* has its origins in *AgentClaimHandling* and *AgentPolicySales* stored in the *MobileAgents* source system.

Formal description as an ontology of both data warehouse architecture and the reference model gives an opportunity to validate their compliance. The ontology defines a number of axioms (total 403) defining constraints using universal,

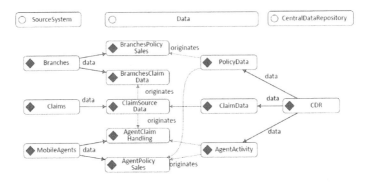

Fig. 5. A partial model of containers and data

existential or cardinality restrictions, eg. process *CDRDataDelivery hasInput exactly 1 CentralDataRepository and hasOutput only ConsumerApplication*. The validation can be achieved with dedicated software tools implementing procedures for consistency checking, or general purpose ontology reasoners, e.g. Pellet or FaCT++. In consequence, a data warehouse designer can be warned about potential flaws in architecture design.

Fig. 6a shows an architecture model, in which the *Delivery* process reads data both from the central repository *CDR* and the data mart *DM*. Such configuration, although possible to implement, is not recognized as valid in the DWARM reference architecture. Implementation of a process reading data from two dependent sources may potentially cause problems with data freshness, data access privileges and data model interpretation. According to formally defined constraints reflecting good practices in data warehouse design, a delivery process should read data from a single container. This is reflected by restrictions on process classes: *DMDataDelivery*, *CDRDataDelivery* and *FDRDataDelivery*, which are to be linked by *hasInput* relation only with containers of type *DataMart*, *CentralDataRepository* and *FederatedDataRepository* respectively.

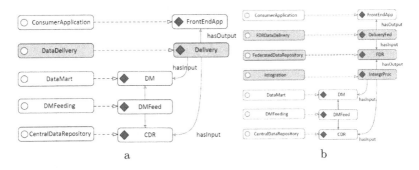

Fig. 6. Examples of models: (a) not compliant with the reference architecture, (b) refactored to the compliant form

The reference architecture model allows for a situation, when it is necessary to deliver data from various repositories. This can be achieved with processes belonging to a data integration layer accompanied by a federated data repository, either in materialized or not materialized form. Fig. 6b presents a refactored architecture complying the reference model. Components marked in Fig. 6 with gray color were replaced by the following elements: *IntegrProc* (an integration process), *FDR* (federated data repository) and *DeliveryFed* (process delivering data from *FDR*).

6   Conclusions

The described here DWARM ontology constitute a formal specification of layered model of data warehouse architecture including basic components: processes, containers, data and metadata. The formal character of the OWL language used to encode the ontology, allows to define constraints, that should be respected within the architectures complying the reference model. This regards both occurrences of elements of particular types, as well as their connections. An advantage of using ontology based architecture models is the possibility to process and analyze them automatically. The ontology was designed to allow future extensions. It is possible to define new component properties and enrich the model wit additional information, e.g. related to performance (number of records, data size, frequency of process executions).

References

1. Alsqour, M., Matouk, K., Owoc, M.L.: A survey of data warehouse architectures - preliminary results. In: 2012 Federated Conference on Computer Science and Information Systems (FedCSIS), pp. 1121–1126. IEEE (2012)
2. Ariyachandra, T., Watson, H.: Key organizational factors in data warehouse architecture selection. Decision Support Systems 49(2), 200–212 (2010)
3. Ariyachandra, T., Watson, H.J.: Which data warehouse architecture is most successful? Business Intelligence Journal 11(1), 4 (2006)
4. Chromiak, M., Stencel, K.: A data model for heterogeneous data integration architecture. In: Proceedings of Beyond Databases, Architectures, and Structures - 10th International Conference, BDAS 2014, Ustron, Poland, May 27-30, pp. 547–556 (2014), http://dx.doi.org/10.1007/978-3-319-06932-6_53
5. Dymek, D., Komnata, W., Kotulski, L., Szwed, P.: Data Warehouse Architectures. Reference model and formal architecture description. In: Polish: Architektury Hurtowni Danych. Model referencyjny i formalny opis architektury. AGH University of Science and Technology Press (2015)
6. Dymek, D., Komnata, W., Szwed, P.: Proposal of a new data warehouse architecture reference model. In: Kozielski, S., Mrozek, D., Kasprowski, P., Malysiak-Mrozek, B., Kostrzewa, D. (eds.) BDAS 2015. CCIS, vol. 521, pp. 210–221. Springer, Heidelberg (2015)
7. Erfanian, A., Aliee, F.S.: An ontology-driven software architecture evaluation method. In: Proceedings of the 3rd International Workshop on Sharing and Reusing Architectural Knowledge, SHARK 2008, pp. 79–86. ACM, New York (2008)

8. Habers, F.: The DW maturity assessment questionnaire. Technical Report UU-CS-2010-021, Institute of Information and Computing Sciences, Utrecht University, 3508 TC, Utrecht, The Netherlands (September 2010), http://computerscience.nl/research/techreps/repo/CS-2010/2010-021.pdf

9. Inmon, W.: Building the Data Warehouse. Wiley (2005), http://books.google.pl/books?id=QFKTmh5IFS4C

10. Kazman, R., Klein, M., Clements, P.: ATAM: Method for architecture evaluation. CMUSEI 4, 83 (2000)

11. Khouri, S., Ladjel, B.: A methodology and tool for conceptual designing a data warehouse from ontology-based sources. In: Proceedings of the ACM 13th International Workshop on Data Warehousing and OLAP, pp. 19–24. ACM (2010)

12. Kimball, R., Ross, M.: The Data Warehouse Toolkit: The Definitive Guide to Dimensional Modeling. Wiley (2013), http://books.google.pl/books?id=WMEqTf2lK84C

13. Kruchten, P.: An ontology of architectural design decisions in software intensive systems, pp. 54–61. Citeseer (2004)

14. Lee, L., Kruchten, P.: Visualizing Software Architectural Design Decisions. In: Morrison, R., Balasubramaniam, D., Falkner, K. (eds.) ECSA 2008. LNCS, vol. 5292, pp. 359–362. Springer, Heidelberg (2008)

15. Nebot, V., Berlanga, R., Pérez, J.M., Aramburu, M.J., Pedersen, T.B.: Multidimensional integrated ontologies: A framework for designing semantic data warehouses. In: Spaccapietra, S., Zimányi, E., Song, I.-Y. (eds.) Journal on Data Semantics XIII. LNCS, vol. 5530, pp. 1–36. Springer, Heidelberg (2009), http://dx.doi.org/10.1007/978-3-642-03098-7_1

16. Shankaranarayanan, G., Even, A.: The metadata enigma. Communications of the ACM 49(2), 88–94 (2006), http://doi.acm.org/10.1145/1113034.1113035

17. Szwed, P., Skrzynski, P.: A new lightweight method for security risk assessment based on Fuzzy Cognitive Maps. Applied Mathematics and Computer Science 24(1), 213–225 (2014), http://dx.doi.org/10.2478/amcs-2014-0016

18. Szwed, P., Skrzynski, P., Rogus, G., Werewka, J.: Ontology of architectural decisions supporting ATAM based assessment of SOA architectures. In: Proceedings of the 2013 Federated Conference on Computer Science and Information Systems, Kraków, Poland, September 8-11, pp. 287–290 (2013)

19. Szwed, P., Skrzynski, P., Rogus, G., Werewka, J.: SOAROAD: An ontology of architectural decisions supporting assessment of service oriented architectures. Informatica (Slovenia) 38(1), 31–42 (2014), http://www.informatica.si/PDF/38-1/05_Szwed%20-%20SOAROAD%20-%20An%20Ontology%20of%20Architectural%20Decisions%20Supporting%20Assessment%20of%20Service%20Oriented%20Architectures.pdf

20. Szwed, P., Wojnicki, I., Ernst, S., Głowacz, A.: Application of new ATAM tools to evaluation of the dynamic map architecture. In: Dziech, A., Czyżewski, A. (eds.) MCSS 2013. CCIS, vol. 368, pp. 248–261. Springer, Heidelberg (2013), http://dx.doi.org/10.1007/978-3-642-38559-9_22

21. Yourdon, E.: Modern Structured Analysis, 2nd edn. Prentice Hall PTR, Upper Saddle River (2000)

Ontologies and Semantic Web

Ontology Learning from Relational Database: How to Label the Relationships Between Concepts?

Bouchra El Idrissi[✉], Salah Baïna, and Karim Baïna

ENSIAS, University Mohammed V Rabat,
BP 713, Rabat Morocco
bouchra.idrissi@um5s.net.ma, {sbaina,baina}@ensias.ma

Abstract. Developing ontology for modeling the universe of a Relational Database (RDB) is a key success for many RDB related domains, including semantic-query of RDB, Linked Data and semantic interoperability of information systems. However, the manual development of ontology is a tedious task, error-prone and requires much time. The research field of ontology learning aims to provide (semi-) automatic approaches for building ontology. However, one big challenge in the automatic transformation, is how to label the relationships between concepts. This challenge depends heavily on the correct extraction of the relationship types. In fact, the RDB model does not store the meaning of relationships between entities, it only indicates the existence of a link between them. This paper suggests a solution consisting of a meta-model for the semantic enrichment of the RDB model and of a classification of relationships. A case study shows the effectiveness of our approach.

Keywords: Ontology · Ontology Learning · Relational Database · Relationships Labeling · Relationships classification · Semantic Enrichment

1 Introduction

The success of many areas related to Relational Databases (RDB), including [13] ontology-based data access, semantic interoperability, mass generation of semantic web data, depends on the proliferation of ontologies (as *formal, explicit specification of shared conceptualization*[7]). However, ontology development was omitted from the life cycle of information systems. In addition, its manual development is difficult, time-consuming and error-prone and requires skills in the ontology formalization languages.

Ontology learning from RDB is a field of research that seeks to provide (semi-) automated approaches for building ontology (see [6] for a classification of ontology learning depending on the type of the input resource). We emphasize that our interest is primarily in the construction of the Terminology Box (TBox) (other term is the schema of the ontology). For the database content, a correlated process, called ontology population [10] is concerned with. Ontology learning is a prerequisite for the ontology population.

© Springer International Publishing Switzerland 2015
S. Kozielski et al. (Eds.): BDAS 2015, CCIS 521, pp. 235–244, 2015.
DOI: 10.1007/978-3-319-18422-7_21

Although the inference of the ontology concepts and data properties from the RDB model seems to be possible and acceptable, it is, however, difficult to believe how the semantics of relationships (the relations between entities of the RDB universe or mini-world) can be directly inferred from the model. In fact, the RDB model does not store the meaning of relationships, it only indicates the existence of a link between entities using foreign-keys. This challenge is based on two key issues: how to discover the type of a relationship?; and how to label this relationship?. By the type of the relationship, we mean its classification either into a well-known relationship pattern like inheritance and fragmentation; or it represents a domain-specific relation between entities from the RDB universe. In this paper, we will see how existing works are based on incorrect rules to extract the types of relationships. For the second issue, the relationships are either labeled with meaningful terms in an automatic process without indicating how it is done; either by the use of foreign-key name; or they are arbitrarily assigned. As a solution, the paper presents our approach that consists in a semantic enrichment process of the RDB model and in the use of a classification of relationships.

The remainder of the paper is structured as follows. The next section gives an overview of related works. The third section presents an use case that illustrates the problem statement. Section 4 describes our solution. In section 5, we present an application of our solution. A conclusion, in section 6, summarizes the whole paper and considers future work.

2 Related Works

In this section, we focus on how the existing approaches have addressed the two introduced issues: the discovery of the type of relationships and relationships labeling.

Discovery of Relationship Types: The majority has been limited to the extraction of taxonomies between concepts. Generally, the corresponding rule is when a primary-key of a relation A is also a foreign-key referencing another relation B, then A is *subclassOf* B (see as examples [2,9]). This is an incorrect rule as we will see in our case study. As an exception, we mention the work of Alalwan et al. [1] that have studied fragmentation (vertical partitioning), unary relationships and have restricted the inference rules by imposing conditions on the domain values of participant relations. However, in their work, the distinction between fragmentation and inheritance is based on the assumption that subtypes can present only partial coverage of the supertype. As stated in [15], subtypes can also be defined as a total coverage of the supertype. In most existing approaches, the default rule is the transformation of each foreign-key to an object property.

Relationships Labeling: A relationship is either labeled with a meaningful term (a term that corresponds to the application semantic of the relationship) in an automatic process without indicating how it is done (see [8,1]); either by the use of the foreign-key name [2]; or they are assigned by the name of the association relation [17]. Squeda et al. [12] have defined a strategy based on the concatenation of the participant relations names and the referenced columns (all separated by a comma). Louhdi et al. in [9] have named the relationships by concatenating the name of the relation that owns the foreign-key with the term *has*. Except [12] and to the best of our knowledge, there is no formal strategy for labeling relationships.

In real-world databases, the names of the database relations, columns and constraints may be abbreviated or codified (see the database of SAP [11]). Thus, they cannot be useful for the labeling of the relationships. We showed in the following section how the existing works failed to answer the questions of the paper.

3   Use Case

The use case example in Fig. 1 shows a simplified physical model of a database for the management of an enterprise products and employees. The enterprise purchases and sales many types of products. Each product has a min and max price. In case of a new product to add to the enterprise catalog, an employee adds a record in the table *product*. The database stores information about employees and additional personal details. Each employee belongs to one department that is managed by one employee. Each department is part of a division. A type of the bought products is the personal computers that are granted to the engineers who are employees of the enterprise. Due to the lack of space, we are limited to the representation of primary and foreign keys in the physical model.

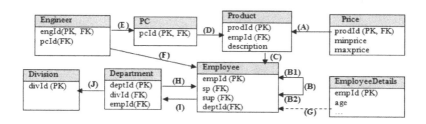

Fig. 1. A simplified physical model of a RDB

From Fig. 1, we can see that the relationship semantics cannot be directly deduced from the model. We reported in table 1 the application semantic of each relationship, the ontology constructs and the resulting labels from the application of some proposed rules from the literature. The inferred relationship (A) resulting in a taxonomy between *price* and *product* is wrong. The correctness of (D) depends on the intended use of the relationship *a-kind-of* (see a discussion in [14] page 462). (B1) and (B2) are inferred as object properties, but without knowing how to label and distinguish them. The implicit relationship (G) is not inferred. For the other cases (C, E, H, I, J), they are all inferred as object properties, but as we can see, the resulting labels failed to cover the semantics of relationships.

Having illustrated some issues for the inference of relationships meaning from a database model, we present in the next section our approach that aims to overcome these limits in order to improve the quality of resulting ontologies.

Table 1. Application Semantics in Comparison to Extracted Semantics

Relationship	Application Semantic	Ontology Constructs	Labeling result
(A)	concerns	$subClassOf[1](C[2]\ (price),\ C(product))$	NA[4]
(B)-(B1)	is the spouse of	$OP[3]\ (C(employee),\ C(employee))$	no rule
(B)-(B2)	is the superior of	$OP(C(employee),\ C(employee))$	no rule
(C)	is created by	$OP(C(product),C(employee))$	pr;em#...[12]
(D)	is a kind of	$subClassOf(C(pc),\ C(product))$	NA
(E)	is granted a	$OP(C(engineer),\ C(pc))$	EnHasPc[6][9]
(F)	is a	$subClassOf(C(engineer),\ C(employee))$	NA
(G)	fragment-of	link not inferred	NA
(H)	is managed by	$OP(C(department),C(employee))$	DeHasEm[9]
(I)	belongs to	$OP(C(employee),C(department))$	EmHasDe[9]
(J)	part-of	$OP(C(department),C(division))$	DeHasDi[9]

[1] $subClassOf(C1,\ C2)$ is the OWL construct for creating a taxonomy between the concepts $C1$ and $C2$.

[2] $C(X)$ is a predicate designing the OWL class of a concept X.

[3] $OP(X,Y)$ designates an *object property* with domain X and range Y.

[4] NA: Not applied.

[5,6] Due to the lack of space, we abbreviated the name of concepts.

4 Proposed Approach

In this section we describe our solution for relationships labeling. We give a short overview of our approach described in [3], before presenting the semantic enrichment model concerning the paper's purpose and the labeling strategy of relationships.

4.1 An Overview

Our ontology learning process is composed of three main phases, which are: (i) **Pre-processing phase**: its goal is to clean the database schema and data from irrelevant, erroneous and incorrect information that may falsify the learned ontology; (ii) **Semantic enrichment phase**: the input RDB model is transformed into an XML document according to a meta-model (an XML schema). The meta-model offers a set of elements for the semantic enrichment of the relational data model. This enrichment can be done manually or automatically through other resources (e.g., mining the semantic of relationships by data mining techniques and augmenting the input database model with the discovered semantics. Then, the user can validate the enriched model before the transformation into OWL). Our enrichment meta-model is motivated and detailed in [4]; (iii) **Transformation into OWL 2**: it transforms the enriched database model to equivalent OWL 2 [19] constructs.

For the paper's objective, the input RDB should be normalized, in at least the third normal form and we consider the manual semantic enrichment.

4.2   The Enrichment Meta-Model

Our meta-model for the semantic enrichment of the database model is an extension of the DDLUtils model [16]. Fig. 2 presents the main elements related to the upgrade of the relationships semantics. The root element is *database* which comprises at least one element *table* and a table may have many foreign-keys. Each foreign-key has at least one reference. We have detailed and motivated, in

Fig. 2. The meta-model elements related to the RDB relationships

the following, the main elements and attributes used to model relationships:

- *ForeignKey*: represents a foreign-key in a table. Its attributes are: *foreignTableName* that identifies the referenced table and *name* is the name of the foreign-key column. A foreign-key may have more than one reference (the case of a composite foreign-key). Instead of our model in [4], we defined the following attributes at the foreign-key and not at the element *reference*, in order to take into consideration the composite foreign-keys.
  - *relType*: it presents the type of the relationship among *RelationshipTypes*. A relationship type is a generic relationship that is not application-specific and it generally presents a pattern of relationships design. Some types of relationships allow the automatic deduction of logical features like transitivity, reflexivity or symmetry (e.g., transitivity holds between *is-a* and *part-of* relationships [14,5]). This attribute may be used for naming the ontology relationships as we will see in the next subsection.
  - *relName*: the domain name of the relationship. When it is difficult to classify a relationship in a type (from *RelationshipTypes*), this term indicates clearly its application-specific meaning.
  - *inverseRel*: unlike most approaches that infer unidirectional relationships (from the relation owner of the foreign-key to the targeted one), this attribute, when fixed, indicates that the inverse relationship (from the targeted relation to the owner of the foreign-key) should be considered in the ontology generation process. Its value represents the name of the

inverse relationship. It can be useful, for example, when it is necessary to query some tables in the inverse direction (e.g., get the min, max price of a product from our case study).

– **Reference**: represents a reference between a column in the local table (that holds the foreign key) and a column in the referenced table. They are useful in the sense that they allow to infer some constraints on relationships. For example, a column with a *not null* constraint indicates a *mincardinality* 1.

– **Relationship Types**: semantic relationships classification has been the purpose of some research like [18,14]. At this step of our work we do not want to go in-depth in this classification, we only retain some types to illustrate the purpose of our work. An initial list is: *has, is-a, part-of* or *component-of, fragment-of.*

We note that in order to clarify the semantics of relationships, the terms used to label concepts, which are the basic elements of an ontology, should also be clear and meaningful. For this, our meta-model defines the attribute *domain name* (dn), which plays an important role in our relationships labeling strategy, as we will see in the next section.

4.3   The Labeling Strategy

In order to publish an ontology, an absolute *Internationalized Resource Identifier (IRI)* [19] should be defined. Each element of the ontology (class or property) is identified by the concatenation of a base *IRI* (for example http://www.fao.org/ontologyName, where *ontologyName* is the name of the ontology, and *fao* is the organization that publishes the ontology), a separator character (generally a #) and the name of the ontology element. The name of each ontology element should be unique within the ontology.

Our strategy for naming object properties is schematized in Fig. 3. It is based on the following principles (*OP* an object property):

– The priority is given to the property *relName*. When this attribute is assigned, the *OP* is named with it. As already stated, we envisage the assignment of this attribute, when the relationship cannot be classified in one of the relationship types.

– If the *relName* is not assigned, we look for the *relType* attribute. If it is not assigned, *op* is then named by the use of the default pattern, which forms the *relName* by concatenating the *dn* of the source table, the word *has* and the *dn* of the targeted table. The user can abbreviate the *dn* of relations in order not to have long names of *OP*.

– If the *relType* is not null, the transformation into OWL is done according to the type of the relationship. If it is an *is-a* a sub-class feature is created. If it is a *fragment-of* a fusion between tables in one concept is done. The default rule is the concatenation of the *dn* of the source table, the *relType* and the *dn* of the targeted table.

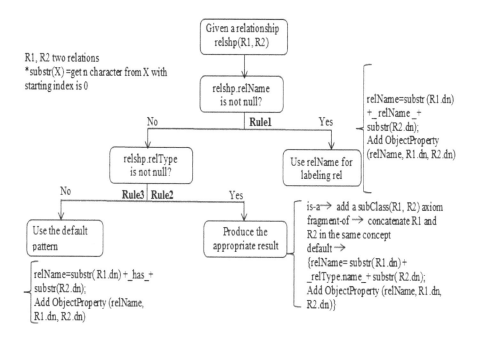

Fig. 3. The strategy followed for labeling relationships in our ontology learning process

In order to show the effectiveness of our approach, we present in the next section, an application of this strategy to the use case that we have given in section 3.

5   Case Study Application

In this section, we reported the result of applying our strategy for the use case from the section 3. First, we illustrate the semantic enrichment phase. Second, we compare the result of the transformation into OWL, based on our labeling strategy with the result shown in table 2.

5.1   The Semantic Enrichment Phase

Due to the lack of space, we shall restrict ourselves to some cases that illustrate the transformation into XML and the enrichment phase performed by a user (a domain expert). Fig. 4 gives some examples. In the transformation into XML, we only show the information about foreign-keys and references. We omitted the description of columns, primary-keys and so on. Line 3 concerns the relationship (C) in the use case. It is enriched by pointing out the meaning of the relationship through the attribute *relName*. For the line 8 that represents the relationship (A), the attribute *inverseRel* is determined in order to output the inverse relationship. We present

```
 1 <database>
 2 <table name="Product" >
 3   <foreign-key foreignTableName="Employee " name="emp_fk"  relName="createdBy" >
 4      <reference local-column="empId" foreign-column="employeeId"/>
 5   </foreign-key>
 6 </table>
 7 <table name="Price" >
 8   <foreign-key foreignTableName=" Product" name="pdt_fk" relName="concerns" inverseRel="has">
 9      <reference local-column="empId" foreign-column="employeeId"/>
10  </foreign-key>
11 </table>
12 <table name="Employee" >
13   <foreign-key foreignTableName="Employee " name="emp_fk1" >
14      <reference local-column="sp"  foreign-column="empId" />
15  </foreign-key>
16 </table>
17 <table name="Engineer " >
18   <foreign-key foreignTableName="Employee " name="emp_fk"  relType="is-a">
19      <reference local-column="engId"   foreign-column="empId"/>
20   </foreign-key>
21 </table>
22</database>
```

Fig. 4. An snippet from an enriched XML document (enriched information are in bold)

on line 13 the relationship (B1), but in this case without any enrichment in order to illustrate the rule 3 from our labeling strategy. Line 18 for the relationship (F) displays a case of the enrichment with the *relType* attribute. In the same way, we can treat the relationships (D, J) by fixing the value of the attribute *relType* respectively at *kind-Of* and *part-Of*. The relationship (G) is handled by adding a virtual foreign-key in the resulting XML document and the setting of the attribute *relType* value to *fragment-of*. For all other relationships (B2, E, H, I) they are enriched by positioning the value of the attribute *relName* in the same manner as 3.

5.2   The Transformation to OWL

In table 2, we outline for each relationship, the resulting OWL constructs and the applied rule of our strategy. For the relationships (B2, C, E, H, I, J), their semantics are now clear through the labels defined in the enrichment phase. For (G), the two classes *EmployeeDetails* and *Employee* are unified in one concept *Employee*. The operation of unification consists of adding all the attributes (non foreign-keys and primary-keys) of *EmployeeDetails* to the concept *Employee*. Unlike the result in table 1, (D) had not resulted in a taxonomy between participant relations. We have complied with the user requirements to model the relationship as a simple binary relation identified by the label *kindOf*. We note also the inference of the inverse relationship of (A), resulting in a binary relation of domain *Product* and range *Price*. For (B1), no information has been added in the semantic enrichment phase. So, the rule that had been applied in this case is the default rule (rule3). However, the label *emp_has_emp* is generic and does not import any semantic. In this case, it was better to set, in the XML document, the *relName* attribute of the relationship.

As we can see, our approach improves the quality of the resulting ontology as it makes clear the semantics of the relationships between concepts. In addition, our approach allows making explicit some hidden and implicit semantics in the input database model (e.g., the relationship (G) in the use case). It also allows the inference of inverse relationships. One of the limits of our approach is the use of the generic relationship *has* in the default rule, which does not appear to be adequate in all cases. This can be avoided by setting the *relType* or *relName* in all foreign-key elements of the XML document.

Table 2. Resulting Semantics from our Approach

Relationship	Resulting OWL Construct	Strategy Rule
(A)	$OP^1(pri\_concerns\_pro, C(price), C(product)) \wedge$ $OP(pro\_has\_pri, C(product), C(price)) \wedge OP\ inverseOf\ OP$	Rule1
(B)-(B1)	$OP(emp\_has\_emp, C(employee), C(employee))$	Rule3
(B)-(B2)	$OP(emp\_spouseOf\_emp, C(employee), C(employee))$	Rule1
(C)	$OP(pro\_createdBy\_emp, C(product), C(employee))$	Rule1
(D)	$OP(pc\_kindOf\_pro, C(PC), C(product))$	Rule1
(E)	$OP(eng\_granted\_pc, C(engineer), C(PC))$	Rule1
(F)	$subClassOf (C(engineer), C(employee))$	Rule2
(G)	$C^2(employeeDetails, employee)$	Rule2
(H)	$OP(dep\_managedBy\_emp, C(department), C(employee))$	Rule1
(I)	$OP(emp\_belongsTo\_dep, C(employee), C(department))$	Rule1
(J)	$OP(dep\_partOf\_div, C(department), C(division))$	Rule1

[1] $OP(N, X, Y)$ designates an OWL *object property* with name N, domain X and range Y.

[2] $C(X, Y)$ designates a concept resulting from the unification of $C(X)$ and $C(Y)$.

6   Conclusion

Automatic approaches for learning ontology from an RDB model fail in discovering the correct semantics of relationships between concepts, due to the fact that these semantics may be implicit, not clear or absent in the RDB model. In this paper, we have proposed an approach based on a semantic enrichment phase to augment the RDB model semantics by the meaning of relationships. Our approach allows to correctly deduce the relationship types and to recover the application-specific meaning of relationships.

We plan to improve this work in different ways. First, we can study how to augment the ontology by logical features (e.g., transitivity, reflexivity, symmetry) either manually by setting them in the *ForeignKey* element or through the relationship types (as we have already discussed). Second, it would be a laborious task for users to perform the manual enrichment of all the RDB foreign-keys, especially for large-scale databases with complex relationships. So, we have to think about a collaborative tool. Finally, this work can be improved by the advances in semantic relationships classification in order to enhance the semantic interoperability of information systems.

References

1. Alalwan, N., Zedan, H., Siewe, F.: Generating owl ontology for database integration. In: Third International Conference on Advances in Semantic Processing, SEMAPRO 2009, pp. 22–31. IEEE (2009)
2. Astrova, I., Korda, N., Kalja, A.: Rule-based transformation of sql relational databases to owl ontologies. In: Proceedings of the 2nd International Conference on Metadata & Semantics Research. Citeseer (2007)
3. El Idrissi, B., Baïna, S., Baïna, K.: A methodology to prepare real-world and large databases to ontology learning. In: Enterprise Interoperability VI, pp. 175–185. Springer (2014)
4. El Idrissi, B., Baïna, S., Baïna, K.: Upgrading the semantics of the relational model for rich owl 2 ontology learning. Journal of Theoretical and Applied Information Technology 68(1) (2014)
5. Evens, M.W.: Relational models of the lexicon: Representing knowledge in semantic networks. Cambridge University Press (2009)
6. Gómez-Pérez, A., Manzano-Macho, D., et al.: A survey of ontology learning methods and techniques. OntoWeb Deliverable D 1, 5 (2003)
7. Gruber, T.R., et al.: A translation approach to portable ontology specifications. Knowledge Acquisition 5(2), 199–220 (1993)
8. Ling, H., Zhou, S.: Mapping relational databases into owl ontology. International Journal of Engineering and Technology (IJET) 5(6), 4735–4740 (2013)
9. Louhdi, M.R.C., Behja, H., El Alaoui, S.O.: Transformation rules for building owl ontologies from relational databases. In: 2nd ICAITA, pp. 271–283 (2013)
10. Petasis, G., Karkaletsis, V., Paliouras, G., Krithara, A., Zavitsanos, E.: Ontology population and enrichment: State of the art. In: Paliouras, G., Spyropoulos, C.D., Tsatsaronis, G. (eds.) Multimedia Information Extraction. LNCS (LNAI), vol. 6050, pp. 134–166. Springer, Heidelberg (2011)
11. SAP ERP Solutions, www.sap.com
12. Sequeda, J.F., Arenas, M., Miranker, D.P.: On directly mapping relational databases to rdf and owl (extended version). arXiv preprint arXiv:1202.3667 (2012)
13. Spanos, D.E., Stavrou, P., Mitrou, N.: Bringing relational databases into the semantic web: A survey. Semantic Web 3(2), 169–209 (2012)
14. Storey, V.C.: Understanding semantic relationships. The VLDB Journal 2(4), 455–488 (1993)
15. Teorey, T.J., Lightstone, S.S., Nadeau, T., Jagadish, H.V.: Database modeling and design: logical design. Elsevier (2011)
16. The Apache DB Project DdlUtils, https://db.apache.org/ddlutils/
17. Tirmizi, S.H., Sequeda, J., Miranker, D.P.: Translating SQL applications to the semantic web. In: Bhowmick, S.S., Küng, J., Wagner, R. (eds.) DEXA 2008. LNCS, vol. 5181, pp. 450–464. Springer, Heidelberg (2008)
18. Ullrich, H., Purao, S., Storey, V.C.: An ontology for classifying the semantics of relationships in database design. In: Bouzeghoub, M., Kedad, Z., Métais, E. (eds.) NLDB 2000. LNCS, vol. 1959, pp. 91–102. Springer, Heidelberg (2001)
19. W3C: Owl 2 web ontology language structural specification and functional-style syntax (second edition). (December 2012), http://www.w3.org/TR/owl2-syntax/

Integration of Facebook Online Social Network User Profiles into a Knowledgebase

Wojciech Kijas[1] and Michał Kozielski[2](✉)

[1] Institute of Informatics, Silesian University of Technology,
Akademicka 16, 44-100 Gliwice, Poland
wojciech.kijas@gmail.com
[2] Institute of Electronics, Silesian University of Technology,
Akademicka 16, 44-100 Gliwice, Poland
michal.kozielski@polsl.pl
http://adaa.polsl.pl

Abstract. The article describes attempts made to integrate variety of user's data available on Facebook online social network. The source of the data are user profiles, publicly available for other Facebook users, which contain data such as visited places, favorite sport teams, TV programs, watched movies, read books and other *likes*. The destination of integrated data is FOAF ontology adopted for integration purposes. The work presents the required FOAF ontology extensions and an approach to Facebook data extraction as a contribution. Also the query and reasoning examples on the created knowledgebase are presented.

Keywords: Online Social Network · Semantic Web · FOAF ontology · Social Network Analysis · Linked Open Data

1 Introduction

During recent years it can be observed that our everyday live is strongly influenced by all kinds of Online Social Networks. A few years ago, only information about known people (politicians, actors, musicians) was available on the web. Now such a data is available for most of people, who have profile on one of the Online Social Networks (OSN). The most known and most widely used for now is Facebook [1] online social network. Facebook users by all their activity, friend's requests, sending messages and sharing *likes*, build a solid database of all the relations. It is not surprising that there is growing interest in the OSN data from various backgrounds.

This paper describes the results of research in the area of integration of Facebook user profiles data to create knowledgebase which can be used by other computer systems. The idea was to prepare ontology which can be used to model variety of data existing in Facebook user's profiles. Using such a model and some Facebook data extraction techniques, it would be possible to create knowledgebase, which can then be easily used by the external systems. In other words, the ontology, together with Facebook extraction tools, can play a role of intermediate layer between Facebook system and external systems, for which the data would be interesting for further analysis.

S. Kozielski et al. (Eds.): BDAS 2015, CCIS 521, pp. 245–255, 2015.
DOI: 10.1007/978-3-319-18422-7_22

The contribution of this work consists of FOAF ontology extensions allowing detailed modeling of user's preferences extracted from Facebook, together with Facebook Crawler - independent of Facebook API system, which allows extracting data from Facebook to the designed knowledgebase.

The structure of this work is as follows. Related work is presented in section 2. In section 3 we present motivation to use Semantic Web technologies. Section 4 summarizes types of data available on Facebook, presents details of ontology preparation and an approach to data extraction from Facebook to a knowledgebase. Examples of using the created knowledgebase are presented in section 5. Conclusions and possible future work are presented in section 6.

2   Related Work

In recent years, with the increasing popularity of new online initiatives such as Online Social Networks, blogs and other user-generated content systems, in natural way new trend showed up for user information modeling. In this new trend the fundamental question that arises is how to bring some benefits from exploring cross-system personalization approaches.

A vision of extracting data from social media websites to make it available following Linked Data principles was presented in [13]. However, extracting and modeling data from Facebook was not considered in this work.

Application of Facebook profile-based user preferences data to recommendation systems was presented in [16]. However, this approach also did not consider any common data model to keep extracted user's data for further analysis.

An example of the ontology application as destination data model was presented in [12]. Ontologies together with Semantic Web technologies become a widely accepted techniques for modeling users data in Social Web [5,9]. A nice survey of so far introduced user information aggregation approaches was presented in [15]. The authors of this work presented also generic model of most frequent user dimensions and attributes available in 17 social applications. However, they did not consider the topic of relationships between users. Other interesting study was presented in [14] where the authors combined user information modeling and profiling with utilizing DBPedia [11] open data. In the papers [19] and [18] the authors besides aggregating user profiles maintain also links between these profiles but they didn't consider any FOAF model extensions to distinguish different types of activities.

In the work [10] our first approach to the problem of parsing Facebook website for collecting data into FOAF (Friend of a Friend) [8] ontology which can be then used in corporate analysis was presented. It was based on a Facebook Crawler system which uses HTML parsing techniques for searching Facebook users for given user email and extracting links between Facebook user's profiles. In the current work we extend this system to extract also user's likes, to distinguish different user activities, introduce new concepts and properties to complement FOAF ontology and employ parsing JSON format to extract basic user data provided by Facebook.

3 Semantic Web Technologies

The idea to establish the standards for describing the content, where the data will exist along with its semantics was presented in [6]. In this way data would be processed not only by human (as in case of traditional web content), but also by other special applications which will be able to process the data according to its meaning. Presented vision required developing a set of standards to allow universal and flexible information storing, defining new concepts, their attributes, as well as new concepts and attributes based on existing ones.

Nowadays Semantic Web is a common initiative led by World Wide Web Consortium (W3C), which is an organization that deals with the establishment of standards for the creation of web content [4]. For now, the main standards that were introduced are: RDF (Resource Description Framework), RDFS (RDF Schema), OWL (Web Ontology Language), SPARQL (SPARQL Protocol And RDF Query Language).

To shortly describe these standards, it can be said that the RDF and RDFS provides abstract data model, OWL adds new vocabulary and advanced modeling capabilities for domain knowledge and ontology development, while SPARQL allows to query the Semantic Web Data.

Due to the nature of presented Semantic Web technologies we used them to solve the problem of Facebook data profiles integration taking advantage of all the facilities offered by the Semantic Web storage. Semantic Web storage is easily extensible and customizable without influence on the already stored data. It also offers SPARQL query language giving a full potential of reasoning over the data that can lead to easy discovery of new connections in a graph data. Moreover, ontologies are increasingly used in modern Web initiatives and the data can be easily linked with other ontologies.

One of the most popular Semantic Web ontology is The Friend of a Friend (FOAF) [8]. Within this project a Web of machine-readable pages describing people, the links between them and the things they create and do is created. The details of how the FOAF ontology was used and extended for this work are described in paragraph 4.2.

4 Integration of Facebook Data

4.1 Types of Data Available on Facebook

The Facebook online social network focuses on user entertainment data. In contrast to professional social networks Facebook maintains mainly entertainment-related data, connected with variety of users free time every day activities, which can be very interesting for corporate analysis while treating Facebook users as potential customers. The fact which stands out Facebook from other similar social networks is categorizing user *likes*. All categories of data which were taken into account are listed in table 1, which also contains percent of analyzed profiles that have any data of the given category and the average number of items of the given category listed in the single profile. The calculations were done for 1396 randomly extracted Facebook profiles.

Table 1. Categories of user *likes* on Facebook

Category of data	Profiles with data filled	Average items count
Favorite sport teams	23.57%	4.05
Favorite sport athletes	34.31%	6.76
Favorite music	56.81%	24.98
Favorite movies	40.26%	10.68
Watched movies	1.43%	30.4
Favorite TV Shows	43.7%	7.18
Watched TV Shows	0.79%	21.73
Favorite books	31.52%	4.11
Read books	1.29%	10.44
Favorite restaurants	24.64%	2.53
Inspiring people	13.04%	1.95
Favorite sports	8.88%	2.02
Favorite clothing	30.01%	4.26
Favorite places	37.39%	6.86
Other likes	74.57%	89.5
User friends	100%	311.54

4.2 Preparing Ontology

The FOAF ontology is one of the most popular projects in Semantic Web world. It is a lightweight model of user's relations, which has become a widely accepted vocabulary for representing Online Social Networks. Besides Social Web, FOAF RDF vocabulary is commonly used by all kinds of Linked Open Data initiatives for describing people and their relations [7].

The FOAF ontology has all the advantages of Semantic Web mentioned in section 3. Besides the core terms which can be used to describe characteristics of people and social groups that are independent of time and technology, it contains also a special group of terms designed especially for describing Internet accounts, address books and other web-based activities.

To model basic information about any new user data extracted from Facebook we use `foaf:Person` class and the properties like `foaf:givenName`, `foaf:familyName`, `foaf:name`, `foaf:gender` and `rdfs:label`. This common set of FOAF/RDFS properties has been enriched with two new properties to store Facebook user name and Facebook id:

```
<rdf:Property rdf:about="http://www.example.com/facebook-crawler#facebookID" rdfs:
    label="Facebook ID" rdfs:comment="A Facebook ID">
  <rdf:type rdf:resource="http://www.w3.org/2002/07/owl#DatatypeProperty"/>
  <rdfs:isDefinedBy rdf:resource="http://www.example.com/facebook-crawler#"/>
  <rdfs:subPropertyOf rdf:resource="http://xmlns.com/foaf/0.1/nick"/>
  <rdfs:domain rdf:resource="http://xmlns.com/foaf/0.1/Person"/>
  <rdfs:range rdf:resource="http://www.w3.org/2000/01/rdf-schema#Literal"/>
</rdf:Property>

<rdf:Property rdf:about="http://www.example.com/facebook-crawler#facebookUserName"
    rdfs:label="Facebook user name" rdfs:comment="A Facebook user name">
```

```
  <rdf:type rdf:resource="http://www.w3.org/2002/07/owl#DatatypeProperty"/>
  <rdfs:isDefinedBy rdf:resource="http://www.example.com/facebook-crawler#"/>
  <rdfs:subPropertyOf rdf:resource="http://xmlns.com/foaf/0.1/nick"/>
  <rdfs:domain rdf:resource="http://xmlns.com/foaf/0.1/Person"/>
  <rdfs:range rdf:resource="http://www.w3.org/2000/01/rdf-schema#Literal"/>
</rdf:Property>
```

To model relationships between users we used `foaf:knows` property which relates `foaf:Person` to another `foaf:Person`. Modeling user *likes* extracted from Facebook would be possible using `foaf:interest` property, which is used to describe interests of the user in a given document, which would be represented by `foaf:Document` class. However, after analyzing the FOAF specification we realized that the natural purpose of using `foaf:interest` property together with `foaf:Document` is to describe some professional interests of the `foaf:Person` and it would be good idea to introduce new class `fc:FacebookDocument` to describe one of the Facebook web sites and `fc:likes` property to allow saying that:

```
<foaf:Person> (subject) <fc:likes>(predicate)
<fc:FacebookDocument>(object)
```

To define the type of FacebookDocument there is additional property `fc:documentType`, which can have one of the literal values: `SportTeam`, `SportAthlete`, `Music`, `Movie`, `TvShow`, `Book`, `Restaurant`, `People`, `Sport`, `Clothing`, `OtherLike` or `Place`. The RDF definition of the mentioned new concepts is the following:

```
<rdf:Property
rdf:about="http://www.example.com/facebook-crawler#likes"
rdfs:label="likes" rdfs:comment="Liked Facebook document">
    <rdf:type rdf:resource="http://www.w3.org/2002/07/owl#ObjectProperty"/>
    <rdfs:isDefinedBy rdf:resource="http://www.example.com/facebook-crawler#"/>
    <rdfs:domain rdf:resource="http://xmlns.com/foaf/0.1/Person"/>
    <rdfs:range rdf:resource="http://www.example.com/facebook-crawler#
        FacebookDocument"/>
</rdf:Property>
<rdfs:Class rdf:about="http://www.example.com/facebook-crawler#FacebookDocument"
    rdfs:label="FacebookDocument" rdfs:comment="Facebook document.">
    <rdf:type rdf:resource="http://www.w3.org/2002/07/owl#Class"/>
    <rdfs:subClassOf>
    <owl:Class rdf:about="http://xmlns.com/foaf/0.1/Document"/></rdfs:subClassOf>
    <rdfs:isDefinedBy rdf:resource="http://www.example.com/facebook-crawler#"/>
</rdfs:Class>
<rdf:Property rdf:about="http://www.example.com/facebook-crawler#documentType" rdfs
    :label="Facebook document type" rdfs:comment="Facebook document type">
    <rdf:type rdf:resource="http://www.w3.org/2002/07/owl#DatatypeProperty"/>
    <rdfs:isDefinedBy rdf:resource="http://www.example.com/facebook-crawler#"/>
    <rdfs:domain rdf:resource="http://www.example.com/facebook-crawler#
        FacebookDocument"/>
    <rdfs:range>
    <rdfs:Datatype>
    <owl:oneOf rdf:parseType="Resource">
    <rdf:first rdf:datatype="http://www.w3.org/2000/01/rdf-schema#Literal">SportTeam
        </rdf:first>
    ...
    <rdf:rest rdf:parseType="Resource">
    <rdf:first rdf:datatype="http://www.w3.org/2000/01/rdf-schema#Literal">Place</rdf
        :first>
    <rdf:rest rdf:resource="http://www.w3.org/1999/02/22-rdf-syntax-ns#nil"/>
    </rdf:rest>
    </rdf:rest>
    </owl:oneOf>
    </rdfs:Datatype>
    </rdfs:range>
</rdf:Property>
```

Besides the `fc:likes` property it was also necessary to introduce similar properties to model the information that a user watched a movie or TV show, read a book or visited some place. For this purpose we introduced the following properties:

```
<rdf:Property
rdf:about="http://www.example.com/facebook-crawler#watched"
rdfs:label="watched" rdfs:comment="Facebook document of watched
movie or tv show">
    <rdf:type rdf:resource="http://www.w3.org/2002/07/owl#ObjectProperty"/>
    <rdfs:isDefinedBy rdf:resource="http://www.example.com/facebook-crawler#"/>
    <rdfs:domain rdf:resource="http://xmlns.com/foaf/0.1/Person"/>
    <rdfs:range rdf:resource="http://www.example.com/facebook-crawler#
        FacebookDocument"/>
  </rdf:Property>

<rdf:Property rdf:about="http://www.example.com/facebook-crawler#read" rdfs:label="
        read" rdfs:comment="Facebook document of read book">
    <rdf:type rdf:resource="http://www.w3.org/2002/07/owl#ObjectProperty"/>
    <rdfs:isDefinedBy rdf:resource="http://www.example.com/facebook-crawler#"/>
    <rdfs:domain rdf:resource="http://xmlns.com/foaf/0.1/Person"/>
    <rdfs:range rdf:resource="http://www.example.com/facebook-crawler#
        FacebookDocument"/>
</rdf:Property>

<rdf:Property rdf:about="http://www.example.com/facebook-crawler#visited" rdfs:
        label="visited" rdfs:comment="Facebook document of visited place">
    <rdf:type rdf:resource="http://www.w3.org/2002/07/owl#ObjectProperty"/>
    <rdfs:isDefinedBy rdf:resource="http://www.example.com/facebook-crawler#"/>
    <rdfs:domain rdf:resource="http://xmlns.com/foaf/0.1/Person"/>
    <rdfs:range rdf:resource="http://www.example.com/facebook-crawler#
        FacebookDocument"/>
</rdf:Property>
```

The Facebook document of some place visited by a user has also latitude and longitude properties extracted from the map, which is presented on a user profile. These localization data can be very helpful for linking an object extracted from Facebook with Linked Open Data (LOD) objects in the future. A sample document fc:FacebookDocument representing Gliwice city has the following form:

```
<fc:FacebookDocument
rdf:about="https://www.facebook.com/pages/Gliwice-Poland/111391415551750">
    <fc:type>Place</fc:type>
    <geo:lat rdf:datatype="&xsd;float">50.2833</geo:lat>
    <geo:long rdf:datatype="&xsd;float">18.6667</geo:long>
</fc:FacebookDocument>
```

Separating fc:likes, fc:watched, fc:read, fc:visited properties from foaf:interest property allows separating user professional interests from some entertainment-related likes. On one hand such an approach allows to perform more accurate analysis of user interests, on the other hand such a user profile based on data extracted from Facebook can be easily linked with the same user professional profile existing on the web without conflicting professional and entertainment-related interests of the user.

4.3 Extracting Data from Facebook to Knowledgebase

The Facebook data in profiles of users, who did not limit the availability of their data using Facebook privacy settings, can be easily available for any user after creating account in the Facebook system. However, these data in most cases are not available for automatic querying by external systems. Although Facebook provides their API, it is not possible to access the whole social graph. Only the part of the data which is directly connected with a logged user is accessible. The only possibility to use Facebook API to query data of a random users would be to prepare Facebook Application asking every user for relevant privileges. The alternative solution is to use HTML parsing techniques to extract the interesting data directly from user's Facebook websites which are easily available, without a need of any additional privileges.

Parsing of user's *likes* web pages and basic user's friends lists was performed by means of the same techniques. The details of the approach were presented in [10]. The employed technologies include the following:

– .NET HttpWebRequest and CookieContainer classes for simulating user logging on Facebook web site,
– .NET WebBrowser Windows Forms control for loading Facebook web pages and simulating user actions,
– Html Agility Pack project [2] libraries for parsing HTML code,
– .NET Linq techniques for querying parsed HTML code.

Facebook JSON data for any Facebook profile is available by manually changing Facebook profile URL. Common form of the user profile URL looks in the following way: `https://www.facebook.com/wojciech.kijas`. Using the URL as input we can generate JSON data URL by replacing "www" substring with "graph" string to get the following sample URL: `https://graph.facebook.com/wojciech.kijas`. Calling the generated URL we can obtain the following sample data in JSON format:

```
{
    "id": "100001184578579",
    "name": "Wojciech Kijas",
    "first_name": "Wojciech",
    "last_name": "Kijas",
    "link": "http://www.facebook.com/wojciech.kijas",
    "gender": "male",
    "locale": "pl_PL",
    "username": "wojciech.kijas"
}
```

To call the URL and get the returned data WebRequest and HttpWebResponse classes are used. Next, the obtained JSON data are parsed by means of Json.NET libraries [3]. Finally, we can complement user's basic data in FOAF knowledgebase. The same technique was used to complement user *likes* knowledgebase resources.

5   Examples of Using the Facebook Data Knowledgebase

The main goal of our work is to bring Facebook data about users, who we are interested in, to knowledgebase to make it easily accessible for external systems. The external systems can take advantage of the data in variety of ways.

5.1   Querying the Facebook Data Knowledgebase

Ability to query the Facebook data knowledgebase can be very useful while doing some analysis of customers. Having all the Facebook data in RDF format (triple store) allows to query it using SPARQL language. Some sample queries include the following:

– friends of some customer:

```
PREFIX foaf:    <http://xmlns.com/foaf/0.1/>
SELECT ?a
WHERE { <https://www.facebook.com/wojciech.kijas> foaf:knows ?a }
```

– common friends of two customers:

```
PREFIX foaf:    <http://xmlns.com/foaf/0.1/>
SELECT ?a
WHERE { <https://www.facebook.com/wojciech.kijas> foaf:knows ?a .
        <https://www.facebook.com/wiola.stanikowska> foaf:knows ?a }
```

– people who like Albert Einstein and visited Gliwice:

```
PREFIX fc: <http://www.example.com/facebook-crawler#>
SELECT ?a
WHERE {?a fc:likes < https://www.facebook.com/AlbertEinstein> .
       ?a fc:visited < https://www.facebook.com/pages/Gliwice-Poland
       /111391415551750 > }
```

Possibility of such an easy access to Facebook data can be useful for both quick queries giving answer for some sample questions regarding corporate customers, but also for retrieving some subset of the data from knowledgebase to use it in further corporate analysis by means of Business Intelligence system. An example of such approach was presented in our previous work [10].

5.2   Taking Advantage of Reasoning

One of the biggest advantages of Semantic Web technologies are reasoning capabilities. Describing the idea of reasoning over Semantic Web data goes beyond the scope of this article, but, saying very simply, reasoning allows discovering of new connections in data graph on the basis of logical consequences of class definitions in ontology and already asserted facts. As the example let's say there are the following facts describing people in the knowledgebase:

```
<foaf:Person rdf:about="https://www.facebook.com/wojciech.kijas">
  <fc:facebookID >10000118457857 9</fc:facebookID>
  <fc:facebookUserName>wojciech.kijas</fc:facebookUserName>
  <fc:likes rdf:resource="https://www.facebook.com/AlbertEinstein" />
  <fc:likes rdf:resource="https://www.facebook.com/
      ZdaniaKtorychGliwiczanieNigdyNieMowia" />
  <fc:visited rdf:resource="https://www.facebook.com/pages/Gliwice-Poland
      /111391415551750" />
</foaf:Person>

<foaf:Person rdf:about="https://www.facebook.com/jankowalski">
  <fc:facebookID >9999999999</fc:facebookID>
  <fc:facebookUserName>jankowalski</fc:facebookUserName>
  <fc:visited rdf:resource="https://www.facebook.com/pages/Wroclaw-Poland
      /110145572342035" />
</foaf:Person>
```

Let's add definition of the following classes to Facebook Crawler ontology:

```
<rdfs:Class
rdf:about="http://www.example.com/facebook-crawler#PersonConnectedWithGliwice">
  <rdf:type rdf:resource="http://www.w3.org/2002/07/owl#Class"/>
  <owl:equivalentClass>
    <owl:Class>
      <owl:intersectionOf rdf:parseType="Collection">
      <owl:Restriction>
        <owl:onProperty rdf:resource="http://www.example.com/facebook-crawler#likes
            " />
        <owl:someValuesFrom rdf:resource="http://www.example.com/facebook-crawler#
            DocumentAboutGliwice" />
      </owl:Restriction>
      <owl:Restriction>
        <owl:onProperty rdf:resource="http://www.example.com/facebook-crawler#
            visited" />
        <owl:hasValue rdf:resource="https://www.facebook.com/pages/Gliwice-Poland
            /111391415551750" />
```

```
        </owl:Restriction>
      </owl:intersectionOf>
    </owl:Class>
  </owl:equivalentClass>
</rdfs:Class>

<rdfs:Class rdf:about="http://www.example.com/facebook-crawler#DocumentAboutGliwice
    ">
  <rdf:type rdf:resource="http://www.w3.org/2002/07/owl#Class"/>
  <owl:equivalentClass>
  <owl:Class>
    <owl:oneOf rdf:parseType="Collection">
      <owl:Thing rdf:about="https://www.facebook.com/
          ZdaniaKtorychGliwiczanieNigdyNieMowia"/>
      <owl:Thing rdf:about="https://www.facebook.com/GliwiceNM"/>
      <owl:Thing rdf:about="https://www.facebook.com/CrossFitGliwice"/>
      <owl:Thing rdf:about="https://www.facebook.com/ClubMardiGras"/>
      <owl:Thing rdf:about="https://www.facebook.com/Gwarek.Gliwice"/>
    </owl:oneOf>
  </owl:Class>
  </owl:equivalentClass>
</rdfs:Class>
```

Now, we can use the following SPARQL query executed in Pellet OWL 2 reasoner [17]:

```
PREFIX rdf: <http://www.w3.org/1999/02/22-rdf-syntax-ns#>
PREFIX fc: <http://www.example.com/facebook-crawler#>
SELECT ?x
WHERE { ?x rdf:type fc:PersonConnectedWithGliwice }
```

By running the query we can easily find people whose profiles fit the definition of classes:

```
Query Results (1 answers):
x
========================
wojciech.kijas
```

If we design such a sample class model, it will work for all the new facts describing new users added to the knowledgebase and the query result will be calculated based on the facts asserted in knowledgebase. This is only a simple example of all the capabilities of OWL/RDFS reasoner.

6   Conclusion and Future Work

In this work we extended the previously created Facebook Crawler [10] which plays a role of intermediate layer between Facebook profiles data and external systems. The new version of the system, besides user's friends extracting capabilities, allows also extracting all kinds of user *likes* from Facebook. Using the data extracted from Facebook, we build FOAF profiles of users after introducing some new vocabulary extending the FOAF ontology. All the extracting of data from Facebook is done by JSON and HTML code parsing to make it independent of Facebook API capabilities.

The general purpose of the work is to allow using Facebook relations of customers in the corporate analysis. After gathering emails of customers, it is possible to use Facebook Crawler to automatically search Facebook users and extract data from their Facebook profiles to knowledgebase. In the last step the data collected in the knowledgebase can be used to perform further analysis or to complement customer's data in corporate Data Warehouse.

Following Tim Berners-Lee's vision [6], employing Linked Open Data capabilities to Facebook knowledgebase is the next step of our work. In further development we plan to introduce algorithm for automatically finding links between

254 W. Kijas and M. Kozielski

Linked Open Data (LOD) objects and knowledgebase objects extracted from Facebook. Incorporating these data can bring new great possibilities for querying and reasoning over the data. We are going to use also Social Network Analysis (SNA) techniques for grouping FOAF profiles and user *likes* for discovering hidden characteristics based on connections between them.

Acknowledgments. This work was supported by Ministry of Science and Higher Education as a Statutory Research Project (decision 8686/E-367/S/2014)

References

1. Facebook online social network, http://facebook.com (accessed: December 01, 2014)
2. Html agility pack project, http://htmlagilitypack.codeplex.com (accessed: December 01, 2014)
3. Json.net library, http://james.newtonking.com/json (accessed: December 01, 2014)
4. W3C consortium, http://www.w3.org (accessed: December 01, 2014)
5. Aroyo, L., Houben, G.J.: User modeling and adaptive semantic web. Semantic Web 1(1), 105–110 (2010)
6. Berners-Lee, T., Hendler, J., Lassila, O.: The semantic web. a new form of web content that is meaningful to computers will unleash a revolution of new possibilities. Scientific American 284(5), 1–5 (2001)
7. Bizer, C., Heath, T., Berners-Lee, T.: Linked data-the story so far. International Journal on Semantic Web and Information Systems 5(3), 1–22 (2009)
8. Brickley, D., Miller, L.: Foaf vocabulary specification 0.98. Namespace Document 9 (2012)
9. Golbeck, J., Rothstein, M.: Linking social networks on the web with foaf: A semantic web case study. In: AAAI, vol. 8, pp. 1138–1143 (2008)
10. Kijas, W.: Facebook crawler as software agent for business intelligence system. Studia Informatica 35(4), 89–110 (2014)
11. Lehmann, J., Isele, R., Jakob, M., Jentzsch, A., Kontokostas, D., Mendes, P.N., Hellmann, S., Morsey, M., van Kleef, P., Auer, S., Bizer, C.: DBpedia - a large-scale, multilingual knowledge base extracted from wikipedia. Semantic Web Journal (2014)
12. Noy, N.F.: Semantic integration: a survey of ontology-based approaches. ACM Sigmod Record 33(4), 65–70 (2004)
13. Orlandi, F.: Multi-source provenance-aware user interest profiling on the social semantic web. In: Masthoff, J., Mobasher, B., Desmarais, M.C., Nkambou, R. (eds.) UMAP 2012. LNCS, vol. 7379, pp. 378–381. Springer, Heidelberg (2012)
14. Orlandi, F., Breslin, J., Passant, A.: Aggregated, interoperable and multi-domain user profiles for the social web. In: Proceedings of the 8th International Conference on Semantic Systems, pp. 41–48. ACM (2012)
15. Plumbaum, T., Wu, S., De Luca, E.W., Albayrak, S.: User modeling for the social semantic web. In: SPIM, pp. 78–89 (2011)
16. Shapira, B., Rokach, L., Freilikhman, S.: Facebook single and cross domain data for recommendation systems. User Modeling and User-Adapted Interaction 23(2-3), 211–247 (2013)

17. Sirin, E., Parsia, B., Grau, B.C., Kalyanpur, A., Katz, Y.: Pellet: A practical owl-dl reasoner. Web Semantics: Science, Services and Agents on the World Wide Web 5(2), 51–53 (2007)
18. Vu, X.T., Morizet-Mahoudeaux, P., Abel, M.H.: Empowering collaborative intelligence by the use of user-centered social network aggregation. In: 2013 IEEE/WIC/ACM International Joint Conferences on Web Intelligence (WI) and Intelligent Agent Technologies (IAT), vol. 1, pp. 425–430. IEEE (2013)
19. Vu, X.T., Morizet-Mahoudeaux, P., Abel, M.H., et al.: User-centered social network profiles integration. In: WEBIST, pp. 473–476 (2013)

RDF Graph Partitions: A Brief Survey

Dominik Tomaszuk[1], Łukasz Skonieczny[2(✉)], and David Wood[3]

[1] Institute of Computer Science, University of Bialystok, Poland
dtomaszuk@ii.uwb.edu.pl
[2] Institute of Computer Science, Warsaw University of Technology, Poland
lskoniec@ii.pw.edu.pl
[3] 3 Round Stones Inc., USA
david@3roundstones.com

Abstract. The paper presents justifications and solutions for RDF graph partitioning. It uses an approach from the classical theory of graphs to deal with this problem. We present four ways to transform an RDF graph to a classical graph. We show how to apply solutions from the theory of graphs to RDF graphs. We also perform an experimental evaluation using the *gpmetis* algorithm (a recognized graph partitioner) on both real and synthetic RDF graphs and prove its practical usability.

1   Introduction

Machines commonly have a need to exchange machine-readable data. One useful approach to facilitate the efficient exchange of such data is to agree upon a common data model under which to structure, represent and store content. This data model should be generic enough to provide a canonical representation for arbitrary content irrespective of its syntax. The data model should also enable processing this content. The core data model chosen for use on the Semantic Web and Linked Data environments is the Resource Description Framework (RDF), an edge-labeled directed graph data model [7].

The size of the RDF graph is often too large to be efficiently managed on a single machine. To deal with this problem many popular RDF graph stores facilitate various ways of distributing RDF data among different nodes of computer cluster. Early techniques came from the adaptation of similar solutions from the RDBMS world e.g. vertical partitioning [1] or horizontal partitioning [19]. These techniques tend to create poor partitions in terms of inter-partition connectivity which leads to poor performance of queries involving many joins. More advanced techniques which take the graph nature of the RDF data, have started to appear only recently [9,18,23,24]. One of the most natural and promising approach involves graph partitioning – a classical problem from theory of graphs.

Section 2 presents a formalized syntax and concept for RDF. Section 3 discusses the graph partition problem in the context of RDF graphs. Section 4 shows the practical relevance of RDF graph partitioning. Finally, section 5 gives some concluding remarks.

© Springer International Publishing Switzerland 2015
S. Kozielski et al. (Eds.): BDAS 2015, CCIS 521, pp. 256–264, 2015.
DOI: 10.1007/978-3-319-18422-7_23

2   Preliminaries

An RDF is used as a general method for the conceptual description or the modeling of information that is available in web resources. It provides the essential foundation and infrastructure to support the description and management of data. In other words, RDF is a very general data model for describing resources and relationships between them.

The RDF data model is based upon the idea of making statements about web resources in the form of subject-predicate-object expressions. These expressions are known as *triples* in the RDF terminology.

An RDF triple consists of a subject, a predicate, and an object. In [6], the meaning of subject, predicate and object is explained.

Definition 2.1 (Subject, Predicate and Object). *The* subject *denotes a resource, the* object *fills the value of the relation, the* predicate *means traits or aspects of the resource, and expresses a relationship between the subject and the object. The predicate denotes a binary relation, also known as a property.* □

Following [6], we provide definitions of RDF triples below.

Definition 2.2 (RDF Triple). *Assume that $\mathcal{I}$ is the set of all Internationalized Resource Identifier (IRI) references, $\mathcal{B}$ an infinite set of blank nodes, $\mathcal{L}$ the set of RDF literals. An RDF triple t is defined as a triple $t = \langle s, p, o \rangle$ where $s \in \mathcal{I} \cup \mathcal{B}$ is called the subject, $p \in \mathcal{I}$ is called the predicate and $o \in \mathcal{I} \cup \mathcal{B} \cup \mathcal{L}$ is called the object.* □

Example 2.1 *The example presents an RDF triple consists of subject, predicate and object.*

```
<http://example.net/me#js>    foaf:name    ''John Smith''.
```

□

The elemental constituents of the RDF data model are RDF terms that can be used in reference to resources: anything with identity. The set of RDF terms is divided into three disjoint subsets:

- IRIs,
- literals,
- blank nodes.

Definition 2.3 (IRIs). IRIs *serve as global identifers that can be used to identify any resource. For example, $<http://dbpedia.org/resource/House>$ is used to identify the house in DBpedia[1].* □

---

[1] http://dbpedia.org/

Note that in RDF 1.0 identifers was RDF URI References. Identifiers in RDF 1.1 are now IRIs, which are a generalization of URIs that permits a wider range of Unicode characters. Every absolute URI and URL is an IRI, but not every IRI is an URI. When IRIs are used in operations that are only defined for URIs, they must first be converted.

Definition 2.4 (Literals). *Literals are a set of lexical values. It can be a set of plain strings, such as "Apple", optionally with an associated language tag, such as "Apple"@en. Literals comprise of a lexical string and a datatype, such as "3.14"^^http://www.w3.org/2001/XMLSchema#float. Datatypes are identified by IRIs, where RDF borrows many of the datatypes defined in XML Schema 1.1 [22]* □

Note that in RDF 1.0 literals with a language tag did not have a datatype URI. In RDF 1.1 literals with language tags have the datatype IRI rdf:langString. Now all literals have datatypes. Implementations might choose to support syntax for simple literals, but only as synonyms for xsd:string literals. Moreover, RDF 1.1 supports the new datatype rdf:HTML. Both rdf:HTML and rdf:XMLLiteral depend on DOM4 (Document Object Model level 4)[2].

Definition 2.5 (Blank Nodes). *Blank nodes are defined as existential variables used to denote the existence of some resource for which an IRI or literal is not given. They are always locally scoped to the file or RDF store, and are not persistent or portable identifiers for blank nodes.* □

Note that RDF 1.0 makes no reference to any internal structure of blank nodes. Given two blank nodes, it is not possible to determine whether or not they are the same. In RDF 1.1 blank node identifiers are local identifiers that are used in some concrete RDF syntaxes or RDF store implementations.

Blank nodes do not have identifiers in the RDF abstract syntax. In situations where stronger identification is needed, some or all of the blank nodes can be replaced with IRIs. Systems wishing to do this should create a globally unique IRI (called a skolem IRI) for each blank node so replaced. This transformation does not appreciably change the meaning of an RDF graph. It permits the possibility of other graphs subsequently using the skolem IRIs, which is not possible for blank nodes. Systems that want skolem IRIs to be recognizable outside of the system boundaries use a well-known IRI [20] with the registered name genid.

A collection of RDF triples intrinsically represents a labeled directed multigraph. The nodes are the subjects and objects of their triples. RDF is often referred to as being *graph structured data* where each $\langle s, p, o \rangle$ triple can be seen as an edge $s \xrightarrow{p} o$.

Definition 2.6 (RDF Graph). *Let $\mathcal{L} = \mathcal{L}_S \cup \mathcal{L}_{\mathcal{L}} \cup \mathcal{L}_{\mathcal{D}}$, $\mathcal{O} = \mathcal{I} \cup \mathcal{B} \cup \mathcal{L}$ and $\mathcal{S} = \mathcal{I} \cup \mathcal{B}$, then $G \subset \mathcal{S} \times \mathcal{I} \times \mathcal{O}$ is a finite subset of RDF triples, which is called RDF graph.* □

---

[2] DOM4 is a way to refer to XML or HTML elements as objects, see http://www.w3.org/TR/dom/

Example 2.2 *The example in Fig. 1 presents an RDF graph of a FOAF [5] profile. This graph includes four RDF triples:*

```
<#js>    rdf:type    foaf:Person  .
<#js>    foaf:name   ''John Smith''.
<#js>    foaf:workplaceHomepage <http://univ.com/> .
<http://univ.com/>  rdfs:label ''University''      .
```

□

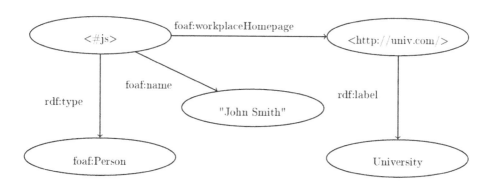

Fig. 1. An RDF graph

When applying classical graph theory on RDF graphs, the RDF graph is usually treated as a directed labeled graph (see definition 2.7) in way that each RDF triple $\langle s, p, o \rangle$ is transformed into corresponding $s' \xrightarrow{p'} o'$ edge, where $s', o' \in V$ and $p' \in L$.

Definition 2.7 (Directed Labeled Graph). Directed labeled graph G *is a quadruple* $G = (V, E, lbl, L)$, *where* V *is a set of vertices,* $E = \{(v_1, v_2)|v_1, v_2 \in V\}$ *is a set of directed edges,* $lbl : E \cup V \to L$ *is a labeling function, and* L *is a set of labels.*

The k-way graph partition problem (see definition 2.8) is in general defined as finding k disjoint subsets of graph vertices. The optimal graph partition is a partition which optimizes some given criteria, e.g. number of edges running between separated components is low (size of the edge cut set, in other words), and the numbers of vertices in every component are close to each other. Please note, that such criteria are especially desirable in case of distributing RDF graphs as it creates highly independent, loosely-coupled partitions, maximizing chances that the RDF query is executed on the minimal number of cluster nodes.

Definition 2.8 (k-way Graph Partition). *Given a graph* $G = (V, E, lbl, L)$, *a k-way graph partitioning, C, is a division of* V *into k partitions* $\{P_1, P_2, ..., P_k\}$

such that $\bigcup_{1 \leq i \leq k} P_i = V$, *and* $P_i \cap P_j = \emptyset$ *for any* $i \neq j$. The edge cut set E_c is the set of edges whose vertices belong to different partitions.

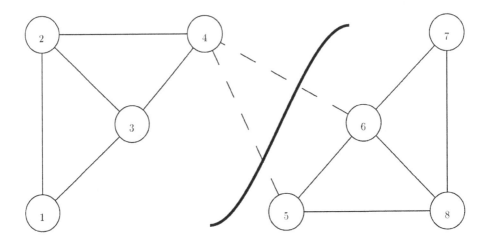

Fig. 2. Graph partition example

An example of graph partition is illustrated in Fig. 2.

3   RDF Graph Partition

3.1   Classical Graph Partitioning

The optimal graph partition problem is known to be NP-complete [8] but a lot of efficient, suboptimal algorithms have been proposed [21,10,11,2,14]. One of the most recognized ones is a *gpmetis* included in the *METIS* software package for partitioning graphs, partitioning meshes and computing fill-reducing orderings of sparse matrices. gpmetis is based on the multilevel graph partitioning paradigm ([13,15]) which consists of three phases: graph coarsening, initial partitioning, graph uncoarsening. The goal of the coarsening phase is to derive a series of smaller graph from the initial input graph by collapsing together a maximal size set of adjacent pairs of vertices. When the resulted graph is low enough (usually a few hundred vertices) it is being partitioned with the Kernighan-Lin algorithm [17] - this is the initial partitioning phase. Then, in the uncoarsening phase, the partitioning is projected to the successively larger graphs. It is done by successive uncollapisng of vertices and assigning vertices that were collapsed together to the same partition. After each projection step, the partitioning is refined by moving vertices between partitions as long as it improves the quality of the partitioning solution. The uncoarsening phase ends when the graph is uncollapsed to the original input graph.

The Kernighan–Lin algorithm which is used in the initial partitioning phase is a $O(n^2 log(n))$ heuristic algorithm for solving the graph partitioning problem. The original algorithm performs 2-way partitioning but can be easily extended to k-way partitioning. The algorithm finds two disjoint, equally sized sets of vertices (namely A and B) which minimizes the sum T of weights of the edges between vertices from A and B. The weights of the edges come from the number of vertices collapsed together in the coarsening phase. Let I_v^V be the sum of the weights of edges between v and vertices in V. The Kernighan-Lin algorithm starts with a random A, B sets and successively interchanges vertices $a \in A$ and $b \in B$ with each other such that the reduction cost $T_{old} - T_{new} = I_a^B - I_a^A + I_b^A - I_b^B - 2c_{a,b}$ is maximized, where $c_{a,b}$ is the cost of the edge between a and b.

3.2   Relationship Between Classical Graphs and RDF Graphs

All graph partitioning algorithms can be applied to RDF graphs if they are transformed to classical graph representation. A typical triple to edge transformation is the simplest one, but as was noted in [12] it is ambiguous and a not one-to-one relationship. A RDF triple $t = \langle s, p, o \rangle$ where $s \in \mathcal{I} \cup \mathcal{B}$, $p \in \mathcal{I}$, $o \in \mathcal{I} \cup \mathcal{B} \cup \mathcal{L}$ (see definition 2.2) is transformed into directed edge $s' \xrightarrow{p'} o'$ edge, where $s', o' \in V$ and $p' \in L$. Notice that the RDF predicate domain intersects with the subject and object domains, so the p predicate might occur as a subject or object in some other triple. In the graph theory, however, E, V and L domains are distinct, i.e. there are no edges coming from/to other edges or labels. There are a few solutions to this problem. **First** is a $\langle s, p, o \rangle$ to $v_1(s') \xrightarrow{p'} v_2(o')$ transformation, where $v_1, v_2 \in V$ and $s', p', o' \in L$, i.e. s' and o' are labels of the vertices. In this solution, predicates may occur either as labels of edges or labels of vertices. The **Second** solution is to make use of hypergraphs instead of simple graphs, that is allowing edges to connect more than two nodes. In this approach all s, p and o are transformed into distinct vertices, and each RDF triple is represented as a directed hyperedge connecting s,p and o with each other. The drawback of this method is that processing hypergraphs requires specialized algorithms (e.g. [16] in case of partitioning), which are in general slower than their simple graphs counterparts. The **Third** solution was presented in [12]. It takes a hypergraph representation as a starting point and transforms it into bipartite graph. In this approach a RDF triple $\langle s, p, o \rangle$ is represented as 4 nodes and 3 edges $s' \xleftarrow{subject} t' \xrightarrow{object} o'$
$\downarrow predicate$
p' . The **Fourth** approach is to transform every RDF triple t into a distinct graph node, and generate edges between those nodes which share subjects, objects and/or predicates. The choice of RDF graph representation is a subject of application. In case of graph partitioning the first and fourth approach seems to be superior but additional research is required. One can notice that the RDF graph is more general than the classical graph. A directed labeled graph

Table 1. Partitioning of elvis, berlin and dbpedia datasets with gpmetis

		elvis	berlin	dbpedia-geo	elvis*
#vertices		742	116728	1637808	774
#edges		774	348106	2104985	41458
k		k-way partitioning			
2	time	252 ms	2617 ms	20954 ms	1792 ms
	$\|E_c\|$	21	44611	226002	1409
3	time	122 ms	2652 ms	14407 ms	224 ms
	$\|E_c\|$	37	79950	330831	2172
4	time	101 ms	3139 ms	16357 ms	180 ms
	$\|E_c\|$	40	96827	415257	3650
5	time	52 ms	3116 ms	14561 ms	194 ms
	$\|E_c\|$	49	119679	396505	11574
6	time	95 ms	3133 ms	17862 ms	92 ms
	$\|E_c\|$	52	131902	473697	12341
7	time	76 ms	3304 ms	14299 ms	137 ms
	$\|E_c\|$	52	139853	456204	16616
8	time	79 ms	3358 ms	16628 ms	291 ms
	$\|E_c\|$	62	145927	504719	15709
9	time	45 ms	3567 ms	13886 ms	87 ms
	$\|E_c\|$	62	155111	515311	17955
10	time	43 ms	3624 ms	16382 ms	154 ms
	$\|E_c\|$	63	162344	520028	21634

can be easily transformed into RDF graph, but the reversed transformation is cumbersome. It means that the complexity of every RDF graph problem is not better than the complexity of the corresponding classical graph problem.

4   Experiments

To examine the practical relevance of RDF graph partitioning we performed an experimental evaluation of the *gpmetis*[3] algorithm from the *grph*[4] library. We chose four RDF data sets: *berlin* - a synthetic dataset generated by Berlin SPARQL Benchmark generator [4], *elvis* – a metadata about Elvis impersonators[5] and *dbpedia-geo* – Geographic Coordinates[6] dataset from DBpedia [3]. Each dataset was partitioned into k partitions with k ranging from 2 to 10. We collected the execution time of the algorithm and the size of the edge cut set $|E_c|$ (see definition 2.8). We were using default parameters of the gpmetis algorithm. The results are presented in table 1. The *elvis** is a graph obtained from elvis dataset with the fourth approach. All other graphs were obtained with the first

---

[3] http://glaros.dtc.umn.edu/gkhome/views/metis
[4] http://www.i3s.unice.fr/~hogie/grph/index.php
[5] http://www.rdfdata.org/
[6] http://wiki.dbpedia.org/Downloads2014

approach. The fourth approach turned out to be not practical as it created very large graphs resulting in out-of-memory errors.

The gpmetis algorithm generally performed very well generating partitions in seconds even for quite big dbpedia graphs. The size of edge cut set depends on the data set, 3%-8% of total number of edges in the case of the elvis graph, which is a very good result, 10%-20% in the case of the dbpedia graph, which is probably acceptable, and 12%-45% in the case of the berlin graph.

5   Conclusions

We outlined a partition of the vertices of an RDF graph into two disjoint subsets. In this paper we presented works from the RDF graph partitions research area. This paper provided insights on classical graph partitioning of RDF graphs. Moreover, we presented formal relationships between classical graphs and RDF graphs. Finally, we presented experiments, which showed a great potential for the presented approaches.

References

1. Abadi, D.J., Marcus, A., Madden, S.R., Hollenbach, K.: Scalable Semantic Web Data Management Using Vertical Partitioning. In: Proceedings of the 33rd International Conference on Very Large Data Bases, VLDB 2007, pp. 411–422. VLDB Endowment (2007)
2. Arora, S., Karger, D., Karpinski, M.: Polynomial Time Approximation Schemes for Dense Instances of NP-Hard Problems. Journal of Computer and System Sciences 58(1), 193–210 (1999)
3. Auer, S., Bizer, C., Kobilarov, G., Lehmann, J., Cyganiak, R., Ives, Z.G.: DBpedia: A Nucleus for a Web of Open Data. In: Aberer, K., et al. (eds.) ASWC/ISWC 2007. LNCS, vol. 4825, pp. 722–735. Springer, Heidelberg (2007)
4. Bizer, C., Schultz, A.: The Berlin SPARQL Benchmark. International Journal on Semantic Web and Information Systems (IJSWIS) 5(2), 1–24 (2009)
5. Brickley, D., Miller, L.: FOAF Vocabulary Specification 0.99. Tech. rep., FOAF Project (January 2014)
6. Cyganiak, R., Lanthaler, M., Wood, D.: RDF 1.1 Concepts and Abstract Syntax. W3C recommendation, World Wide Web Consortium (February 2014)
7. Czajkowski, K., Trela, T.: Semantic web - standard, narzędzia, implementacje. Studia Informatica 33(2A), 379–393 (2012)
8. Garey, M.R., Johnson, D.S., Stockmeyer, L.: Some simplified NP-complete graph problems. Theoretical Computer Science 1(3), 237–267 (1976)
9. Goczyła, K., Waloszek, A., Waloszek, W.: Techniki modularyzacji ontologii. Bazy danych. Rozwój metod i technologii–Architektura, metody formalne i zaawansowana analiza danych, red.: Kozielski, S. and Małysiak, B. and Kasprowski, P. and Mrozek, D. and WKŁ s 309, 322 (2008)
10. Goldschmidt, O., Hochbaum, D.S.: Polynomial algorithm for the k-cut problem. In: 29th Annual Symposium on Foundations of Computer Science, pp. 444–451 (October 1988)

11. Guttmann-Beck, N., Hassin, R.: Approximation Algorithms for Minimum K -Cut. Algorithmica 27(2), 198–207 (2000)
12. Hayes, J., Gutierrez, C.: Bipartite Graphs as Intermediate Model for RDF. In: McIlraith, S.A., Plexousakis, D., van Harmelen, F. (eds.) ISWC 2004. LNCS, vol. 3298, pp. 47–61. Springer, Heidelberg (2004)
13. Karypis, G., Kumar, V.: A Fast and High Quality Multilevel Scheme for Partitioning Irregular Graphs. SIAM J. Sci. Comput. 20(1), 359–392 (1998)
14. Karypis, G., Kumar, V.: Multilevel Algorithms for Multi-constraint Graph Partitioning. In: Proceedings of the 1998 ACM/IEEE Conference on Supercomputing, pp. 1–13. IEEE Computer Society, Washington (1998)
15. Karypis, G., Kumar, V.: Multilevel K-way Partitioning Scheme for Irregular Graphs. J. Parallel Distrib. Comput. 48(1), 96–129 (1998)
16. Karypis, G., Kumar, V.: Multilevel K-way Hypergraph Partitioning. In: Proceedings of the 36th Annual ACM/IEEE Design Automation Conference, DAC 1999, pp. 343–348. ACM, New York (1999)
17. Kernighan, B.W., Lin, S.: An Efficient Heuristic Procedure for Partitioning Graphs. The Bell System Technical Journal 49(1), 291–307 (1970)
18. Lee, K., Liu, L.: Scaling Queries over Big RDF Graphs with Semantic Hash Partitioning. Proc. VLDB Endow. 6(14), 1894–1905 (2013)
19. Mulay, K., Kumar, P.S.: SPOVC: A Scalable RDF Store Using Horizontal Partitioning and Column Oriented DBMS. In: Proceedings of the 4th International Workshop on Semantic Web Information Management, SWIM 2012, pp. 8:1–8:8. ACM, New York (2012)
20. Nottingham, M., Hammer-Lahav, E.: Defining Well-Known Uniform Resource Identifiers (URIs). RFC 5785, Internet Engineering Task Force (April 2010)
21. Saran, H., Vazirani, V.V.: Finding k-cuts within Twice the Optimal (1995)
22. Sperberg-McQueen, M., Thompson, H., Peterson, D., Malhotra, A., Biron, P.V., Gao, S.: W3C XML Schema Definition Language (XSD) 1.1 Part 2: Datatypes. W3C recommendation, World Wide Web Consortium (April 2012)
23. Wang, R., Chiu, K.: A Graph Partitioning Approach to Distributed RDF Stores. In: Proceedings of the 2012 IEEE 10th International Symposium on Parallel and Distributed Processing with Applications, ISPA 2010, pp. 411–418. IEEE Computer Society, Washington, DC (2012)
24. Yan, Y., Wang, C., Zhou, A., Qian, W., Ma, L., Pan, Y.: Efficient Indices Using Graph Partitioning in RDF Triple Stores. In: Proceedings of the 2009 IEEE International Conference on Data Engineering, ICDE 2009, pp. 1263–1266. IEEE Computer Society, Washington (2009)

Artificial Intelligence, Data Mining and Knowledge Discovery

Optimization of Inhibitory Decision Rules Relative to Coverage – Comparative Study

Beata Zielosko[✉]

Institute of Computer Science, University of Silesia
39, Będzińska St., 41-200 Sosnowiec, Poland
beata.zielosko@us.edu.pl

Abstract. In the paper, a modification of a dynamic programming algorithm for optimization of inhibitory decision rules relative to coverage is proposed. The aim of the paper is to study the coverage of inhibitory decision rules constructed by the proposed algorithm and comparison of coverage of inhibitory rules constructed by a dynamic programming algorithm and greedy algorithm.

Keywords: Inhibitory decision rules · Coverage · Dynamic programming algorithm · Greedy algorithm

1 Introduction

Decision rules are known and popular form of knowledge representation. They are used in many areas connected with data mining [6,10,11].

Inhibitory rules, in contrast to ordinary decision rules, on the right-hand side have a relation "attribute $\neq$ value". It was shown in [14] that, for some information systems, ordinary rules cannot describe the whole information contained in the system. However, inhibitory rules describe the whole information for every information system [9]. It means that inhibitory rules can express more information encoded in information system than ordinary rules. Classifiers based on inhibitory rules have often better accuracy than classifiers based on ordinary rules [7,8].

There are different rule quality measures that are used for induction or classification tasks [12,13]. In the paper, the coverage of inhibitory rules is studied. It is a rule's evaluation measure that allows to discover major patterns in the data. Construction and optimization of inhibitory rules relative to coverage can be considered as important task for knowledge representation and knowledge discovery.

In the paper, a modification of a dynamic programming algorithm for optimization of inhibitory decision rules relative to coverage is presented. The paper is a continuation of research connected with a problem of scalability for decision rule optimization relative to coverage based on the dynamic programming approach [15]. However in [15], approximate deterministic rules were studied, now exact inhibitory rules are considered. Motivation for usage of inhibitory rules is

© Springer International Publishing Switzerland 2015
S. Kozielski et al. (Eds.): BDAS 2015, CCIS 521, pp. 267–276, 2015.
DOI: 10.1007/978-3-319-18422-7_24

that coverage of such rules is often greater than the coverage of ordinary rules. Dynamic programming approach allows to obtain optimal inhibitory decision rules, i.e., rules with the maximum coverage. Proposed method of rule induction is based on the analysis of the directed acyclic graph constructed for a given decision table. Such graph can be huge for larger data sets (see in [15], comparison of the number of nodes and edges in the graph). The aim of the paper, is to find a heuristic, modification of a dynamic programming algorithm that allows to obtain values of coverage of inhibitory decision rules close to optimal ones [3], and the size of the graph should be smaller than in case of dynamic programming approach.

In [9], it was shown that under some natural assumptions on the class NP, the greedy algorithm is close to the best polynomial approximate algorithms for the minimization of length of inhibitory decision rules. There is an intuition, that in case of coverage we can have similar situation, so greedy algorithm is considered also. It is obvious, that greedy approach is simpler than dynamic programming approach, the aim is to compare how close is proposed and greedy solution to optimal solution. To evaluate such a difference between proposed approach and greedy approach, a Wilcoxon test was applied.

Presented algorithm is based on a dynamic programming algorithm for inhibitory decision rules optimization relative to coverage [3]. For a given decision table T a directed acyclic graph $\Lambda(T)$ is constructed. Nodes of this graph are subtables of a decision table T described by descriptors (pairs attribute = value). The partitioning of a subtable is finished when it has less different decisions than T. In comparison with algorithm presented in [3], subtables of the graph $\Lambda(T)$ are constructed for one attribute with the minimum number of values, and for the rest of attributes from T - the most frequent value of each attribute (value of an attribute attached to the maximum number of rows) is selected. So, the size of the graph $\Lambda(T)$ is smaller than the size of the graph constructed by the dynamic programming algorithm. This fact is important from the point of view of scalability. Based on the graph $\Lambda(T)$ it is possible to describe sets of inhibitory decision rules for rows of table T. Then, using procedure of optimization of the graph $\Lambda(T)$ relative to coverage it is possible to find for each row r of T an inhibitory rule with the maximum coverage.

In [4], a dynamic programming algorithm for optimization of inhibitory rules relative to length was considered. In [3], algorithms for optimization relative to coverage and sequential optimization relative to length and coverage were studied, also a notion of a totaly optimal rule was presented. In [2], different kinds of greedy algorithms for inhibitory rules construction were studied.

The paper consists of six sections. Section 2 contains main notions connected with a decision table and inhibitory decision rules. In section 3, proposed algorithm for construction of a directed acyclic graph is presented. Section 4 contains a description of a procedure of optimization relative to coverage. In section 5, a greedy algorithm for inhibitory decision rules construction is presented. Section 6 contains experimental results with decision tables from UCI Machine Learning Repository, and section 7 - conclusions.

2   Main Notions

In this section, notions corresponding to decision tables and inhibitory decision rules are presented.

A *decision table* T is a rectangular table with n columns labeled with conditional attributes $f_1, \ldots, f_n$. Rows of this table are filled with nonnegative integers that are interpreted as values of conditional attributes. Rows of T are pairwise different and each row is labeled with a nonnegative integer (decision) that is interpreted as a value of a decision attribute. The set of decisions attached to rows of the table T is denoted by $D(T)$. The number of rows in the table T is denoted by $N(T)$.

A table obtained from T by the removal of some rows is called a *subtable* of the table T. A subtable T' of the table T is called *reduced* if $|D(T')| < |D(T)|$, and *unreduced* otherwise when $|D(T')| = |D(T)|$.

Let T be nonempty, $f_{i_1}, \ldots, f_{i_m} \in \{f_1, \ldots, f_n\}$ and $a_1, \ldots, a_m$ be nonnegative integers. The subtable of the table T which contains only rows that have numbers $a_1, \ldots, a_m$ at the intersection with columns $f_{i_1}, \ldots, f_{i_m}$ is denoted by $T(f_{i_1}, a_1) \ldots (f_{i_m}, a_m)$. Such nonempty subtables (including the table T) are called *separable subtables* of T.

An attribute $f_i \in \{f_1, \ldots, f_n\}$ is *not constant* on T if it has at least two different values. For the attribute that is not constant on T it is possible to find *the most frequent value*. It is an attribute's value attached to the maximum number of rows in T.

The set of attributes from $\{f_1, \ldots, f_n\}$ which are not constant on T is denoted by $E(T)$. For any $f_i \in E(T)$, the set of values of the attribute f_i in T is denoted by $E(T, f_i)$. If $f_i \in E(T)$ is the attribute with the most frequent value then $E(T, f_i)$ contains only one element.

The expression

$$f_{i_1} = a_1 \wedge \ldots \wedge f_{i_m} = a_m \rightarrow d \neq k \tag{1}$$

is called an *inhibitory rule over* T if $f_{i_1}, \ldots, f_{i_m} \in \{f_1, \ldots, f_n\}$, $a_1, \ldots a_m$ are nonnegative integers, and $k \in D(T)$. It's possible that $m = 0$. In this case (1) is equal to the rule

$$\rightarrow d \neq k. \tag{2}$$

Let Θ be a subtable of T and $r = (b_1, \ldots, b_n)$ be a row of Θ. The rule (1) is called *realizable for* r, if $a_1 = b_{i_1}, \ldots, a_m = b_{i_m}$. The rule (2) is realizable for any row from Θ.

The rule (1) is called *true for* Θ if each row of Θ for which the rule (1) is realizable has the decision attached to it that is different from k. The rule (2) is true for Θ if and only if each row of Θ is labeled with the decision different from k. If the rule (1) is an inhibitory rule over T which is true for Θ and realizable for r, then (1) is an *inhibitory rule for* Θ *and* r *over* T.

Let Θ be a subtable of T, τ be an inhibitory rule over T, and τ be equal to (1).

The *coverage* of τ relative to Θ is the number of rows in Θ for which τ is realizable and which are labeled with decisions other than k. It is denoted by

$c(\tau)$. The coverage of inhibitory rule (2) relative to Θ is equal to the number of rows in Θ which are labeled with the decisions other than k. If τ is true for Θ then $c(\tau) = N(\Theta(f_{i_1}, a_1) \ldots (f_{i_m}, a_m))$.

3   Algorithm for Directed Acyclic Graph Construction

In this section, a modification of a dynamic programming algorithm that construct, for a given decision table T, a *directed acyclic graph* $\Lambda(T)$ is presented. Based on this graph it is possible to describe the set of inhibitory decision rules for T and each row r of T. Nodes of the graph are separable subtables of the table T. During each step, the algorithm processes one node and marks it with the symbol *. At the first step, the algorithm constructs a graph containing a single node T that is not marked with *.

Let us assume that the algorithm has already performed p steps. Now, the step $(p+1)$ will be described. If all nodes are marked with the symbol * as processed, the algorithm finishes its work and presents the resulting graph as $\Lambda(T)$. Otherwise, choose a node (table) Θ, that has not been processed yet. If Θ is reduced, then mark Θ with the symbol * and go to the step $(p+2)$. Otherwise, for each $f_i \in E(\Theta)$, draw a bundle of edges from the node Θ, if f_i is the attribute with the minimum number of values. If f_i is the attribute with the most frequent value, draw one edge from the node Θ. Let f_i be the attribute with the minimum number of values and $E(\Theta, f_i) = \{b_1, \ldots, b_t\}$. Then draw t edges from Θ and label them with pairs $(f_i, b_1) \ldots (f_i, b_t)$ respectively. These edges enter to nodes $\Theta(f_i, b_1), \ldots, \Theta(f_i, b_t)$. For the rest of attributes from $E(\Theta)$ draw one edge, for each attribute, from the node Θ and label it with pair (f_i, b_1), where b_1 is the most frequent value of the attribute f_i. This edge enters to a node $\Theta(f_i, b_1)$. If some of nodes $\Theta(f_i, b_1), \ldots, \Theta(f_i, b_t)$ are absent in the graph then add these nodes to the graph. Each row r of Θ is labeled with the set of attributes $E_{\Lambda(T)}(\Theta, r) \subseteq E(\Theta)$. The node Θ is marked with the symbol * and proceed to the step $(p+2)$.

The graph $\Lambda(T)$ is a directed acyclic graph. A node of this graph will be called *terminal* if there are no edges leaving this node. Note that a node Θ of $\Lambda(T)$ is terminal if and only if Θ is reduced.

In the next section, a procedure of optimization of the graph $\Lambda(T)$ relative to the coverage will be described. As a result a graph Γ is obtained, with the same sets of nodes and edges as in $\Lambda(T)$. The only difference is that any row r of each unreduced table Θ from Γ is labeled with a set of attributes $E_\Gamma(\Theta, r) \subseteq E(\Theta, r)$.

Let G be the graph $\Lambda(T)$ obtained from $\Lambda(T)$ by procedure of optimization relative to coverage.

Now, for each node Θ of G and for each row r of Θ a set of inhibitory rules $Rul_G(\Theta, r)$ over T will be described.

Let Θ be a terminal node of G, i.e., Θ is a reduced table. Then

$$Rul_G(\Theta, r) = \{\rightarrow d \neq k : k \in D(T) \setminus D(\Theta)\}.$$

Let now Θ be a nonterminal node of G such that for each child Θ' of Θ and for each row r' of Θ' the set of rules $Rul_G(\Theta', r')$ is already defined. Let $r = (b_1, \ldots, b_n)$ be a row of Θ. For any $f_i \in E_G(\Theta, r)$, the set of rules $Rul_G(\Theta, r, f_i)$ is defined as follows:

$$Rul_G(\Theta, r, f_i) = \{f_i = b_i \wedge \alpha \to d \neq k : \alpha \to d \neq k \in Rul_G(\Theta(f_i, b_i), r)\}.$$

Then $Rul_G(\Theta, r) = \bigcup_{f_i \in E_G(\Theta, r)} Rul_G(\Theta, r, f_i)$.

To illustrate the presented algorithm a decision table T depicted on the top of Fig. 1 is considered. The graph $\Lambda(T)$ is denoted by G.

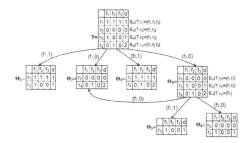

Fig. 1. Directed acyclic graph for decision table T

Now, for each node Θ of the graph G and for each row r of Θ the set $Rul_G(\Theta, r)$ of inhibitory rules for Θ and r over T will be presented, starting from terminal nodes. Terminal nodes of the graph G are Θ_1, Θ_2, Θ_3, Θ_5 and Θ_6. For these nodes, $Rul_G(\Theta_1, r_1) = Rul_G(\Theta_1, r_3) = \{\to \neq 0, \to \neq 2\}$; $Rul_G(\Theta_2, r_2) = Rul_G(\Theta_2, r_4) = \{\to \neq 1\}$; $Rul_G(\Theta_3, r_1) = Rul_G(\Theta_3, r_4 = \{\to \neq 0\}$; $Rul_G(\Theta_5, r_3) = \{\to \neq 0, \to \neq 2\}$; $Rul_G(\Theta_6, r_2) = Rul_G(\Theta_6, r_3) = \{\to \neq 2\}$. Now, it is possible to describe the sets of rules attached to rows of nontermi-nal node Θ_4. For this subtable children (subtables Θ_2, Θ_5 and Θ_6) are already treated, and we have:

$Rul_G(\Theta_4, r_2) = \{f_1 = 0 \to \neq 1, f_2 = 0 \to \neq 2\}$;
$Rul_G(\Theta_4, r_3) = \{f_1 = 1 \to \neq 0, f_1 = 1 \to \neq 2, f_2 = 0 \to \neq 2\}$;
$Rul_G(\Theta_4, r_4) = \{f_1 = 0 \to \neq 1\}$;
Finally, it is possible to describe the sets of rules attached to rows of T:
$Rul_G(T, r_1) = \{f_1 = 1 \to \neq 0, f_1 = 1 \to \neq 2, f_2 = 1 \to \neq 0\}$;
$Rul_G(T, r_2) = \{f_1 = 0 \to \neq 1, f_3 = 0 \wedge f_1 = 0 \to \neq 1, f_3 = 0 \wedge f_2 = 0 \to \neq 2\}$;
$Rul_G(T, r_3) = \{f_1 = 1 \to \neq 0, f_1 = 1 \to \neq 2, f_3 = 0 \wedge f_1 = 1 \to \neq 0, f_3 = 0 \wedge f_1 = 1 \to \neq 2, f_3 = 0 \wedge f_2 = 0 \to \neq 2\}$;
$Rul_G(T, r_4) = \{f_1 = 0 \to \neq 1, f_2 = 1 \to \neq 0, f_3 = 0 \wedge f_1 = 0 \to \neq 1\}$.

4   Procedure of Optimization Relative to Coverage

In this section, a procedure of optimization of the graph G relative to the cov-erage c is presented.

The algorithm moves from the terminal nodes of the graph G which are re-duced subtables to the node T. It will assign to each row r of each table Θ the number $Opt_G^c(\Theta, r)$ – the maximum coverage of an inhibitory rule from $Rul_G(\Theta, r)$, and it will change the set $E_G(\Theta, r)$ attached to the row r in the nonterminal table Θ. The obtained graph is denoted by $G(c)$.

Let Θ be a terminal node of G. Then the number

$$Opt_G^c(\Theta, r) = N(\Theta)$$

is assigned to each row r of Θ.

Let Θ be a nonterminal node and all children of Θ have already been treated. Let $r = (b_1, \ldots, b_n)$ be a row of Θ. The number

$$Opt_G^c(\Theta, r) = \max\{Opt_G^c(\Theta(f_i, b_i), r) : f_i \in E_G(\Theta, r)\}$$

is assigned to the row r in the table Θ and set

$$E_{G(c)}(\Theta, r) = \{f_i : f_i \in E_G(\Theta, r), Opt_G^c(\Theta(f_i, b_i), r) = Opt_G^c(\Theta, r)\}.$$

One can show that, for each node Θ of the graph $G(c)$ and for each row r of Θ, the set of rules $Rul_{G(c)}(\Theta, r)$ coincides with the set of all rules with the maximum coverage from $Rul_G(\Theta, r)$.

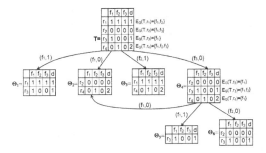

Fig. 2. Graph G^c

Fig. 2 presents the directed acyclic graph $G(c)$ obtained from the graph G (see Fig. 1) by the procedure of optimization relative to the coverage. As a result each row r_i, $i = 1, \ldots, 4$ of the table T has assigned the set $Rul_{G(c)}(T, r)$ of inhibitory rules for T and r_i with the maximum coverage. The value $Opt_G^c(T, r_i)$ is equal to the maximum coverage of an inhibitory rule for T and r_i.

Using the graph $G(c)$ it is possible to describe for each row r_i, $i = 1, \ldots, 4$, of the table T the set $Rul_{G(c)}(T, r_i)$ of inhibitory decision rules for T and r_i with the maximum coverage. The value $Opt_G^c(T, r_i)$ which is equal to the maximum coverage of an inhibitory decision rule for T and r_i is also obtained. We have

$Rul_G(T, r_1) = \{f_1 = 1 \rightarrow\neq 0, f_1 = 1 \rightarrow\neq 2, f_2 = 1 \rightarrow\neq 0\}$, $Opt_G^c(T, r_1) = 2$;
$Rul_G(T, r_2) = \{f_1 = 0 \rightarrow\neq 1, f_3 = 0 \wedge f_1 = 0 \rightarrow\neq 1, f_3 = 0 \wedge f_2 = 0 \rightarrow\neq 2\}$,
$Opt_G^c(T, r_2) = 2$;

$Rul_G(T, r_3) = \{f_1 = 1 \rightarrow\neq 0, f_1 = 1 \rightarrow\neq 2, f_3 = 0 \wedge f_2 = 0 \rightarrow\neq 2\}$,
$Opt_G^c(T, r_3) = 2$
$Rul_G(T, r_4) = \{f_1 = 0 \rightarrow\neq 1, f_2 = 1 \rightarrow\neq 0, f_3 = 0 \wedge f_1 = 0 \rightarrow\neq 1\}$,
$Opt_G^c(T, r_4) = 2$.

5   Greedy Algorithm

In this section, a greedy algorithm for inhibitory rule construction is presented. The algorithm uses an uncertainity measure for decision table $R(T')$ which is the number of unordered pairs of rows with different decisions in T'. At each iteration an attribute $f_i \in \{f_1, \ldots, f_n\}$ is selected, such that uncertainty of corresponding subtable is minimum. Algorithm constructs, for the row r labeled with decision q an inhibitory rule for each decision $k \in D(T) \setminus \{q\}$. As a result, for each row r of T, a set of inhibitory rules is constructed. Among them, for the row r, an inhibitory rule with the maximum coverage is selected.

Algorithm 1. Greedy algorithm for inhibitory decision rule construction

---

Require: Decision table T with conditional attributes $f_1, \ldots, f_n$, row $r = (b_1, \ldots, b_n)$
labeled with a decision q, decision $k \in D(T) \setminus \{q\}$.
Ensure: Inhibitory decision rule for T, r and k.
  $Q \leftarrow \emptyset$;
  $T' \leftarrow T$;
  while T' contains rows labeled with decision k **do**
    select $f_i \in \{f_1, \ldots, f_n\}$ such that $R(T'(f_i, b_i), k)$ is minimum;
    $T' \leftarrow T'(f_i, b_i)$;
    $Q \leftarrow Q \cup \{f_i\}$;
  end while
  $\bigwedge_{f_i \in Q}(f_i = b_i) \rightarrow\neq k$.

---

6   Experimental Results

Experiments were done on decision tables from UCI Machine Learning Repository [5]. Some decision tables contain conditional attributes that take unique value for each row. Such attributes were removed. In some tables there were equal rows with, possibly, different decisions. In this case each group of identical rows was replaced with a single row from the group with the most common decision for this group. In some tables there were missing values. Each such value was replaced with the most common value of the corresponding attribute.

For each such decision table T, using modified dynamic programming algorithm, the directed acyclic graph $\Lambda(T)$ was constructed. Then, optimization relative to coverage was applied. For each row r of T, the maximum coverage of an inhibitory decision rule for T and r was obtained. After that, for rows of T the average coverage of rules with the maximum coverage - one for each row, was calculated.

Table 1. Average coverage of inhibitory rules

Decision table	attr	rows	avg_cov	[s]	dp_cov	[s]	greedy_cov
adult-stretch	4	16	6.25	1	7.00	1	7.00
balance-scale	4	625	10.89	0	11.94	1	11.66
breast-cancer	9	266	6.15	4	9.53	19	4.09
cars	6	1728	470.33	1	543.74	8	419.06
lymphography	18	148	141.00	91	141.00	275	20.84
monks-2-test	6	432	12.11	1	12.36	4	5.29
monks-2-train	6	169	4.32	1	6.38	2	6.25
nursery	8	12960	4997.17	18	5400.00	193	3084.01
shuttle-landing-control	6	15	1.80	0	2.13	0	1.87
soybean-small	35	47	37.00	16	37.00	32	8.89
teeth	8	23	16.22	0	16.22	0	3.74
zoo-data	16	59	50.46	16	50.46	27	13.37

Table 1 presents the average coverage of inhibitory decision rules. Column *attr* contains the number of attributes in T, column *rows* - the number of rows in T. Values of the average coverage of inhibitory rules constructed by the proposed algorithm are contained in a column *avg_cov*, values of inhibitory rules constructed by the dynamic programming algorithm are contained in the column *dp_cov*, and values of inhibitory rules constructed by the greedy algorithm are contained in the column *greedy_cov*. Time execution of performed experiments, in seconds, is contained in a column *[s]*. The value 0 denotes the time execution below one second.

Comparison with optimal values (obtained by the dynamic programming algorithm) is presented in table 2. Based on the average relative difference it is possible to see how close on average coverage is proposed and greedy solution to optimal solution. Columns *rel_diff* and *greedy-rel_diff* contain a relative difference. It is equal to $(Opt\_Coverage - Coverage)/Opt\_Coverage$, where $Opt\_Coverage$ denotes the average coverage of inhibitory decision rules constructed by the dynamic programming algorithm, *Coverage* denotes the average coverage of inhibitory decision rules constructed by the proposed algorithm (in a case of column *rel_diff*) and greedy algorithm (in a case of column *greedy-rel_diff*). The last row in Table 2 presents the average value of the relative difference for considered decision tables. These values show, that on average, proposed approach allows to obtain values of coverage of constructed rules closer to optimal ones than greedy approach.

Table 2. Relative difference on average coverage of inhibitory rules

Decision table	attr	rows	rel_diff	greedy-rel_diff
adult-stretch	4	16	0.11	0.00
balance-scale	4	625	0.09	0.02
breast-cancer	9	266	0.35	0.57
cars	6	1728	0.14	0.23
lymphography	18	148	0.00	0.85
monks-2-test	6	432	0.02	0.57
monks-2-train	6	169	0.32	0.02
nursery	8	12960	0.07	0.43
shuttle-landing-control	6	15	0.15	0.12
soybean-small	35	47	0.00	0.76
teeth	8	23	0.00	0.77
zoo-data	16	59	0.00	0.74
average			0.10	0.42

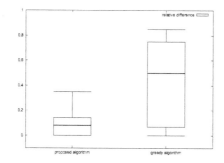

Fig. 3. Comparison of relative difference for proposed algorithm and greedy algorithm

Fig. 3 presents a relative difference of coverage of inhibitory rules constructed by the proposed algorithm and greedy algorithm.

Experimental results of the proposed algorithm and greedy algorithm were compared also using Wilcoxon test (available among others, on web-site vassarstats.net). It shows that the average coverage of rules is significantly affected by the proposed algorithm ($W(12) = 10$, $p > 0.05$, two-tailed test).

Experiments were done using software system Dagger [1] which is implemented in C++ and uses Pthreads and MPI libraries for managing threads and processes respectively, on computer with i5-3230M processor and 8 GB of RAM.

7   Conclusions

A modification of the dynamic programming algorithm for optimization of inhibitory decision rules relative to the coverage was presented. Using relative difference on average coverage of inhibitory rules it was possible to see how close is proposed and greedy solution to optimal solution. Experimental results of the proposed algorithm and greedy algorithm were compared using Wilcoxon test. It showed that the average coverage of rules is significantly affected by the proposed algorithm.

In the future works, accuracy of rule based classifiers using considered algorithms will be compared.

Acknowledgement. The author wishes to thanks the anonymous reviewers for useful comments. Thank you to Prof. Moshkov and Dr. Chikalov for possibility to work with Dagger software system.

References

1. Alkhalid, A., Amin, T., Chikalov, I., Hussain, S., Moshkov, M., Zielosko, B.: Dagger: A tool for analysis and optimization of decision trees and rules. In: Computational Informatics, Social Factors and New Information Technologies: Hypermedia Perspectives and Avant-Garde Experiences in the Era of Communicability Expansion, pp. 29–39. Blue Herons (2011)

2. Alsolami, F., Chikalov, I., Moshkov, M.: Comparison of heuristics for inhibitory rule optimization. In: Jędrzejowicz, P., Jain, L.C., Howlett, R.J., Czarnowski, I. (eds.) KES 2014. Procedia Computer Science, vol. 35, pp. 378–387. Elsevier (2014)

3. Alsolami, F., Chikalov, I., Moshkov, M., Zielosko, B.M.: Length and coverage of inhibitory decision rules. In: Nguyen, N.T., Hoang, K., Jędrzejowicz, P. (eds.) ICCCI 2012, Part II. LNCS, vol. 7654, pp. 325–334. Springer, Heidelberg (2012)

4. Alsolami, F., Chikalov, I., Moshkov, M., Zielosko, B.: Optimization of inhibitory decision rules relative to length. Studia Informatica 33(2A(105)), 395–406 (2012)

5. Asuncion, A., Newman, D.J.: UCI Machine Learning Repository (2007), http://wwwicsuciedu/~mlearn/, http://www.ics.uci.edu/~mlearn/

6. Błaszczyński, J., Słowiński, R., Szeląg, M.: Sequential covering rule induction algorithm for variable consistency rough set approaches. Inf. Sci. 181(5), 987–1002 (2011)

7. Delimata, P., Moshkov, M.J., Skowron, A., Suraj, Z.: Two families of classification algorithms. In: An, A., Stefanowski, J., Ramanna, S., Butz, C.J., Pedrycz, W., Wang, G. (eds.) RSFDGrC 2007. LNCS (LNAI), vol. 4482, pp. 297–304. Springer, Heidelberg (2007)

8. Delimata, P., Moshkov, M., Skowron, A., Suraj, Z.: Lazy classification algorithms based on deterministic and inhibitory rules. In: Magdalena, L., Ojeda-Aciego, M., Verdegay, J.L. (eds.) Information Processing and Management of Uncertainty in Knowledge-Based Systems, pp. 1773–1778 (2008)

9. Delimata, P., Moshkov, M., Skowron, A., Suraj, Z.: Inhibitory Rules in Data Analysis. SCI, vol. 163. Springer, Heidelberg (2009)

10. Moshkov, M., Zielosko, B.: Combinatorial Machine Learning - A Rough Set Approach. SCI, vol. 360. Springer, Heidelberg (2011)

11. Nguyen, H.S.: Approximate boolean reasoning: foundations and applications in data mining. In: Peters, J.F., Skowron, A. (eds.) Transactions on Rough Sets V. LNCS, vol. 4100, pp. 334–506. Springer, Heidelberg (2006)

12. Sikora, M., Wróbel, Ł.: Data-driven adaptive selection of rule quality measures for improving rule induction and filtration algorithms. Int. J. General Systems 42(6), 594–613 (2013)

13. Stańczyk, U.: Decision rule length as a basis for evaluation of attribute relevance. Journal of Intelligent and Fuzzy Systems 24(3), 429–445 (2013)

14. Suraj, Z.: Some remarks on extensions and restrictions of information systems. In: Ziarko, W.P., Yao, Y. (eds.) RSCTC 2000. LNCS (LNAI), vol. 2005, pp. 204–211. Springer, Heidelberg (2001)

15. Zielosko, B.: Optimization of approximate decision rules relative to coverage. In: Kozielski, S., Mrózek, D., Kasprowski, P., Małysiak-Mrózek, B. (eds.) BDAS 2014. CCIS, vol. 424, pp. 170–179. Springer, Heidelberg (2014)

Application of the Shapley-Shubik Power Index in the Process of Decision Making on the Basis of Dispersed Medical Data

Małgorzata Przybyła-Kasperek[✉]

Institute of Computer Science, University of Silesia,
Będzińska 39, 41-200 Sosnowiec, Poland
malgorzata.przybyla-kasperek@us.edu.pl
http://www.us.edu.pl

Abstract. The paper considers the issues that are related to the process of decision-making that is based on dispersed knowledge. In previous papers the author proposed a dispersed decision support system with a dynamic structure, which is the approach that is used. The novelty, that is analyzed in this paper is the application of a power index in this system. Together with the Shapley-Shubik index, a simple method of determining local decisions has been applied. The purpose of this was to reduce the computational complexity in comparison with the approach that was proposed in the earlier papers. In experiments the situation is considered in which medical data from one domain are collected in many medical centers. We want to use all of the collected data at the same time in order to make a global decisions.

Keywords: Knowledge-based systems · Group decisions and negotiations · Global decision · Dispersed knowledge · Shapley-Shubik power index

1 Introduction

In many real-life situations, there is a need separate entities - agents - to make joint decisions. In many circumstances in which agents need to make a joint decision, voting is used to aggregate the agents' preferences. The weight assigned to a player is not equal to its actual influence on the outcome of the decisions that are made using the weighted voting game. Consider, for example, a weighted voting game in which the threshold value is equal to the sum of the weights of all of the players. In the literature we can find various ways of calculating the power of agents in the weighted voting game. Some of these proposals are given in the papers [1,3,14]. In this paper we describe the application of the Shapley-Shubik power index in a dispersed decision support system. Issues related to the use of dispersed knowledge were considered by the author in earlier papers [10,11,12,9]. In the paper [11] a decision-making system with a dynamic structure that is created using a negotiations stage was proposed. A system with such a structure is used in this paper. A description of the creation of this structure is as follows. A vector that describes the classification of a test object that is made on

© Springer International Publishing Switzerland 2015
S. Kozielski et al. (Eds.): BDAS 2015, CCIS 521, pp. 277–287, 2015.
DOI: 10.1007/978-3-319-18422-7_25

the basis of the local base is generated for each local knowledge base. Among the local knowledge bases three types of relations - friendship, conflict and neutrality - are defined. In the first step of a process of connecting the local knowledge bases in coalitions, the bases that remain in a friendship relation are combined into groups. The second step consists of re-examining the relations between the initial coalitions that were created. A negotiation process is implemented, in which in addition to the initial coalitions, the local knowledge bases that remain in a neutrality relation are included. In this paper, after the creation of the system structure a very simple method of generating local decisions within a cluster are used. The main aim of this paper is to describe the application of the Shapley-Shubik index in the process of determining global decisions.

Power indexes have been proposed to measure the ability of agents to influence the outcome of a vote. The most popular power indexes have been proposed by L. Shapley and M. Shubik in the paper [14] and J. Baznhaf in the paper [1]. Both indexes measure the probability that after an agent joins to a coalition, the status of the coalition will change from losing to winning.

The issue of making decisions based on dispersed knowledge is widely considered in the literature. For example, this issue is considered in the multiple model approach [4,5]. The concept of distributed decision-making is widely discussed in the paper [13]. In addition, the problem of using distributed knowledge is considered in many other papers [2,16,17]. A very important issue that is discussed in this paper is formulating the coalition and the process of negotiations. An approach to the issue of coalition formation was considered in the papers of Z. Pawlak [7,8]. The concept of inference is commonly considered in the literature. Examples of applying the inference in rule knowledge bases are descibed in [15]. The article [6] presents the problem of outlier detection in rule-based knowledge bases.

2   Notations and Definitions

We assume that the set of local knowledge bases that contain dispersed medical data from one domain is pre-specified. We assume that each local knowledge base is managed by one agent, which is called a resource agent.

Definition 1. *We call ag in $Ag = \{ag_1, \ldots, ag_n\}$ a resource agent if it has access to resources represented by a decision table $D_{ag} := (U_{ag}, A_{ag}, d_{ag})$, where U_{ag} is the universe; A_{ag} is a set of conditional attributes, V_{ag}^a is a set of values of the attribute a; d_{ag} is a decision attribute, V_{ag}^d is called the value set of d_{ag}.*

The purpose of this paper is to present the application of the Shapley-Shubik index in the global decision-making process. This index will be used to evaluate the strength of an agent in the decision making process. In the proposed approach, agents that take similar decisions are connected into group. The power index will be calculated for the synthesis agents. A synthesis agent is a superordinate agent with respect to a group of resource agents. This agent will be defined later in the paper.

2.1   The Process of Creating Clusters

The method of generating clusters with the stage of negotiations was proposed in the paper [11]. The method consists in defining groups of resource agents who agree on the classification of the test object. Each resource agent $ag \in Ag$, based on the decision table D_{ag}, can independently determine the value of the decision for a test object for which the values of the set of attributes A_{ag} are defined. Let there be given a test object $\bar{x}$ for which we want to generate a global decision. In order to determine groups of agents, from each decision table of a resource agent $D_{ag_i}, i \in \{1, \ldots, n\}$ and from each decision class $X_v^{ag_i}, v \in V^{d_{ag_i}}$, the smallest set containing at least m_1 objects is chosen, for which the values of conditional attributes bear the greatest similarity to the test object. The value of the parameter m_1 is selected experimentally. The subset of relevant objects is the union of the sets of objects selected from all decision classes. The next stage in the process of generating groups of agents is to determine the vectors of values specifying the classification of the test object made by the agents. Each coordinate of the vector is determined on the basis of relevant objects that were previously selected from the decision table of the resource agent. Thus, for each resource agent ag_i, $i \in \{1, \ldots, n\}$, a c-dimensional vector $[\bar{\mu}_{i,1}(\bar{x}), \ldots, \bar{\mu}_{i,c}(\bar{x})]$ is generated, where the value $\bar{\mu}_{i,j}(\bar{x})$ means the certainty with which the decision $v_j \in V^d, j \in \{1, \ldots, c\}, c = card\{V^d\}$ is made about the object $\bar{x}$ by the resource agent ag_i. The value $\bar{\mu}_{i,j}(\bar{x})$ is defined as follows:

$$\bar{\mu}_{i,j}(\bar{x}) = \frac{\sum_{y \in U_{ag_i}^{rel} \cap X_{v_j}^{ag_i}} s(\bar{x}, y)}{card\{U_{ag_i}^{rel} \cap X_{v_j}^{ag_i}\}}, i \in \{1, \ldots, n\}, j \in \{1, \ldots, c\}, \tag{1}$$

where $c = card\{V^d\}$, $U_{ag_i}^{rel}$ is the subset of relevant objects selected from the decision table D_{ag_i} of a resource agent ag_i and $X_{v_j}^{ag_i}$ is the decision class of the decision table of resource agent ag_i; $s(x, y)$ is the measure of similarity between objects x and y. On the basis of the vector of values defined above a vector of rank assigned to the values of the decision attribute is specified. The vector of rank is defined as follows: rank 1 is assigned to the values of the decision attribute which are taken with the maximum level of certainty, rank 2 is assigned to the second best decisions etc. Proceeding in this way for each resource agent $ag_i, i \in \{1, \ldots, n\}$, the vector of rank $[r_{i,1}(\bar{x}), \ldots, r_{i,c}(\bar{x})]$ will be defined. The definitions of friendship relation, conflict relation and neutrality relation are given next. Definitions of the relations are based on the concepts that were provided by Pawlak [7]. We define the function $\phi_{v_j}^x$ for the test object x and each value of the decision attribute $v_j \in V^d$; $\phi_{v_j}^x : Ag \times Ag \to \{0, 1\}$;

$$\phi_{v_j}^x(ag_i, ag_k) = \begin{cases} 0 \text{ if } r_{i,j}(x) = r_{k,j}(x) \\ 1 \text{ if } r_{i,j}(x) \neq r_{k,j}(x) \end{cases} \text{ where } ag_i, ag_k \in Ag.$$

Definition 2. *Agents* $ag_i, ag_k \in Ag$ *are in a friendship relation due to the object* x *and decision class* $v_j \in V^d$, *which is written* $R_{v_j}^+(ag_i, ag_k)$, *if and only if* $\phi_{v_j}^x(ag_i, ag_k) = 0$. *Agents* $ag_i, ag_k \in Ag$ *are in a conflict relation due to the object* x *and decision class* $v_j \in V^d$, *which is written* $R_{v_j}^-(ag_i, ag_k)$, *if and only if* $\phi_{v_j}^x(ag_i, ag_k) = 1$.

We also define the intensity of conflict between agents using a function of the distance between agents. We define the distance between agents ρ^x for the test object x: $\rho^x(ag_i, ag_k) = \frac{\sum_{v_j \in V^d} \phi^x_{v_j}(ag_i, ag_k)}{card\{V^d\}}$, where $ag_i, ag_k \in Ag$.

Definition 3. *Let p be a real number, which belongs to the interval $[0, 0.5)$. We say that agents $ag_i, ag_k \in Ag$ are in a friendship relation due to the object x, which is written $R^+(ag_i, ag_k)$, if and only if $\rho^x(ag_i, ag_k) < 0.5 - p$. Agents $ag_i, ag_k \in Ag$ are in a conflict relation due to the object x, which is written $R^-(ag_i, ag_k)$, if and only if $\rho^x(ag_i, ag_k) > 0.5 + p$. Agents $ag_i, ag_k \in Ag$ are in a neutrality relation due to the object x, which is written $R^0(ag_i, ag_k)$, if and only if $0.5 - p \leq \rho^x(ag_i, ag_k) \leq 0.5 + p$.*

The first step in the process of clusters creating is to define the initial group of agents remaining in the friendship relation. Let Ag be the set of resource agents. The initial cluster due to the classification of object x is the maximum, due to the inclusion relation, subset of resource agents $X \subseteq Ag$ such that $\forall_{ag_i, ag_k \in X} \ R^+(ag_i, ag_k)$. After the first stage of clusters creating we obtain a set of initial clusters and a set of agents which are not included in any cluster. For each agent, which is not attached to any clusters, relations between this agent and the generated initial clusters and other agents without coalition are analyzed. In the second stage, agents without coalition are connected to each initial cluster, with which his relations will be good enough. Also new clusters, consisting of agents without coalition, which are in a sufficiently good relations, are creating. Let $C_1, \ldots, C_k$ be a set of initial clusters and $Ag \setminus \bigcup_{i=1}^{k} C_i$ be a set of agents which are not attached to any clusters. In the second stage of clustering process a generalized distance function between agents is defined. This definition assumes that during the negotiation, agents put the greatest emphasis on compatibility of the ranks assigned to the decisions with the highest ranks. That is the values of the decisions that are most significant for the agent. Compatibility of the ranks assigned to the less meaningful decision is omitted during the second stage of clustering process.

We define the function ϕ^x_G for the test object x; $\phi^x_G : Ag \times Ag \to [0, \infty)$; $\phi^x_G(ag_i, ag_j) = \frac{\sum_{v_l \in Sign_{i,j}} |r_{i,l}(x) - r_{j,l}(x)|}{card\{Sign_{i,j}\}}$ where $ag_i, ag_j \in Ag$ and $Sign_{i,j} \subseteq V^d$ is the set of significant decision values for the pair of agents ag_i, ag_j. In the set $Sign_{i,j}$ there are the values of the decision, which the agent ag_i or agent ag_j gave the highest rank. During the second stage of the clusters creating process - the negotiation process, the intensity of the conflict between the two groups of agents is determined by using the generalized distance. We define the generalized distance between agents ρ^x_G for the test object x; $\rho^x_G : 2^{Ag} \times 2^{Ag} \to [0, \infty)$

$$\rho^x_G(X, Y) = \begin{cases} 0 & \text{if } card\{X \cup Y\} \leq 1 \\ \frac{\sum_{ag, ag' \in X \cup Y} \phi^x_G(ag, ag')}{card\{X \cup Y\} \cdot (card\{X \cup Y\} - 1)} & \text{else} \end{cases} \quad \text{where } X, Y \subseteq Ag.$$

Then the agent ag is included to all initial clusters, for which the generalized distance does not exceed a certain threshold, which is set by the system's user.

Also agents without coalition, for which the value of the generalized distance function does not exceed the threshold, are combined into a new cluster. After completion of the second stage of the process of clustering we get the final form of clusters.

The proposed decision-making system has a hierarchical structure. For each cluster that contains at least two resource agents, a superordinate agent is defined, which is called a synthesis agent, as_j, where j is the number of cluster. The synthesis agent, as_j, has access to knowledge that is the result of the process of inference carried out by the resource agents that belong to its subordinate group. As is a finite set of synthesis agents. Now, we can provide a formal definition of a dispersed decision-making system. By a dispersed decision-making system (multi-agent system) with dynamically generated clusters we mean $WSD_{Ag}^{dyn} = \langle Ag, \{D_{ag} : ag \in Ag\}, \{As_x : x \text{ is a classified object}\}, \{\delta_x : x \text{ is a classified object}\}\rangle$ where Ag is a finite set of resource agents; $\{D_{ag} : ag \in Ag\}$ is a set of decision tables of resource agents; As_x is a set of synthesis agents, $\delta_x : As_x \to 2^{Ag}$ is a injective function that each synthesis agent assigns a cluster.

The paper [12] presents an example of creating a dynamic structure in the manner described above.

2.2   Application of the Shapley-Shubik Index

The aggregation of vectors is a method for determining the local decisions made by one cluster. This method consists in simple arithmetic operations that are performed on the vectors that are assigned to the resource agents. The vectors that are assigned to the resource agents were defined earlier in the paper at the stage of defining the relations between agents. Each resource agent ag_i, $i \in \{1, \ldots, n\}$, a c-dimensional vector $[\bar{\mu}_{i,1}(x), \ldots, \bar{\mu}_{i,c}(x)]$ is assigned, where $c = card\{V^d\}$, given by formula 1. In the proposed method of creating the system's structure, inseparable clusters are generated. When an agent belongs to several clusters, it is not fully committed in any of them. This means that the partial participation of the agent in the creation of the cluster should be considered. A coefficient of agent's membership in clusters is defined for each resource agent $ag \in Ag$ and given test object x, $m_{ag}^x = \frac{1}{card\{as \in As_x : ag \in \delta_x(as)\}}$. The value of the agent's membership in clusters is inversely proportional to the number of clusters to which the agent belongs. Then, the vector of values is determined for each cluster, according to the formula $[\mu_{j,1}(x), \ldots, \mu_{j,c}(x)] = \sum_{i \in \delta_x(as_j)} m_{ag_i}^x \cdot [\bar{\mu}_{i,1}(x), \ldots, \bar{\mu}_{i,c}(x)]$. This vector is the weighted average of the vectors that were assigned to the resource agents that belong to the cluster that is subordinate to the j-th synthesis agent. This method for generating local decisions of one cluster has very low computational complexity; significantly lower than the approximated method of the aggregation of decision tables, used in the paper [9]. The Shapley-Shubik index is widely used when decisions are taken by voting. To apply this index we have to deal with a simple game. The definition of a simple game is given below.

Definition 4. *The simple game is an ordered pair* (N, W), *where* N *is the set of players and* $W \subseteq 2^N$ *is the set of winning coalitions, such that:* $\emptyset \notin W$ - *an empty set is a losing coalition,* $N \in W$ - *a coalition of all players is a winning coalition, if* $S \in W$ *and* $S \subseteq T$, *then* $T \in W$ - *if* S *is a winning coalition, then each coalition containing* S *is also a winning coalition.*

In the proposed approach, the power of each synthesis agents is calculated. In the situation that is considered in this paper, the set of players is the set of synthesis agents $N = As$. A separate simple game is considered for each decision value $v_l \in V^d$. We assume that the j-th synthesis agent has a number of votes equal to the coordinate $\mu_{j,l}(x)$. Based on preliminary experiments, it was found that the best results are achieved when the threshold for the winning coalition is equal to the average value of the coordinate vectors of resource agents. Let them $q = \frac{1}{card\{As\}} \sum_{as_j \in As} \mu_{j,l}(x)$. We say that a coalition of synthesis agents $S \subseteq As$ is a winning coalition when the total number of votes at its disposal is more than q: $\sum_{as_j \in S} \mu_{j,l}(x) > q$.

Definition 5. *Let* $S \subseteq As$ *be a coalition and* $j \notin S$ *be a player. The player* j *is called a decisive player for coalition* S *if and only if the coalition* S *is a losing coalition, but adding the player* j *to coalition* S *will change the status of the coalition to a winning coalition.*

The value of the Shapley-Shubik power index is equal to the probability that the player will be decisive for the coalition, assuming that all of the arrangements are equally likely. The higher the value of the player is, the more he has to say when the group makes decisions, according to the rule for making decisions that is determined by the game. According to the above definition, the Shapley-Shubik index for the decision $v_l \in V^d$ and the agent $as_j \in As$ is defined as follows:

$$\varphi_j^l(x) = \frac{1}{(card\{As\})!} \sum_{\substack{S \subseteq As \\ \text{agent } j \text{ is decisive player for} \\ \text{the coalition } S}} \Big(card\{S\}\Big)! \Big(card\{N\} - card\{S\} - 1\Big)!$$

A general power of a synthesis agent is defined by the sum of the Shapley-Shubik index for all of the values of the decision $\varphi_j(x) = \sum_{v_l \in V^d} \varphi_j^l(x)$. This value is used for the conversion of the vector that is assigned to a given synthesis agent. The vectors are calculated by the following formula $\varphi_j(x) \cdot [\mu_{j,1}(x), \ldots, \mu_{j,c}(x)]$, for $j \in \{1, \ldots, card\{As\}\}$. Because of this transformation, agents who have greater power, as expressed by the value of Shapley-Shubik index, will also have a greater impact on global decisions. In the last stage of taking global decisions the vectors will be used to determine the global decisions that are taken by all of the agents. In order to determine the level of certainty with which individual decisions are taken by all of the agents, the sum of the vectors that are assigned to particular clusters should be calculated. The set of global decisions is determined using the method of ε -neighborhood. In the first stage of this method, decisions that are taken with the maximum level of certainty are selected. Next, we define a set of decisions that are in the ε -neighborhood of decisions with the maximum level of certainty. The set of global decisions for test object x that are generated by a dispersed decision-making system is defined as follows $\hat{d}_{WSD_{Ag}^{dyn}}(x) = \left\{ v_i \in V^d : \left| \sum_{j=1}^{card\{As\}} \varphi_j(x) \cdot \mu_{j,max}(x) - \sum_{j=1}^{card\{As\}} \varphi_j(x) \cdot \mu_{j,i}(x) \right| \leq \varepsilon \right\}$, where $v_{max}(x)$ is a decision that is made with the maximum level of certainty for test object x.

3 Experiments

The aim of the experiments is to examine the quality of the classification made on the basis of dispersed medical data by the decision-making system with using the Shapley-Shubik power index. An additional objective is to compare the effectiveness in terms of accuracy and execution time of this system with the results obtained in the paper [9] (where the Shapley-Shubik power index was not used). For the experiments the following data, which are in the UCI repository (archive.ics.uci.edu/ml/), were used: Lymphography data set, Primary Tumor data set. Both sets of data was obtained from the University Medical Centre, Institute of Oncology, Ljubljana, Yugoslavia (M. Zwitter and M. Soklic provided this data). In order to determine the efficiency of inference of the proposed decision-making system with respect to the analyzed data, each data set was divided into two disjoint subsets: a training set and a test set. A numerical summary of the data sets is as follows: Lymphography: # The training set - 104; # The test set - 44; # Conditional - 18; # Decision - 4; Primary Tumor: # The training set - 237; # The test set - 102; # Conditional - 17; # Decision - 22.

We will consider a situation in which medical data from one domain are collected in different medical centers. We want to use all of the collected data at the same time in order to make a global decisions. This approach not only allows the use of all available knowledge, but also should improve the efficiency of inference. In order to consider the discussed situation it is necessary to provide the knowledge stored in the form of a set of decision tables. Therefore, the training set was divided into a set of decision tables. For each of the data sets used, the decision-making system with five different versions were considered: WSD_{Ag1}^{dyn} - 3 resource agents; WSD_{Ag2}^{dyn} - 5 resource agents; WSD_{Ag3}^{dyn} - 7 resource agents; WSD_{Ag4}^{dyn} - 9 resource agents; WSD_{Ag5}^{dyn} - 11 resource agents. Note that the division of the data set was not made in order to improve the quality of the decisions taken by the decision-making system, but in order to store the knowledge in a distributed form. The measures of determining the quality of the classification are: *estimator of classification error e* in which an object is considered to be properly classified if the decision class used for the object belonged to the set of global decisions generated by the system; *estimator of classification ambiguity error e_{ONE}* in which object is considered to be properly classified if only one, correct value of the decision was generated to this object; *the average size of the global decisions sets $\overline{d}_{WSD_{Ag}^{dyn}}$* generated for a test set. For clarity, some designations for algorithms have been adopted in the description of the results of the experiments: S - the aggregation of the vectors that determine the level of certainty with which resource agents make decisions. This method involves calculating the weighted average of the vectors that are assigned to the resource agents, $N(\varepsilon)$ - the ε - neighborhood method, where ε determines the radius of the neighborhood, $A(m_2)$ - the approximated method of the aggregation of decision tables; used in the paper [9], $G(\varepsilon, MinPts)$ - the method of a density-based algorithm; used in the paper [9]. The results of the experiments with the proposed approach and the Lymphography data set are presented in the first part of table 1. In the table the following information is given: the name of multi-agent decision-making system (System); the optimal parameter values (Parameters); the algorithm's symbol (Algorithm); the three measures discussed earlier $e, e_{ONE}, \overline{d}_{WSD_{Ag}^{dyn}}$; the time t needed to analyse a test set expressed in milliseconds. Such a unit for expressing the execution time was chosen because in this article it is important to compare the complexity of the considered methods. For comparison, in the second part of table 1, the results of the experiments that are presented in the paper [9] are given. These results were obtained by using a dispersed decision-making system, wherein the method of forming the system structure is the same as that presented in this paper. However, in this approach the Shapley-Shubik power index was not used. The approximated method of the aggregation of decision tables with the method of a density-based algorithm have been used. The results of the experiments with the Primary Tumor data set are presented in table 2. In the first part of the table the results for the proposed approach are described. For comparison, in the second part of table 2, the results of the experiments that are presented in the paper [9] are given. In tables the best results in terms of the measures e and $\overline{d}_{WSD_{Ag}^{dyn}}$ are bold.

Based on the results of the experiments given in tables 1 and 2 the following conclusions can be drawn. The use of the Shapley-Shubik power index instead of the approximated method of the aggregation of decision tables significantly reduces the execution time. This is very important if we want to apply the method to large distributed data. For the Lymphography data set, in the case of systems with 5 (WSD_{Ag4}^{dyn}), 7 (WSD_{Ag3}^{dyn}) and 9 (WSD_{Ag4}^{dyn}) resource agents the aggregation of the vectors of the values with the Shapley-Shubik power index gave better results. But in the case of systems with 3 (WSD_{Ag1}^{dyn}) and 11 (WSD_{Ag5}^{dyn}) resource agents the approximated method of the aggregation of decision tables provided better results. For the Primary Tumor data set the proposed approach generates comparable results with the approach proposed in the paper [9]. In the paper [9] the results of the experiments using other existing approaches, with the Lymphography and the Primary Tumor data set, are listed. Comparing these results we can say that the results presented in this paper may be considered as a quite good. The author checked that the proposed approach can be applied to the data from another field than medicine, but these results will be presented in another paper.

Table 1. Experiments results with the Lymphography data set

Using the Shapley-Shubik index and the vector aggregation

System	Parameters	Algorithm	e	e_{ONE}	$\overline{d}_{WSD^{dyn}_{Ag}}$	t
WSD^{dyn}_{Ag1}	$m_1 = 10,\ p = 0.05$	$SN(0.216)$	0.068	0.591	1.545	62
	$m_1 = 10,\ p = 0.05$	$SN(0.126)$	0.136	0.409	1.295	62
WSD^{dyn}_{Ag2}	$m_1 = 5,\ p = 0.05$	$SN(0.315)$	**0.068**	0.568	**1.568**	78
	$m_1 = 5,\ p = 0.05$	$SN(0.084)$	0.136	0.250	1.159	78
WSD^{dyn}_{Ag3}	$m_1 = 3,\ p = 0.05$	$SN(0.081)$	**0.045**	0.409	**1.386**	172
	$m_1 = 3,\ p = 0.05$	$SN(0.036)$	0.136	0.273	1.159	172
WSD^{dyn}_{Ag4}	$m_1 = 4,\ p = 0.05$	$SN(0.078)$	**0.045**	0.500	**1.455**	1 550
	$m_1 = 4,\ p = 0.05$	$SN(0.063)$	0.114	0.477	1.364	1 550
WSD^{dyn}_{Ag5}	$m_1 = 5,\ p = 0.05$	$SN(0.03)$	0.182	0.614	1.432	19 670
	$m_1 = 5,\ p = 0.05$	$SN(0.048)$	0.205	0.545	1.341	19 670

Results presented in the paper [9]

System	Parameters	Algorithm	e	e_{ONE}	$\overline{d}_{WSD^{dyn}_{Ag}}$	t
WSD^{dyn}_{Ag1}	$m_1 = 2,\ p = 0.3$	$A(1)G(0.0268;2)$	**0.068**	0.568	**1.523**	78
	$m_1 = 2,\ p = 0.3$	$A(1)G(0.0004;2)$	0.159	0.182	1.023	78
WSD^{dyn}_{Ag2}	$m_1 = 2,\ p = 0.3$	$A(3)G(0.042;2)$	0.091	0.682	1.591	560
	$m_1 = 2,\ p = 0.3$	$A(3)G(0.0128;2)$	0.136	0.318	1.182	560
WSD^{dyn}_{Ag3}	$m_1 = 1,\ p = 0.05$	$A(1)G(0.0515;2)$	0.114	0.523	1.409	203
	$m_1 = 1,\ p = 0.05$	$A(1)G(0.0005;2)$	0.159	0.273	1.114	203
WSD^{dyn}_{Ag4}	$m_1 = 1,\ p = 0.05$	$A(1)G(0.0625;2)$	0.114	0.591	1.477	1 156
	$m_1 = 1,\ p = 0.05$	$A(1)G(0.052;2)$	0.136	0.5	1.364	1 156
WSD^{dyn}_{Ag5}	$m_1 = 5,\ p = 0.3$	$A(2)G(0.058;2)$	**0.159**	0.568	**1.409**	23 175
	$m_1 = 5,\ p = 0.3$	$A(2)G(0.0292;2)$	0.182	0.545	1.364	23 175

Table 2. Experiments results with the Primary Tumor data set

Using the Shapley-Shubik index and the vector aggregation

System	Parameters	Algorithm	e	e_{ONE}	$\overline{d}_{WSD^{dyn}_{Ag}}$	t
WSD_{Ag1}	$m_1 = 4,\ p = 0.2$	$SN(0.072)$	**0.363**	0.814	**3.010**	500
WSD_{Ag2}	$m_1 = 1,\ p = 0.1$	$SN(0.0003)$	0.363	0.853	3.284	390
WSD_{Ag3}	$m_1 = 1,\ p = 0.15$	$SN(0.0009)$	**0.363**	0.892	**3.706**	670
WSD_{Ag4}	$m_1 = 1,\ p = 0.05$	$SN(0.0003)$	0.333	0.902	4.147	2 280
WSD_{Ag5}	$m_1 = 1\ p = 0.05$	$SN(0.0003)$	0.324	0.912	4.245	25 300

Results presented in the paper [9]

System	Parameters	Algorithm	e	e_{ONE}	$\overline{d}_{WSD^{dyn}_{Ag}}$	t
WSD_{Ag1}	$m_1 = 5,\ p = 0.05$	$A(1)G(0.00546;2)$	0.373	0.814	3.020	780
WSD_{Ag2}	$m_1 = 3,\ p = 0.1$	$A(2)G(0.00001;2)$	**0.343**	0.814	**3.029**	1 219
WSD_{Ag3}	$m_1 = 2,\ p = 0.05$	$A(1)G(0.00001;2)$	0.373	0.902	3.745	1 640
WSD_{Ag4}	$m_1 = 4,\ p = 0.1$	$A(2)G(0.00001;2)$	0.353	0.882	3.686	4 720
WSD_{Ag5}	$m_1 = 2\ p = 0.2$	$A(3)G(0.00001;2)$	**0.314**	0.892	**4.245**	38 450

4   Conclusion

In this paper, using the Shapley-Shubik power index in a dispersed decision-making system is proposed. In cases in which global decisions are taken based local decisions, the voting method is used very often. When decisions are taken by a vote, the Shapley-Shubik power index is often used to evaluate the actual ability of agents to influence the outcome of the vote. Therefore, using the Shapley Shubik index in a dispersed decision system appears to be very natural and obvious approach. In the experiments, which are presented in the article, dispersed medical data have been used: Lymphography data set, Primary Tumor data set. The usage of dispersed medical data is very important, because in many medical centers, information from one domain, are collected. Thus, these data are in the dispersed form. Based on the presented results of experiments it can be concluded that the use of the Shapley-Shubik power index in decision-making system provides good results for dispersed medical data. And what is the most important significantly reduces the execution time.

References

1. Banzhaf, J.F.: Weighted voting doesn't work: A mathematical analysis. Rutgers Law Review 19(2), 317–343 (1965)
2. Delimata, P., Suraj, Z.: Feature selection algorithm for multiple classifier systems: A hybrid approach. Fundam. Inform. 85(1-4), 97–110 (2008)
3. Dubey, P., Shapley, L.S.: Mathematical properties of the banzhaf power index. Mathematics of Operations Research 4(2), 99–131 (1979)
4. Gatnar, E.: Multiple-model approach to classification and regression. PWN, Warsaw (2008)
5. Kuncheva, L.I.: Combining pattern classifiers methods and algorithms. John Wiley & Sons (2004)
6. Nowak-Brzezińska, A.: Outlier mining in rule-based knowledge bases. In: Yao, J., Yang, Y., Słowiński, R., Greco, S., Li, H., Mitra, S., Polkowski, L. (eds.) RSCTC 2012. LNCS, vol. 7413, pp. 206–211. Springer, Heidelberg (2012), http://dx.doi.org/10.1007/978-3-642-32115-3_24
7. Pawlak, Z.: On conflicts. Int. J. of Man-Machine Studies 21(2), 127–134 (1984)
8. Pawlak, Z.: An inquiry into anatomy of conflicts. Inf. Sci. 109(1-4), 65–78 (1998)
9. Przybyła-Kasperek, M.: Global decisions taking process, including the stage of negotiation, on the basis of dispersed medical data. In: Proceedings of BDAS 2014, Ustron, Poland, May 27-30, pp. 290–299 (2014), http://dx.doi.org/10.1007/978-3-319-06932-6_28
10. Przybyła-Kasperek, M., Wakulicz-Deja, A.: Application of reduction of the set of conditional attributes in the process of global decision-making. Fundam. Inform. 122(4), 327–355 (2013)
11. Przybyła-Kasperek, M., Wakulicz-Deja, A.: A dispersed decision-making system - the use of negotiations during the dynamic generation of a system's structure. Inf. Sci. 288, 194–219 (2014), http://dx.doi.org/10.1016/j.ins.2014.07.032
12. Przybyła-Kasperek, M., Wakulicz-Deja, A.: Global decision-making system with dynamically generated clusters. Inf. Sci. 270, 172–191 (2014), http://dx.doi.org/10.1016/j.ins.2014.02.076

13. Schneeweiss, C.: Distributed decision making. Springer, Berlin (2003)
14. Shapley, L.S., Shubik, M.: A method for evaluating the distribution of power in a committee system. American Political Science Review 48(3), 787–792 (1954)
15. Siminski, R.: Extraction of rules dependencies for optimization of backward inference algorithm. In: Proceedings of the BDAS 2014, Ustron, Poland, May 27-30, pp. 191–200 (2014), http://dx.doi.org/10.1007/978-3-319-06932-6_19
16. Skowron, A., Wang, H., Wojna, A., Bazan, J.: Multimodal classification: Case studies. In: Peters, J.F., Skowron, A. (eds.) Transactions on Rough Sets V. LNCS, vol. 4100, pp. 224–239. Springer, Heidelberg (2006)
17. Suraj, Z., El Gayar, N., Delimata, P.: A rough set approach to multiple classifier systems. Fundam. Inform. 72(1-3), 393–406 (2006), http://iospress.metapress.com/content/02xh0rh5u0am8g1t/

Inference in Expert Systems Using Natural Language Processing

Tomasz Jach[(✉)] and Tomasz Xięski

Institute of Informatics,
University of Silesia, Poland
{tomasz.jach,tomasz.xieski}@us.edu.pl
http://www.ii.us.edu.pl

Abstract. The authors show the real life application of an expert system using queries submitted by the user using natural language. The system is based on polish language. The two stage process (involving data preparation and the inference itself) is proposed in order to complete the inference.

Keywords: Expert systems · Inference · NLP · Natural language processing · Morphological analysis

1 Introduction to Decision Support Systems

Under the classical definition of the Decision Support System (DSS) the authors mean the combination of a knowledge base and inference algorithms. Both rely on rules, where every one of it consists of two parts: decisional and conditional. Formally, the Decision Support System is given by[16]:

$DSS =< U, A, V, f >$
U − Nonempty, finitive set of rules;
A − Nonempty, finitive set of atributes; A_i - the attribute set of i-th rule
V − Nonempty, finitive set of values of atributes
$C \cup D = A; C \cap D = \emptyset;$ C − conditional attributes; D − decision
$V = \cup_{a \in A} V_a$ V_a − the domain of attribute a
$f : U \times A \to V$ − information function

The authors in previous works ([19,11,10]) proposed some additions to DSSs which were able to speed-up the inference process. In this paper however, we are aiming to properly start the inference process using the query stated in natural language by the user.

Each DSS consists of modules (shown on Fig. 1) which cooperate with each other and will be described in more detail.

Knowledge Base is where all the user knowledge is stored. The authors use a rule based system, therefore rules and facts are the basis of the system. The properties of a knowledge base, especially the way how it's organised,

© Springer International Publishing Switzerland 2015
S. Kozielski et al. (Eds.): BDAS 2015, CCIS 521, pp. 288–298, 2015.
DOI: 10.1007/978-3-319-18422-7_26

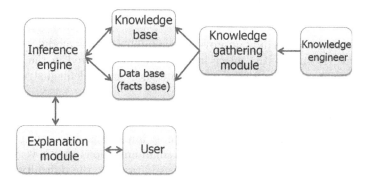

Fig. 1. The modules of Decision Support System

are crucial to an efficient inference process. The authors during the past few years have developed a variety of methods in order to boost both the speed and the quality of inference [12,14,13].

Data Base (Facts Base) is a part of DSS where user submitted facts are gathered. As it will be shown further in this paper, the user provides data using natural language, which is then processed and matched to existing rules in the system. Afterwards, the proposed approach asks questions about the values of attributes of which the rules are formed. Those descriptors (attribute-value pairs) are then stored in the facts base and are used to conduct the inference process.

Inference Engine is divided by the authors into two subsystems: the inference engine itself and the database crawler. Both are essential parts of the proposed system and are explained in detail further on. The inference engine is responsible for the generation of conclusions (previously unknown facts) by activating rules which all the premisses are satisfied.

Knowledge Gathering Module is a part of DSS responsible for development and extension of the knowledge base. In the proposed approach it is realised by a web page where an user is able to submit new rules which are then processed by the system. Newly added rules are then available to be taken into consideration during the inference phase.

Explanation Module is responsible for explaining to the user how the inference was performed. During the inference the user may wonder 'why' the system asked for a particular attribute and 'how' it has reached the conclusion. Both of these questions are answered and explained by this module.

The Communications Module which implementation is the novel approach described in this paper. The authors used the morphological analysis system ([8]) in order to process the user-submitted query and to identify meaningful keywords. Those keywords are cross-referenced to previously stored ones inside the knowledge base. The process is very fast and completely transparent to the end-user. Prior to the inference process, the knowledge base is limited only to the set of possibly relevant rules. The process is explained in detail further on.

The presented work mainly improves the communications module, so the user does not have to submit the query in the exact form stored in the knowledge base and presents the new way of mixed inference on the defined structure.

1.1 Communication with the User

The communication module is one of the most important aspects of DSS. Most of the modern systems do not allow direct communications as it is very difficult to process the informations from the user. Additionally, the difficulty of Polish language processing makes this process very hard.

Most of the Decision Support Systems focus only on one particular subject [1,4]. In this case, usually identified by a single decision attribute, DSS can start asking the users questions which afterwards can reach one of the conclusions stored in the knowledge base. The interaction with the user is limited to displaying closed questions (which are backed-up by the rules from the knowledge base) and performing the backward or forward inference [5].

The more difficult task is to provide multiple decision attributes and therefore - allowing to reach multiple conclusions. This kind of DSSs, prior to performing the inference, asks the user about the subject of inference. For example, one can develop a system which diagnoses multiple medical diseases. Before the process of inference, user should pick the one which is being considered.

The most difficult situation provides the extensive knowledge base with multiple conclusions, usually about a loosely connected subject. By using the user submitted query, DSS chooses the probable inference path and tries to prove it. In case of failure, it automatically switches to an another one (or starts the inference process from the beginning). The system proposed in this paper belongs to this group. By using natural language processing, along with a decision support system and a modified inference algorithm we were able to provide real-time inference with couple of thousands of rules.

There are many difficulties when developing such a system. The end-user quite frequently supplies the query with spelling errors, however we were able to compensate them in a limited way.

The proposed system is visualised by a chatterbot [2]. This way, the user is faced with a more natural way of communication.

2 The Basics of Natural Language Processing

Natural Language Processing (NLP) is the computerized approach to analyzing text. The analysis can be performed both on written and spoken texts. NLP problems are very hard to be dealt with, especially when it comes to languages which are hard to be automatically analysed (such as Polish).

Natural Language Processing is a theoretically motivated range of computational techniques for analysing and representing naturally occurring texts at one or more levels of linguistic analysis for the purpose of achieving human-like language processing for a range of tasks or applications [7].

The goal of NLP as stated above is "to accomplish human-like language processing". The choice of the word "processing" is very deliberate, and should not be replaced with "understanding".

A full NLU System should be able to [7]:

- Paraphrase an input text.
- Translate the text into another language.
- Answer questions about the contents of the text.
- Draw inferences from the text.

In the presented system, we were able to paraphrase the input text by using the Morfologik stemming engine[8].

Today, the leading polish NLP research is held on The Institute of Computer Science, Polish Academy of Sciences. The research focuses on different aspects of NLP including building the corpora of different Polish texts [15], cross-language comparative analysis [6], building adaptive systems to support problem-solving on the basis of document collections in the Internet [17] and much more. In 2011 the National Corpus of Polish Language was built [15].

2.1   Polish Language

Polish language belongs to the group of West Slavic languages. Its alphabet consists of 32 letters. There are two number classes (singular and plural) and deprecated and unclassified dual form (eg. for the word hand, singular: "ręka", plural: "rękami", dual: "rękoma").

Polish retains the Old Slavic system of cases for nouns, pronouns, and adjectives. There are seven cases: nominative (mianownik), genitive (dopełniacz), dative (celownik), accusative (biernik), instrumental (narzędnik), locative (miejscownik), and vocative (wołacz). It's still a modest figure compared to the Hungarian language – which is a representative of a class of conglomerate languages – having 29 cases. In comparison in English there are only two cases which are usually distinguished only for pronouns.

There are three main genders: masculine, feminine and neuter. Masculine nouns are then divided into animate and inanimate (which are distinguished in the singular), and personal and non-personal (being relevant in the plural). Each case is characterized by complex rules of complying, for which there are many exceptions.

Verbs in Polish can be either imperfective (denoting continous or habitual events) or perfective (meaning single, fixed in time completed events). The majority of polish verbs have both forms, e.g. "pić" (to drink, activity not completed) and "wypić" (to drink, activity completed).

Imperfective verbs have three tenses: present, past and future, which is a compound tense with some exceptions. Perfective verbs don't have present tense and are generally easier to conjugate. Both types also have imperative and conditional forms.

The present tense of imperfective verbs (and future tense of perfective verbs) consists of three persons and two numbers giving the total number of six forms. For example, the present imperfective tense of "jeść" (to eat) is "jem", "jesz", "je", "jemy", "jecie", "jedzą".

All the above information is only a brief outline, but in a perfect way reflects the complexity of Polish compared to English. In contrast to the Polish language, the English inflection has virtually disappeared. The declination is limited to residual cases that are gradually being replaced. It is worth to mention that the lack of inflection in English (which due to the limited possibilities for creating correct forms of words) greatly facilitates the natural language processing.

3   The Proposed System

As it was stated before, traditional Decision Support Systems only allow to perform inference about one decision variable. Additionally, one of the major drawbacks of these systems is the lack of a natural way of communication with the user.

To overcome this issue, we have developed our own version of decision support system with natural language processing called RATT. It is currently implemented in one of the builder companies where the number of rules (in a knowledge base) has already exceeded four and a half thousand. Additionally, there are nearly three thousand attributes in the system, of which about one thousand is a part of decision. It is worth noticing, that some of the decisional attributes from one rule can be part of conditional attributes from another rule.

RATT consists of two modules: dbCrawler and dbInferer. The first one prepares the knowledge base by automatically generating the set of keywords to every rule, where the second performs the inference process itself. By the demands of the client, the whole system communicates with user in a real time and matches the submitted query to the stored rules.

In the picture 2 one can see the view presented to the user used to perform the inference process.

3.1   Database Crawler

The purpose of this part of the system is to prepare the keywords table. It is used further on to compare the query submitted by the user to the rules stored in knowledge base. After several tests of different morphological analysis systems (i.a. Hunspell [9], Stempel [3] and Morfologik [8]) the authors chose the last one because of the best quality of generated stems. All of these systems take words as input and give the so-called "stems" as output. The term "stem" is used to denote the primary form of a word or a root of the word. For example, words "pies" and "psa" (both meaning dog in different cases) have the same stem "pies'. It is worth noting, that to obtain the similarity of two words, one can use the Levenstein distance [20]. However in case of developing the expert system using Polish natural language, the Levenstein distance is not sufficient and error-prone.

Fig. 2. The main window of chatterbot

The algorithm itself is given in the following listing:

Algorithm 1. dbCralwer - computing the keywords for the rules

---
Data: $U = \{r_1 \ldots r_n\}$ rules from knowledge base; m - number of
    keywords to be stored for every rule; $STOPWORDS$ - the set of
    common words meaningless to the process of inference;
Result: $K = \cup_{i=1}^{n} K_i$ - n arrays of keywords ordered by keyword
    significance to a particular rule
begin
    Read rules from knowledge base;
    Populate the $STOPWORDS$ set;
    `/* For every rule`   `*/`
    foreach r_i *in* U **do**
        foreach *word in the description of* r_i **do**
            $K[r_i] \leftarrow$ stem($word$);
        end
    end
    `/* Sort keywords table by the incidence of keywords` `*/`
    sortByValues(K);
    `/* Limit the number of keywords to` m `*/`
    limitKeywords(K,m);
end

The algorithm uses the *stem()* function in order to obtain the roots of words given as the input. Each rule is characterised by:

1. Rule's description.
2. Description of attributes belonging to the rule.
3. Attributes' names and synonyms of attributes' names.

During the preliminary phase of RATT implementation we used the description of all the attributes belonging to the rule. However, after several tests, we started to only use the conclusion attributes to compute the keywords describing the rule. Most of the time, the user asks for help of a particular problem. The problem can be solved by the expert system only when it has rules which conclusions are similar to the problem stated by the user. The user is not interested in conditional attributes, but only in those belonging to the conclusion.

The *STOPWORDS* table was given by the linguistic experts and is formed by words which do not provide extra knowledge. These words include conjunctions ("and", "or", etc.), common words ("analysis", "how", "but", etc.) and meaningless verbs ("to be", "to like", "to belong", etc.). The synonyms are also given by the domain experts vastly improving the quality of inference.

The limit of m keywords per rule was enforced to improve performance of the system. Currently, we store 15 keywords per rule and this number seems to be sufficient.

3.2   The Inference Engine

The second vital part of RATT is the inference engine. It was developed as an Internet REST service able to respond with JSON data format. The inference engine receives either a query in natural language (in the preliminary phase of the inference) or a descriptor (attribute-value pair).

During the process, the user is given a question connected with the appropriate attribute along with its set of values. He or she can then choose one of the given options (to answer it).

The result of the inference is either success with the information about the value of the decision attribute or the indication of failure meaning that the system was not able to perform inference under given conditions.

The main algorithm is given on listing 2. It proceeds in two phases. First one results in limiting the knowledge base only to the rules which are relevant to user's query. By performing this step we were able to limit the number of further analysed rules tenfold. The *STOPWORDS* table is used here as well so that the user query if filtered only to those words which can be meaningful for the morphological analyser. The stemming process sometimes gives multiple stems to a single word. This case is also handled by our system by utilising all those given stems.

RATT uses mixed inference in order to perform quick and efficient process of decision support. At each step of the algorithm, the user is being asked about the value of the first unconfirmed attribute of the most relevant rule. Then, all the rules which have the same attribute, but different value in their description, are excluded from the inference process. In this way, only potentially relevant rules are considered at each stage of inference.

If during the process all of the rules' attributes' are confirmed (all of them belong to the facts set), the rule is considered confirmed and the conclusion is given to the user. Otherwise, when the set of rules U becomes empty, the inference is considered as unsuccessful.

Algorithm 2. dbInfern - inference engine

Data: $U = \{r_1 \ldots r_n\}$ rules from knowledge base; $STOPWORDS$ - the set of common words meaningless to the process of inference; $query$ - user submitted query in natural language; $K = \cup_{i=1}^{n} K_i$ - n arrays of keywords ordered by keyword significance to a particular rule;

Result: $S = [f_1, \ldots, f_i, \ldots, f_n]$ - the values of similarity between every rule and user submitted query; $d \in D$, $d := (a_j, v_j)$ - the decision given by the value of one of the conclusion attributes; F - the set of facts given as descriptors

begin

  /* First phase: limiting the knowledge base only to rules relevant to the query   */

  Read rules from knowledge base;

  Populate the $STOPWORDS$ set;

  foreach *word in query* **do**

    foreach $word_s \in stem(word)$ **do**

      if $word_s \notin STOPWORDS$ **then**

        foreach r_i *in* U **do**

          if $word_s \in K[r_i]$ **then**

            | $S[i] := S[i] + 1$;

          end

        end

      end

    end

  end

  /* Remove irrelevant rules   */

  foreach $r_i \in U$ **do**

    if $S[i] = 0$ **then**

      | $U = U \backslash r_i$;

    end

  end

  /* Sort rules by value of similarity to the query   */

  $sort(U, S)$;

  /* Second phase - perform inference on a limited set of rules   */

  while $card(U) \geq 1$ **do**

    /* Ask the user about the value of the first unconfirmed attribute in the most relevant rule   */

    $(a_u, v_u) \leftarrow ask(U[1], a_u \notin F)$;

    /* Erase the rules which have the same attribute, but a different value in the premiss part   */

    foreach $r_i \in U$ **do**

      if $\exists_{a_x \in A_i} f(r_i, a_u) \neq v_u$ **then**

        | $U = U \backslash r_i$;

      else

        if $\forall_{a_x \in A_i} a_x \in F$ **then**

          /* Success - rule r_i is confirmed   */

          return *Conclusion of $i - th$ rule;*

        end

      end

    end

  end

  if $card(U) = 0$ **then**

    | **return** *Failure*

  end

end

---

4 Summary

During research and implementation the authors faced some difficulties. The major one involved relatively small number of rules in system. Because of it's nature, RATT provides a comprehensive way to support multiple decision problems within the scope of building and decorating houses. Because of the very broad aspect, the number of rules should be further increased. These rules, which are in the knowledge base were created by the external experts (from the architecture field) who did not have any knowledge about computer decision systems. Therefore, at the beginning the vast majority of rules had not been prepared in an optimal way (e.g. experts did not understood the concept of multi-level inference). At this point most of the rules were repaired or formed from scratch.

Because of the fact, that knowledge came from multiple experts the problems of synonyms and spelling errors was observed. To cope this obstacle the experts were being presented with rules similar to those just submitted. In this way, some level of consistency was obtained. Additionally, Morfologik was able to compensate most of the common spelling errors without a major impact on performance and efficiency.

In the future, the authors would like to extend the knowledge base. Furthermore, we would like to implement the previously shown methods of inference with incomplete knowledge, especially the IF method [18].

In order to better understand and adapt to the user's needs, the authors would like not only to analyse and compare keywords, but also conduct the semantics analysis of queries.

Currently, the presented system is being implemented as one of the modules in a bigger project. The preliminary tests proved that for most of the queries, the answer was found in about 80% of times, provided that the system had knowledge about the subject. As it was stated before, the scope of the project is to support the user in every aspect connected with building and decorating houses. The specific and detailed evaluation will be performed after gathering enough data from the users and after publicly announcing the system.

The computation times of the proposed approach are satisfactory. The full processing of 4718 rules by the dbCrawler takes about 3 minutes. This step has to be done only after modifying the knowledge base and is more than satisfying for the purposes of the project. There are currently 76 words in the STOPWORDS table. The result of a single user query is returned within a few seconds and is gradually reduced during the inference process (because of the smaller group of potentially relevant rules).

Acknowledgments. This work is a part of the project *"Exploration of rule knowledge bases"* founded by the Polish National Science Centre (2011/03/D/ ST6/03027).

References

1. Bolc, L., Coombs, M.J.: Expert System Applications. Springer (1988)
2. De Angeli, A.: To the rescue of a lost identity: Social perception in human-chatterbot interaction. In: Virtual Agents Symposium, pp. 7–14 (2005)
3. Galambos, L.: Stempel (2004), http://www.getopt.org/stempel/
4. Ignizio, J.P.: An introduction to Expert Systems. McGraw-Hill (1991)
5. Jackson, P.: Introduction to Expert Systems. Addison Wesley (1999)
6. Kacprzak, S., Ziółko, M., Mąsior, M., Igras, M., Ruszkiewicz, K.: Statisical analysis of phonemic diversity in languages across the world. In: XIX Krajowa Konferencja Zastosowań Matematyki w Biologii i Medycynie (2013)
7. Liddy, E.D.: Natural Language Processing. Syracuse University Libraries (2001)
8. Miłkowski, M., Weiss, D.: Morfologik (2014), http://morfologik.blogspot.com
9. Németh, L.: Hunspell (2010), http://hunspell.sourceforge.net/
10. Nowak-Brzezińska, A., Jach, T.: Wnioskowanie w systemach z wiedzą niepeĆną (*Inference with incomplete knowledge*). Studia Informatica, Zeszyty Naukowe Politechniki śląskiej, 377–389 (2011)
11. Nowak-Brzezińska, A., Jach, T.: Wybrane aspekty wnioskowania w systemach z wiedzą niepeĆną (*Inference processes using incomplete knowledge in Decision Support Systems - chosen aspects*). Studia Informatica, Zeszyty Naukowe Politechniki śląskiej, 465–477 (2012)
12. Nowak-Brzezińska, A., Jach, T.: Metoda wspĆczynnikw niepeĆności wiedzy w systemach wspomagania decyzji (*The incompletness factor method in decision support systems*). Studia Informatica, Zeszyty Naukowe Politechniki śląskiej, 227–238 (2013)
13. Nowak-Brzezińska, A., Jach, T., Xięski, T.: Analiza hierarchicznych i niehierarchicznych algorytmw grupowania dla dokumentw tekstowych (The analysis of hierarchical and partitional clustering algorithms applied to text documents). Studia Informatica, Zeszyty Naukowe Politechniki śląskiej, 245–258 (2009)
14. Nowak-Brzezińska, A., Jach, T., Xięski, T.: Wybr algorytmu grupowania a efektywność wyszukiwania dokumentw (*Choosing the effective documents clustering algorithm*). Studia Informatica, Zeszyty Naukowe Politechniki śląskiej, 147–162 (2010)
15. Przepiórkowski, A., Bańko, M., Górski, R.L., Lewandowska-Tomaszczyk, B.: Narodowy Korpus Języka Polskiego (*The National Corpus of Polish Language*). Wydawnictwo Naukowe PWN (2012)
16. Simiński, R., Nowak-Brzezińska, A., Jach, T., Xięski, T.: Towards a Practical Approach to Discover Internal Dependencies in Rule-Based Knowledge Bases. In: Yao, J., Ramanna, S., Wang, G., Suraj, Z. (eds.) RSKT 2011. LNCS, vol. 6954, pp. 232–237. Springer, Heidelberg (2011)
17. Sydow, M., Ciesielski, K., Wajda, J.: Introducing diversity to log-based query suggestions to deal with underspecified user queries. In: Bouvry, P., Kłopotek, M.A., Leprévost, F., Marciniak, M., Mykowiecka, A., Rybiński, H. (eds.) SIIS 2011. LNCS, vol. 7053, pp. 251–264. Springer, Heidelberg (2012)
18. Wakulicz-Deja, A., Nowak-Brzezińska, A., Jach, T.: Inference Processes in Decision Support Systems with Incomplete Knowledge. In: Yao, J., Ramanna, S., Wang, G., Suraj, Z. (eds.) RSKT 2011. LNCS, vol. 6954, pp. 616–625. Springer, Heidelberg (2011)

19. Nowak-Brzezińska, A., Jach, T., Wakulicz-Deja, A.: Inference processes using incomplete knowledge in decision support systems-chosen aspects. In: Yao, J., Yang, Y., Słowiński, R., Greco, S., Li, H., Mitra, S., Polkowski, L. (eds.) RSCTC 2012. LNCS, vol. 7413, pp. 150–155. Springer, Heidelberg (2012)
20. Yujian, L., Bo, L.: A normalized levenshtein distance metric. IEEE Transactions on Pattern Analysis and Machine Intelligence 29(6), 1091–1095 (2007)

Impact of Parallel Memetic Algorithm Parameters on Its Efficacy

Miroslaw Blocho[1,2] and Jakub Nalepa[1,3(✉)]

[1] Institute of Informatics,
Silesian University of Technology, Gliwice, Poland
jakub.nalepa@polsl.pl
[2] ABB ISDC, Krakow, Poland
miroslaw.blocho@pl.abb.com
[3] Future Processing, Gliwice, Poland
jnalepa@future-processing.com

Abstract. The vehicle routing problem with time windows (VRPTW) is an NP-hard discrete optimization problem with two objectives—to minimize a number of vehicles serving a set of dispersed customers, and to minimize the total travel distance. Since real-life, commercially-available road network and address databases are very large and complex, approximate methods to tackle the VRPTW became a main stream of development. In this paper, we investigate the impact of selecting two crucial parameters of our parallel memetic algorithm—the population size and the number of children generated for each pair of parents—on its efficacy. Our experimental study performed on selected benchmark problems indicates that the improper selection of the parameters can easily jeopardize the search. We show that larger populations converge to high-quality solutions in a smaller number of consecutive generations, and creating more children helps exploit parents as best as possible.

Keywords: Parallel memetic algorithm · Island model · Population size · Number of children · VRPTW

1 Introduction

The vehicle routing problem with time windows (VRPTW) is an important NP-hard discrete optimization problem, consisting in determining the minimum cost routing plan to deliver goods from a single depot to geographically dispersed customers. Its first objective is to minimize the fleet size, and the secondary one is to minimize the total distance traveled by the vehicles.

There exist two development streams of tackling the VRPTW. First, exact approaches aim at solving the VRPTW to the optimality [10,1]. Since their computation time may become enormously large for many scheduling circumstances, heuristics are being developed. They do not guarantee obtaining optimal solutions but usually execute very fast and converge to very high-quality (nearly-optimal) solutions in a time acceptable in real-life applications. In construction heuristics, customers are iteratively inserted into a partial solution [21],

© Springer International Publishing Switzerland 2015
S. Kozielski et al. (Eds.): BDAS 2015, CCIS 521, pp. 299–308, 2015.
DOI: 10.1007/978-3-319-18422-7_27

whereas improvement heuristics modify an initial solution [5,14]. Other approximate algorithms include simulated annealing [6], tabu searches [8], evolutionary algorithms (EAs) (both sequential and parallel) [22], and more [5,2,9]. Genetic algorithms (GAs) evolve populations of solutions, which are improved in the biologically-inspired manner. Memetic algorithms (MAs) combine an EA for the global exploration of a solution space with a local search algorithm for its exploitation. They are very effective in solving the VRPTW [15,18,23,4,17], and a plethora of other optimization and pattern recognition problems [13,11,20,3].

An important shortcoming of the mentioned EAs is an unclear selection of their parameters. Since these methods use *static* values (they do not change during the execution), this decision significantly affects the performance. The state-of-the-art EAs for VRPTW must be executed multiple times to determine an acceptable set of appropriate parameters (usually not independent). The other approach is to control the algorithm parameters on the fly [7]. It enables finding "optimal" parameters (which may vary in time) for different steps of execution.

In this work, we investigate the impact of two crucial parameters of our parallel MA for solving the VRPTW (PMA–VRPTW) [16]—the population size and the number of children generated for each pair of parents—on its efficacy and convergence. PMA–VRPTW is an island-model parallel MA in which processes co-operate in order to exchange solutions found so far. This co-operation (its topology, migration frequency, and handling of migrants) is defined by a co-operation scheme [16]. We performed an extensive experimental study on selected Gehring and Homberger's benchmark problems (of various sizes and structures) in order to investigate the impact of the mentioned PMA–VRPTW parameters. Finally, we show that improper selection of the parameters can jeopardize the search and affect the algorithm convergence capabilities.

The remainder of the paper is organized as follows. Section 2 formulates the VRPTW. In section 3 we discuss PMA–VRPTW. Section 4 contains the analysis of the computational experiments. Section 5 concludes the paper.

2   Vehicle Routing Problem with Time Windows

The VRPTW is defined on a complete graph $G = (V, E)$ with a set of vertices $V = \{v_0, v_1, ..., v_C\}$, and edges $E = \{(v_i, v_j): v_i, v_j \in V, i \neq j\}$. The node v_0 is a depot, and the set of nodes $\{v_1, v_2, ..., v_C\}$ represents the customers to be served. With each $v_i \in V$, there are associated a load q_i ($q_0 = 0$), a service time s_i ($s_0 = 0$), and a time window $[e_i, l_i]$. Every edge (v_i, v_j) has a travel distance d_{ij}, and a non-negative travel time c_{ij}. A feasible solution to the VRPTW is a set of K routes such that: (i) each route starts and ends at the depot, (ii) every v_i belongs to one route, (iii) the load of each route does not exceed the vehicle capacity Q, (iv) the service at each customer v_i begins between e_i and l_i.

A desired feasible solution of the NP-hard VRPTW incorporates the minimum K (primary objective), and the minimum total travel distance T (secondary objective). Let (K_α, T_α) represent a solution α. The solution β, represented by (K_β, T_β), is of a higher quality, if $(K_\beta < K_\alpha)$ or $(K_\beta = K_\alpha$ and $T_\beta < T_\alpha)$.

3   Parallel Memetic Algorithm

In PMA–VRPTW, a population of N solutions evolves in time to optimize T (Alg. 1). The initial population is generated using the parallel guided ejection search (P–GES) (line 1), which is applied to minimize K at first, and then to generate N solutions with K routes each (see [16] for details). Next, the population is subject to transformations using standard genetic operators (*selection*, *crossover*, and *mutation*), complemented with a memetic operator, called *education* (lines 3–15). Since PMA–VRPTW is an island-model parallel MA, each process (out of π) (an *island*) executes the same MA. Then, the islands co-operate to exchange the knowledge already acquired during the search process.

1: Generate initial population of N solutions with K routes each;   ▷ P–GES
2: **for** $P_i \leftarrow P_1$ **to** P_π **do in parallel**
3:    **while not** *finished* **do**
4:       Select N pairs of (σ_A, σ_B);
5:       **for all** (σ_A, σ_B) **do**
6:          $\sigma_C^B \leftarrow \sigma_A$;
7:          **for** $j \leftarrow 1$ **to** N_c **do**
8:             Generate a child σ_C;   ▷ See Fig. 2
9:             Update σ_C^B if it is necessary;
10:          **end for**
11:       **end for**
12:       Create the next generation;
13:       Co-operate (if it is necessary) and handle immigrants/emmigrants;
14:       Verify termination conditions and update *finished*;
15:    **end while**
16: **end for**
17: **return** the best solution σ^B among all islands;

Alg. 1. Parallel memetic algorithm for the VRPTW (PMA–VRPTW)

3.1   Genetic and Memetic Operators

At first, N pairs of parents (σ_A, σ_B) are determined using the AB-selection scheme [12] (line 4). Then, for each (σ_A, σ_B), N_c children are generated (see Fig. 2). It involves crossing over the selected individuals using the edge-assembly operator (EAX) [15], and restoring the feasibility of a child (if it is necessary) by applying local search moves[1], which decrease the time window and capacity penalties. These penalties define the severity of violating the constraints by an infeasible solution [15]. If the child is feasible, it is then enhanced by additional moves in the education procedure (only moves which decrease T are applied).

---

[1] These moves include 2–Opt*, Out-exchange, Out-relocate, In-relocate, In-exchange, and GENIUS-exchange moves [16].

Afterwards, σ_C is mutated by feasible moves (not necessarily improving its fitness) to diversify the search[2]. Finally, σ_C replaces the best child generated for a given (σ_A, σ_B), if it is of a higher quality than σ_C^B. In PMA–VRPTW, we generate N_c, $N_c > 1$, children for each pair of selected individuals to exploit parents as best as possible. Intuitively, if N_c is large, then the possibility of ending up with a valuable child (i.e., with a decreased T) grows. On the other hand, it was shown that creating a child takes $O(C^2)$ time, where C denotes the number of customers in each solution. Thus, too large values of N_c may significantly slow down the process of generating consecutive populations.

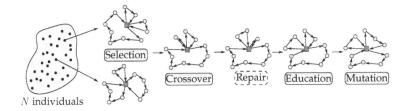

Fig. 2. Generating a child in PMA–VRPTW

After the recombination, a new generation containing N best children is formed (Alg. 1, line 12) (note that the elitist approach is intrinsically applied due to the AB-selection scheme, and best solutions survive [12]). Finally, the islands exchange the best solutions found up to date (line 13). In this paper, we utilized the R–EAX co-operation scheme (a randomized ring topology with an additional crossover of immigrant(s) and best island solutions [16]). It is worth mentioning that PMA–VRPTW can be terminated if (i) its execution time exceeds an assumed time limit, (ii) a solution of a required quality has been found, or (iii) the optimization process is unlikely to converge to better solutions (e.g., because of the diversity crisis) (line 14). Here, we impose the maximum execution time of PMA–VRPTW, τ. Finally, the best solution among all parallel islands (σ^B) is returned (line 17). In this study, we do not introduce any new genetic material (e.g., in the re-generation process) to verify how PMA–VRPTW copes with possible saturation of populations with similar individuals.

4   Experimental Results

4.1   Settings

PMA–VRPTW was implemented in C++ using the Message Passing Interface (MPI) library (MVAPICH1 v. 0.9.9), and the source code was compiled using Intel 10.1 compiler. The experiments were conducted on 64 processors[3] of an

---

[2] The repair, education, and mutation are hill-climbing procedures based on standard VRPTW neighborhoods.

[3] Either a uniprocessor or a core of a multicore processor.

SMP cluster (Galera)[4]. The maximum execution time of PMA–VRPTW was $\tau = 180$ min. Note that τ does not include the P–GES execution time (which is neglectable compared to τ). The experiments showed that 10 minutes were sufficient to find the best-known K for all benchmark tests.

The algorithm was run on selected Gehring and Homberger's (GH) benchmark instances to minimize T (note that for all tests we obtained $K = K_{WB}$, where K_{WB} is the current world's best K[5]). Six test groups highlight several real-life scheduling factors. The C1 and C2 groups include clustered customers, while in the R1 and R2 groups the locations are random. The RC1 and RC2 groups contain a mix of random and clustered customers. C1, R1 and RC1 tests are characterized by smaller vehicle capacities and shorter time windows than instances in the other groups. The GH tests consist of five sets containing $C = 200$, 400, 600, 800 and 1000, with 60 instances in each set. Each test is distinguished by its name, α_β_γ, where α denotes the class (C1, C2, R1, R2, RC1, or RC2), β is the problem size (2 for 200 customers, 4 for 400, and so forth), and γ is the test's identifier (from 1 to 10). In this paper, we investigated instances from various GH groups which reflect different scheduling scenarios.

4.2   Impact of the Population Size

The influence of the population size on the total travel distance of solutions obtained by the PMA–VRPTW for $\pi = 64$ processors was examined on tests C2_6_2 and R2_4_9. The following population sizes were investigated: $N = 10$, 50, 100, and 200, and each configuration was executed 100 times (thus, there were 800 PMA–VRPTW runs in total). Here, we show the total travel distance averaged for 100 independent executions of each PMA–VRPTW configuration. The number of child solutions was fixed to $N_c = 20$ for both GH tests (as suggested in [15]). The experimental results shown in Figs. 3–4 (for C2_6_2 and R2_4_9, respectively) present the achieved averaged total travel distance T for different values of N, matched against the generation number (g). The maximum value of g was fixed to 60 (C2_6_2), and to 50 (R2_4_9).

In both tests, the best results were obtained with the largest population of $N = 200$ solutions. This indicates that larger populations give better diversification of solutions. However, the larger number of solutions entails the larger computational effort which grows linearly with N. Therefore, in order to complete the computation of the same number of generations, PMA–VRPTW with the population size of $N = 200$ requires $20\times$ more time than with $N = 10$.

For test C2_6_2 (Fig. 3), PMA–VRPTW with $N = 200$ reached the previous world's best result $T = 8380.49$ in 19 generations, while the configurations with $N = 100$ and $N = 50$ individuals required, respectively, 40 and 43 generations to achieve the same goal. The small population ($N = 10$) did not reach the previous

---

[4] See http://task.gda.pl/hpc-en/ for details; reference date: October 23, 2014.

[5] For the instance definitions and world's best results see http://www.sintef.no/Projectweb/TOP/VRPTW/Homberger-benchmark/; reference date: October 23, 2014.

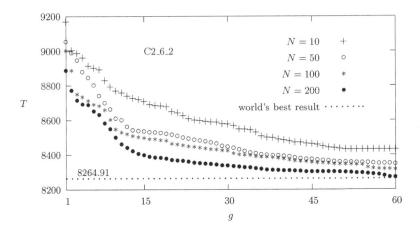

Fig. 3. The total travel distance T obtained using PMA–VRPTW for C2_6_2 (averaged for 100 independent executions) for various population sizes N

world's best result in 60 generations, and the steady state of computation for such a population began in the 52nd generation. However, these generations were executed very fast, and the maximum number of generations can be increased for small populations without affecting the execution time of PMA–VRPTW.

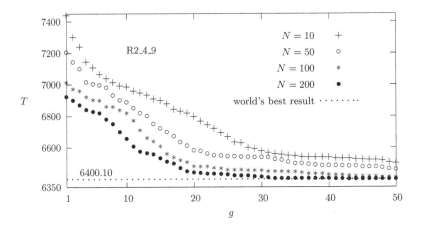

Fig. 4. The total travel distance T obtained using PMA–VRPTW for R2_4_9 (averaged for 100 independent executions) for various population sizes N

Considering the test R2_4_9 (Fig. 4), PMA–VRPTW with $N = 200$ individuals improved the previous world's best result ($T = 6493.14$) in only 18 generations, while the configurations containing $N = 100$ and $N = 50$ solutions required, accordingly, 20 and 38 generations to converge to best solutions. Similarly to the previous test case, the population of $N = 10$ solutions was insufficient to achieve the previous world's best result within the first 50 generations.

4.3   Impact of the Number of Children

The influence of the number of child solutions N_c on the total travel distance obtained using PMA–VRPTW was evaluated on two GH tests—C2_8_8 and RC1_8_2. For the comparative analysis, the following numbers of children were selected: $N_c = 10, 20, 50, 100$. Each PMA–VRPTW configuration was run 80 times (640 tests were executed in total). The population size was fixed for all tests to $N = 50$, and the maximum value of g was fixed to 60 (C2_8_8), and to 130 (RC1_8_2). Similarly to the previous analyses, the averaged T values (over 80 independent runs) were plotted against the generation number g. The results of the experiments are presented in Figs. 5–6. It is worth noting that in both cases the best results were achieved with the largest number of N_c. This indicates that exploiting the selected parents as best as possible is beneficial and allows for obtaining very high-quality children.

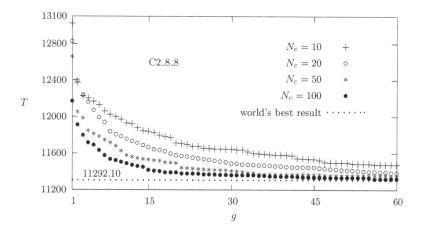

Fig. 5. The total travel distance T obtained using PMA–VRPTW for C2_8_8 (averaged for 80 independent executions) for various numbers of children N_c

Considering the test C2_8_8 (Fig. 5), it was possible to improve the previous world's best result ($T = 12927.45$) in only 2 generations (with all values of N_c). The optimization process with $N_c = 50$ and $N_c = 100$ converged relatively fast (in about 20 generations) to the quite similar results. It suggests that any further effort to increase the number of child solutions N_c will not lead to improvements of solutions quality. Contrary to the test C2_8_8, the instance RC1_8_2 (Fig. 6) appeared much more challenging, and the best results were obtained while generating 100, 50 and 20 children. It is worth pointing out that a very large N_c

increases the time necessary for generating a new population (as already mentioned, creating a child individual takes $O(C^2)$ time). Thus, the time necessary to create a consecutive generation grows linearly with the number of children produced during the recombination process.

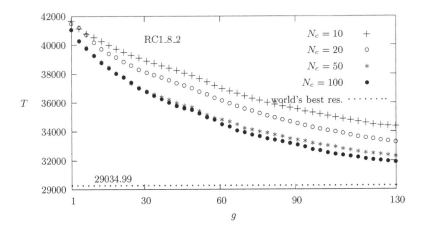

Fig. 6. The total travel distance T obtained using PMA–VRPTW for RC1_8_2 (averaged for 80 independent executions) for various numbers of children N_c

5   Conclusions and Future Work

In this paper, we investigated the impact of the population size and the number of children generated during the recombination of each pair of parents on the efficacy of our parallel MA. Since these parameters are crucial to ensure proper convergence capabilities of any EA, they should be selected carefully in order to guide the optimization efficiently, and to keep the algorithm computation time as low as possible. Our experimental study performed on selected benchmark problem instances indicated that generating a relatively large number of children can help exploit the parents very intensively and lead to creating high-quality individuals. On the other hand, the execution time increases linearly with the number of children. We showed that the optimization of larger populations converges to very well-fitted individuals in a significantly smaller number of consecutive generations. However, the execution time of a single generation is larger. Although this induces increasing the computation time of PMA–VRPTW, this issue can be easily solved by an additional parallelization (e.g., using the OpenMP interface [19]) of the recombination process (generating children for a given pair of selected parents is independent from generating children for other pairs). It is worth noting that evolving small populations (with relatively small number of created children) is very fast and applicable in many real-time scheduling applications (even using the sequential version of PMA–VRPTW). This, in turn, enables analyzing realistic road network databases (usually available via Geographic Information Systems) in a reasonable time.

Our ongoing research includes designing adaptive and self-adaptive (which evolve parameters) MAs for the VRPTW and the pickup and delivery problem with time windows (PDPTW). These adaptive techniques will control parameters (including the population size and the number of children) on the fly. This will mitigate the necessity of conducting a time-consuming tuning process to select a set of acceptable values before the execution. Also, we aim at investigating mutual dependencies between parameters. Finally, we plan to solve other variants of complex routing problems using PMA–VRPTW, especially the PDPTW.

Acknowledgments. This work was supported by the National Science Centre under research grant no. DEC-2013/09/N/ST6/03461. The research was performed using the infrastructure supported by POIG.02.03.01-24-099/13 grant: "GeCONiI–Upper Silesian Center for Computational Science and Engineering". We thank the Academic Computer Centre in Gdańsk TASK, where the computations were carried out. We thank Zbigniew J. Czech for fruitful discussions and suggestions.

References

1. Baldacci, R., Mingozzi, A., Roberti, R.: Recent exact algorithms for solving the vehicle routing problem under capacity and time window constraints. European Journal of Operational Research 218(1), 1–6 (2012)
2. Banos, R., Ortega, J., Gil, C., Márquez, A.L., de Toro, F.: A hybrid meta-heuristic for multi-objective vehicle routing problems with time windows. Compuetrs & Industrial Engineering 65(2), 286–296 (2013)
3. Benlic, U., Hao, J.K.: Memetic search for the quadratic assignment problem. Expert Systems with Applications 42(1), 584–595 (2015)
4. Blocho, M., Czech, Z.J.: A parallel memetic algorithm for the vehicle routing problem with time windows. In: Proceedings of the 2013 Eighth International Conference on P2P, Parallel, Grid, Cloud and Internet Computing, 3PGCIC 2013, pp. 144–151 (2013)
5. Bräysy, O., Gendreau, M.: Vehicle routing problem with time windows, part II: Metaheuristics. Transportation Science 39(1), 119–139 (2005)
6. Chiang, W.C., Russell, R.A.: Simulated annealing metaheuristics for the vehicle routing problem with time windows. Annals of Operations Research 63(1), 3–27 (1996)
7. Eiben, A.E., Hinterding, R., Michalewicz, Z.: Parameter control in evolutionary algorithms. IEEE Transactions on Evolutionary Computation 3(2), 124–141 (1999)
8. Ho, S.C., Haugland, D.: A tabu search heuristic for the vehicle routing problem with time windows and split deliveries. Computers & Operations Research 31(12), 1947–1964 (2004)
9. Hu, W., Liang, H., Peng, C., Du, B., Hu, Q.: A hybrid chaos-particle swarm optimization algorithm for the vehicle routing problem with time window. Entropy 15(4), 1247–1270 (2013)
10. Irnich, S., Villeneuve, D.: The shortest-path problem with resource constraints and k-cycle elimination for $k \leq 3$. INFORMS Journal on Computing 18(3), 391–406 (2006)
11. Jin, Y., Hao, J.K., Hamiez, J.P.: A memetic algorithm for the minimum sum coloring problem. Computers & Operations Research 43, 318–327 (2014)

12. Kawulok, M., Nalepa, J.: Support vector machines training data selection using a genetic algorithm. In: Gimel'farb, G., Hancock, E., Imiya, A., Kuijper, A., Kudo, M., Omachi, S., Windeatt, T., Yamada, K. (eds.) SSPR & SPR 2012. LNCS, vol. 7626, pp. 557–565. Springer, Heidelberg (2012), http://dx.doi.org/10.1007/978-3-642-34166-3_61

13. Li, Y., Li, P., Wu, B., Jiao, L., Shang, R.: Kernel clustering using a hybrid memetic algorithm. Natural Computing 12(4), 605–615 (2013)

14. Nagata, Y., Bräysy, O.: A powerful route minimization heuristic for the vehicle routing problem with time windows. Operations Research Letters 37(5), 333–338 (2009)

15. Nagata, Y., Bräysy, O., Dullaert, W.: A penalty-based edge assembly memetic algorithm for the vehicle routing problem with time windows. Computers & Operations Research 37(4), 724–737 (2010)

16. Nalepa, J., Blocho, M.: Co-operation in the parallel memetic algorithm. International Journal of Parallel Programming, 1–28 (2014), http://dx.doi.org/10.1007/s10766-014-0343-4

17. Nalepa, J., Blocho, M., Czech, Z.J.: Co-operation schemes for the parallel memetic algorithm. In: Wyrzykowski, R., Dongarra, J., Karczewski, K., Waśniewski, J. (eds.) PPAM 2013, Part I. LNCS, vol. 8384, pp. 191–201. Springer, Heidelberg (2014)

18. Nalepa, J., Czech, Z.J.: New selection schemes in a memetic algorithm for the vehicle routing problem with time windows. In: Tomassini, M., Antonioni, A., Daolio, F., Buesser, P. (eds.) ICANNGA 2013. LNCS, vol. 7824, pp. 396–405. Springer, Heidelberg (2013), http://dx.doi.org/10.1007/978-3-642-37213-1_41

19. Nalepa, J., Kawulok, M.: Fast and accurate hand shape classification. In: Kozielski, S., Mrozek, D., Kasprowski, P., Małysiak-Mrozek, B. z. (eds.) BDAS 2014. CCIS, vol. 424, pp. 364–373. Springer, Heidelberg (2014), http://dx.doi.org/10.1007/978-3-319-06932-6_35

20. Nalepa, J., Kawulok, M.: A memetic algorithm to select training data for support vector machines. In: Proceedings of the 2014 Conference on Genetic and Evolutionary Computation, GECCO 2014, pp. 573–580. ACM, New York (2014)

21. Pang, K.W.: An adaptive parallel route construction heuristic for the vehicle routing problem with time windows constraints. Expert Systems with Applications 38(9), 11939–11946 (2011)

22. Repoussis, P.P., Tarantilis, C.D., Ioannou, G.: Arc-guided evolutionary algorithm for the vehicle routing problem with time windows. IEEE Transactions on Evolutionary Computation 13(3), 624–647 (2009)

23. Vidal, T., Crainic, T.G., Gendreau, M., Prins, C.: A hybrid genetic algorithm with adaptive diversity management for a large class of vehicle routing problems with time-windows. Computers & Operations Research 40(1), 475–489 (2013)

Data Processing in Immune Optimization of the Structure

Arkadiusz Poteralski$^{(\boxtimes)}$

Institute of Computational and Mechanical Engineering,
Faculty of Mechanical Engineering, Silesian University of Technology,
ul. Konarskiego 18a, 44-100 Gliwice, Poland
arkadiusz.poteralski@polsl.pl

Abstract. The paper is devoted data processing in immune optimization (using artificial immune system - AIS) to selected optimization problems of the structures. The Procedure for the Exchange of Data - PED in immune optimization is presented. During this procedure important information about design variables and the objective function are saved in specific files. Additional commercial software *MSCPatran* and *Nastran* to analyze of mechanical structures is used.

Keywords: Data processing · Artificial immune system (AIS) · Optimization · Computational intelligence · Artificial immune algorithm

1 Introduction

Data transfer in immune optimization is very important stage of immune algorithm. One of the primary goals defined by creating software based on artificial immune system was its universality and simplicity in applying it with any programs to solve problem on the basis of the changing design variables. The papers is devoted to application of the Procedure for the Exchange of Data - PED in artificial immune algorithm. During this procedure any program (*solver*) is used to calculate the objective function. One of the possibilities to use this solver is commercial software MSC Patran and Nastran to analyze of mechanical structures. It requires preparation in an appropriate manner input and output files. Additionally a short descriptions of biological aspect of natural immune systems [1] is described in the context of optimization procedures. The clonal selection algorithm which represents one of the main features of the artificial immune system is described. The main innovation of this paper is standard and modified versions of artificial immune systems and its applications in different optimization problems of mechanical structures were widely presented by the authors [2,14,13,18]. In the present paper also the application of this algorithm to topology optimization problems of structures is demonstrated [4,11,17].

2 Artificial Immune Systems

The artificial immune systems (AIS) are developed on the basis of a mechanism discovered in biological immune systems [15]. An immune system is a complex

© Springer International Publishing Switzerland 2015
S. Kozielski et al. (Eds.): BDAS 2015, CCIS 521, pp. 309–319, 2015.
DOI: 10.1007/978-3-319-18422-7_28

system which contains distributed groups of specialized cells and organs. The main purpose of the immune system is to recognize and destroy pathogens - funguses, viruses, bacteria and improper functioning cells. The lymphocytes cells play a very important role in the immune system. The lymphocytes are divided into several groups of cells. There are two main groups B and T cells, both contains some subgroups (like B-T dependent or B-T independent). The B cells contain antibodies, which could neutralize pathogens and are also used to recognize pathogens. There is a big diversity between antibodies of the B cells, allowing recognition and neutralization of many different pathogens. The B cells are produced in the bone marrow in long bones. A B cell undergoes a mutation process to achieve big diversity of antibodies. The T cells mature in thymus, only T cells recognizing non self cells are released to the lymphatic and the blood systems. There are also other cells like macrophages with presenting properties, the pathogens are processed by a cell and presented by using MHC (Major Histocompatibility Complex) proteins. The recognition of a pathogen is performed in a few steps. First, the B cells or macrophages present the pathogen to a T cell using MHC, the T cell decides if the presented antigen is a pathogen. The T cell gives a chemical signal to B cells to release antibodies. A part of stimulated B cells goes to a lymph node and proliferate (clone). A part of the B cells changes into memory cells, the rest of them secrete antibodies into blood. The secondary response of the immunology system in the presence of known pathogens is faster because of memory cells. The memory cells created during primary response, proliferate and the antibodies are secreted to blood. The antibodies bind to pathogens and neutralize them. Other cells like macrophages destroy pathogens. The number of lymphocytes in the organism changes, while the presence of pathogens increases, but after attacks a part of the lymphocytes is removed from the organism. The artificial immune systems [11] take only a few elements from the biological immune systems. The most frequently used are the mutation of the B cells, proliferation, memory cells, and recognition by using the B and T cells. The artificial immune systems have been used to optimization problems in classification and also computer viruses recognition. The cloning algorithm presented by von Zuben and de Castro [5] uses some mechanisms similar to biological immune systems to global optimization problems. The unknown global optimum is the searched pathogen. The memory cells contain design variables and proliferate during the optimization process. The B cells created from memory cells undergo mutation. The B cells evaluate and better ones exchange memory cells. In Wierzchoń S. T. [20] version of Clonalg the crowding mechanism is used - the diverse between memory cells is forced. A new memory cell is randomly created and substitutes the old one, if two memory cells have similar design variables. The crowding mechanism allows finding not only the global optimum but also other local ones. The presented approach is based on the Wierzchoń S. T. algorithm [20], but the mutation operator is changed. The Gaussian mutation is used instead of the nonuniform mutation in the presented approach . The Fig. 1 presents the flowchart of an artificial immune system.

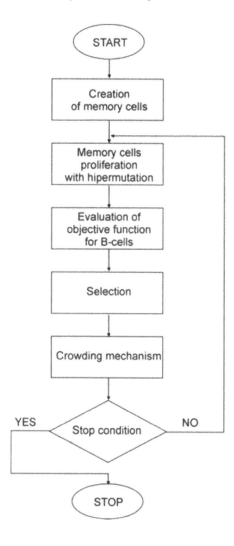

Fig. 1. The algorithm of an artificial immune system

The memory cells are created randomly. They proliferate and mutate creating B cells. The number of clones created by each memory cell is determined by the memory cells objective function value. The objective functions for B cells are evaluated. The selection process exchanges some memory cells for better B cells. The selection is performed on the basis of the geometrical distance between each memory cell and B cells (measured by using design variables) (Fig. 2).

For example compared two memory cells "A" and "C" the better one (memory cell "C" - because the objective function is better) goes to next iteration. The crowding mechanism removes similar memory cells. The similarity is also determined as the geometrical distance between memory cells (Fig. 3) measured using parameter min_{dis} (1).

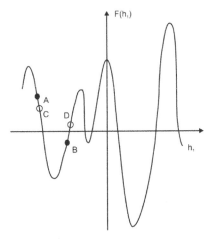

Fig. 2. The idea of the selection mechanism

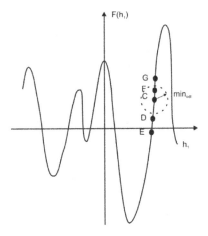

Fig. 3. The idea of crowding mechanism

For example for memory cell "C" only memory cell "F" is in the similarity area. The better one (memory cell "C") stay in population and a worse (memory cell "F") is eliminated. Finally new memory cell is generated in randomly way. The process is iteratively repeated until the stop condition is fulfilled. The stop condition can be expressed as the maximum number of iterations. The unknown global optimum is represented by the searched pathogen. The memory cells contain design variables and proliferate during the optimization process.

Minimal geometrical distance between two memory cells in the search space:

$$\min_{dis} = 0.1 \min_{dom} \left(\sqrt{\sum_{i=1}^{n} ([h_i^j]_{\max} - [h_i^j]_{\min})^2} \right) \qquad (1)$$

where: $h_{i\,\min}^j$ - i-th minimal value of the parameter for j-th memory cell, $h_{i\,\max}^j$ - i-th maximal value of the parameter for j-th memory cell, n - number of design variables, min_{dom} - parameter deciding about size of search space of similarities of the memory cells.

3 Data Processing in Immune Optimization Process (The Procedure for the Exchange of Data - PED)

One of the primary goals defined by creating software based on artificial immune system was its universality and simplicity in applying it with any programs to solve problem on the basis of the changing design variables. Therefore, the Procedure for the Exchange of Data (PED) between an artificial immune system (AIS), and any program that solves the problem used to calculate the objective function is created. The optimize software is implemented procedure to record information about the design variables to the special file, on the basis of any program which solves a specific task. After performing the calculations the value of the objective function is written to the special file. A user who wants to apply optimization algorithm based on artificial immune systems for optimization or identification problems must prepare the procedure, which will be on the basis of data downloaded and prepared by AIS calculate the value of the objective function, and save it in the output file. Immune algorithm uses the above mentioned procedure in iterative way (at the beginning of each iteration), for each B-cell, until the stop condition of optimization process is fulfilled. The block diagram of this procedure (PED) in the Fig. 4 is presented.

In a first stage, an artificial immune system generates (at the first iteration usu-ally in randomly way) design variables and sends them to the *Data.in* file. After each generation of the data the specific problem for each B-cell is solved (Solver – any program that solves given problem). In the second step the data from *Data.in* (Fig. 5a) file are downloaded, calculations are performed and then the value of the objective function is sent to the file *Data.out* (Fig. 5b). In the third step the artificial immune system takes data from the *Data.out* file. In this way, we have related to each other: the design variables and the corresponding value of the objective function (Fig. 6).

At this moment we have a database that includes information about every possible solution. At any time we can on the basis of design variables to generate interesting us solution (Fig. 6).

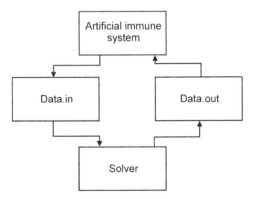

Fig. 4. Scheme of procedure for the exchange of data (PED)

a)   b)

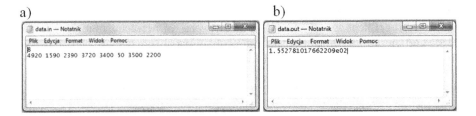

Fig. 5. Example of files: a) input – *data.in*, output – *data.out*

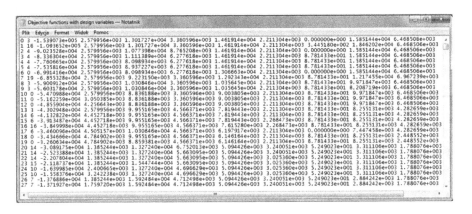

Fig. 6. Database of the best B-cell in optimization process (objective function with design variables)

4   The Use of Commercial Codes MSC Software During Immune Optimization

This chapter is devoted to applications of the artificial immune system to selected shape and topology optimization problems of structures analyzed by finite element method FEM [21] using Procedure for the Exchange of Data - PED. One of the most popular software to analyze of mechanical structures like: $MSCPatran$ and $MSCNastran$ is used. Use of this package allows to obtain a proper geometric structure and the corresponding finite element mesh. $MSCPatran$ software has been used to create a mechanical system and then the discrete model creating finite element mesh. Next the structure was optimized using immune algorithm. During immune optimization structures were analyzed many times using a $MSCNstran$. The adequate prepare input files to the program $MSCNastran$ is the condition for cooperation between these two programs. The construction of these files was specific and required knowledge of the structure of $MSCNastran$ files. The input file to the program $MSCNastran$ is a file with the extension .bdf with the structure shown in Fig. 7. This is an example to describe the geometry, ie. elements, nodes, forces and the boundary conditions (Fig. 7).

Next, the .bdf file was imported into the program $MSCNastran$, and after using this solver output file .$f06$ was generated (Fig. 8). From this file needed information were downloaded ie. resultant displacements and equivalent stresses von Huber-Misses.

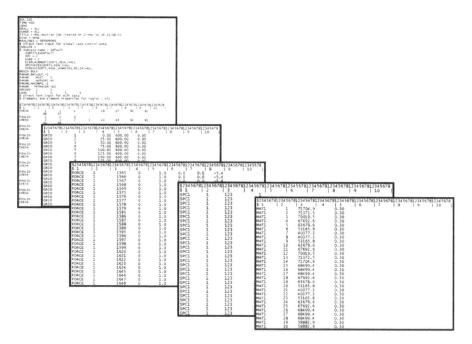

Fig. 7. The part of the input file containing the: elements, nodes, forces and the boundary conditions of the MSC Nastran program

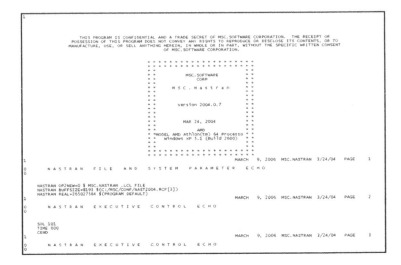

Fig. 8. Part of the output file (.f06) containing the information about stresses and displacements of the *MSCNastran* program

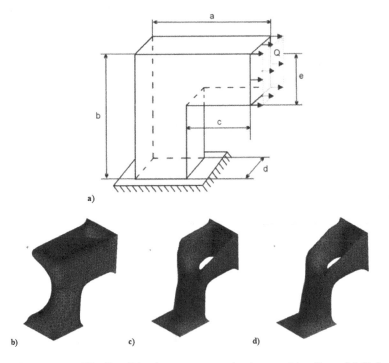

Fig. 9. 3-D structure like L solid: a) geometry and scheme of loading of 3-D L solid, b) optimization result after 1st iteration, c) optimization result after 21st iteration, d) the best optimization result after 51st iteration

Table 1. The input data to optimization task of a 3-D structure

a x b x c d x e	Maximal displacement [mm]	Maximal stress [MPa]	Q [kN]	range of ρ [g/cm3] existence or elimination of the finite element
48 x 48 x 24 24 x 24	0.8	600	8450	$0 \leq \rho_e < 3.14$ elimination $3.14 \leq \rho_e < 7.85$ existence

Table 2. Parameters of AIS

No. of memory cells=10
No. of the clones=10
Crowding factor=0.5
The probability of Gaussian mutation=50%

Table 3. Values of design variables for three variants: first, 21st iteration and the best one

Objective	function	(1st iteration):	3.291500e-002	
Design variables: 2.422791e-001	3.652954e-002	1.391907e-001	5.190582e-001	4.583054e-001
7.247925e-002	7.681885e-001	6.423950e-002	5.240021e-001	1.080933e-001	9.439087e-001
7.381287e-001	7.530212e-001	1.089172e-001	1.089172e-001	9.928894e-001	8.199158e-001
1.025391e-002	7.623596e-001	2.444458e-002	7.196655e-001	5.795441e-001	4.930878e-001
	7.840271e-001	7.827759e-001	7.730103e-001	5.396576e-001

Objective	function	(21st iteration):	7.256000e-003	
Design variables: 4.372101e-001	3.777148e-001	1.175842e-001	8.760719e-001	1.715419e-001
1.776670e-001	2.096558e-001	7.195562e-001	7.948608e-001	3.488617e-001	4.173164e-001
9.500427e-001	6.262982e-001	5.446218e-001	1.881930e-001	1.408386e-001	3.354034e-001
3.496857e-001	3.989143e-001	7.241325e-001	2.746582e-003	7.650757e-002	6.576106e-001
	7.091675e-001	1.617432e-003	6.178640e-001	1.229858e-002

Objective	function	(51st iteration):	3.721000e-003	
Design variables: 4.372101e-001	4.965159e-001	1.096166e-001	7.678363e-001	4.822388e-001
0.000000e+000	2.420731e-001	7.398847e-001	6.988843e-001	3.383611e-001	4.410858e-001
6.650124e-001	5.353661e-001	6.202469e-001	1.460584e-001	6.004333e-003	3.354034e-001
1.213570e-001	3.989143e-001	6.118825e-001	0.000000e+000	0.000000e+000	5.715176e-002
	8.758202e-001	1.617432e-003	9.001376e-001	1.229858e-002

5   Immune Optimization Example Using Procedure for the Exchange of Data

Topology optimization of 3D structure like L-solid (Fig. 9a) by the minimization of the mass of the structure and with imposed stress or displacement constraints [3]. The structures are considered in the framework of the theory of elasticity. The input data to the optimization task are included in table 1. Numerical results are presented in the Fig. 9b. The parameters of the algorithms are presented in the table 2. Three structures generated from database after 1st, 21st and 51st iteration in the table 3 and Fig. 9 are presented.

6 Conclusions

In the paper, a description of the Procedure for the Exchange of Data – PED and the algorithm of artificial immune approach is presented and applied to optimization of structures. Data transfer in immune optimization is very important stage of immune algorithm therefore the procedure of exchange data was applied. During this procedure information about the design variables are record to the input file, on the basis of any program which solves a specific task. After performing the calculations the value of the objective function is written to the output file. During optimization process commercial software MSC Patran and Nastran to analyze of mechanical structures was used. It requires preparation in an appropriate manner input and output files. Artificial immune system belong to methods based on population of solutions and they have some interesting features which can be considered as alternative to evolutionary algorithms [8] or particle swarm optimizers [6,7,16,19]. Comparison between AIS, PSO and EA in the paper [12] is presented. In the paper [12], the formulation and application of the finite element method and the artificial immune system and particle swarm optimizer to optimization of stacking sequence of plies in composites is presented. Described approach has applied to simultaneous shape, topology and material optimization of 3D structures. There are possibilities of further efficiency improvement of the proposed method, e.g. by the application of adjoint variable method in the sensitivity analysis. Also, the application of another hybridized global optimization algorithms, like hybrid artificial immune system, would be interesting. Moreover, the use of fuzzy approach in the optimization process, like the one presented in works by Mrozek et al. [9,10] may bring some improvements when working with uncertainties. Using this approach the multi-objective optimization is also possible to use with artificial immune system. It is the next step of improve this algorithm.

References

1. Balthrop, J., Esponda, F., Forrest, S., Glickman, M.: Coverage and generalization in an artificial immune system. In: Proceedings of the Genetic and Evolutionary Computation Conference, GECCO 2002, pp. 3–10. Morgan Kaufmann, New York (2002)
2. Burczyński, T., Długosz, A., Kuś, W., Orantek, P., Poteralski, A., Szczepanik, M.: Intelligent computing in evolutionary optimal shaping of solids. In: 3rd International Conference on Computing, Communications and Control Technologies, Proceedings, vol. 3, pp. 294–298 (2005)
3. Burczyński, T., Kuś, W., Długosz, A., Poteralski, A., Szczepanik, M.: Sequential and distributed evolutionary computations in structural optimization. In: Rutkowski, L., Siekmann, J.H., Tadeusiewicz, R., Zadeh, L.A. (eds.) ICAISC 2004. LNCS (LNAI), vol. 3070, pp. 1069–1074. Springer, Heidelberg (2004)
4. Burczyński, T., Szczepanik, M.: Intelligent optimal design of spatial structures. Computer and Structures. Elsevier (2013)
5. de Castro, L.N., Timmis, J.: Artificial immune systems as a novel soft computing paradigm. Soft Computing 7(8), 526–544 (2003)

6. Heppner, F., Grenander, U.: A stochastic nonlinear model for coordinated bird flocks. The Ubiquity of Chaos. AAAS Publications, Washington, DC (1990)
7. Kennedy, J., Eberhart, R.C.: Particle Swarm Optimisation. In: Proceedings of IEEE Int. Conf. on Neural Networks, Piscataway, NJ (1995)
8. Michalewicz, Z.: Genetic algorithms + data structures = evolutionary programs. Springer, Berlin (1992)
9. Mrozek, D., Małysiak, B., Kozielski, S.: An optimal alignment of proteins energy characteristics with crisp and fuzzy similarity awards. In: 2007 IEEE International Conference on Fuzzy Systems, FUZZ-IEEE 2007, London, England, pp. 1513–1518 (2007)
10. Mrozek, D., Małysiak, B., Kozielski, S.: Alignment of protein structure energy patterns represented as sequences of fuzzy numbers. In: 2009 Annual Meeting of the North-American-Fuzzy-Information-Processing-Society, Cincinnati, OH, USA, pp. 35–40 (2009)
11. Poteralski, A.: Optimization of mechanical structures using artificial immune algorithm. In: Kozielski, S., Mrozek, D., Kasprowski, P., Małysiak-Mrozek, B. z. (eds.) BDAS 2014. CCIS, vol. 424, pp. 280–289. Springer, Heidelberg (2014)
12. Poteralski, A., Szczepanik, M., Beluch, W., Burczyński, T.: Optimization of composite structures using bio-inspired methods. In: Rutkowski, L., Korytkowski, M., Scherer, R., Tadeusiewicz, R., Zadeh, L.A., Zurada, J.M. (eds.) ICAISC 2014, Part II. LNCS, vol. 8468, pp. 385–395. Springer, Heidelberg (2014)
13. Poteralski, A., Szczepanik, M., Dziatkiewicz, G., Kuś, W., Burczyński, T.: Comparison between PSO and AIS on the basis of identification of material constants in piezoelectrics. In: Rutkowski, L., Korytkowski, M., Scherer, R., Tadeusiewicz, R., Zadeh, L.A., Zurada, J.M. (eds.) ICAISC 2013, Part II. LNCS, vol. 7895, pp. 569–581. Springer, Heidelberg (2013)
14. Poteralski, A., Szczepanik, M., Ptaszny, J., Kuś, W., Burczyński, T.: Hybrid artificial immune system in identification of room acoustic properties. Inverse Problems in Science and Engineering. Taylor and Francis (2013)
15. Ptak, M., Ptak, W.: Basics of immunology. Jagiellonian University Press, Cracow (2000) (in Polish)
16. Reynolds, C.W.: Flocks, herds, and schools, a distributed behavioral model. Computer Graphics 21, 25–34 (1987)
17. Szczepanik, M., Burczyński, T.: Swarm optimization of stiffeners locations in 2-d structures. Bulletin of the Polish Academy of Sciences, Technical Sciences 60(2), 241–246 (2012)
18. Szczepanik, M., Poteralski, A., Długosz, A., Kuś, W., Burczyński, T.: Bio-inspired optimization of thermomechanical structures. In: Rutkowski, L., Korytkowski, M., Scherer, R., Tadeusiewicz, R., Zadeh, L.A., Zurada, J.M. (eds.) ICAISC 2013, Part II. LNCS, vol. 7895, pp. 79–90. Springer, Heidelberg (2013)
19. Szczepanik, M., Poteralski, A., Ptaszny, J., Burczyński, T.: Hybrid particle swarm optimizer and its application in identification of room acoustic properties. In: Rutkowski, L., Korytkowski, M., Scherer, R., Tadeusiewicz, R., Zadeh, L.A., Zurada, J.M. (eds.) EC 2012 and SIDE 2012. LNCS, vol. 7269, pp. 386–394. Springer, Heidelberg (2012)
20. Wierzchoń, S.T.: Artificial Immune Systems, Theory and Applications. EXIT, Warsaw (2001) (in Polish)
21. Zienkiewicz, O.C., Taylor, R.L.: The finite element method, vol. I., II. McGraw Hill (1989, 1991)

A Prudent Based Approach for Customer Churn Prediction

Adnan Amin[1][(✉)], Faisal Rahim[1], Muhammad Ramzan[2], and Sajid Anwar[1]

[1] Institute of Management Sciences Peshawar, Khyber Pukhtunkhwa, Pakistan
adnan.amin@live.co.uk, faisal.rahimpk@gmail.com,
sajid.anwar@imsciences.edu.pk
[2] Saudi Electronic University, Riyadh, Saudi Arabia
m.ramzan@seu.edu.sa

Abstract. This study contributes to formalize a three phase customer churn prediction technique. In the first phase, a supervised feature selection procedure is adopted to select the most relevant subset of features by laying-off the redundancy and increasing the relevance that leads to reduced and highly correlated features set. In the second phase, a knowledge based system (KBS) is built through Ripple Down Rule (RDR) learner which acquires knowledge about seen customer churn behavior and handles the problem of brittle in churn KBS through prudence analysis that will issue a prompt to the decision maker whenever a case is beyond the maintained knowledge in the knowledge database. In the final phase, a technique for Simulated Expert (SE) is proposed to evaluate the Knowledge Acquisition (KA) in KB system. Moreover, by applying the proposed approach on publicly available dataset, the results show that the proposed approach can be a worthy alternate for churn prediction in telecommunication industry.

Keywords: Churn Prediction · Classification · Simulated Expert · Prudence Analysis · Ripple Down Rules

1 Introduction

Customer Churn-Shifting from one service provider to the next competitor in the market is an alarming issue for various service based industries and particularly for telecommunication industry [29]. The prediction of such customer churn is highly important for project managers because losing a customer is a low cost opportunity for competitors to gain customer [17,10]. It has been reported that the associated cost with acquisition of new customers is ten times more than retaining the existing customers [14]. Retaining existing customers also leads to significant increase in sales and reduced marketing cost.

These facts have focused on customer churn prediction as an indispensable part of telecom companies strategic decision making and planning process which ultimately is primary objective of customer relationship management (CRM) as well. The importance of this emerging issue has led to the development of several

© Springer International Publishing Switzerland 2015
S. Kozielski et al. (Eds.): BDAS 2015, CCIS 521, pp. 320–332, 2015.
DOI: 10.1007/978-3-319-18422-7_29

predictive tools that support some vital tasks in predictive modeling and classification process. In the recent decades, the data explosion is a challenging task for getting valuable information that is contained in this data. On the other hand, data mining entails the overall process of knowledge extraction from the data bank. There are many data mining applications which have been successfully applied to various knowledge discovery techniques [29,24,15] and [12] to extract hidden and meaningful relationships between entities and attributes. These facts led competitive companies to invest in CRM to maintain customer information. Data maintained in such CRM System can be converted into meaningful information in order to identify the customer's churn behavior before they are lost, which increases customer strength [24].

Customer churn prediction modeling has been extensively studied in various domains such as banking, online gaming, airlines, telecommunication, financial services and social network services [30]. Various machine learning and knowledge acquisition techniques are applied; however, these techniques have been criticized certain reasons such as (a) bottleneck problem in KA, (b) training classifier and transfer of knowledge to system whenever a new nature of instance or case appears which do not render the old knowledge or rule changes that is very costly, (c) human expert or knowledge engineer is required for maintenance and handling latest changes which is also very costly. However, these factors can be overcome by efficient implementation of suitable prudence system and SE based technique for churn prediction in telecommunication sector. The prudence system will produce a warning prompt whenever new changes or a case occurs that is beyond the current knowledge in KB. On the other hand, SE will update and maintain the knowledge base just like virtual human expert or knowledge engineer.

The remainder of this paper is organized as follows; the paper comprises of five sections; in section 2, the domain of customer churn and churn prediction modeling is briefly described by means of broad literature review. Section 3 presents a detailed evaluation setup that comprises of dataset, different evaluation measures used in this study, proposed supervised feature selection process, RDR based classification and classifier evaluation. Section 4 introduces and summarize the key finding of a novel technique for recognition of unseen instances using prudence analysis with simulated expert, and the paper is concluded with the direction of future work in section 5.

2   Churn Prediction Modeling

Churn prediction has been widely studied in the past decade by researchers from academia and industry. Several methodologies and approaches have been proposed which mainly leverage both static and dynamic analysis for churn prediction modeling. Although, Churn analysis problem is an alarming issue for various domains such as Credit cards accounts [17], Banks & Financial Services [27], Human resource management [22], Insurance & subscription services [25], games [26] and social networks [28], this section represents various related studies about customer churn prediction modeling in telecommunication sector.

Ahn et al. [2] conducted a study on proprietary dataset of Korean mobile telecom services and revealed that the service quality factor is highly influenced on customer churn. Kang et al [13] introduced support vector machine recursive feature elimination (SVM-RFE) approach to identify the key attributes for customer churn and rule out the related and repeated attributes which reduced the dimensions of data. We also used similar approach to eliminate the redundant attributes from the dataset to select best features set for churn prediction using supervised feature selection technique.

Sharma et al [24] proposed a neural network (NN) approach and identified the importance of those attributes which are highly important for predicting customer churn in telecom's subscription services. Burez and Poel [4] developed a churn prediction model for pay-TV European company by using random forest and Markov chains. They also indicated two types of targeted approaches (reactive and proactive) for managing customer churn in telecommunication sector. Once the company identifies those customers who may likely to churn from the service providers before customers do it, the decision makers can provide targeted proactive offers and promotions to retain them.

Clemet Kirui et al [15] investigated new set of features and evaluate it through Naive Bayes, Bayesian Network and C4.5 for improving the churn recognition rate to identify the possible customer churn beforehand to retain as many as possible.

As we noticed in the above discussion that many classification techniques have been used for churn prediction but, an appropriate use of technique for classification is still an open research problem. Some experts from research community have investigated that SVM is one the state-of-the-art approaches for classification in machine learning due to its ability of model nonlinearities [8]. On other hand, a couple of studies [16,20] have reported that artificial neural networks can outperform as compared to other traditional machine learning algorithms.

It is clear from the literature that different researchers have approached the customer churn problem using different algorithms and techniques. This study is another attempt to propose a benchmarking and empirical study with the aim to produce further contribution in the said domain. The proposed study, introduces a novel procedure for useful utilization of prudence analysis which will produce a warning prompt every time for unusual cases and also introduces more efficient SE which will map the unseen cases to the conclusion used to test the KBS being built and update accordingly.

3   Evaluation Setup

3.1   Data Set

In this study, we have used publicly available dataset which can be obtained from URL [1]. The data set contains total 3333 instances with 21 attributes including one decision attribute "Churn?" and one unique attribute "Phone". Some considerations are required on data preparation such as removing of unique attributes "Phone" from the data set. Otherwise, the system will generate alert of unmatched case because the value of phone number will also be considered in the process. The attribute "state" is some sort of categorical attribute which contains 51 distinct string values but we represent them by numerical ID for simplicity. Detail attribute's description can be obtained from URL [1].

3.2   Evaluation Measures

It is nearly impossible to build a perfect classifier or a model that could perfectly characterize all instances of the test set [5], the following measures were used for the evaluation of proposed study. The detail of evaluation measures which are used in proposed study can be found in [3].

$$Sensitivity\ (REC) = \frac{TP}{TP + FN} \tag{1}$$

$$Precision\ (PRE) = \frac{TP}{TP + FP} \tag{2}$$

$$Accuracy = \frac{TP + TN}{TP + FN + TN + FP} \tag{3}$$

$$Misclassification\ Error\ (MisErr) = 1 - Accuracy \tag{4}$$

$$F - Measure\ (F - M) = 2\left(\frac{Precision.Recall}{Precision + Recall}\right) \tag{5}$$

3.3   Proposed Supervised Based Features Selection Process

In the proposed study, we have determine best subset of more relevant features based on different measures and procedures as follows; (1) To find the final best subset of features, we have used supervised feature selection through base-line algorithm "Ridor" using open-source data mining toolkit WEKA [11], (2) three set of features list were prepared. The original set of features list without any ranking of features, (3) we used ranked filter methods such as Information Gain Attribute Evaluator (IGAE) and Correlation Attribute Evaluator (CAE), (4) we have identified tightly correlated attributes from original set of features, subset of CAE and IGAE and removed one of these correlated attributes arbitrarily for the reason that it just increases the computational cost and effects on performance as well.

So ultimately we have obtained three subsets of features i.e. Subset 1 contains attributes obtained from Original Set (Original Set without Correlated attributes), Subset 2 contains those attributes that were obtained from IGAE (IGAE without Correlated attributes) and Subset 3 holds those attributes in area which we have obtained from CAE (CAE without Correlated attributes). (5) The weighted average values of ROC and simple accuracy for the selection of best feature set were computed and finally the best subset of features and subset of rejected attributes were identified. Fig. 1 reflects selection of optimum subset of features. The following Fig. 2 reflects the average performance of targeted combination of features subset to avoid the biasness in this study. Based on these results (i.e. Fig. 2) we have obtained the best result of each method as shown in table 1. Table 2 reflect, the best subset of features that were selected based on CAE as the CAE has the highest average performance as compared to other feature evaluation methods.

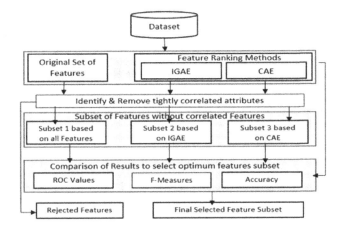

Fig. 1. Supervised Features Selection

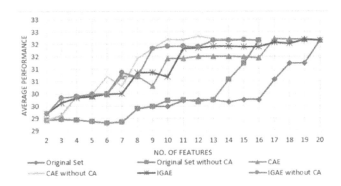

Fig. 2. Average Performance of targeted combination of features

Table 1. Summary of optimal obtained results

Methods	# of features	ROC Value	F-Measures Value	Accuracy
Original Set	20	0.839	0.944	0.9472
Original Set without Correlated Attributes	16	0.839	0.944	0.9473
CAE	18	0.846	0.945	0.9483
IGAE	19	0.841	0.944	0.9477
CAE without Correlated Attributes	12	0.857	0.949	0.9516
IGAE without Correlated Attributes	11	0.841	0.944	0.9477

Table 2. Final Optimal Subset Of Features

Attribute Name	Distinct Count	Means	StdDev	Categorical Values
State	51	25.269	14.737	-
Intl_Plan	2	0.097	0.296	Y=323, N=3010
VMail_Message	46	8.099	13.688	-
Day_Mins	1667	179.77	54.467	-
Day_Calls	119	100.43	20.069	-
Eve_Mins	1611	200.98	50.741	-
Night_Mins	1591	200.87	50.574	-
Intl_Mins	162	10.237	10.792	-
Intl_Calls	21	4.479	2.461	-
CutServ_Calls	10	1.563	1.315	-
Churn?	2	-	-	Class Label

3.4 Knowledge Acquisition and Ripple Down Rules Based Classification

RDR (Ripple Down Rules) was originally introduced by Compton and Jansen [6] in 1990. They proposed the RDR technique as a suitable methodology for Knowledge Acquisition (KA) as well as maintenance of large scale rule based system [9]. RDR was developed to deal with the contextual nature of knowledge expert [6,21]. Basically RDR is a list of rules where each rule can be linked to another list of rules that is called exceptions. If the exception next general rule is applicable then an exception is applicable [23]. Details about RDR can be found in studies [9,23].

RDR can be used for knowledge acquisition and classification. To extract the rules with exceptions and cases for building KBS, we have applied machine learning (ML) algorithm "Ripple Down Rules Learner" [9] on the prepared dataset (see Table 2) using open source toolkit WEKA [11]. 10-fold cross validation

Table 3. Describes the decision rules

# of Cases	Rules
Case 0	If TRUE then Churn? = NC //Default Rules
Case 1	Except (Day_Mins >224.6) and (VMail_Plan <= 0.5) and (Eve_Mins >183.75) and (Night_Mins >161.8) THEN Churn? = C
Case 2	Except (CustServ_Calls >3.5) and (Day_Mins <= 160.2) THEN Churn? = C
Case 3	Except (Intl_Plan >0.5) and (Intl_Mins >13.1) THEN Churn? = C
Case 4	Except (Day_Mins >221.85) and (Eve_Mins >241) and (VMail_Plan <= 0.5) and THEN Churn? = C
Case 5	Except (Day_Mins >244.95) and (VMail_Plan <= 0.5) and (Eve_Mins >223.75) and (Night_Mins >169.8) THEN Churn? = C
Case 6	Except (Day_Mins >236.45) and (VMail_Plan <= 0.5) and (Eve_Mins >144.35) THEN Churn? = C
Case 7	Except (Intl_Plan >0.5) and (Intl_Calls <= 2.5) =>Churn? = C
Case 8	Except (CustServ_Calls >3.5) and (Eve_Mins <= 191.85) and (Day_Mins <= 175.35) THEN Churn? = C

method is used to evaluate the performance of classifier (RDR) at initial stage. Rules and exceptions were transformed into decision rules list for easy interpretation and understanding. Based on these rules, the classification and prediction of decision class can be performed easily. Table 3 describes the decision rules with conditions and conclusions.

3.5   Evaluation of the Classifier

As discussed earlier, the evaluation of the classifier is performed using 10-fold cross validation on the dataset. Table 4 describes the weighted average evaluation measures used for calculating predictive power of the classifier.

Table 4. Performance of Classifier

CLASS	TPR	FPR	PRE	REC	F-M
NC	0.991	0.277	0.955	0.991	0.972
C	0.723	0.009	0.928	0.723	0.813
Weighted Avg.	0.952	0.239	0.951	0.952	0.949
MisErr	4.84%				
Accuracy	95.169%				

It is observed that correctly classified instances rate is above 95% (or 3172 individual instances) while incorrectly classified instances rate is above 4.8%

(or 161 individual instances). Therefore, to strengthen the ultimate results, the following section describes a prudent based approach to handle unseen instances and simulation expert approach to handle incorrectly classified instances. Such a prudent based approach would be very much interesting when it recognizes those situation that are beyond the available KBS and prompt for acquiring some new knowledge for KBS as well.

4   Proposed Prudent Based Recognition of Unseen Instances Using Simulated Expert

The term prudence means to describe the behavior of such an approach which will fire a warning every time when a case is in some way unusual [7]. Prudence Analysis was an RDR technique that was discovered as an alternate to deal with the brittleness of KBS [19]. In the literature [7,18], different mechanisms have been developed for prudent based KBS. Compton and Preston [7] used a technique in which a set of seen attribute's values associated with each rules and conclusion were maintained in a list. If the attribute's value for any case does not already exist in the prepared list, then an alert is generated for reviewing the case. Another study [18] presented ripple down model (RDM) which has two main function for detecting the knowledge boundaries based on range probabilities. One function is used for observing ranges values of continuous attribute and another is used for observing the already maintained values of categorical attribute. For this study, the proposed prudent based recognition of unseen instances using SE consists of the following steps;

- Develop a table with all those attributes which were extracted through RDR for building KBS with three lists titled as churn, non-churn and Both (churn, non-churn or complete dataset) as shown in table 5.
- Calculates minimum and maximum values of each attribute into the three specified lists. Table 5 represents the attributes range's information.

Once the table is maintained with required information, the following algorithms can be used for required purpose. For understanding purpose and to maintain the continuity, RDR based classification is also represented as algorithmic steps.

(See Algorithm: RDR based procedure for recognition of unseen instances using SE).

Initially, every case that is to be processed will be tested with RDR rules in KBS. If the case could not satisfy any rules in the KBS or reach to the wrong conclusion (Class Label) then it will be processed through the attributes' range values table. The list of attributes and values of attributes of instance will be compared with the ranges in table. If one of them is not already included in the list of attributes or list of values' range then a warning prompt will be generated to check the conclusion for new case and a call will be sent to the simulated expert (SE) with a parameter 1 (SE is discussed in next section).

328   A. Amin et al.

Table 5. List of Attributes Value's Ranges

Attribute	Both		Churn		Non-Churn	
	Min	Max	Min	Max	Min	Max
State	1	51	1	51	1	51
Intl_Plan	0	1	0	1	0	1
VMail_Plan	0	1	0	1	0	1
VMail_Message	0	51	0	48	0	51
Day_Mins	0	350.8	0	350.8	0	315.6
Day_Calls	0	165	0	165	0	163
Eve_Mins	0	363.7	70.9	365.7	0	361.8
Night_Mins	23.2	395	47.4	354.9	23.2	395
Intl_Mins	0	20	2	20	0	18.9
Intl_Calls	0	20	1	20	0	19
Cust_Serv_Calls	0	9	0	9	0	8

4.1   Simulated Expert

A SE is mapping of the unseen cases to the conclusion used to test the KBS being built [18]. In the proposed study, the technique of SE is applied as a procedure that will receive a call with parameter from prudent prompt. The SE will easily determine the nature of the assigned task based on the parameter value (e.g. 1, 2). If the parameter value is 1 then SE will add a new rule into KBS which satisfies the new case after the last evaluated rule in KBS based on domain expert decision. RDR is useful for adding new rule without effecting previous KBS. If the SE received a parameter value 2 then SE will determine from table 5 that whether it belong to churn list or non-churn list. Once it detect that an instance belong to churn list or non-churn list, SE will just simply test the corresponding rules and will try to reach to the correct conclusion. There may occur a rough instance case where an instance may belong to more than one list. In such rough instance case, SE will count the match cases and class of the majority matched cases will be assigned to that instance.

4.2   Evaluation of Simulated Expert

We have performed an experiment for approaching prudent based churn prediction. The data set is prepared from the misclassified cases which are given in table 4 (see section 3.5). The New or each misclassified instance which are figure out in table 4, was selected and passed through prudent check and SE. By doing this the initial KBS was also maintained up-to-date. There were 161 cases which were misclassified by simple RDR based classification while SE correctly predicted the conclusion for these cases through the procedure which is explained in section 4. Ultimately, the proposed technique correctly concluded 27 cases as churn and 134 cases as non-churn.

Algorithm: RDR based Procedure for recognition of unseen instances using SE

//Preparation of KBS and range table

A1. START

A2. Load prepared dataset (i.e table 2) to an RDR based classifier and evaluate the performance of classifier by applying 10-fold cross validation.

    A2.1. Extract the rules and exception

    A2.2. Create list of decision rules from extracted rules and exception i.e. table 3 //Set majority class set as default rule

    A2.3. IF default rule then Non-Churn except (list of decision rules with class label as Churn)

    A2.4. Keep track of misclassified instances. //False Positives

A3. Build a KBS from the list of decision rules set given in step A2.3

A4. Initialized a range table which keep track of the following; (i.e. table 5)

    A4.1. Add list of attributes from prepared dataset (i.e. table 2).

    A4.2. Create three lists for class labels (e.g. Churn, Non-churn, Both) and find the minimum and maximum value of each attribute for each class label.

//Prudence checkpoint prompt alert

//Read Instances

B1. IF instance belong to A2.4 THEN

    B1.2. Call to SE<instance, parameter 2 >

B2. ELSE IF value beyond the competency of KBS by comparing with both list given in step **A4.2** THEN

// For new instance

    B2.1. generate a prudent prompt AND Call to SE<instance, parameter 1>

B3. Repeat Step B1 until instances exists

B4. Stop

//Simulated Expert (SE)

C1. Start

C2. Initialized Counters INC=0, INNC=0 //increment for churn and for non-churn

C3. SE <instance ref, parameter value >

C4. IF parameter value is 2 THEN

    C4.1 Read attribute value until attributes exists for each instance received from Step **B1.2** //To keep track of attributes values for each instance

        C4.1.1. IF attribute value belong to range of churn list THEN INC+=1

        C4.1.2. ELSE IF value belong to range of non-churn list THEN INNC+=1

    C4.2. Repeat Step **C4.1**

    C4.3. IF INC >INNC THEN Assign Class label to instance Churn

    C4.4. IF INC <INNC THEN Assign Class Label to instance Non-Churn

    C4.5. IF INC==INNC THEN Assign Class Label to instance which have Maximum Count of matched In Step **A3**

C5. ELSE parameter value 1 THEN add a new rule into KBS which satisfy the new case after the last evaluated rule in KBS based on domain expert decision.

C6. Stop

5   Conclusion

The presented approach appears to be a worthy alternate for churn prediction in the telecommunication industry where prudent and simulated expert based approach in combination with RDR classifier has resulted in promising results. The proposed approach has the ability to handle the maintenance problem of KB and KBS brittleness. Furthermore, the proposed technique has also shown that all the misclassified and new cases can be successfully classified by updating rules in KB and assigning correct conclusion that were wrongly classified initially, before applying the proposed technique. So far this work presents that building KBS with RDR prudence is potentially useful in customer churn prediction in telecommunication industry. The future work is related to developing generic prudent base customer churn prediction tool for various domains.

References

1. Data Set (2015), http://www.sgi.com/tech/mlc/db/ (accessed December 20, 2014)
2. Ahn, J.H., Han, S.P., Lee, Y.S.: Customer churn analysis: Churn determinants and mediation effects of partial defection in the Korean mobile telecommunications service industry. Telecommunications Policy 30(10-11), 552–568 (2006), http://www.sciencedirect.com/science/article/pii/S0308596106000760
3. Amin, A., Shehzad, S., Khan, C., Ali, I., Anwar, S.: Churn prediction in telecommunication industry using rough set approach. In: Camacho, D., Kim, S.W., Trawiski, B. (eds.) New Trends in Computational Collective Intelligence. SCI, vol. 572, pp. 83–95. Springer, Heidelberg (2015), http://dx.doi.org/10.1007/978-3-319-10774-5_8
4. Burez, J., Van den Poel, D.: CRM at a pay-TV company: Using analytical models to reduce customer attrition by targeted marketing for subscription services. Expert Systems with Applications 32(2), 277–288 (2007), http://www.sciencedirect.com/science/article/pii/S0957417405003374
5. Burez, J., Van den Poel, D.: Handling class imbalance in customer churn prediction. Expert Systems with Applications 36(3), 4626–4636 (2009), http://linkinghub.elsevier.com/retrieve/pii/S0957417408002121
6. Compton, P., Jansen, R.: Knowledge in context: a strategy for expert system maintenance, pp. 292–306 (March 1990), http://dl.acm.org/citation.cfm?id=89411.89756
7. Compton, P., Preston, P., Edwards, G., Kang, B.: Knowledge based systems that have some idea of their limits. In: Tenth Knowledge Acquisition for Knowledge-Based Systems Workshop (1996)
8. Farquad, M.A.H., Ravi, V., Raju, S.B.: Churn prediction using comprehensible support vector machine: An analytical CRM application. Applied Soft Computing 19, 31–40 (2014), http://www.sciencedirect.com/science/article/pii/S1568494614000507
9. Gaines, B.R., Compton, P.: Induction of ripple-down rules applied to modeling large databases. Journal of Intelligent Information Systems 5(3), 211–228 (1995), http://dl.acm.org/citation.cfm?id=218246.218250
10. Hadden, J., Tiwari, A., Roy, R., Ruta, D.: Computer assisted customer churn management: State-of-the-art and future trends. Computers & Operations Research 34(10), 2902–2917 (2007), http://www.sciencedirect.com/science/article/pii/S0305054805003503

11. Holmes, G., Donkin, A., Witten, I.H.: WEKA: a machine learning workbench. In: Proceedings of ANZIIS 1994 - Australian New Zealnd Intelligent Information Systems Conference, pp. 357–361. IEEE (1994), http://ieeexplore.ieee.org/articleDetails.jsp?arnumber=396988
12. Huang, B., Kechadi, M.T., Buckley, B.: Customer churn prediction in telecommunications. Expert Systems with Applications 39(1), 1414–1425 (2012), http://dl.acm.org/citation.cfm?id=2038068.2038213
13. Kang, C., Pei-ji, S.: Customer Churn Prediction Based on SVM-RFE. In: 2008 International Seminar on Business and Information Management, vol. 1, pp. 306–309. IEEE (December 2008), http://ieeexplore.ieee.org/articleDetails.jsp?arnumber=5117490
14. Khan, I., Usman, I., Usman, T., Rehman, G.U., Rehman, A.U.: Intelligent Churn prediction for Telecommunication Industry. International Journal of Innovation and Applied Studies 4(1), 165–170 (2013)
15. Kirui, C., Hong, L., Cheruiyot, W., Kirui, H.: Predicting Customer Churn in Mobile Telephony Industry Using Probabilistic Classifiers in Data Mining. IJCSI International Journal of Computer Science Issues 10(2), 165–172 (2013)
16. Lazarov, V., Capota, M.: Churn Prediction. Business Analytics Course. TUM Computer Science (2007), http://home.in.tum.de/~lazarov/files/research/papers/churn-prediction.pdf
17. Lin, C.S., Tzeng, G.H., Chin, Y.C.: Combined rough set theory and flow network graph to predict customer churn in credit card accounts. Expert Systems with Applications 38(1), 8–15 (2011), http://linkinghub.elsevier.com/retrieve/pii/S0957417410004501
18. Maruatona, O., Vamplew, P., Dazeley, R.: Knowledge Management and Acquisition for Intelligent Systems. Springer, Heidelberg (2012)
19. Maruatona, O.O., Vamplew, P., Dazeley, R.: Prudent Fraud Detection in Internet Banking. In: 2012 Third Cybercrime and Trustworthy Computing Workshop, pp. 60–65. IEEE (October 2012), http://ieeexplore.ieee.org/articleDetails.jsp?arnumber=6498429
20. Mozer, M.C., Wolniewicz, R., Grimes, D.B., Johnson, E., Kaushansky, H.: Predicting subscriber dissatisfaction and improving retention in the wireless telecommunications industry. IEEE Transactions on Neural Networks / A Publication of the IEEE Neural Networks Council 11(3), 690–696 (2000), http://ieeexplore.ieee.org/articleDetails.jsp?arnumber=846740
21. Richards, D., Compton, P.: Taking up the situated cognition challenge with ripple down rules. International Journal of Human-Computer Studies 49(6), 895–926 (1998), http://www.sciencedirect.com/science/article/pii/S1071581998902312
22. Saradhi, V.V., Palshikar, G.K.: Employee churn prediction. Expert Systems with Applications 38(3), 1999–2006 (2011), http://www.sciencedirect.com/science/article/pii/S0957417410007621
23. Scheffer, T.: Algebraic Foundation and Improved Methods of Induction of Ripple Down Rules. In: Pasific Knowledge Acquisition Workshop, Sydney, pp. 23–25 (1996)
24. Sharma, A., Prabin Kumar, P.: A Neural Network based Approach for Predicting Customer Churn in Cellular Network Services. International Journal of Computer Applications 27(11), 26–31 (2011)
25. Soeini, R.A., Rodpysh, K.V.: Applying Data Mining to Insurance Customer Churn Management 30, 82–92 (2012)
26. Suznjevic, M., Stupar, I., Matijasevic, M.: MMORPG Player Behavior Model based on Player Action Categories. IEEE (2011)

27. Van den Poel, D., Larivière, B.: Customer attrition analysis for financial services using proportional hazard models. European Journal of Operational Research 157(1), 196–217 (2004),
    http://www.sciencedirect.com/science/article/pii/S0377221703000699
28. Verbeke, W., Martens, D., Baesens, B.: Social network analysis for customer churn prediction. Applied Soft Computing 14, 431–446 (2014),
    http://linkinghub.elsevier.com/retrieve/pii/S1568494613003116
29. Verbeke, W., Martens, D., Mues, C., Baesens, B.: Building comprehensible customer churn prediction models with advanced rule induction techniques. Expert Systems with Applications 38(3), 2354–2364 (2011),
    http://www.sciencedirect.com/science/article/pii/S0957417410008067
30. Wolniewicz, R.H., Dodier, R.: Predicting customer behavior in telecommunications. IEEE Intelligent Systems 19(2), 50–58 (2004),
    http://ieeexplore.ieee.org/articleDetails.jsp?arnumber=1274911

Forecasting Daily Urban Water Demand Using Dynamic Gaussian Bayesian Network

Wojciech Froelich[✉]

Institute of Computer Science, University of Silesia, Sosnowiec, Poland
wojciech.froelich@us.edu.pl

Abstract. The objective of the presented research is to create effective forecasting system for daily urban water demand. The addressed problem is crucial for cost-effective, sustainable management and optimization of water distribution systems. In this paper, a dynamic Gaussian Bayesian network (DGBN) predictive model is proposed to be applied for the forecasting of a hydrological time series. Different types of DGBNs are compared with respect to their structure and the corresponding effectiveness of prediction. First, it has been found that models based on the automatic learning of network structure are not the most effective, and they are outperformed by models with the designed structure. Second, this paper proposes a simple but effective structure of DGBN. The presented comparative experiments provide evidence for the superiority of the designed model, which outperforms not only other DGBNs but also other state-of-the-art forecasting models.

Keywords: Forecasting time series · Water demand · Bayesian networks

1 Introduction

Forecasting of time series and especially forecasting of urban water demand is an important problem considered by many researches [3,16,8,10]. Dependent on the length of the prediction horizon, forecasting water demand can be long-, medium-, short- or very-short-term [3]. Long-term predictions are made in decades or in yearly scale, and they are crucial for planning and design of water distribution systems [21]. Medium-term forecasting made yearly is important for making improvements to already existing water distribution systems [3]. Short-term forecasting is made in a time scale of months [9,5], days [19,2] or hours [12,18]. For example, daily predictions are required for cost-effective, sustainable management and optimization of urban water supply systems [3].

This paper addresses the problem of daily water demand. Existing literature on the topic involves many solutions. An analytic predictive model based on a set of arbitrary designed equations was proposed in [25]. The other approach is to learn predictive models using historical data gathered from water distribution systems. Linear regression and artificial neural networks were applied in [20]. A hybrid approach combining feed-forward artificial neural networks, fuzzy logic, and a genetic algorithm was proposed in [19]. An extensive overview of papers

© Springer International Publishing Switzerland 2015
S. Kozielski et al. (Eds.): BDAS 2015, CCIS 521, pp. 333–342, 2015.
DOI: 10.1007/978-3-319-18422-7_30

devoted to water demand forecasting was made in [21] and [8]. One of the first attempts and maybe the best known example of using Bayesian networks for forecasting is [1]. Bayesian networks were used for forecasting aggregated end-use water heater load in residential areas [24]. Recently, an overview of Bayesian forecasting systems has been done in [4].

The main deficiency of existing approaches is the still limited forecasting accuracy. In this paper, an effort is undertaken to design a new predictive model for more effective forecasting of daily water demand. For that purpose, a Bayesian network approach is proposed. To the best of our knowledge, Bayesian networks and especially their dynamic Gaussian version have been never applied for the prediction of water demand.

The effort undertaken in this study is to create the structure and learn the parameters of a Gaussian Bayessian network, thus aiming to achieve the best possible forecasting accuracy of daily urban water demand. After numerous comparative experiments, it has been found that the models based on automatic learning of network structures are not the most effective. They can be out-performed by models with manually designed structures. For that reason, this study investigates and compares different simple DGBNs. The experiments provide evidence for the superiority of the simple DGBN model that outperforms other selected state-of-the-art forecasting methods.

The remainder of this paper is organized as follows. Section 2 introduces background knowledge related to the presented research, i.e., basic notions related to Gaussian Bayesian networks and their application to forecasting. Section 2 also presents a selected forecasting accuracy measure and a statistical test applied to the comparison of forecasting errors. Section 3 presents the contribution of this paper, after specifying the addressed problem, in which the dynamic Gaussian Bayesian network approach is adapted to the forecasting of daily water demand. The experimental section 4 contains the presentation of comparative experiments to provide evidence for high effectiveness of the proposed DGBN approach when applied to water demand forecasting.

2   Background Knowledge

Numerous probabilistic models can be applied to the forecasting of time series. One of them is a Bayesian network.

2.1   Bayesian Networks

In general two types of Bayesian networks can be distinguished: 1) discrete Bayesian networks 2) Gaussian Bayesian networks.

A discrete Bayesian network is a triple $BN = (X, DAG, P)$, where X is a set of random variables, $DAG = (V, E)$ is a directed acyclic graph and P is a set of conditional probability distributions [14]. Each node of the graph $v \in V$ corresponds to a discrete random variable $X_i \in X$ with a finite set of mutually exclusive states. Directed edges $E \subseteq V \times V$ of DAG express conditional dependence (and independence) between random variables. For every child node X_i, there is $P(X_i|X_{pa(i)})$, which is the conditional probability distribution for each $X_i \in X$, where: $pa(i)$ is set of parent (conditioning) variables of the variable X_i.

A Gaussian Bayesian network (GBN) is a model that specifies a probabilistic distribution over a mixture of continuous and discrete variables [14]. The set X is partitioned to the subset of continuous variables X_Γ and the set of discrete variables X_Δ. In comparison with the discrete Bayesian networks, GBNs contain continuous random variables with the linear conditional Gaussian distribution. Those Gaussian density functions are specified by their means and variances. Formally, GBN is a tuple $GBN = (X, DAG, P, F)$, where X is a set of discrete or continuous random variables, and $DAG = (V, E)$ is a directed acyclic graph, P denotes a set of conditional probability distributions, and F is a set of Gaussian density functions $\aleph(\mu, \sigma^2)$ assigned to continuous random variables, where μ denotes the mean and σ^2 is the variance. Every Gaussian probability density function that is assigned to continuous random variable X_i is conditional on the configuration of the parent variables. In case the parents of the continuous random variable are discrete variables, the variable possesses a single Gaussian distribution function for each combination of values of the parents. If the parents are continuous, the mean of density function assigned to the variable depends linearly on the values of its continuous parents.

For both types of Bayesian networks, when a temporal aspect of data is analyzed, a dynamic version is used. The Dynamic Bayesian network (DBN) is a simple extension of standard BN. In DBN, a discrete time scale is used, and selected random variables are related to parent variables of the previous time points. DBN contains arcs that are related to time delays. The directed arcs of the DBN flow forward in accordance with the time flow. This way DBN is an art of network that relates probabilistically random variables assuming that some of them are lagged in time.

Using the above described typology, it is possible to distinguish specific dynamic Gaussian Bayesian networks that combine features of Gaussian and dynamic extensions of Bayesian networks.

Bayesian networks can be designed by an expert or can be learned from historical data. The learning of Bayesian network consists of two steps: 1)structure learning 2) parameters learning.

Specifically for Gaussian Bayesian networks, the learning of their structure can be performed using several learning algorithms [22]:

- Hill-Climbing is a score-based greedy search on the space of the directed graphs,
- Tabu Search is modified Hill-Climbing that escapes local optima by selecting a network that minimally decreases the score function,
- Max-Min Hill-Climbing is a hybrid algorithm combining the Max-Min Parents and Children algorithm (to restrict the search space) and the Hill-Climbing algorithm,

- Restricted Maximization is a generalized implementation of Max-Min Hill-Climbing using a combination of constraint- and score-based algorithms.

It is worth emphasizing that all of the mentioned learning algorithms are more general, they are not dedicated to the dynamic version of Gaussian Bayesian networks. For that reason the direction of the arcs within the trained structure (probabilistic dependency) may be contradictory with the direction of time flow that is assigned to them. However, in spite of it, the prediction can be performed using the appropriate backward reasoning algorithms. The other learning algorithms that produce networks with undirected edges were not considered. To the best of our knowledge there are no publicly available implementations of learning algorithms that are dedicated to learn the structure of DGBN.

Assuming that the structure of the DGBN is already learned or given by an expert, the only suitable method for learning parameters is available in 'bnlearn' library of R, it is the Maximum Likelihood Estimation (MLE) [22]. Moreover, to the best of our knowledge, at the moment of writing this paper, only that method was publicly available.

2.2 Measuring the Accuracy of Forecasting

Independent of the applied predictive model, the error for any single forecast is calculated as $e(t) = X'(t) - X(t)$, where $X'(t)$, $X(t)$ denote the predicted and actual values of the series respectively. There are numerous methods for measuring the cumulative accuracy of a time series. However, for the purpose of this paper, we decided to apply only mean absolute percentage error (MAPE) given as formula (1). The reason of such selection is that the MAPE is a simple, easy-to-interpret scale-independent error measure. By exploiting it, it is easy to compare the efficiency of forecasts independent of the absolute values of the considered time series.

$$MAPE = \frac{1}{n} \sum_{t=1}^{n} |\frac{e_t}{x_t}| \times 100\% \tag{1}$$

Besides cumulative estimation of prediction errors, for the purposes of the comparison of residuals (series of forecasting errors) generated by different models, a Diebold-Mariano (DM) statistical test is commonly used [7]. The null hypothesis of the DM test H_0: Model 1/Model 2 is that Model 1 is more accurate than Model 2. The alternative hypothesis is that Model 2 is better than Model 1.

3 Dynamic Gaussian Bayesian Network for the Forecasting of Water Demand

Let $X \in \Re$ be a real-valued variable and let $t \in [1, 2, \ldots, n]$ be a discrete time scale, where $n \in \aleph$ is its length. The values of $X(t)$ are observed over time. A time series is denoted as a sequence $\{X(t)\} = \{X(1), X(2), \ldots, X(n)\}$.

For the purpose of this paper it is assumed that $X(t)$ corresponds to the continuous random variable whose value is to be predicted. The lagged versions $X(t-1), X(t-2), \ldots, X(t-n)$ also correspond to random variables. We assume that their values are known from historical data.

Let us also assume that $K_i(t) \in K, i \in [1, \ldots, m]$ are the other known i^{th} time series and K is the set of such series. We assume here that the random variables corresponding to the series from K together with $X(t-1), X(t-2), \ldots X(t-n)$ constitute the set E of the so-called evidence variables, i.e., $E = K \bigcup \{X(t-1), X(t-2), \ldots X(t-n)\}$.

A Bayesian network model can be used to calculate the posterior probability of unknown $X(t)$ given the set of evidence variables E. Formally, the objective of Bayesian forecasting is to calculate the probability distribution $P(X'(t)|E)$, where $X'(t)$ denotes the predicted value and E is the observed evidence. The evidence variables from E are also called explanatory variables. The first issue during the construction of our DGBN model is their proper selection.

The predicted variable $X(t)$ is the water flow denoted for the sake of simplicity as $flow$. To denote the time lag of any variable, the lower index is used, e.g., $flow_1$ denotes $flow(t-1)$ and $flow_7$ denotes $flow(t-7)$. Similar notation is used for the rest of the time series and related variables.

As the existing studies show, the lower precipitation and higher temperature may lead to higher water demand [3]. For that reason we decided to check the influence of weather-related variables on water demand. Variables $rain$ and $rain_1$ related to precipitation are selected. Also the temperature variables T and T_1 are taken into account.

In the case of urban water demand, weekly seasonality is expected [3]. Weekly seasonality is the phenomenon related to the division of a week into working days when water demand is usually higher and to the days of Saturday and Sunday with lower water demand [3]. For that reason, the time lag of 7 of water demand is considered. In this paper this fact is taken into account by including random variable $flow_7$ into the set of evidence variables.

Also to reflect the expected weekly seasonality, the variable day related to the day of the week is considered. It can be expected that the variable $month$ may help to capture the potential monthly seasonality related to summer or winter holidays [3]. The main problem encountered while using the calendar-related variables was related to the practical implementation. To the best of our knowledge, at the moment of writing the paper there was no free implementation of learning DGBN and forecasting on that basis in the case of using a mixture of continuous and symbolic types of variables.

The feature is not implemented in the 'bn.learn' package of R. Although such models can be learned using the 'deal' packages, the inference is not implemented. The use of commercial packages was not taken into account due to their limited availability. To overcome the problem we decided to exploit the fact that the variables day and $month$ are, in fact, numeric integer values. This enabled us to use the 'bn.learn' package. This way the numeric values of day and $month$ were used as the parameters of the mean of the related probabilistic density functions.

Table 1. The set of evidence variables

Variable	Description
$flow_1$	water demand on previous day
$flow_7$	water demand in previous week
T	temperature
T_1	temperature on previous day
$rain$	rain
$rain_1$	rain on previous day
day	day of week
$month$	month

The final set of the evidence variables is given in table 1.

4   Experiments

For the validation of the proposed approach real-world data were used. Historical water demand data (water flow) were acquired from the urban water distribution network of Sosnowiec, Poland. Weather data were gathered and prepared at the University of Silesia weather station. All data were collected from the period of 16 June 2007 to 31 December 2013 containing 2386 records.

For the following learn-and-test trials, the time series had to be partitioned into learning and testing data. For that purpose, the idea of a growing learning window was applied, see Fig. 1. It assumes that the learning period begins at the first available observation of data and finishes at time $t-1$. As the time flows and the amount of available data grows, also the learning window grows till the last available data at time $t-1$. The length of the growing window is $t-1$. Because the prediction is made one-step ahead the DGBN model is updated every day. According to discussion regarding the minimal data samples [13], the minimum length of the growing learning window was set to 25, i.e., $t > 24$.

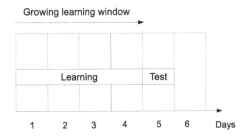

Fig. 1. Growing window

In the first set of experiments the temperature T and the *rain* variables were explicitly used in the network structure. As they were unknown during the day t they had to be predicted. This way, the problem was related to weather forecasting [11,15]. First, the structure of DGBN was learned automatically using all the algorithms given in section 2. The obtained prediction efficiency turned out to be very poor. Second, an attempt was undertaken to design the network structure by trial and error; however, the results were also very poor. For that reason we do not provide those results here.

In the next set of experiments the variables T and *rain* were not used. As a result the structure of DGBN has been simplified containing besides evidence variables only a single variable *flow* to be predicted at time t.

Similarly as before, two trails were made. First, the structure of DGBN1-DGBN4 was learned automatically using the available algorithms (see section 2). The obtained results are given in table 2. Second, the structure of DGBN5 was designed. It quickly became clear that better results can be obtained while assuming a direct influence of explanatory variables on the predicted water demand *flow*.

In the following experiments we decided to test how the elimination of explanatory variables influences the accuracy of forecasting. Due to space restrictions, we can provide here only the selected results of those experiments. In DGBN6 the *month* and *day* variables were not used. In DGBN7 the weather variables T_1 and $rain_1$ were neglected. After further step-wise elimination of explanatory variables their influence on the prediction accuracy was examined. The results are given in table 2.

Table 2. Results

Model	Evidence variables considered	MAPE
naive	univariate time series	9.517191
ES	univariate time series	9.624532
ARIMA	univariate time series	14.56436
LR	all from table 1	10.30014
DGBN 1	all - Hill-Climbing	7.858135
DGBN 2	all - Tabu Search	15.32959
DGBN 3	all- Max-Min Hill-Climbing	7.649636
DGBN 4	all - Restricted Maximization	7.654599
DGBN 5	$flow_1$, $flow_7$, T_1, $rain_1$, $month$, day	7.599925
DGBN 6	$flow_1$, $flow_7$, T_1, $rain_1$	7.663248
DGBN 7	$flow_1$, $flow_7$, $month$, day	7.656192
DGBN 8	$month$, day	38.85962
DGBN 9	$flow_1$, $flow_7$	7.660317
DGBN 10	$flow_1$	7.688957
DGBN 11	$flow_7$	13.57425

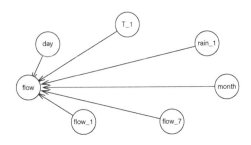

Fig. 2. Structure of the winning DGBN5

To compare the forecasting accuracy of the Bayesian approach with the other state-of-the art models, numerous comparative experiments were performed. On the basis of literature study, those models were selected for comparisons, for which the lowest forecasting errors were reported. Also the naive model, for which $X(t) = X(t-1)$, was used, traditionally applied for benchmarking of time series forecasting. The set of competitive to DGBN predictive models contained naive, exponential smoothing (ES), auto-regressive integrated moving average (ARIMA), and linear regression (LR). The description of those state-of-the-art methods fell beyond the scope of the paper but can be found in [6,23]. In every case, the parameters of the applied models were adjusted by numerous try-and-test experiments. All coefficients of ES model were learned automatically using the function 'ets' that is available in the 'forecast' library of the R package [17]. For fitting the ARIMA model to the data and adjusting the parameters automatically, we used the function 'auto.arima' from the package 'forecast.' The applied 'auto.arima' recognize also the seasonality in data and implements in fact the seasonal version of ARIMA. Also in case of linear regression the training process was performed using R package. The details on the adjustment of parameters are described in the documentation of R package.

As can be noted in table 2, the values of MAPE show that DGBN5 model was the best. The structure of DGBN5 is shown in Fig. 2. It is worth noting that the DGBN5 is equivalent to continuous version of naive Bayesian classifier.

Table 3. Diebold-Mariano test

Model1/Model2	p-value
DGBN5/DGBN3	0.9987
DGBN5/DGBN4	0.9994
DGBN5/DGBN6	0.9916
DGBN5/DGBN7	0.4912
DGBN5/DGBN9	0.9087
DGBN5/DGBN10	0.9999

Furthermore, for the most competitive models the Diebold-Mariano statistical test was performed. The results are given in table 3. Only the DGBN7 with the explanatory variables $flow_1$, $flow_7$, $month$, day can be interpreted as competitive to DGBN5; however, the null hypothesis that DGBN5 is better was not rejected. It leads to the conclusion that the use of weather variables does not bring much benefit while forecasting the considered time series.

5 Final Remarks

The paper proposed to apply a dynamic Gaussian Bayesian network predictive model to forecast daily water demand. The experimental studies have shown that DGBNs with the designed structure outperform those with an automatic learned structure. Moreover, comparative experiments show that the winning DGBN that is equivalent to continuous version of naive Bayes classifier outperformed selected state-of-the-art forecasting models. In spite of the obtained improvement in terms of forecast accuracy, the MAPE for the winning DGBN reaches 7.6%, which, in our opinion, motivates further research on the addressed problem.

Acknowledgments. The work was supported by ISS-EWATUS project which has received funding from the European Union's Seventh Framework Programme for research, technological development and demonstration under grant agreement no. 619228.

The author would like also to thank the water distribution company in Sosnowiec (Poland) for gathering water demand data and the personal of the weather station of the University of Silesia for collecting and preparing meteorological data.

References

1. Abramson, B., Brown, J., Edwards, W., Murphy, A., Winkler, R.L.: Hailfinder: A Bayesian system for forecasting severe weather. International Journal of Forecasting 12(1), 57–71 (1996)
2. Adamowski, J., Adamowski, K., Prokoph, A.: A spectral analysis based methodology to detect climatological influences on daily urban water demand. Mathematical Geosciences 45(1), 49–68 (2013)
3. Billings, R.B., Jones, C.V.: Forecasting urban Water Demand. American Water Works Association (2008)
4. Biondi, D., De Luca, D.L.: Performance assessment of a bayesian forecasting system (bfs) for real-time flood forecasting. Journal of Hydrology 479, 51–63 (2013)
5. Bougadis, J., Adamowski, K., Diduch, R.: Short-term municipal water demand forecasting. Hydrological Processes 19(1), 137–148 (2005)
6. Cowpertwait, P.S.P., Metcalfe, A.V.: Introductory Time Series with R, 1st edn. Springer Publishing Company, Incorporated (2009)
7. Diebold, F.X., Mariano, R.S.: Comparing predictive accuracy. Journal of Business and Economic Statistics 13(3), 253–263 (1995)

8. Donkor, E., Mazzuchi, T., Soyer, R., Roberson, A.J.: Urban water demand forecasting: Review of methods and models. Journal of Water Resources Planning and Management 140(2), 146–159 (2014)
9. Firat, M., Turan, M.E., Yurdusev, M.A.: Comparative analysis of neural network techniques for predicting water consumption time series. Journal of Hydrology 384(1-2), 46–51 (2010)
10. Froelich, W., Papageorgiou, E.I.: Extended evolutionary learning of fuzzy cognitive maps for the prediction of multivariate time-series. In: Papageorgiou, E.I. (ed.) Fuzzy Cognitive Maps for Applied Sciences and Engineering. ISRL, vol. 54, pp. 121–131. Springer, Heidelberg (2014)
11. Froelich, W., Salmeron, J.L.: Evolutionary learning of fuzzy grey cognitive maps for the forecasting of multivariate, interval-valued time series. International Journal of Approximate Reasoning 55(6), 1319–1335 (2014)
12. Herrera, M., Torgo, L., Izquierdo, J., Pérez-García, R.: Predictive models for forecasting hourly urban water demand. Journal of Hydrology 387(1-2), 141–150 (2010)
13. Hyndman, R.J., Kostenko, A.V.: Minimum sample size requirements for seasonal forecasting models. The International Journal of Applied Forecasting 6, 12–15 (2007)
14. Jensen, F.V.: Bayesian Networks and Decision Graphs. Springer (2001)
15. Juszczuk, P., Froelich, W.: Learning fuzzy cognitive maps using a differential evolution algorithm. Polish Journal of Environmental Studies 12(3B), 108–112 (2009)
16. Machiwal, D., Jha, M.K.: Hydrologic Time Series Analysis: Theory and Practice. Springer (2012)
17. package, R.: http://www.r-project.org
18. Preis, A., Whittle, A.J., Ostfeld, A., Perelman, L.: Efficient hydraulic state estimation technique using reduced models of urban water networks. Journal of Water Resources Planning and Management (2011)
19. Pulido-Calvo, I., Gutiérrez-Estrada, J.C.: Improvedfrigation water demand forecasting using a soft-computing hybrid model. Biosystems Engineering 102(2), 202–218 (2009)
20. Pulido-Calvo, I., Montesinos, P., Roldán, J., Ruiz-Navarro, F.: Linear regressions and neural approaches to water demand forecasting in irrigation districts with telemetry systems. Biosystems Engineering 97(2), 283–293 (2007)
21. Qi, C., Chang, N.B.: System dynamics modeling for municipal water demand estimation in an urban region under uncertain economic impacts. Journal of Environmental Management 92(6), 1628–1641 (2011)
22. Scutari, M.: Bayesian network structure learning, parameter learning and inferenceg (2014), http://www.bnlearn.com/
23. Shumway, R.H., Stoffer, D.S.: Time Series Analysis and Its Applications. Springer (2000)
24. Vlachopoulou, M., Chin, G., Fuller, J.C., Lu, S., Kalsi, K.: Model for aggregated water heater load using dynamic bayesian networks. In: Proceedings of the DMIN 2012 International Conference on Data Mining, pp. 1–7 (2012)
25. Zhou, S.L., McMahon, T.A., Walton, A., Lewis, J.: Forecasting daily urban water demand: a case study of melbourne. Journal of Hydrology 236(3-4), 153–164 (2000)

Parallel Density-Based Stream Clustering Using a Multi-user GPU Scheduler

Ayman Tarakji[1](✉), Marwan Hassani[2], Lyubomir Georgiev, Thomas Seidl[2], and Rainer Leupers[3]

[1] Research Group for Operating Systems, Faculty of Electrical Engineering, RWTH Aachen University, Germany
ayman.tarakji@rwth-aachen.de
[2] Data Management and Data Exploration Group, RWTH Aachen University, Germany
{hassani,seidl}@cs.rwth-aachen.de
[3] Institute for Communication Technologies and Embedded Systems, Faculty of Electrical Engineering, RWTH Aachen University, Germany
rainer.leupers@ice.rwth-aachen.de

Abstract. With the emergence of advanced stream computing architectures, their deployment to accelerate long-running data mining applications is becoming a matter of course. This work presents a novel design concept of the stream clustering algorithm *DenStream*, based on a previously presented scheduling framework for GPUs. By means of our scheduler *OCLSched*, *DenStream* runs together with general computation tasks in a multi-user computing environment, sharing the GPU resources. A major point of concern throughout this paper has been to disclose the functionality and purposes of the applied scheduling methods, and to demonstrate the *OCLSched*'s ability of managing highly complex applications in a multi-task GPU environment. Also in terms of performance, our tests show reasonable improvements when comparing the proposed parallel concept of *DenStream* with a single-threaded CPU version.

Keywords: GPGPU · OpenCL · *DenStream* · Data Mining · Stream clustering · Task scheduling

1 Introduction

Over the last few years, data mining applications have grown in size and complexity. Because of their long runtime on conventional processors, researchers are looking at running such applications on computational accelerators. The deployment of GPUs, for instance, can lead to enormous improvement in their performance as shown in a related work [16]. However, providing high-performance solutions for highly complex applications might be a time-consuming task [20], especially, if an efficient use of shared devices should be considered during the development. Further, numerous studies have shown that resource utilization constitutes a serious problem when considering the typically available computation power of modern accelerators [6,11,17]. For these reasons, an autonomous

© Springer International Publishing Switzerland 2015
S. Kozielski et al. (Eds.): BDAS 2015, CCIS 521, pp. 343–360, 2015.
DOI: 10.1007/978-3-319-18422-7_31

scheduling unit with built-in libraries, that provide the functionality for certain type of applications, is becoming a necessity to facilitate the access to shared resources and to exploit the computation power of available accelerators. A multi-user functionality would improve the overall throughput rate in accelerator-based systems, if multiple tasks are issued to be run simultaneously. These and other challenging issues have been pursued by our research during the last few years. We introduced *OCLSched*, a scheduling framework for common computation tasks on heterogeneous computing systems in a previous work [18]. Our scheduler provides for a multi-user functionality by means of the well established server-client model. It runs in background and manages the distribution and execution of tasks centrally, exhausting the available computing units.

This is, however, the first work, in which we introduce *OCLSched* in combination with a complex data mining algorithm. We propose a well thought-out OpenCL-based implementation concept of a recent data stream clustering application *DenStream* [2], with the intention of providing a library-like environment for this kind of applications as a supplement to *OCLSched*. Serving as a starting point for customizing typical data mining strategies for HPC accelerators, we focus in this paper on the cluster and outlier analysis [3]. Within the execution model of *OCLSched*, the clustering functionality is described as a single user process, achieving its defined targets asynchronous to other general purpose computations. In order to overcome these challenges, the proposed *OCLSched*'s implementation of *DenStream* combines very recent techniques from two research areas; Heterogeneous parallel computing and data mining. We mainly discuss different important aspects and identify a lot of trade-offs of running highly complex data mining applications on the GPU. Besides proving the ability of the *OCLSched*'s task-concept to handle complex problems from the new application domain of data mining, this work opens up new avenues to a successful development of innovative GPU-programing designs for data mining applications in general, and for modern stream clustering algorithms in particular.

As a major modification towards providing an optimized built-in library for stream clustering algorithms, the task-concept of *OCLSched* is improved and extended increasing the efficiency of the considered algorithms. For instance, a special memory allocation and transfer functionality has been necessary to be established, in order to achieve an efficient memory management related to the density-based clustering algorithm. In the context of this work, *DenStream* is completely redesigned to meet the characteristics and special design requirements of the parallel GPU architecture. A number of *OCLSched* tasks describing *DenStream* are designed in such a way that they could be alternatively combined and separately launched on different processing devices, following the stream computing paradigm. Using a variety of real data sets, we conduct numerous experiments to expose the influence of certain parameters on *DenStream*'s performance on the one side, and to discuss the GPU-resource utilization on the other.

The reminder of this paper is organized as follows: After providing a short overview about the scheduling framework, which is used in the implementation of *DenStream*, we briefly discuss the key concepts of the stream clustering algorithm. In section 3, we present the strategies followed in the task-design of the parallel *DenStream* algorithm. Later in section 4, we discuss the implementation details of the parallel algorithm. A thorough evaluation of the presented strategy is introduced in section 5, and section 7 concludes the work and discusses future directions immediately after the related work in section 6.

1.1   Why OpenCL?

A variety of programing frameworks have been recently released, by means of which software developers can design and implement their applications on high performance accelerators. The most well-established frameworks in this context seem to be *OpenCL* (Open Computing Language), *OpenMP* (Open Multi-Processing) and *CUDA* (Compute Unified Device Architecture). In terms of performance, it might be argued that *CUDA* can be more efficient on NVIDIA devices. However, Komatsu et al. [10] demonstrated that the performance gap between *CUDA* and *OpenCL* can be overcome by performing special purpose and hardware-specific optimizations on OpenCL programs.

Besides the performance issue, the advantage of *OpenCL* is represented in its portability [12]. To be successful, an accelerator programing model must be low-level enough to express potential parallelism in existing sequential code, but high-level and portable enough to allow efficient implementation-mappings to a variety of accelerators. For these reasons, we believe that *OpenCL* offers a promising programing framework for future applications, when it comes to heterogeneous computing.

1.2   *OCLSched*

OCLSched stands for OpenCL-based scheduling framework, by means of which only basic skills relating to the OpenCL programming model would be sufficient to issue any given application to be run efficiently, sharing OpenCL computing devices with other users. It manages the processing of common computation tasks on accelerator-based heterogeneous systems, exhausting the resources of available accelerators. Multiple tasks written in OpenCL can be issued by means of a C++ API that relies on the OpenCL C++ wrapper. At this point, a daemon takes over the control immediately and performs load scheduling. The major purpose of our scheduler is to manage the execution of multiple tasks issued by multiple users on different shared OpenCL devices. It acts as a server process, accepting user client connections and providing for a multi-user functionality by means of the well established server-client model. Based on preemption and context funneling, the core functionality of our scheduler specifically encompasses the establishment of shared multi-GPU environment to maximize the utilization of available resources. Due to its implementation, *OCLSched* can be easily applicable to a common OS. In Fig. 1, an overview of the *OCLSched*'s execution model is provided. On this basis, the clustering functionality discussed in this paper is described as a single user process within the execution model of *OCLSched*.

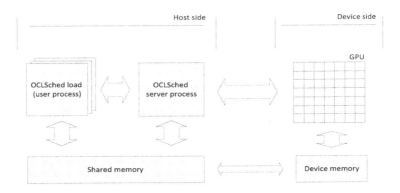

Fig. 1. The Execution Model of *OCLSched*

2   DenStream

A data stream is a massive, continuous and rapid sequence of data elements without length limitation, containing (d, t)-*tuples* where d is a data record, and t is a time-stamp. Each reading process of such a sequence is achieved through a linear scan. The data stream model is widely used for modeling tasks such as telephone records, financial transaction statistics and multimedia data lists. For the purpose of statistical data analysis, clustering is used in many fields and represents the process of grouping similar data objects into so-called clusters. Since a variety of clustering strategies exists currently [13,8], there are different formal definitions of *cluster* depending on the desired clustering algorithm. *DenStream* is one of the data stream clustering approaches, as it was presented in [2].

2.1   Strategy of Clustering

DenStream uses micro clusters to form the final clustering of the data stream at time t. Based on an exponentially fading function $f(t) = 2^{-\lambda t}, 1 > \lambda > 0$, whereby λ is an input parameter and describes the density level of the clusters (see Algorithm. 1), the weight of old data records is reduced over time. The importance of the historical data becomes lower if we select a higher value of the input parameter λ. The micro clusters in *DenStream* (as shown in Fig. 2) build a summary representation of groups of close data points. This representation defines a spherical region in the data set with center c, radius $r < \epsilon$ and weight w. The micro clusters evolve with the time by the new data, changing w, c and r. At time t, the weight of the micro cluster is defined as the sum over the weights of the included points. A threshold $\beta.minPts$ separates the micro clusters into two classes (types):

- Micro clusters with $w \geq \beta.minPts$ are potential micro clusters and are grouped to build the final clusters using DBSCAN.
- Micro clusters with $w < \beta.minPts$ are outlier micro clusters, which are still not dense enough at the current point of time.

Data: $DataStream, \epsilon, \beta, MinPts, \lambda$
Result: Clustering

forall the $p \in DataStream$ **do**
    Merge(p);

    if $t \bmod Tp == 0$ **then**
        | Check all micro clusters and perform downgrade if necessary;
    end
    if *clustering request* **then**
        | // offline phase
        | Generate clusters from microclusters using DBSCAN;
    end
end

Algorithm 1. DenStream

A very important property of the micro clusters is that they can be maintained incrementally. Adding a new point to a micro cluster or fading another one by the fading function $f(t) = 2^{-\lambda t}$ for an input constant parameter λ, does not require re-computation of that micro cluster. Hence, the stream data is processed only once, extracting the required features without the need to save each point in the memory for further use.

While incrementally maintaining the micro clusters, the weighted linear and the weighted squared sum of the points are stored in the micro cluster. For a micro cluster summarizing n points $p_1, p_2, \ldots, p_n$, with time stamps $T_1, T_2, \ldots, T_n$ and considering the fading function $f(t)$ for the points' weights, the linear sum at time t is $\overline{CF1} = \sum_{i=1}^{n} f(t - T_i)p_i$. The weighted squared sum for the same micro cluster is $\overline{CF2} = \sum_{i=1}^{n} f(t - T_i)p_i^2$.

Using $\overline{CF1}, \overline{CF2}$ and w as features defining a micro cluster $mc = \{\overline{CF1}, \overline{CF2}, w\}$, a merge of a new point p (considered in Algorithm 1) is realized modifying mc to $mc = \{\overline{CF1}+p, \overline{CF2}+p^2, w+1\}$. Fading out a micro cluster for an interval δt is performed using the transformation: $mc = \{\overline{CF1}.2^{-\lambda \delta t}, \overline{CF2}.2^{-\lambda \delta t}, w.2^{-\lambda \delta t}\}$.

The radius and center of each micro cluster, which are used in the *DenStream* algorithm, will be computed each time from the features $\overline{CF1}, \overline{CF2}$ and w. For $mc = \{\overline{CF1}, \overline{CF2}, w\}$, it applies that $c = \overline{CF1}/w$ and r is the maximal standard deviation of the data over all dimensions of the micro cluster. The outlier micro clusters save also their creation time t_0 as an additional feature.

The *DenStream* algorithm allows a micro cluster to change its type or fade completely out and disappear. The period $Tp = \lceil \frac{1}{\lambda} \log\left(\frac{\beta\mu}{\beta\mu-1}\right) \rceil$ determines how often a micro cluster should be checked whether it has faded out being outlier. Thus, if the weight w of a given outlier micro cluster is defined as: $w < \xi(t, t_0) = \frac{2^{-\lambda(t-t_0+Tp))}-1}{2^{-\lambda Tp}-1}$ at time $t \geq k.Tp$ for $k \geq 1$, this micro cluster can be discarded.

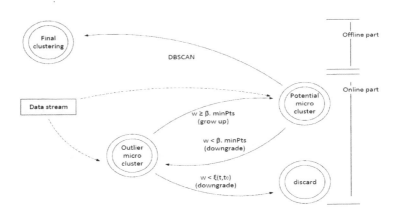

Fig. 2. *DenStream* and Micro Cluster Types

2.2   Algorithm

The *DenStream* clustering algorithm basically consists of two major parts; Online part and offline part. The complete algorithm can be described in a series of steps as depicted in Fig. 3.

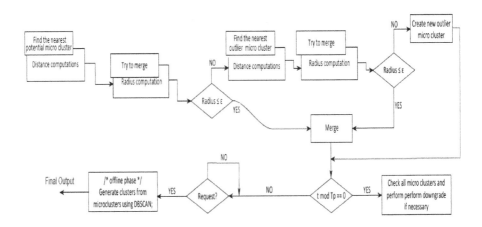

Fig. 3. A sequential model of *DenStream*

Online Part. This part of the algorithm is responsible for the micro cluster maintenance. The maintenance includes: Merging new data, creating new micro clusters and fading existing micro clusters. For each newly arriving point from the data stream, *DenStream* tries to merge to the nearest potential micro cluster without increasing its radius over the value of ϵ, which is an input parameter of the algorithm (shown in Algorithm 1) and strongly associates with the density level of the micro cluster. If this is not possible, the algorithm tries to merge it to the nearest outlier micro cluster, considering the maximum radius constraint upper bounded by ϵ. If this is also not possible, the new data point is considered as a center for a new outlier micro cluster. Before the algorithm can take the decision how to merge the new data to some existing micro cluster, or start a new one, the distance between the newly arriving data point and the center of each of the existing micro clusters must be computed. In the case of *merge*, the radius of the merged micro cluster must be re-computed. The online part is also responsible for fading out the old micro clusters, assuming that historical data is losing importance exponentially over time. During this phase, the micro clusters are checked for the minimum weight bound to keep them as potential or outlier micro clusters, which is considered as the grade down strategy of *DenStream*. According to this strategy, potential micro clusters are changed to outliers, and the outliers with weight that is lower than the threshold can be deleted.

Offline Part. The offline part runs DBSCAN (Density-Based Spatial Clustering of Applications with Noise) over the currently existing micro clusters. In this work, we follow the definition of DBSCAN that was presented in [14]. DBSCAN is characterized by detecting the clusters in the data set as dense region of data points. It defines the neighborhood of some point p, from a data set D as $N_\epsilon(p) = \{q \in D | dist(p,q) \leq \epsilon\}$ for a given distance function $dist$. The ϵ is the first input parameter of DBSCAN. The second, called $MinPts$, describes a number of points. The so called core-object is the point q, where $|N_\epsilon(q)| \geq MinPts$. Hence, a point p is directly density-reachable from a point q, if $p \in N_\epsilon(q)$ and q is core-object. Furthermore, density reachability is defined as a relation of two points in D, connected by directly density-reachable chains of points, implicating that density reachable points belong to the same dense area, thus, building a cluster.

3   *DenStream*'s Task-Design for *OCLSched*

An efficient implementation of *DenStream* by means of *OCLSched* demands a well-thought deployment of its task-concept [18]. In this section, the main *OCLSched* tasks used for *DenStream* are discussed. A major goal of this work has been establishing a challenging test case of application for our scheduling strategy. Implementing such a complex application as *DenStream* with *OCLSched* allows to investigate the adaptivity of the scheduler's task-concept to different computation patterns. For instance, since the *DenStream* algorithm needs a huge number of task starts, it has been necessary to improve the task-model of

OCLSched during the development, completely modifying the task's reset and reuse methods towards avoiding the recreation of tasks, in order to decrease the host-side time overhead. Further, motivated by the implementation of memory allocation by pages used in our *DenStream* implementation (as will be described later), additional offset were added to the copy tasks, allowing to copy a given data element to a specific place in the device memory buffer. Generally, the online part of *DenStream* is run for each new point arriving from the data stream. This implies three *OCLSched* tasks, which are responsible for merging the new data and actualizing the existing micro clusters. These tasks are designed in such a way that they could be alternatively combined and separately launched on different processing devices. These are represented as follows:

- *Candidate Task*: It supports the data aggregation decision. *candidate* collects the data related to the merge decision (potential micro cluster), this includes the distances between the new data point and the existing micro clusters' centers. In this phase, as many instances of the distance computation kernel (threads) as available micro clusters will be issued to execution. For efficiency reasons, the functionality of this task (OpenCL kernel) is extended to compute also the new radius of a micro cluster caused by merging the new point. Thus, in this phase all the data required for the merge decision is delivered in form of two vectors of length n, where n is the current number of the micro clusters.
- *Merge Task*: This task is responsible for merging the new data points to existing micro clusters and fading out the old micro clusters to give less importance to old data at an exponential rate. In practice, this kind of fading out (implicitly fading out) is realized by multiplying the micro cluster's features $CF1$, $CF2$ and w with the fading factor described in a previous section. For the micro cluster that will include the current point, the multiplication is executed whenever the point is merged to it. This ensures that the micro cluster fades out for the time period between the last and the current *merge* actions. However, only one micro cluster is updated with the currently managed point. The *merge* operation is characterized by a divergent program flow and a concurrent memory access. For this reason, the *merge* task will not be as optimally parallelized as the *candidate* task, as discussed later.
- *Fade Task*: This task ensures a proper fading out of each micro cluster at most after a period Tp. The *fade* task also checks all the micro clusters for the minimum weight bound to keep them as potential or outlier micro clusters, following a special method of *DenStream* (called *downgrade*). According to this strategy, potential micro clusters change into outlier, and outliers whose weight is below the lower bound can be deleted. The *fade* task can be executed in parallel for all micro clusters by means of a single one dimensional OpenCL kernel (one thread for each existing micro cluster).

In contrast to the online part, a single *OCLSched* task called *Intersects* is responsible for the final clustering during the offline part of *DenStream*. This task launches a two dimensional OpenCL kernel and calculates the centers of the final micro clusters while concluding their overlapping areas. In this work, we adopt a slightly modified version of the DBSCAN implementation from [19] for the offline part of *DenStream*, to generate the final request output.

4 Implementation

GPGPU programming follows the stream computing paradigm, based on which massive sets of data are processed. Basically, there would exist three major problems in a native GPU implementation of *DenStream*: The high memory transfer volume that becomes a real bottleneck when it comes to GPUs [4], the high number of task starts and also the processing of only one point per iteration and time unit in the main program loop. Since only one point is merged per program loop, the application would need a high number of loops to manage the whole data stream. In particular, the achievement of several tasks for each iteration would cause high scheduling overhead. For an efficient implementation of *DenStream* by means of *OCLSched*, a completely new design of *DenStream* tasks is required, in order to face all the mentioned challenges.

4.1 Task Design

In the presented *OCLSched*'s implementation of *DenStream*, the data stream is divided into packages of points. A data-point package contains a number of individual data points, all these points are assumed to arrive with the same time stamp. As a result, the time stamp is generated by the number of the packages being processed and not by the points' number from the input.

We apply a *merge* method that is similar to the approach from [7], in order to reduce the number of task starts, joins and resets. In total, three OpenCL kernels are implemented, while no more than two of them are started in one loop iteration, processing one data-point package each time. Compared with a native GPU implementation, our strategy saves on task starts, and thus, it reduces overhead.

The online part of the *OCLSched*'s *DenStream* implementation applies a functional *merge* of tasks. Two tasks are prepared during the initialization phase and recreated only in case of memory buffer expansion on the device. In the first task, both the *candidate* and *merge* parts of *DenStream* are combined in one OpenCL kernel, briefly called *CM* (*candidate* and *merge*). The second task is an extended version of *CM*, in which the fading routine for all micro clusters is also considered, briefly called *FCM* (*fade, candidate* and *merge*). Thus, in total three tasks are deployed when considering the online and offline parts of *DenStream*. One of four task combinations is started in each host program loop:

C1 Start, join and reset the CM-task.
C2 Start, join and reset the FCM-task.
C3 Start, join and reset the CM-task, followed by start, join and reset of the offline task.
C4 Start, join and reset the FCM-task, followed by start, join and reset of the offline task.

Data: $Rp, DataStream, \lambda, \epsilon, \beta, MinPts, pps$
Result: Clustering
initialization();
forall the $p \in DataStream$ **do**
    build packageOfPoints();
    if $packageOfPoints.size() == PpS$ **then**
        if $request(Rp)$ **then**
            if $fadePhase()$ **then**
                | C4−>run();
            else
                | C3−>run();
            end
            Request−>DBSCAN();
            Request−>output();
        else
            if $fadePhase()$ **then**
                | C2−>run();
            else
                | C1−>run();
            end
        end
    end
    if $MicroClusters \; \xi= MemRange$ **then** expandMemRange();
    ;
end

Algorithm 2. Host program in Pseudocode

First of all, the host program prepares a data-point package in a host buffer and decides which task combination to start. Then according to the Algorithm 2, the data package is managed. The decision about which task combination should be started is taken on the basis of the request period (Rp) and the size of the package (PpS), that is a priori defined in the configuration file. On each Rp data-point package, a request has to be started. Finally, the fading procedure is applied according to the *DenStream* parameter λ, thus affecting Tp.

The *CM* kernel consists of two main parts: The *candidate* part and the *merge* part. The *candidate* procedure executes in an optimally efficient parallel way, as can be seen from the code snippet in Listing 1.1. However, when adding the *merge* part to the kernel, only one thread will perform the *merge* procedure. The FCM kernel works on the same way, executing also the fading for all micro clusters in parallel, before the single thread executes the sequential *merge*. This kernel part achieves the calculation of the distances, centers and also the fading process for each micro cluster. The other kernel part is responsible for the *merge* procedure, it implements the sequential program flow described in the original *DenStream* (see section 2). The *candidate* part of the kernel should be completed for all threads, before starting both *scan* and *merge* procedures, which are also part of the same kernel. For this purpose, a counting semaphore (depends on the number of running threads) is applied at the point between the *candidate* computation and the scan of results.

```
1     ...
2         private size_t Id = get_global_id(0);
3         private float distance=0;
4         private float radius=0;
5
6         for(private size_t i=0; i < *dim ; i++){
7
8             CF1[(*dim)*Id+i] *= fade;
9             CF2[(*dim)*Id+i] *= fade;
10
11            distance += pow( point - CF... ,2);
12            radius = max(radius , sqrt( CF... ));
13
14            }
15
16            dists[Id]=sqrt(distance);
17            rads[Id]=radius * (*radExp);
18    ...
```

Listing 1.1. The *candidate* part of the FCM kernel

At first, both CM and FCM tasks transfer a set of points P that are wrapped as a data-point package ($|P| = k$) to the device memory. Then, either CM or FCM is launched for k iterations, each iteration processes one point of the data-point package. With *OCLSched*, a specialized *iteration* parameter can be passed to start a number of compute iterations, instead of recalling the same task many times. By means of the *iteration* parameter, the same kernel is launched for all threads, considering the data dimensionality *dim* (a constant parameter) and accessing $point[(iteration) * (*dim) + i], \forall i \in \{0 \ldots * dim - 1\}$, where the point array is filled like $point = \{p_1^\top, p_2^\top \ldots p_k^\top\}$.

Finally, a copy task returns the free memory space on the device after finishing the *merge* procedure. This information is used to decide if a memory expansion is required at this point. A possible optimization in this context would be to process all points in the data-point package in parallel. However, a parallel *merge* would change the logic of the *DenStream* algorithm and also cause concurrent access to the data structures holding the micro clusters data.

4.2 Memory Allocation and Transfer

The stream data is transferred to the GPU by building data-point packages. The memory transfer from the GPU to host includes the micro cluster intersections matrix. In order to decrease the amount of memory transfers between the host and the device, the micro-cluster data is completely stored on the device memory. Features of a given micro cluster are stored in individual memory buffers. For n micro clusters with dimensionality of d, the scalar valued attributes w_i and

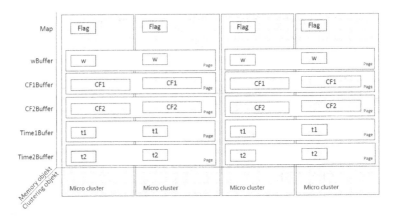

Fig. 4. Micro cluster representation in the memory

the *timestamps* t_i are stored in the same order of the micro clusters. The vector features $CF1$ and $CF2$ follow the same logic.

On the host side, the buffers are organized in pages. All host buffers consist of $p \geq 1$ pages of size $k > 1$, where k is the number of features in the page. One page set is the set of pages from different memory buffers, holding the micro cluster's features for a given micro cluster. A page set consists of five pages from the five different memory buffers, as illustrated in Fig. 4. All host buffers are placed in the shared host memory and managed by *OCLSched*. Contrary, the device buffers are directly allocated with size: $p.sizeof(page)$ (the summed up size of the allocated memory on the host side). When memory expansion is necessary, the device buffers are freed and reallocated with bigger size copying the complete micro clusters' data from the host to the device.

5   Experiments

In this section, we evaluate our proposal of *DenStream* on the basis of *OCLSched*. All the time measurements represent the total execution time, including the initialization phase, the stream processing and the request response for each Rp point. This design allows input memory transfers in big data portions, reducing the time overhead for small copy tasks. A major improvement of the *OCLSched*'s design of *DenStream* against its supposed native GPU implementation is represented in merging multiple kernels, saving on task starts, which might be caused if too many small tasks are scheduled on the GPU individually. It will be shown in the evaluation that both design changes have significant impacts on the measured execution time when enough points per package PpS are involved in the computation.

Table 1. Data Sets Used

data set	number of data objects	type of data objects
DE8K	60000	German cites
ICML	720792	sensor data

5.1 Hardware Environment

All tests are performed on a Fedora 17 Linux PC with a Quad-Core-CPU Intel(R) Core(TM) i5-3550 CPU @ 3.30GHz and 4 GB RAM. The test platform is equipped with an AMD Radeon HD 7870 GPU, with the following technical specifications:1000MHz Engine Clock, 2GB GDDR5 Memory, 1200MHz Memory Clock (4.8 Gbps GDDR5), 153.6 GB/s maximal memory bandwidth, 2.56 TFLOPS in Single Precision, 20 Compute Units (1280 Stream Processors), $256 - bit$ GDDR5 memory interface, PCI Express 3.0x16 bus interface and an OpenCL(TM) 1.2 support.

5.2 Test Data Sets

Two real data sets (listed in table 1) are used to test the quality and performance of the clustering algorithm with *OCLSched*, compared with a CPU version. The first data set *DE8K* contains geographical positions. The data is provided by the OpenStreetMap project under the Open Data Commons Open Database License (ODbL). The other one represents the Physiological Data Modeling Contest at ICML 2004 data set, which is a collection of activities collected by test subjects with wearable sensors over several months. Each data object exhibits 15 attributes and consists of 55 different labels for the activities and one additional label if no activity was recorded. We picked 9 numerical attributes.

5.3 Evaluation

In this section, we discuss the experiments that are designed to expose the influence of certain parameters on *DenStream*'s performance on the one side, and to discuss the GPU-resource utilization on the other. Since the request period Rp constitutes an important parameter in *DenStream*, the first test focuses on the impact of this parameter on performance, when comparing the *OCLSched*'s *DenStream* implementation with a CPU version. In general, the user can adjust the period of output requests. For example, when setting the request period of 1000, the clustering output will be generated every 1000 data records, revealing the changes of the data stream over time.

As can be seen from Fig. 5(a), when clustering is performed using the *DE8K* data set with different request periods, the shorter the request periods are, the smaller is the performance gap between the CPU and the *OCLSched*'s implementation of *DenStream*. When the same experiment is repeated using the high dimensional data set *ICML*, as can bee clearly seen in Fig. 5(b) the runtime results keep the same tendency, but only for a small request period ($Rp = 100$). This test shows that in case of often requests and high dimensional data set, *OCLSched*'s implementation of *DenStream* achieves a slight improvement in performance. Combining multiple kernels to reduce the task starts has a negative impact on the total runtime. This is caused by the *merge* procedure, that performs sequentially during the kernel execution. The responsible thread for this procedure takes longer than the other kernel instances, thus, leading to a threads-synchronization delay.

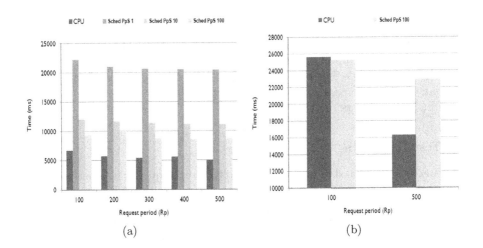

Fig. 5. A performance comparison between the *OCLSched*'s implementation and a conventional CPU version of *DenStream*: (a) for different Request Periods Rp and different points number PpS while using the $DE8K$ data set, (b) for a certain number of points per package PpS while using the ICML data set.

Resource utilization, generally, constitutes an important point of concern when programming GPUs. In our case, considering the *DenStream* algorithm, the resource utilization rate depends on two parameters of the data set: The dimensionality on the one hand, and both the data distribution and the speed of data fading defined by the input parameter λ on the other hand. A small number of micro clusters, which might be caused by a higher value of λ or by data distribution that is concentrated in a small data range, can not achieve high utilization levels of available GPU resources. For a small number of micro clusters, the offline part of *DenStream* performs also well on the host, and no benefit of the highly parallel GPU architecture can be obtained in this case. The AMD APP profiler also indicated that the NDRange size (problem size in OpenCL) is relatively small in such cases, where less than 512 micro clusters are examined during the *candidate* phase. These effects are clearly shown when comparing the results in Fig. 6(a) and Fig. 6(b). The $DE8K$ data set in the former generates a relatively small number of micro clusters for the selected λ, if it is compared to the high dimensional data set (ICML Physiological data set) used in the posterior. The mean number of micro clusters for the $DE8K$ data set equals 288, while it equals 681 for the ICML Physiological data set. Further, using the high dimensional data set with request period $Rp = 100$, the scheduler version of *DenStream* performs mostly better than the sequential version. For such a big data set, the offline part of the algorithm generates a significant CPU load in the conventional CPU version of *DenStream*. This explains the advantage of the GPU implementation over the CPU one.

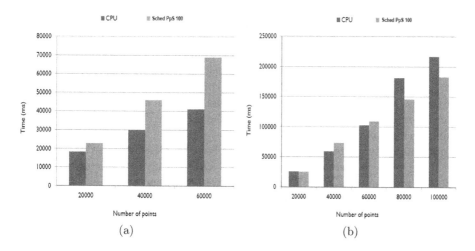

Fig. 6. A performance comparison between the *OCLSched*'s implementation and a traditional CPU version of *DenStream* when varying the number of streaming points: (a) using the $DE8K$ data set, (b) using the ICML data set.

In summary, we propose parallelization of both parts of *DenStream*. However, due to the sequential data *merge* logic, further optimization of the parallel *DenStream* using *OCLSched* would be necessary to achieve better resource utilization. This would require a data *merge* algorithm that is completely different from the originally used method in *DenStream*.

The proposed *OCLSched*'s design of *DenStream* offers a solution to reduce the CPU load and the computation time using the GPU co-processor. The main contribution has been to provide a sustainable development of *DenStream* to run on GPUs by means of our OpenCL-based scheduler. Actually, better performance of *DenStream* on the basis of *OCLSched* will be expected, when certain limitations in *OCLSched* are overcome soon. Especially, when it comes to the support of concurrent execution, future work might show major potential for improvement by boosting the *OCLSched*'s performance, if OpenCL device partitioning [5] becomes available. Currently, this OpenCL feature is only supported by CPUs and Cell devices.

6   Related Work

Numerous GPU implementation strategies have been introduced in data mining approaches, achieving high performance and presenting new ideas in the field of big data processing. In a similar contribution to ours, a data stream solution was presented in [4], focusing on the performance tuning in OpenCL. It discussed a variety of methods to optimize the device memory management and the selection of the OpenCL work-group size as well, based on a k-means clustering algorithm. To save and reuse the device memory, the presented solution processes the problem domain in portions, optimizing the memory rakes' size. However, in our approach we consider the density-based stream clustering problem, which is much more complex and accurate than the clustering algorithm managed in [4].

Another strategy for Clustering huge data sets using CUDA was implemented and discussed in [21]. The presented method applied data partitioning on large data sets in such a way that each data block could fit into the GPU's dedicated memory. Also included in the related work, Shalom et al. [15] focused on accelerators' speedups against the CPU clustering version, performing multiple distance computations that are required in the k-means clustering algorithm (a popular technique for cluster analysis).

Additionally, two interesting publications handled DBSCAN in two different ways [1,19]. The former strategy was based on creating multiple instances of DBSCAN and summarizing the results. The idea behind that was the use of density chains that are defined as connected, instead of treating high-density regions. This method caused, however, relatively high memory consumption, and additional overhead due to particularly divergent control flow.

To the best of our knowledge, there is no available work that performs both the online and offline parts of *DenStream* in a parallel manner. Our approach relies on considering DBSCAN as a part of *DenStream* and follows the strategy from [19]. In a separate approach [9], we discussed several strategies to design a GPU-based parallel *DenStream* algorithm. Now we focus on a solution, that is designed to be run with general computation tasks, sharing the GPU by means of a GPU-scheduler *OCLSched* [18].

7   Conclusion and Future Prospects

This work presented an OpenCL implementation of a complex problem from the field of data mining, which involves several tasks running together at nonstandard terms. A redesign of the stream clustering algorithm *DenStream* to meet the characteristics and the special design requirements of the GPU architecture was discussed in the context of a GPU scheduler. We showed that the presented *OCLSched'* implementation of *DenStream* performs well and saves on task starts and memory transfers. Through dividing the transferred data in portions and starting multiple loops of one thread over the streaming data, better performance of the application could be achieved. In future work, we are planning to apply the proposed strategy to anytime stream clustering, which is applicable to modern and emerging data stream scenarios.

References

1. Böhm, C., Noll, R., Plant, C., Wackersreuther, B.: Density-based clustering using graphics processors. In: Proceedings of the 18th ACM Conference on Information and Knowledge Management, CIKM 2009, pp. 661–670. ACM (2009)
2. Cao, F., Ester, M., Qian, W., Zhou, A.: Density-based clustering over an evolving data stream with noise. In: Proceedings of the SIAM Conference on Data Mining (2006)
3. Ester, M., Sander, J.: Knowledge Discovery in Databases. Springer (2000)
4. Fang, J., Varbanescu, A.L., Sips, H.: An auto-tuning solution to data streams clustering in opencl. In: 2011 IEEE 14th International Conference on Proceedings of Computational Science and Engineering (CSE), pp. 587–594 (2011)
5. Gaster, B.R.: OpenCL Device Fission (March 2011), http://www.khronos.org/assets/uploads/developers/library/2011_GDC_OpenCL/AMD-OpenCL-Device-Fission_GDC-Mar11.pdf
6. Gregg, C., Dorn, J., Hazelwood, K., Skadron, K.: Fine-grained resource sharing for concurrent GPGPU kernels. In: Proceedings of the 4th USENIX Conference on Hot Topics in Parallelism, pp. 10. USENIX Association (2012)
7. Gunarathne, T., Salpitikorala, B., Chauhan, A., Fox, G.: Iterative statistical kernels on contemporary gpus. Int. J. Comput. Sci. Eng. 58–77 (2013)
8. Hassani, M., Kranen, P., Saini, R., Seidl, T.: Subspace Anytime Stream Clustering. In: Proc. of the 26th International Conference on Scientific and Statistical Database Management (SSDBM 2014). ACM, Aalborg (2014)
9. Hassani, M., Tarakji, A., Georgiev, L., Seidl, T.: Parallel Implementation of a Density-Based Stream Clustering Algorithm Over a GPU Scheduling System. In: Peng, W.-C., Wang, H., Bailey, J., Tseng, V.S., Ho, T.B., Zhou, Z.-H., Chen, A.L.P. (eds.) PAKDD 2014 Workshops. LNCS(LNAI), vol. 8643, pp. 441–454. Springer, Heidelberg (2014)
10. Komatsu, K., Sato, K., Arai, Y., Koyama, K., Takizawa, H., Kobayashi, H.: Evaluating performance and portability of OpenCL programs. In: The fifth International Workshop on Automatic Performance Tuning, p. 7 (2010)
11. Pai, S., Thazhuthaveetil, M.J., Govindarajan, R.: Improving GPGPU Concurrency with Elastic Kernels. In: Proceedings of the Eighteenth International Conference on Architectural Support for Programming Languages and Operating Systems, ASPLOS 2013, pp. 407–418. ACM (2013)
12. Pennycook, S.J., Hammond, S.D., Wright, S.A., Herdman, J.A., Miller, I., Jarvis, S.A.: An investigation of the performance portability of OpenCL. Journal of Parallel and Distributed Computing 73, 1439–1450 (2013)
13. RazaviZadegan, S.G., RazaviZadegan, S.M.: A Novel Clustering Approach: Simple Swarm Clustering. In: Kozielski, S., Mrozek, D., Kasprowski, P., Małysiak-Mrozek, B. z. (eds.) BDAS 2014. CCIS, vol. 424, pp. 222–237. Springer, Heidelberg (2014)
14. Sander, J., Ester, M., Kriegel, H.P., Xu, X.: Density-based clustering in spatial databases: The algorithm gdbscan and its applications. Data Min. Knowl. Discov. 2, 169–194 (1998)
15. Shalom, S.A., Dash, M., Tue, M.: Efficient k-means clustering using accelerated graphics processors. In: Proceedings of the 10th International Conference on Data Warehousing and Knowledge Discovery, pp. 166–175 (2008)

16. Tarakji, A., Hassani, M., Lankes, S., Seidl, T.: Using a Multitasking GPU Environment for Content-Based Similarity Measures of Big Data. In: Murgante, B., Misra, S., Carlini, M., Torre, C.M., Nguyen, H.-Q., Taniar, D., Apduhan, B.O., Gervasi, O. (eds.) ICCSA 2013, Part V. LNCS, vol. 7975, pp. 181–196. Springer, Heidelberg (2013)
17. Tarakji, A., Marx, M., Lankes, S.: The Development of a Scheduling System *GPUSched* for Graphics Processing Units. In: Proceedings of the International Conference on High Performance Computing Simulation (HPCS 2013), pp. 566–557. ACM / IEEE (2013)
18. Tarakji, A., Salscheider, N.O., Hebbeker, D.: OS Support four Load Scheduling on Accelerator-based Heterogeneous Systems. In: Proceedings of the 2014 International Conference on Computational Science. Procedia Computer Science (2014)
19. Thapa, R.J., Trefftz, C., Wolffe, G.: Memory-efficient implementation of a graphics processor-based cluster detection algorithm for large spatial databases. In: 2010 IEEE International Conference on Proceedings of Electro/Information Technology (EIT), pp. 1–5 (2010)
20. Wienke, S., Plotnikov, D., an mey, D., Bischof, C., Hardjosuwito, A., Gorgels, C., Brecher, C.: Simulation of bevel gear cutting with gpgpus–performance and productivity. Computer Science - Research and Development 26, 165–174 (2011)
21. Wu, R., Zhang, B., Hsu, M.: Clustering billions of data points using gpus. In: Proceedings of the Combined Workshops on UnConventional High Performance Computing Workshop Plus Memory Access Workshop, UCHPC-MAW 2009, pp. 1–6. ACM (2009)

Image Analysis
and Multimedia Mining

Lossless Compression of Medical and Natural High Bit Depth Sparse Histogram Images

Roman Starosolski[✉]

Institute of Informatics, Silesian University of Technology,
Akademicka 16, 44-100 Gliwice, Poland
roman.starosolski@polsl.pl

Abstract. In this paper we overview histogram packing methods and focus on an off-line packing method, which requires encoding the original histogram along with the compressed image. For a diverse set containing medical MR, CR and CT images as well as various natural 16-bit images, we report histogram packing effects obtained for several histogram encoding methods. The histogram packing improves significantly JPEG2000 and JPEG-LS lossless compression ratios of high bit depth sparse histogram images. In case of certain medical image modalities the improvement may exceed a factor of two, which indicates that histogram packing should be exploited in medical image databases as well as in medical picture archiving and communication systems in general as it is both highly advantageous and easy to apply.

Keywords: Image processing · Lossless image compression · High bit depth images · Medical images · Sparse histogram · Histogram packing · Histogram encoding · Image coding standards · JPEG2000 · JPEG-LS · DICOM

1 Introduction

Most single-frame single-band medical images, like MR, CR and CT and are of a high nominal bit depth, which usually varies from 12 to 16 bits per pixel. The number of active levels, i.e., intensity levels actually used by image pixels, may be smaller, than implied by the nominal bit depth, by an order of magnitude or even more. Furthermore, active levels are distributed throughout almost all the entire nominal intensity range, i.e., the images have sparse histograms of intensity levels. Also, the continuous tone natural (photographic) images of high bit depths may have sparse histograms. The image histogram is sparse, when the acquisition device quantizes analog image intensities to a number of levels, which is smaller than the nominally possible, and then distributes the quantized levels over a wider range. In case of multiple band (e.g., color or multispectral) images each band may use a different set of active levels. This makes viewing images easier–the brightest actual level gets closer to the brightest nominally possible. On the other hand, numeric values of originally consecutive quantization levels cease to be consecutive integers making compressing of the images less effective [4]. The acquisition device

S. Kozielski et al. (Eds.): BDAS 2015, CCIS 521, pp. 363–376, 2015.
DOI: 10.1007/978-3-319-18422-7_32

characteristic is not the only reason for image histogram sparseness. Some of the routine image processing methods, e.g., gamma correction or contrast adjustment, may make the histogram sparse. Histograms of some images are inherently sparse. Although this observation probably won't lead to improving the compression ratios for such images, we note the obvious fact: regardless of the nominal bit depth, the number of active levels cannot be greater, than the number of pixels. All in all, sparse histograms occur frequently in high bit depth images. The impact of histogram sparseness on image compression ratios is well known–applying to sparse histogram images, prior to regular compression, a histogram packing [23,12] leads to significant ratio improvement.

Image compression algorithms are based on sophisticated assumptions as to characteristics of images they process. Sparse histogram is clearly different from what is expected by most lossless image compression algorithms. An initial step of processing image data by these algorithms is aimed at making the data easier to compress. Predictive image compression algorithms (e.g., JPEG-LS [7]) use the predictor function to guess the pixel intensities and then the prediction errors, i.e., differences between actual and predicted pixel intensities, are encoded instead of pixel intensities. Even using extremely simple predictors, such as one that predicts that pixel's intensity is identical to its left-hand side neighbor, improves the resulting compression ratio. For typical single-band images, the pixel intensity distribution is roughly close to uniform. Prediction error distribution is close to Laplacian, i.e., symmetrically exponential. Therefore entropy of prediction errors is significantly smaller than entropy of pixel intensities and the resulting compression ratio is improved accordingly. However, since in sparse histogram images the pixel intensity distribution is not uniform the prediction may (and usually does) increase the entropy. Transformation image compression algorithms (e.g., JPEG2000 [5]) instead of pixel intensities encode a matrix of transformation coefficients (cosine or wavelet transformation), making probability distribution peaked and reducing the entropy of the data. In case of multiple-band color image compression, since bands are correlated, the first step is to apply a color space transformation that removes correlation. Most such transformations for color images [19] produce 3 bands: luminance band of entropy similar to original bands (provided that they were not sparse) and 2 chrominance bands of peaked distributions and reduced entropies. Again, if histograms of original bands are sparse, then the color space transformation increases entropy of all 3 bands. In the above mentioned cases, histogram packing should be applied before regular compression. Histogram packing does not reduce the entropy of the data (actually it may cause a small increase), but it allows entropy reduction in the first step of the regular algorithm (prediction, band transformation or color space transformation).

The simplest method of histogram packing is called the off-line histogram packing. This method simply maps all the active levels to the lowest part of the nominal intensity range (order-preserving one-to-one mapping). The off-line packing requires the information, describing how to expand the histogram after decompressing an image, to be encoded along with the compressed image–along with the compressed image we have to encode the original histogram. Below we briefly characterize other methods targeted at the sparse histogram image compression.

- The on-line preprocessing technique [14]–the histogram is built on-line, yet still prior to actual compression. The technique may be used as a preprocessing step for any image compression algorithm. The compression ratio improvement is reported to be about the same as for the off-line packing.
- The integrated on-line packing [12] is a variant of the on-line preprocessing technique integrated into the JPEG-LS algorithm. The compression ratio improvement is reported to be about the same as for the off-line packing.
- The extended prediction mode of the 2nd part of the JPEG-LS standard [8] is a technique designed for sparse histogram images. The compression ratio improvement reported for this technique [13] is several times smaller compared with the improvement obtained using the off-line packing.
- Embedded Image Domain Adaptive Compression (EIDAC) [24]–relatively complex, progressive compression algorithm targeted at the sparse histogram images.
- Piecewise Constant Image Model algorithm (PWC) [1] is designed for palette images having low number of colors (for images of up to 16 and up to 256 colors), for binary images, and for grayscale images of 8-bit depth. For low bit depth sparse histogram grayscale images it outperforms the EIDAC algorithm.

In the case of the off-line histogram packing, on-line preprocessing, integrated on-line packing, and EIDAC the histogram has to be encoded and transmitted explicitly as the side information to the decoder. In the 2nd part of the JPEG-LS, the extended prediction mode prohibits predicting intensity levels not found in the already processed part of the image. This way histogram is built on-line by the decoder based on the already decoded pixels. In the PWC algorithm, when the pixel intensity is different from intensities of it's neighbors, then the limited length LRU chain of active levels is used to guess the pixel intensity, and if this fails, then the intensity level is encoded in the predictive way–the difference between actual intensity and intensity predicted from the pixels neighborhood (using the standard predictor of JPEG-LS [7]) is encoded and transmitted to the decoder.

Standard algorithms, which along with the compressed image store the image palette (PNG [20]) or the level mapping table (JPEG-LS [7]), may be practically useful. For these algorithms, we use the off-line histogram packing to improve compression ratios. The decompression reconstructs image and its original histogram solely by means of the algorithm, i.e., no additional step of histogram expanding is required after decompressing images encoded using these algorithms. The support for histogram transformations is included in the 2nd part (annex K) of the JPEG2000 standard [6]. The standard describes two non-linear transformations, which may be applied to decoded pixel intensity levels: the piecewise linear function and the gamma-style function. These transformations are probably aimed mainly for lossy compression and for systems combining image transformations and compression rather than for lossless compression of images having sparse histograms. Fortunately, the level mapping table is a special case of the piece-wise linear function.

In the case of high bit depth sparse histogram images, the direct use of the above standards is not as straightforward as it seems–storing the 16-bit mapping table may be not supported by the algorithm itself, as in the case of PNG, or by the algorithm implementation. Popular implementations of JPEG2000 core coding system [5] may not support non-linear transformations, since these transformations are extensions defined in the 2nd part of the standard.

When the histogram of an image is not sparse globally, but the image contains sparse histogram areas, then the compression ratio may be improved by exploiting the local image characteristics [15]. Also, the histogram of an image containing uniform intensity areas or, after the initial prediction step, containing runs of equal prediction errors, may be considered locally (highly) sparse. Several modern algorithms use special mode of processing such data, usually improving this way both the compression ratio and the speed. In the JPEG-LS algorithm, instead of encoding each pixel separately, we encode, with a single codeword, the number of consecutive pixels of equal intensity. In the PWC, a more sophisticated method, the Skip-Innovation model [1], is used for encoding runs of equal intensity pixels. In the CALIC algorithm [22], we encode in a special way sequences of pixels that are of at most two levels.

Histogram packing techniques are also used in compression methods, which may be generally described as near lossless. In [10], for given grayscale image and for given tolerable error threshold, a minimum entropy sparse histogram is computed. Then image is transformed into sparse-histogram image, which is subject to histogram packing followed by lossless compression. For small error threshold, the scheme outperforms most of other tested near-lossless algorithms in terms of compression ratio. In [2], for biological micrographs, a model of noise being a function of signal in the imaging system is constructed using the measured acquisition device characteristics. Statistically insignificant intensity levels are discarded making the histogram of image sparse, then histogram packing and lossless compression is used. The scheme preserves image information content better than standard lossy algorithms; compared to lossless algorithms it obtains significantly better compression ratios.

First successful attempts to use histogram packing method in lossy image coding were recently reported. In [9] it was shown, that applying histogram packing and lossless compression to original image quantized using weighted median cut results in better reconstructed image quality (in terms of PSNR) than in case of JPEG2000; the method also allows for fine rate control for high bit rates.

Methods of image compression exploiting the histogram sparseness were found to be effective for low bit depth images. To our best knowledge, except for our previous work (see [16] and [18], of which this paper is a revised version) and the lossy method mentioned above [9], the compression of high bit depth images having sparse histograms has not been investigated. In this paper, we report effects of the off-line histogram packing in the case of high bit depth images. High bit depth images require the histogram to be encoded efficiently–simple off-line methods that are suitable for the 8-bit images, like encoding the histogram using the level mapping table, for 16-bit images may cause the data expansion. We analyze efficient methods of encoding this information.

The reminder of this paper is organized as follows. In section 2, we discuss methods of encoding histograms of high bit depth images. In section 3, for medical MR, CR and CT images as well as natural high bit depth grayscale images, we report effects of these methods and of applying the off-line histogram packing to JPEG2000, and JPEG-LS algorithms. Section 4 summarizes the research.

2   Histogram Encoding Methods

The off-line histogram packing method actually is an image transformation; we apply it to an image before the compression. It transforms sparse histogram image into the packed histogram image. The transformation is reversible if, along with the compressed image, we encode the original histogram. For the histogram expanding, it is enough to encode which of intensity levels are active–we do not need to know how many times the active level was used. The size of the encoded histogram, for some high bit depth images, is not negligible.

2.1   Mapping Table

For of 8-bit images, we may simply encode binary all the active levels. Following the JPEG-LS terminology, we call this method of histogram encoding the Mapping Table (MT). Actually, for the Mapping Table method, we have to store both the number of active levels and the levels, so for a histogram of an N-bit image containing L active levels we need $(L + 1)N$ bits. This way, encoding the histogram of an 8-bit image requires not more than about a quarter of kilobyte. Typical size of an image after compression varies from a couple of dozens of kilobytes to several megabytes. So, even in the case of small images, the size of encoded histogram is a negligible factor in the overall compression ratio.

For 16-bit images, encoding the histogram using the Mapping Table method may lead to significant worsening of the compression ratio. In the worst case, when the histogram is not sparse, we would need 128 kilobytes to encode the histogram–the nominal number of image intensity levels is 216 and we need two bytes to encode binary the specific level. Therefore we need more efficient methods of encoding histograms of high bit depth images. Below we describe a couple of them.

2.2   Bit-Array

Instead of encoding the intensity level of each active level, we encode, for all nominally available levels, the information whether the specific level is active. Therefore, we need 2^N bits to encode the histogram of an N-bit image, regardless of the number of active levels. This method of histogram encoding was used for 8-bit images in the EIDAC algorithm starting from its first version [23]. We call a histogram encoded in this manner the Bit-Array of the histogram. For a 16-bit image, the Bit-Array requires 8 kilobytes, for an 8-bit image, 32 bytes only.

2.3   Run Length Encoding

Some images, like MR images used for experiments in this paper, use below 1% of all the nominally possible levels. A histogram of such image, encoded using the Bit-Array method, contains long runs of 0s separated by single 1s. Such histogram could be represented more compactly if we encoded lengths of runs of 0s. If, on the other hand, the histogram is not sparse, then it contains long runs (or just one long run) of 1s. Therefore we encode the Bit-Array of the histogram using the Run Length Encoding (RLE) variant described in the table 1 [16]. Note, that the RLE variant cannot be used if the last run of bits in the Bit-Array is not followed by single bit of value opposite to bits of run. In such a case we assume that a single bit being negation of the last bit in the Bit-Array follows the array–we encode this extra bit, but do not decode any bits after having decoded 2^N bits. Encoding the histogram using the RLE method is most efficient when the number of levels is close to 0 or close to 2^N. In the worst case, i.e., when every second level is used, we need 2^{N+2} bits for the RLE encoded histogram–32 kilobytes for the worst case histogram of a 16-bit image.

Table 1. Run Length Encoding of histograms of images of bit depths up to 16 bits

RLE codeword	Sequence
$0\ b_6\ b_5\ b_4\ b_3\ b_2\ b_1\ b_0$	run of $r + 1$ 0s followed by single 1, $r = b_6...b_0, r < 126$
$0\ 1\ 1\ 1\ 1\ 1\ 1\ 0\ b_7...b_0$	run of $r + 127$ 0s followed by single 1, $r = b_7...b_0$
$0\ 1\ 1\ 1\ 1\ 1\ 1\ 1\ b_{15}...b_0$	run of $r + 383$ 0s followed by single 1, $r = b_{15}...b_0$
$1\ b_6\ b_5\ b_4\ b_3\ b_2\ b_1\ b_0$	run of $r + 1$ 1s followed by single 0, $r = b_6...b_0, r < 126$
$1\ 1\ 1\ 1\ 1\ 1\ 1\ 0\ b_7...b_0$	run of $r + 127$ 1s followed by single 0, $r = b_7...b_0$
$1\ 1\ 1\ 1\ 1\ 1\ 1\ 1\ b_{15}...b_0$	run of $r + 383$ 1s followed by single 0, $r = b_{15}...b_0$

2.4   Further Compression of the Encoded Histogram

The Bit-Array is inefficient when the number of active levels is low; the RLE may be inefficient for certain numbers of intensity levels. Fortunately, both the Bit-Array of the histogram and the RLE encoded histogram may be further compressed. In the cases, when the above methods are most inefficient, the histograms encoded using them are likely to contain multiple repetitions of long sequences of symbols (bits or RLE codewords). For compressing such data we may use a universal compression algorithm capable of capturing long contexts, like the LZ77 universal dictionary compression algorithm [25].

3   Experimental Results and Discussion

We have compared experimentally the presented methods of histogram encoding. We also evaluated effects of packing histograms of high bit depth images on compression ratios of standard image compression algorithms. In experiments, we used MR, CR, and CT medical images as well as various 16-bit natural images, i.e., images acquired from scenes available for the human eye (photographic images).

In order to evaluate the impact of histogram sparseness on compression ratio for typical medical image of a certain modality, we used all the MR, CR, and CT medical images from a test image set described in another study [17]. There were 12 images of each of the modalities. Not all the medical images are of 16-bit depth and not every medical image has sparse histogram. Obviously, for the 10- or 12-bit images the method of histogram encoding gets less important for the overall compression ratio. Natural continuous tone grayscale images of 16-bit depth were included in experiments to evaluate effects of histogram packing on various non-medical images. These images included unprocessed images of various sizes as well as processed ones. Following groups of non-medical images were evaluated, each containing 4 images:

- Medium–natural (photographic) images of 16-bit depth classified in the above-mentioned set [17] as medium-sized;
- Contrast–Medium images with contrast increased by 25%;
- Gamma–Medium images with gamma (value 1.25) correction applied;
- Small–small images, which are reduced size (ninefold) Medium images.

The characteristics of images are reported in the table 2 (for brevity we report averaged results only). To characterize numerically image sparseness, we define the image level utilization $U = L/(1 + l_{hi} - l_{lo})$, where l_{lo} and l_{hi} are respectively the lowest and the highest active level, and L is number of active levels. In the table, images are characterized by the image name, size (number of pixels), nominal depth (N), nominal (2^N) and actual (L) number of intensity levels, and by the level utilization (U). Results of encoding histograms, expressed as sizes (in bytes) of encoded histograms averaged for image groups, are reported in the table 3 for the following methods:

- MT–Mapping Table method;
- BA–Bit-Array method;
- BA+LZ77–Bit-Array of the histogram, compressed using LZ77;
- RLE–RLE method;
- RLE+LZ77–histogram encoded using RLE method followed by LZ77.

For the LZ77 compression we used the popular gzip compression utility, in the case of the RLE method it was applied directly to the encoded histogram. For the Bit-Array, since gzip is byte-wise, prior to compression, each bit was expanded to a byte.

Table 2. Characteristics of test images (averages for groups)

Images	Pixels	N	2^N	L	U
MR	196608	16.0	65536	1104	1.7%
CR	3527076	12.5	23296	7878	59.5%
CT	257569	14.7	45056	1951	17.3%
Medium	440746	16.0	65536	55839	87.1%
Contrast	440746	16.0	65536	23737	36.4%
Gamma	440746	16.0	65536	28076	44.4%
Small	48776	16.0	65536	25174	39.7%

Table 3. Encoded histogram size [B] (averages for groups)

Images	MT	BA	BA+LZ77	RLE	RLE+LZ77
MR	2210	8192	550	1127	102
CR	15184	2912	285	7071	179
CT	3592	5632	541	1852	219
Medium	111681	8192	4358	6231	3528
Contrast	47475	8192	1251	23737	909
Gamma	56154	8192	1546	28080	1314
Small	50350	8192	8447	13195	7288

The RLE+LZ77 method appears to be the most efficient. It obtains the shortest encoded histogram length for nearly all tested images. In the case of medical images, on average, it results in the encoded histogram length about 3 times shorter, than the second best BA+LZ77 method, for non-medical images the difference in favor of RLE+LZ77 is about 20%. Therefore, for evaluating effects of histogram packing on compression ratios of standard image compression algorithms, we use the RLE+LZ77 method.

The compression ratios obtained for images before histogram packing (Norm.), after packing (Pack.), and the ratio improvements due to histogram packing are reported in the table 4. The compression ratio is expressed in bits per pixel [bpp]: $8e/n$, where n is the number of pixels in the image, e–the size in bytes of the compressed image (including the size of the histogram encoded using the RLE+LZ77 method in the case of ratio after packing). We performed experiments for JPEG-LS [7,21] and JPEG2000 [5,3] standard image compression algorithms.

We notice, that effects of packing histograms on the compression ratios of tested algorithms are, for both algorithms, highly similar (see Fig. 1). Therefore we discuss results obtained for the more frequently used JPEG2000 algorithm only. As expected, the histogram packing does not improve compression ratios for Small images. For these images we observe noticeable worsening of compression ratios. Histograms of Small images are sparse ($U = 39.7\%$) and actually the packed histogram images would compress better by about 7% if we did not consider the encoded histogram size. For these images, the histogram sparseness is caused by the image size and also because of the image size the encoded

Table 4. Effects of histogram packing on compression ratios of JPEG-LS, and JPEG2000; results obtained for histograms encoded using RLE+LZ77 method (averages for groups)

Images	U	JPEG-LS			JPEG2000		
		Norm. [bpp]	Pack. [bpp]	Impro-vement	Norm. [bpp]	Pack. [bpp]	Impro-vement
MR	1.7%	10.009	4.944	50.6%	10.024	4.849	51.6%
CR	59.5%	6.343	5.398	14.9%	6.394	5.426	15.1%
CT	17.3%	7.838	4.557	41.9%	8.044	4.630	42.4%
Medium	87.1%	11.829	11.844	-0.1%	12.058	12.082	-0.2%
Contrast	36.4%	11.416	9.992	12.5%	11.951	10.558	11.7%
Gamma	44.4%	11.950	10.676	10.7%	12.183	10.965	10.0%
Small	39.7%	12.414	12.813	-3.2%	12.712	13.180	-3.7%

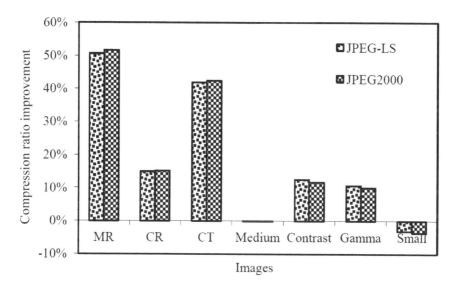

Fig. 1. Average compression ratio improvement due to histogram packing (RLE+LZ77)

histogram size is an important factor in the overall compression ratio–if compression ratios of such images could be improved by histogram packing, then we could try to improve the compression ratio for any image by splitting it into several smaller ones and compressing them separately.

Histograms of most Medium images are non-sparse. Only in the case of one of these images the histogram may be considered sparse, since for this image $U = 70.6\%$. The impact of histogram packing on compression ratios is similar for all Medium images–it negligibly worsens the compression ratios. Based on the image level utilization we'd rather expect compression ratio improvement for the above mentioned image. Analyzing histogram of this image we found, that almost all image pixels are of low intensities and the histogram is sparse

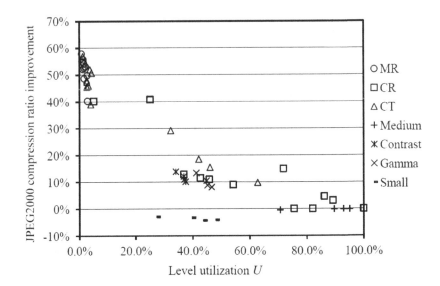

Fig. 2. JPEG2000 improvements for individual images due to histogram packing (RLE+LZ77)

only in the high intensity range. Similar image characteristics and similar lack of impact of histogram packing on compression ratio may be observed for two CR medical images. These observations show, that image level utilization is not a perfect histogram sparseness measure. Except for the cases described above, the histogram packing improves compression ratios for high bit depth sparse histogram images. The improvement varies depending on the image level utilization U, which we use as a measure of the histogram sparseness (see Fig. 2). For $U < 1/4$ the compression ratio improvement is roughly 50%, i.e., the size of the compressed image gets halved by applying the histogram packing method. For $U \approx 1/2$ we get the compression ratio improvement of about 10–20%; this level of improvement is not negligible for lossless image compression algorithm–the difference in compression ratio between algorithms obtaining best ratios and algorithms obtaining best speeds usually does not exceed 10% for the images used [17]. For $U > 3/4$ the histogram packing improves ratios for some images only, however, it does not worsen noticeably ratios for the remaining ones.

The RLE+LZ77 histogram encoding method outperforms others. It is interesting, however, whether it is practically justified to use it instead of some simpler one when we consider the overall image compression ratio, not the encoded histogram size alone. In the table 5, we report the JPEG2000 compression ratios for packed histogram images (calculated assuming, that the histogram is encoded using 0 bytes) and the cost of encoding histogram for the described histogram encoding methods. The histogram encoding cost is expressed in bits per image pixel and as relative to the above compression ratio. For brevity we report averaged results only.

Table 5. The cost of histogram encoding per image pixel [bpp] and relative to JPEG2000 compression ratio of packed histogram images. JPEG2000 ratio calculated excluding the encoded histogram size.

Images	JPEG2000 ratio [bpp]	MT [bpp]	relative	BA [bpp]	relative
MR	4.938	0.141	2.8%	0.500	10.1%
CR	5.398	0.044	0.8%	0.017	0.3%
CT	4.550	0.121	2.7%	0.168	3.7%
Medium	12.018	2.027	16.9%	0.149	1.2%
Contrast	10.542	0.862	8.2%	0.149	1.4%
Gamma	10.941	1.019	9.3%	0.149	1.4%
Small	11.618	8.258	71.1%	1.344	11.6%

Images	BA+LZ77 [bpp]	relative	RLE [bpp]	relative	RLE+LZ77 [bpp]	relative
MR	0.032	0.6%	0.072	1.5%	0.006	0.1%
CR	0.001	0.0%	0.021	0.4%	0.001	0.0%
CT	0.017	0.4%	0.061	1.3%	0.007	0.2%
Medium	0.079	0.7%	0.113	0.9%	0.064	0.5%
Contrast	0.023	0.2%	0.431	4.1%	0.017	0.2%
Gamma	0.028	0.3%	0.510	4.7%	0.024	0.2%
Small	1.385	11.9%	2.163	18.6%	1.195	10.3%

In the case of some images, the effects of histogram packing vary significantly depending on the method of encoding the histogram. The best method for all image groups is the RLE+LZ77. The BA+LZ77 method also obtains good ratios, however in the case of certain modalities, namely MR and CT, the cost of encoding histograms using this method is several times greater than the cost of RLE+LZ77. Compared to the RLE+LZ77 method, the simplest methods, which were successfully used for low bit depth images (MT and BA) are, for some modalities, highly inefficient.

The RLE code is constructed ad-hoc; the algorithm for encoding the RLE sequence was selected without thorough analysis of the RLE sequence structure. Also, we did not analyze other algorithms that could be used instead of LZ77. Experiments were done using gzip, a popular general purpose compression utility not adjusted to characteristics of data produced by our RLE variant–it expands the shortest RLE sequences. Therefore there is certainly a possibility to encode the histogram more efficiently, than using our RLE+LZ77 method. From practical point of view, however, there is no need to encode the histogram more efficiently. Except for the Medium and Small images, which do not benefit from histogram packing, by finding a better method we could get further ratio improvement of no more than about 0.2% only.

For medical images, the cost of encoding histograms using the MT method is a small factor in the ratio improvement obtained due to histogram packing. Decoding of images with packed histograms encoded using the MT method is already supported by the JPEG-LS [7] standard and by the 2nd part (annex K) of the JPEG2000 standard [6]. The JPEG-LS standard is included in the

DICOM standard [11] commonly used in medical picture archiving and communication systems. Therefore the MT method may in practice be very useful for medical devices acquiring MR, CT, and CR images. Provided that the 16-bit JPEG-LS mapping table is supported by the decompression software, using the off-line histogram packing and the MT method of histogram encoding we may significantly improve compression ratios while maintaining compatibility with current standards. We note that the annex K of the 2nd part of the JPEG2000 standard is not included in the the DICOM standard, as opposed to JPEG2000 core coding system.

4   Conclusions

In this paper we overview histogram packing methods and report effects of packing histograms of high bit depth images on compression ratios obtained by lossless image compression algorithms. Experiments were performed for a diverse set of test images, including medical MR, CR, and CT images as well as unprocessed and processed (gamma and contrast adjustment) natural 16-bit images. We focused on the off-line packing method. The off-line packing requires the information, describing how to expand the histogram after decompressing an image, to be encoded along with the compressed image–along with the compressed image we have to encode the original histogram. The size of the encoded histogram, for some high bit depth images, is not negligible. One of the histogram encoding methods (RLE+LZ77) obtains the shortest encoded histogram length for nearly all tested images and in practice is sufficiently good for encoding histograms of wide range of images. A simpler method (MT) may be useful for medical images. For these images, its use results in improvements of the compression ratio little worse compared to RLE+LZ77, but decoding of images with packed histograms encoded using the MT method is already supported by JPEG-LS (included in DICOM) and JPEG2000 (part 2) standards. The effects of packing histograms on the compression ratios of JPEG2000 and JPEG-LS are, for both tested algorithms, very similar–histogram packing improves significantly lossless compression ratios for high bit depth sparse histogram images. The ratio improvement due to histogram packing may approach or exceed a factor of two, as in case of CT and MR medical images, respectively. Though effects of histogram packing are known for over a dozen years, the technique is not considered a routine step of reversible image compression algorithms. The results presented in this paper indicate, that at least in case of MR and CT modalities, histogram packing should be exploited in medical image databases as well as in medical picture archiving and communication systems in general as it is both highly advantageous and easy to apply.

Acknowledgments. This work was supported by BK-266/RAu2/2014 grant from the Institute of Informatics, Silesian University of Technology and by POIG.02.03.01-24-099/13 grant: GeCONiI–Upper Silesian Center for Computational Science and Engineering.

References

1. Ausbeck Jr., P.J.: The piecewice-constant image model. Proc. of the IEEE 88(11), 1779–1789 (2000)
2. Bernas, T., Starosolski, R., Robinson, J.P., Rajwa, B.: Application of detector precision characteristics and histogram packing for compression of biological fluorescence micrographs. Computer Methods and Programs in Biomedicine 108(2), 511–523 (2012), http://dx.doi.org/10.1016/j.cmpb.2011.03.012
3. Christopoulos, C., Skodras, A., Ebrahimi, T.: The JPEG2000 still image coding system an overview. IEEE Trans. on Consumer Electronics 46(4), 1103–1127 (2000)
4. Ferreira, P.J., Pinho, A.J.: Why does histogram packing improve lossless compression rates? IEEE Signal Processing Letters 9(8), 259–261 (2002)
5. INCITS/ISO/IEC, ITU-T: Information technology – JPEG 2000 image coding system: Core coding system, INCITS/ISO/IEC International Standard 15444-1 and ITU-T Recommendation T.800 (2004)
6. INCITS/ISO/IEC, ITU-T: Information technology – JPEG 2000 image coding system: Extensions, INCITS/ISO/IEC International Standard 15444-2 and ITU-T Recommendation T.801 (2008)
7. INCITS/ISO/IEC, ITU-T: Information technology – Lossless and near-lossless compression of continuous-tone still images – Baseline, INCITS/ISO/IEC International Standard 14495-1 and ITU-T Recommendation T.87 (2006)
8. INCITS/ISO/IEC, ITU-T: Information technology – Lossless and near-lossless compression of continuous-tone still images: Extensions, INCITS/ISO/IEC International Standard 14495-2 and ITU-T Recommendation T.870 (2003)
9. Iwahashi, M., Kobayashi, H., Kiya, H.: Lossy compression of sparse histogram image. In: Proc. ICASSP 2012, pp. 1361–1364 (2012), http://dx.doi.org/10.1109/ICASSP.2012.6288143
10. Nasr-Esfahani, E., Samavi, S., Karimi, N., Shiran, S.: Near lossless image compression by local packing of histogram. In: Proc. ICASSP 2008, pp. 1197–1200 (2008), http://dx.doi.org/10.1109/ICASSP.2008.4517830
11. NEMA: Digital imaging and communications in medicine (DICOM) part 5: Data structures and encoding, National Electrical Manufacturers Association Standard DICOM PS3.5 2014c (2014)
12. Pinho, A.J.: On the impact of histogram sparseness on some lossless image compression techniques. In: Proc. ICIP 2001, vol. II, pp. 442–445 (2001)
13. Pinho, A.J.: A comparison of methods for improving the lossless compression of images with sparse histograms. In: Proc. ICIP 2002, vol. 2, pp. 673–676 (2002)
14. Pinho, A.J.: An online preprocessing technique for improving the lossless compression of images with sparse histograms. IEEE Signal Processing Letters 9(1), 5–7 (2002)
15. Pinho, A.J.: Preprocessing techniques for improving the lossless compression of images with quasi-sparse and locally sparse histograms. In: Proc. ICME 2002, vol. 1, pp. 633–636 (2002)
16. Starosolski, R.: Compressing images of sparse histograms. In: Kowalczyk, A., Fercher, A.F., Tuchin, V.V. (eds.) Proc. SPIE: Medical Imaging, vol. 5959, pp. 209–217 CID:595912. The International Society for Optical Engineering, Bellingham (2005), http://dx.doi.org/10.1117/12.624489

17. Starosolski, R.: Performance evaluation of lossless medical and natural continuous tone image compression algorithms. In: Kowalczyk, A., Fercher, A.F., Tuchin, V.V. (eds.) Proc. SPIE: Medical Imaging, vol. 5959, pp. 116–127 CID:59590L. The International Society for Optical Engineering, Bellingham (2005), http://dx.doi.org/10.1117/12.624136

18. Starosolski, R.: Compressing high bit depth images of sparse histograms. In: Simos, T.E., Psihoyios, G. (eds.) International Electronic Conference on Computer Science, AIP Conference Proceedings, vol. 1060, pp. 269–272. American Institute of Physics, USA (2008), http://dx.doi.org/10.1063/1.3037069

19. Starosolski, R.: New simple and efficient color space transformations for lossless image compression. Journal of Visual Communication and Image Representation 25(5), 1056–1063 (2014), http://dx.doi.org/10.1016/j.jvcir.2014.03.003

20. W3C, ISO/IEC: Information technology – Computer graphics and image processing – Portable Network Graphics (PNG): Functional specification, World Wide Web Consortium Recommendation and ISO/IEC International Standard 15948 (2004)

21. Weinberger, M.J., Seroussi, G., Sapiro, G.: The LOCO-I lossless image compression algorithm: Principles and standardization into JPEG-LS. IEEE Trans. Image Proc. 9(8), 1309–1324 (2000)

22. Wu, X., Memon, N.: Context-based, adaptive, lossless image codec. IEEE Trans. on Communications 45(4), 437–444 (1997)

23. Yoo, Y., Kwon, Y.G., Ortega, A.: Embedded image-domain compression of simple images. In: Proc. of the 32nd Asilomar Conf. on Signals, Systems, and Computers, vol. 2, pp. 1256–1260 (1998)

24. Yoo, Y., Kwon, Y.G., Ortega, A.: Embedded image domain compression using context models. In: Proc. ICIP 1999, pp. 477–481 (1999)

25. Ziv, J., Lempel, A.: A universal algorithm for sequential data compression. IEEE Trans. on Information Theory 32(3), 337–343 (1997)

Query by Shape for Image Retrieval from Multimedia Databases

Stanisław Deniziak and T. Michno[✉]

Kielce University of Technology, Poland
{s.deniziak,t.michno}@tu.kielce.pl

Abstract. Efficient methods of image retrieval is one of the most important challenges in the scope of the management of large multimedia databases. Existing methods for querying, based on a textual description e.g. keywords or based on image content, are not sufficient for the most applications. Methods based on semantic features are more suitable. In this paper we propose a new query by shape (QS) method for image retrieval from multimedia databases. Each image in the database is represented as a set of graphical objects, which are specified using graphical primitives like lines, circles, polygons etc. To retrieve images containing the given object, the object shape should be provided. Next, the efficient algorithm for testing the similarity of shapes is applied. The preliminary results showed the high effectiveness of the QS method.[1]

Keywords: Query by shape · Image retrieval · Multimedia database · Graphs

1 Introduction

Nowadays, in the world of a large scale multimedia data stored in databases, one of the most important aims is retrieving precise results of queries. The classical textual methods are suitable for texts or numbers, but not for images with many details. More appriopriate are methods which uses a sample image as a query. The main problems of such methods are the algorithm determining the similarity of images and the representation of images in the database.

We propose a new Query by Shape (QS) method for image retrieval from multimedia databases. The method is based on object decomposition into features. Each feature may consist of shape, color, texture or other attributes and can be connected to others, creating a graph. The query is specified by the shape of the requested object. Then, all images containing the given object are retrieved from the database. In this paper we only focused on shape features, specified by primitives, i.e. lines and circles.

The paper is organized as follows: section 2 presents the related works in the area of image storage and multimedia database queries. In the section 3 a motivation for the research is given. The section 4 describes the main idea of our

---

[1] The research used equipment funded by the European Union in the Innovative Economy Programme, MOLAB - Kielce University of Technology.

S. Kozielski et al. (Eds.): BDAS 2015, CCIS 521, pp. 377–386, 2015.
DOI: 10.1007/978-3-319-18422-7_33

approach. The experimental results are presented in the section 5. The section 6 describes conclusions and further direction of research.

2   Related Works

The problem of image searching in a multimedia database has been a subject of many researches. Among the algorithms, we can distinguish two groups: Keyword Based Image Retrieval (KBIR) and Content Based Image Retrieval (CBIR) [10,16]. The first group is not very efficient because it relies on an image description (i.e. keywords), manually assigned by a user. This results in poor scalability and very often does not contain the complete information about the image. Also the query in such database needs the involvement of a user, in order to create the textual description [10,16]. The second group is based on querying the database by an image or a video file. Because of the fact that most often input image contains almost all searched information, the results are more precise. Low level and high level CBIR algorithms may be distinguished [16].

The low level algorithms are based on certain features of images, like lightness or color. For example in [12] a normalized color histogram is used with both recursive and non-recursive algorithms, in [13] a spatial domain representation is applied. Kriegel et al. [7] described similarity search in large multimedia databases which is focused on the video. A weighted approach was based on image and audio representation of a video. For each frame, a color histogram, a contrast, an entropy and an inverse difference moment are used. Lalos et al. [8] described a new technique for indexing multimedia content in pervasive environments. Their algorithm is based on uniform content representation using M-hyper rectangle. For image representation in a video file, a shape, color and texture descriptors are used. Images with similar content are grouped into a cell topology in order to reduce computational complexity and searching time. The query is compared with cells and the closest images within them are retrieved.

All of the above CBIR algorithms compute features for the whole image or frame. This approach may be not sufficient for querying a database for the specific object. In order to partially overcame this problem, a region-based CBIR algorithms were created. In this type of algorithms, each image is divided into smaller parts, called regions. Each region involves pixels with the same or very similar features. Then, a region correspondence is computed which is the most often represented as a graph with attributes. During the database querying, an inexact graph matching is performed, using Maximum Likelihood estimation [10]. iPURE [1] uses intra-query modification and learning of user perception at the client-site. In this approach, an input image query is segmented into color regions, which provides first iteration of results for a client. Then a user selects segments which are the most important (segment(s) of interest), generating new results based on the first iteration without communication with the server. The system can learn initial user perception, providing more suitable results. Additionally, a new color image segmentation algorithm, based on region-growing approach, was developed. This incorporates elimination or modification of region edges as a result of testing contrast, gradient and shape of the region boundary. K.S. Fu et al. in [9] proposed an algorithm which seems similar to our approach. The algorithm uses ellipses, parallelograms, symmetric arcs and corners detection in order to extract regions from fixed resolution grayscale images. For each region, links between regions on its left and right sides are stored. In our method we use shape primitives in order to create an object's skeleton without region extraction, we use also another representation of the detected primitives and another query matching.

Above CBIR algorithms needs an image as a query. This can be a problem for a user, because often there is no available image. One of the most promising algorithms which tries to overcome this problem was presented by Kato et al. [6]. It is based on the idea that people most often do not remember all details, but can draw a rough sketch of the image which they are looking for. The algorithm creates a sketch from images in the database, converting them to 64x64 images. During the query, a sketch is provided as an input. Next, the input and the database images are divided into blocks of 8x8 pixels and compared.

Some pattern recognition methods also may be applied in the scope of the image retrieval system. As an example fuzzy IE graphs representation for syntactic recognition of fuzzy patterns [2] or moment-based local operators [15] could be given. Some works of Jakubowski (e.g. [5]) are also interesting because they consider object's contour as a composition of primitives which are grouped into a set.

3 Motivation

The goal of searching the multimedia database is to retrieve images or videos with the content specified by a query. The main problem encountered in the content based image retrieval research is the semantic gap between the representation of low-level features and high-level semantics in the images [16]. Since "a picture is worth a thousand words", the KBIR methods are usually ambiguous and inadequate. As far as CBIR algorithms are concerned, the visual retrieval methods are the most appropriate to human perception. But existing methods are based on exact matching of complex sketches [4,3,6] or images. Sometimes it is difficult to draw such a shape. Moreover, the methods are not efficient for images containing rotated or resized objects.

We believe that the most applicable CBIR method should fulfill the following requirements:

- queries should be specified as geometrical shapes, consisting of primitives like lines and circles, and other attributes like color and texture,
- shapes specifying the query should be easily obtainable, e.g. drawn by a user or extracted from the real images,
- queries should concern any part of the image,
- the matching should recognize transformed shapes,
- since some objects may be more or less recognizable, the matching should be controlled by the level of similarity.

4   The Shape Matching Algorithm

The main idea of the Query by Shape method is based on decomposition of an image into features. Each feature consists of the shape, color and texture. The approach presented in this paper is limited to the shape features, consisting of lines and circles. But it can be easily extended to comply other primitives and other features.

Any shape may be specified as a set of correlated primitives. Our method is limited to lines and circles because there exists well known precise algorithms to detect them, e.g. Circular and Line Hough Transform. Each shape is specified by the following properties:

- bounding box which covers all detected features in the object,
- for circle, a ratio between its diameter to the diagonal of the bounding box (relative size),
- for line, a ratio between its length and height (line slope).

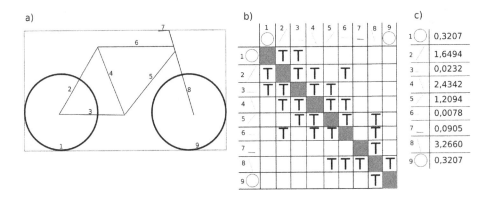

Fig. 1. A bicycle object stored in the database, a) the object after shape detection (the dotted rectangle shows object's bounding box), b) shape graph neighbourhood matrix (note that node 5 is connected with 6 because of minimal distance threshold), c) parameters for each primitive

Relations between primitives are represented by a shape graph, where nodes correspond to primitives and edges are between the connected components. To take into consideration some inaccuracy of shapes, the minimal distance is used as a threshold. The comparison between graphs, corresponding to tested shapes, is made on two levels: node level and feature level. The node level consists of testing if the number of links for each node is the same and if all links have the same type. If the similarity is equal or greater than ϵ_N, the feature level test is performed on the same subgraph. For circles the algorithm checks if difference between relative sizes is equal or smaller than ϵ_F. For lines, the slopes are compared in the same way. If both tests are successful, the subgraphs are considered as the same. The similarity coefficient is defined as a ratio of matched nodes to all nodes. To find the similar objects two thresholds are used:

− ϵ_H - above this level the objects are treated as the same,
− ϵ_L - under this level the objects are treated as different.

We assume that the database contains graphs corresponding to all images. The object matching algorithm is as follows:

1. Find all circles and lines in the query image (or shape).
2. Create shape graph G, for each node store its parameter value:
   (a) for the circle node - store its radius in relation to object's bounding box diagonal
   (b) for the line node - store its slope as a relation of height to width (*height/width*)
3. Compare query graph G with the shape graph M from the database:
   (a) For each node g_i in G:
       i. Find in the graph M the most similar node m_j of the same class (circle or line), omit already found nodes.
       ii. Check the connections of g_i with other nodes:
           A. Compute the node level similarity coefficient
           $$ns = \frac{the\ number\ of\ the\ links\ with\ the\ same\ type\ for\ g_i}{the\ number\ of\ links\ for\ m_j}$$
           B. If $ns < \epsilon_N$, go to 3.1. and test another node from G.
           C. Compare the parameter's value of g_i and m_j,
              if $|g_i.value - m_j.value| > \epsilon_F$, go to 3.1. and test another node from G.
           D. Map the nodes connected to g_i into nodes connected to m_j, choosing the smallest difference between parameter's values. If the number of links connected to m_j is smaller than the number of links connected to g_i, choose the best matching and omit the rest nodes connected with g_i.
           E. compute P as the number of pairs with the difference between parameter's values equal or smaller than ϵ_F
           F. compute the feature level similarity coefficient
           $$fs = \frac{P+1}{the\ number\ of\ links\ for\ m_j+1}$$
           G. If $fs < \epsilon_N$, go to 3.1. and test another node from G.
           H. Mark g_i node as similar to m_j and m_j as found.
   (b) Sum up the number of similar nodes in G.
       Compute the graph's similarity coefficient sim:
       $$sim = \frac{number\ of\ similar\ nodes}{number\ of\ nodes\ in\ M}$$
   (c) If $sim >= \epsilon_H$ the graph G is the same as M
       If $sim < \epsilon_H$ AND $sim > \epsilon_L$ the graph G is similar to M
       If $sim <= \epsilon_L$ the graph G is different than M

The main idea of the algorithm will be presented using the following example. Assume that the database contains an image with a bicycle (Fig. 1). The database contains also the neighbourhood matrix (Fig. 1b) and parameters of all primitives defining the shape of the bicycle (Fig. 1c). Assume that, as a query, an image containing a car is given, and that the parameters ϵ_H, ϵ_L, ϵ_N and ϵ_F are equal to 0.7,

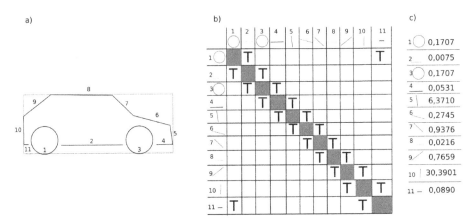

Fig. 2. A car, a) the object after shape detection (the dotted rectangle shows object's bounding box), b) shape graph neighbourhood matrix, c) parameters for each primitive

0.3, 0.7 and 0.2. During the first step the shape detection is performed(Fig. 2a) and the object's bounding box is defined. Next, the shape graph is being build (Fig. 2b shows graph's neighbourhood matrix), using a minimal distance threshold for establishing connections between nodes. It is defined as a value relative to bounding rectangle's diagonal. After this step, a comparison between graphs is performed on the node level, starting from circle nodes. First, a car's node no. 1 (with 2 connections to lines) is compared with bicycle's node no. 1 (also with 2 connections to lines). The similarity is equal to 1, so the test on the feature level has to be performed. First, the algorithm computes the difference between parameters of the car's node no. 1 and bicycle's node no. 1: $|0.1707 - 0.3207| = 0.15$, $0.15 < \epsilon_F$ (0.15 < 0.2). This means that circles are very similar, so the connections should be also tested. During the tests, the nodes are compared with each others and the most similar ones are chosen. This results in: for car's node no. 2 the most similar is bicycle's node no. 3 (similarity=0.0157, $< \epsilon_F$) and no. 11 is the most similar to node no. 2 (similarity=1.5604 $> \epsilon_F$). Summarizing, the feature level test ended up with 2 of 3 matched nodes, which is equal to ≈ 0.6 and less than ϵ_N, therefore these subgraphs are not similar. Next, another node is chosen and compared in the same way. When all nodes are tested, the number of matched ones is summed up and divided by the number of all nodes of the object from the database. If the result will be greater than ϵ_H, then the objects may be assumed as the same. The problem of different objects orientation and perspective deformations will be considered during future research.

5   Experimental Results

First, the Query by Shape algorithm was examined using the algorithm described in section 4 and the following set of real life images: 2 bicycles, 1 motorbike and 2 cars (Fig. 3). The tests were performed using the bicycle (Fig. 3 a) as the

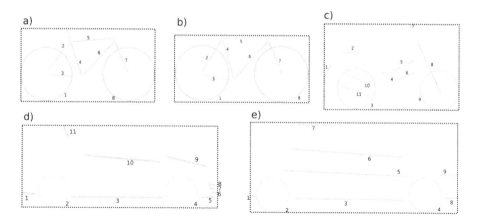

Fig. 3. Objects used for tests after shape detection a) and b) bicycles, c) motorbike, d) and e) cars. By dotted rectangles a bounding box of each object is marked.

query object. The following values were used as algorithm parameters: $\epsilon_H = 0.6$, $\epsilon_L = 0.4$, $\epsilon_N = 0.7$, $\epsilon_F = 0.4$. The test showed (table 1) that for bicycle query, the algorithm gave only one positive result, where the similarity coefficient was equal to 1 (for bicycle from Fig. 3 b)), this means that objects were detected as strictly the same. When comparing the bicycle with cars, the similarity coefficient was equal or close to 0. This means that these objects are completely different.

To verify our approach for more real life images the algorithm was implemented in C++, using OpenCV library for image operations. For line and circles detection, the Line and Circle Hough Transform algorithms were used. As a requirement for images, the uniform objects background and side view was used.

As a next test, eight images were examined. The used images are shown in Fig. 4 and results in table 2. When choosing as a query images c) and f), all results were produced properly with similarity values greater than ϵ_H. Additionally, all lines and circles were detected correctly.

Finally, to evaluate the reliability of the algorithm, defined by the number of properly and improperly matched objects and to detect possible matching problems, the next test was performed. Six object classes were used: bicycles, cars, motorbikes, scooters, tanks and chairs. The last two classes were used to determine the similarity coefficient for objects which are completely different. As database and query images, random real life photos were used (only for the tank a drawing was used). For each class, at least one image was present in the database, the total number of images was 102 (including images shown in Fig. 4). The summarized results are presented in table 3. It may be observed, that the objects which are not similar to others, have high number of correctly matched objects with similarity equal or higher than ϵ_H. If objects are similar to others, like a motorbike and a bicycle, they may have high *sim* values for both classes or even higher for improper class. This problem will be solved as a result of a future research.

Table 1. The test results for a query using a bicycle image (*bicycle a*)). The number for each node is the ns or fs coefficient value.

node	bicycle b)	motor c)	car d)	car e)
1	1	0	0	0,25
2	1	0	0,6666666667	0,6666666667
3	1	0,6666666667	0	0
4	1	0,5	0,3333333333	0,5
5	1	0,5	0,5	0
6	1	0,5	0,25	0
7	1	0	0,25	0
8	1	0,3333333333	0,25	0,25
9		1	0	0
10		0,5	0	
11		0,5	0	
$sim =$	1	0,125	0	0
add to query results	yes	no	no	no

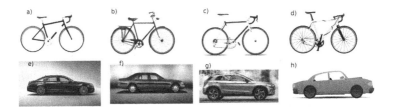

Fig. 4. Images used for examining the influence of model image details to detection results. Source: a), b), c), d), h) from OpenClipart.org, e), f), g) from mercedes-benz.pl

Table 2. The similarity coefficient values after algorithm execution for images shown in Fig. 4 and choosing as a model images c) and f) ($\epsilon_H = 0.7$, $\epsilon_L = 0.5$).

	Database models	
	c (bicycle)	f (car)
a (bicycle)	0,880933	0,345491
b (bicycle)	0,853214	0,419496
c (bicycle)	0,979681	0,110489
d (bicycle)	0,843183	0,824929
e (car)	0,184325	0,968342
f (car)	0,105222	0,924459
g (car)	0,135336	0,929266
h (car)	0,119824	0,940007

Some number of improperly classified objects was caused by improperly shape detection by Hough Transform algorithm. This problem appeared especially for a car and a scooter images, where not all circles were detected. Applying better algorithms should highly improve the quality of results. Also other values of the algorithm parameters should be examined in order to find the optimal ones for most images.

Table 3. The part of the results table (due to limited space) for the reliability test. The upper part contains similarity values, the lower part summarized number of correct results ($\epsilon_H = 0.75$, $\epsilon_L = 0.35$).

	Query image					
results	bicycle	car	motorbike	scooter	tank	chair
tank	0	0	0	0.0487013	0.98358	0
scooter	0.00806452	0.0882931	0	0.976903	0.172414	0.00454545
scooter	0.517998	0.181527	0.0348797	0.0728819	0	0.0181818
chair	0.0107527	0.242543	0	0	0	0.982277
motorbike	0.725014	0.323048	0.890854	0.111008	0.172414	0.0136364
motorbike	0.534222	0.0889066	0.148858	0.111008	0.172414	0.0181818
	(...)					
bicycle	0.880933	0.345491	0.181099	0.608081	0.413793	0.0181818
bicycle	0.843183	0.824929	0.868368	0.0564935	0	0.00909091
car	0.105222	0.924459	0.0270768	0.098021	0.103448	0.0136364
car	0.119824	0.940007	0.0119153	0.111008	0.103448	0.0181818
	(...)					
The % of correct matches for:						
$>= \epsilon_H$	93.94	66.67	28.57	20	100	100
$< \epsilon_H$ **and** $>= \epsilon_L$	30	60	30	0	0	0
$< \epsilon_L$	6.9	31.03	5.19	1.1	0	0

6   Conclusions

In this paper a new method of Content Based Image Retrieval was proposed. The main idea of our approach is based on the new representation of images in the database. Objects are decomposed into features. Each feature consist of its type and a parameter. In our approach only shape features were examined. The tests showed that our method properly recognizes different objects. To distinguish similar objects more detailed specification of shapes is required.

In the future research we consider applying other types of primitives for specification of shapes (e.g. arcs, polygons) as well as other types of features, e.g. colors or textures. Furthermore, some advanced algorithms for shape detection will be used, e.g. based on decision-making problem [14]. Another direction of research will concern developing the database structure for storing images according to shape characteristics, e.g. a tree-based. In order to improve the performance of the system, a distributed database structure will be taken into account [11]. The goal of our research is to develop efficient query by shape method that efficiently retrieves images containing multiple objects from large multimedia databases.

References

1. Aggarwal, G., Ashwin, T.V., Ghosal, S.: An image retrieval system with automatic query modification. IEEE Transactions on Multimedia 4(2), 201–214 (2002)
2. Bielecka, M., Skomorowski, M.: Fuzzy-aided parsing for pattern recognition. In: Kurzynski, M., Puchala, E., Wozniak, M., Zolnierek, A. (eds.) Computer Recognition Systems 2. ASC, vol. 45, pp. 313–318. Springer, Heidelberg (2007)
3. Daoudi, M., Matusiak, S.: Visual image retrieval by multiscale description of user sketches. J. Vis. Lang. Comput. 11(3), 287–301 (2000)
4. Del Bimbo, A., Pala, P.: Visual image retrieval by elastic matching of user sketches. IEEE Trans. on Pattern Analysis and Machine Intelligence 19(2), 121–132 (1997)
5. Jakubowski, R.: Extraction of shape features for syntactic recognition of mechanical parts. IEEE Trans. on Systems, Man and Cybernetics SMC 15(5), 642–651 (1985)
6. Kato, T., Kurita, T., Otsu, N., Hirata, K.: A sketch retrieval method for full color image database-query by visual example. In: 11th IAPR International Conference on Pattern Recognition, Vol. I. Conference A: Computer Vision and Applications, pp. 530–533 (August 1992)
7. Kriegel, H.P., Kroger, P., Kunath, P., Pryakhin, A.: Effective similarity search in multimedia databases using multiple representations. In: 12th International Multi-Media Modelling Conference Proceedings, p. 4 (2006)
8. Lalos, C., Doulamis, A., Konstanteli, K., Dellias, P., Varvarigou, T.: An innovative content-based indexing technique with linear response suitable for pervasive environments. In: International Workshop on Content-Based Multimedia Indexing, pp. 462–469 (June 2008)
9. Lee, H.C., Fu, K.S.: Generating object descriptions for model retrieval. IEEE Trans. on Pattern Analysis and Machine Intelligence PAMI 5(5), 462–471 (1983)
10. Li, C.Y., Hsu, C.T.: Image retrieval with relevance feedback based on graph-theoretic region correspondence estimation. IEEE Transactions on Multimedia 10(3), 447–456 (2008)
11. Lukawski, G., Sapiecha, K.: Balancing workloads of servers maintaining scalable distributed data structures. In: 19th Euromicro International Conference on Parallel, Distributed and Network-Based Processing, pp. 80–84 (February 2011)
12. Mocofan, M., Ermalai, I., Bucos, M., Onita, M., Dragulescu, B.: Supervised tree content based search algorithm for multimedia image databases. In: 6th IEEE International Symposium on Applied Computational Intelligence and Informatics, pp. 469–472 (May 2011)
13. Shih, T.K.: Distributed Multimedia Databases. In: Distributed Multimedia Databases, pp. 2–12. IGI Global, Hershey (2002)
14. Sitek, P., Wikarek, J.: A hybrid framework for the modelling and optimisation of decision problems in sustainable supply chain management. International Journal of Production Research (2015)
15. Sluzek, A.: On moment-based local operators for detecting image patterns. Image and Vision Computing 23(3), 287–298 (2005)
16. Wang, H.H., Mohamad, D., Ismail, N.A.: Approaches, challenges and future direction of image retrieval. CoRR abs/1006.4568 (2010)

Real-Time People Counting from Depth Images

Jakub Nalepa[1,2(✉)], Janusz Szymanek[1], and Michal Kawulok[1,2]

[1] Future Processing, Gliwice, Poland
{jnalepa,jszymanek,mkawulok}@future-processing.com
[2] Institute of Informatics, Silesian University of Technology, Gliwice, Poland
{jakub.nalepa,michal.kawulok}@polsl.pl

Abstract. In this paper, we propose a real-time algorithm for counting people from depth image sequences acquired using the Kinect sensor. Counting people in public vehicles became a vital research topic. Information on the passenger flow plays a pivotal role in transportation databases. It helps the transport operators to optimize their operational costs, providing that the data are acquired automatically and with sufficient accuracy. We show that our algorithm is accurate and fast as it allows 16 frames per second to be processed. Thus, it can be used either in real-time to process traffic information on the fly, or in the batch mode for analyzing very large databases of previously acquired image data.

Keywords: People counting · Object detection · Object tracking · Depth image

1 Introduction

The analysis of image sequences acquired in public areas, including public transport vehicles, became an important problem tackled by researchers over last years. Since the cost of surveillance cameras and sensors keeps decreasing, they are used not only for security applications, but can be applied for estimating passenger flow and predicting overcrowding. To achieve this, images must be automatically processed to provide a meaningful input to transportation databases. This helps improve the quality of public transport services by proper allocation of vehicles and humans. It results in decreasing the operational costs of transport operators by optimizing routing schedules. This also affects the traffic congestion and environmental pollution which are very important concerns nowadays [15].

Many approaches to estimating people flows utilize mechanical devices (gates, turnstiles, and other), and require passenger interaction and additional attention [2]. Although these devices are very robust and straightforward to install, they are of a high cost. Since each interaction with such devices takes time, these

J. Szymanek—This work has been supported by the European Regional Development Fund under Operational Programme Innovative Economy 2007–2013, based on the Agreement No. UDA-POIG.01.04.00-24-138/11-01.

M. Kawulok—This work was supported by the Polish Ministry of Science and Higher Education under research grant no. IP2012 026372 from the Science Budget 2013–2015.

© Springer International Publishing Switzerland 2015
S. Kozielski et al. (Eds.): BDAS 2015, CCIS 521, pp. 387–397, 2015.
DOI: 10.1007/978-3-319-18422-7_34

systems may lead to overcrowding in rush hours. Also, they need continuous maintenance as they are built from mechanical parts. Finally, the information captured by such systems is very simple, and usually it is hard to extract other useful insights about the traffic nature from that data.

To overcome the drawbacks of device-based systems, image and video analysis algorithms emerged to fully automate the counting process. It is a challenging task due to the dynamic nature of the underlying problem, varying lighting conditions, uncontrolled operation environments, and other real-life circumstances. Importantly, passengers may get on/off the bus (alone or in groups) simultaneously so the crowd moves in two opposite directions. These complex image frames must be analyzed very fast to ensure real-time traffic service support. Alternatively, images may be processed in a batch mode allowing for the analysis of large image databases. The main drawback of the latter approach is the size of data which can become enormously large during the continuous acquisition.

There exist a plethora of approaches for counting people in uncontrolled environments [9,2,8]. In a bunch of techniques, the problem is tackled by using multiple cameras to resolve the issues concerned with illumination changes, overcrowding, and occlusions in a scene by analyzing images acquired at different views [22,7,14]. Another stream of development encompasses the analysis of image sequences captured by a single camera. Zhao et al. proposed an algorithm for counting people in public places (including public vehicles) based on face detection and tracking [24]. However, this system requires relatively stable lighting conditions (face detection is very computationally-intensive and extremely challenging in uncontrolled environments [12,11]). Other techniques include those based on separating background and foreground (moving) objects [17], extracting features from segmented frames [3], and many others [1,20,23,21,4,19,5,6].

In this paper, we propose a fast and effective algorithm to count people from depth images acquired using a single Kinect sensor installed inside the bus over an entrance. In the algorithm, standard morphology operators are followed by an efficient segmenting and grouping of image regions. Our extensive experimental study performed on images acquired in various conditions showed that the algorithm can process up to 16 frames per second. Thus, it can be executed in the embedded environment to count people in real-time even in varying lighting conditions of public vehicles. Alternatively, it can be applied for analyzing very large video surveillance databases. The sensitivity analysis demonstrates how various components influence the algorithm efficacy and execution time.

This paper is organized as follows. Our real-time algorithm for counting people is outlined in detail in section 2. The discussion on the experimental results along with the description of generated datasets are given in section 3. Section 4 concludes the paper and highlights the directions of our future work.

2   Real-Time People Counting

In this section, we discuss in detail our real-time algorithm for counting people from depth image sequences. The approach is visualized in the flowchart shown in Fig. 1. It is complemented with exemplary images showing the output of each step of the proposed algorithm (Fig. 2). It is worth pointing out that we applied an additional lookup table to transform an input image (the second step in Fig. 2) for a better visualization of the consecutive operations of our technique (the original image is always fed into the pipeline – see Fig. 1).

Fig. 1. Flowchart of the proposed algorithm

2.1 Dilation and Erosion

At first, an input image is subject to standard morphology operators (dilation and erosion), in order to remove small "noisy" structures from the image (see Figs. 2C–D). Instead of running multiple iterations of dilation/erosion, we have proposed two-pass counterparts which make it possible to process an image in only two iterations. An input image is analyzed from top to bottom at first, and then reversely – from bottom to top, similarly as in [13]. It is worth noting that we analyzed both versions of each operation here – clearly, they give the same resulting intermediate images. Once the image is dilated and eroded, it undergoes the segmentation procedure (Fig. 1).

2.2 Segmentation

Here, we implemented a simple segmentation routine, which we extended so that it benefits from the information about the mutual distance between pixels. In the basic segmentation process (referred to as the *simple* segmentation), we determine local minima of pixel depth values at first. We start from the top-left image pixel (and trace towards the bottom-right pixel), and compare its depth value (v) with the neighboring pixels depth values. If a smaller value is found, then the visited pixels are excluded from the further analysis. Alternatively, if larger values are found, v is appended to the vector containing the minima determined so far. Similarly, the visited pixels are marked as processed. The analysis is then restarted from the first not visited pixel to find other minima.

These values are sorted and constitute a vector of minimum values in an analyzed depth image. A modified flood fill algorithm is applied to append neighboring pixels to the found minima. The pixel is appended to the region if the difference (θ) between the considered pixel depth value and the one taken from the vector of local minima (m) is less than an assumed threshold (ϵ): $\theta < \epsilon$, where $\theta = |v - m|$. It is easy to see that the position of a given pixel within an image with respect to the currently considered minimum depth value pixel is not taken into account during this segmentation procedure. It is therefore possible

(A: Input image)   (B: Apply LUT)   (C: Dilate)

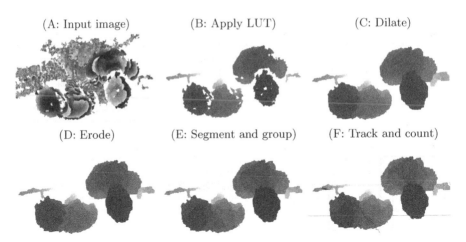

(D: Erode)   (E: Segment and group)   (F: Track and count)

Fig. 2. Images obtained at each step of the algorithm

to group distant pixels with similar depth values which results in "leakages" near region boundaries. This is a significant drawback of the *simple* segmentation.

To mitigate the shortcoming of the *simple* segmentation, we employed the distance transform (DT) to consider not only the depth value of each pixel while segmenting an image, but also its position in the spatial domain with respect to other pixels [13]. In the DT segmentation, the DT is performed from each local minimum at first. Then, the procedure executes similarly to the *simple* segmentation – the flood fill is applied. Once pixels are clustered and form blobs near the pixels determined as local minima, the DT is performed to verify if appended pixels are "close" to the local minimum. If the added pixels are far from the local minimum (based on the DT values), then they are removed from the segment and added to the one containing the closest minimum. Thus, the depth information is complemented by the spatial information extracted using the DT performed in the depth map. The final step of the DT segmentation involves merging neighboring segments if their depth values are similar.

2.3   Grouping

The next algorithm step comprises clustering regions into larger blobs (Fig. 1). This operation is performed in order to group segments representing various parts of a human body (head, shoulders, torso and possibly legs) into a single segment (which will be tracked and counted afterwards). It also allows for distinguishing between two persons getting on/off the bus close to each other. We proposed two variants of grouping: *fast* grouping, and the one based on the analysis of depth local minima (LM). In the fast grouping, two segments are merged if (i) the area of their overlapping part is at least 20% of either segment (the overlapping part is relatively large, and indicates that the neighboring segments should be merged to form one segment), or (ii) if the length of any edge of the overlapping part of their bounding boxes is greater than a half of the edge length of either bounding box. In the latter case, the segments are placed very close to each other (so that their bounding boxes overlap), but the actual overlapping area of these segments is not necessarily significant.

In the local minimum based grouping (LM grouping), the local minima considered in the segmentation step are sorted at first. Then, they are annotated with identifiers (the first minimum gets 0 – the identifiers increase towards the end of the vector with minima). Furthermore, we determine which segments are overlapping. The analysis is started from the last segment in the sorted (by the depth values) vector of minima, and it is verified which segment was the most overlapping with the considered one. Then, they are merged and put into the vector to replace the last vector. We proceed with the same procedure for each segment in the vector. Finally, the grouped segments are tracked and counted once the ground-truth door line is crossed in either direction (see Fig. 1 and Fig. 2F – the bus door line is rendered in light pink).

3 Experimental Results

3.1 Setup

We performed an extensive experimental study to investigate the efficacy of the proposed algorithm, and to verify how each algorithm component contributes to its performance and execution time. The investigated algorithm variants are given in table 1. The algorithm was implemented in the C++ programming language and executed on a computer equipped with an Intel Core i7 2.3 GHz (16 GB RAM) processor. The segmentation threshold was set to $\epsilon = 1000$ (see section 2 for more details). The proposed algorithm operates asynchronously (the next frame is taken from the stream once the current frame has been analyzed).

3.2 Datasets

The proposed algorithm was tested on five depth image sequences acquired in the real-world environment. The Kinect sensor was placed above the bus door, and people flows were recorded to capture different scenarios. The sequence durations and the ground-truth data, are summarized in table 2. Exemplary frames captured in each video are given in Fig. 3. These images show (a) at least two people in the region of interest, (b) one person, and (c) an empty region of interest. It is easy to note that the people coming in and out the bus are of a different height (the taller person the darker image region showing his head – indicating smaller depth values, since the head was closer to the sensor).

3.3 Analysis and Discussion

The proposed algorithm was verified on several depth image sequences acquired in the real-life bus environment. In order to compare various algorithm variants, we executed each variant on the entire dataset. The results (cumulated for all

Table 1. The investigated variants of the proposed algorithm

ID	Dilation	Erosion	Segmentation	Grouping
(1)	Standard	Standard	Simple	Fast
(2)	Standard	Standard	Simple	LM
(3)	Standard	Standard	DT	Fast
(4)	Standard	Standard	DT	LM
(5)	Standard	Two-pass	Simple	Fast
(6)	Standard	Two-pass	Simple	LM
(7)	Standard	Two-pass	DT	Fast
(8)	Standard	Two-pass	DT	LM
(9)	Two-pass	Standard	Simple	Fast
(10)	Two-pass	Standard	Simple	LM
(11)	Two-pass	Standard	DT	Fast
(12)	Two-pass	Standard	DT	LM
(13)	Two-pass	Two-pass	Simple	Fast
(14)	Two-pass	Two-pass	Simple	LM
(15)	Two-pass	Two-pass	DT	Fast
(16)	Two-pass	Two-pass	DT	LM

Table 2. Settings of depth image sequences along with the ground-truth data used in this study (GT_{In} and GT_{Out} denotes the number of people in the incoming and outcoming flows, respectively). The frame rate of each video is 30 frames/second.

ID	Duration (in sec.)	GT_{In}	GT_{Out}
I.	71	13	15
II.	58	11	11
III.	49	5	5
IV.	18	2	2
V.	72	16	15
Total	268	47	48

sequences) are given in table 3: In and Out denote the number of passengers coming in/out the bus determined by the corresponding algorithm variant (1–16), $r_{In} = In/GT_{In}$, and $r_{Out} = Out/GT_{Out}$ are the ratios of In/Out and the ground-truth values, and δ denotes the error given as $\delta = |1 - r_{In}| + |1 - r_{Out}|$. It is easy to note that $r < 1$ indicates that the corresponding variant ended up with a lower number of passengers getting on/off the bus than it was expected (analogously, if $r > 1$, then this value was larger than expected). The error δ is an absolute error for both incoming and outcoming people movements. Additionally, we present the number of frames analyzed per second using each variant (fps).

The results confirm that the two-pass versions of standard morphology operators significantly speed up the algorithm execution (see (5) and (9) compared with (1), along with (12) and (8) in comparison with (4) in table 3). It is worth noting that the algorithm components are *not* independent from each other (especially in the case of segmenting and grouping). Also, it is worth pointing out

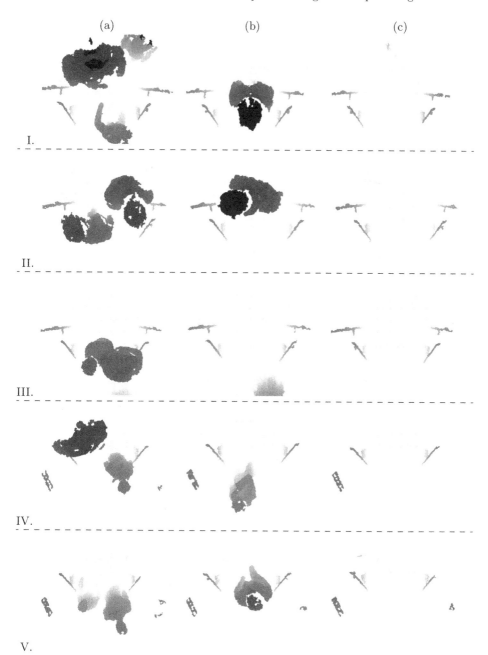

Fig. 3. Examplary frames captured in each depth image sequence (I–V) showing (a) at least two people, (b) one person, (c) empty region of interest

Table 3. The results obtained using different variants of the proposed algorithm. The smallest error δ and the largest fps rate are rendered in boldface.

Alg.	In	Out	r_{In}	r_{Out}	δ	fps
(1)	30	30	0.64	0.63	0.74	11.60
(2)	47	60	1.00	1.25	0.25	11.19
(3)	49	46	1.04	0.96	0.08	12.39
(4)	42	46	0.89	0.96	0.15	12.54
(5)	30	31	0.64	0.65	0.72	11.85
(6)	55	72	1.17	1.50	0.67	11.76
(7)	49	45	1.04	0.94	0.11	12.69
(8)	40	47	0.85	0.98	0.17	12.85
(9)	30	31	0.64	0.65	0.72	12.50
(10)	50	65	1.06	1.35	0.42	11.40
(11)	48	47	1.02	0.98	**0.04**	14.70
(12)	43	46	0.91	0.96	0.13	14.74
(13)	30	31	0.64	0.65	0.72	12.44
(14)	55	72	1.17	1.50	0.67	13.12
(15)	49	45	1.04	0.94	0.11	16.04
(16)	41	46	0.87	0.96	0.17	**16.34**

that although both versions of dilation and erosion operators give the same intermediate images in the pipeline, the final algorithm outcome may slightly differ across the variants with different implementations of morphology operators (see the results for e.g., (1) and (13)). Since the algorithm executes asynchronously, analyzing a larger number of frames may introduce an error if passengers move close to the door line (the tracked point "oscillates" near this line and may be counted as an entering/exiting passenger erroneously). This shortcoming has been mitigated by an additional analysis of the length of the tracked path (a threshold hysteresis has been applied). It stabilizes the results, however a more sophisticated approach may improve them even further.

It is easy to see that the applied segmentation and grouping procedures drastically influence the counting error δ. Exploiting the mutual information between pixels in an input image (using a DT) helps improve the quality of final results (see e.g., (13) compared with (15), and (1) compared with (3) in table 3 – the error decreased from $\delta = 0.72$ to $\delta = 0.11$ in the first case, and from $\delta = 0.74$ to $\delta = 0.08$ in the latter one). Since the DT executes very fast, it does not slow down the image analysis. Additionally, more accurate segments extracted at this algorithm step constitute a better input for the final grouping. This leads to not only better-quality counting but also to speeding up the execution (the version (15) processes at least 3.5 frames more than (13) on average). The results clearly confirm that the grouping step is crucial and significantly affects the error values. Since it is dependent on the segmenting step outcome, the choice of the grouping version should be undertaken based on the characteristics of segments constructed during the segmentation. In general, the LM grouping performs much better when coupled with the *Simple* segmentation. On the other hand, the *Fast* grouping is better when the DT segmentation is applied.

As already mentioned, the algorithm components influence one another. It is worth noting that the algorithm version with the two-pass morphology operators, DT-based segmentation, and the LM grouping appeared to be the fastest among the investigated ones (see table 1). It allowed analyzing more than 16 frames per second (11.6 frames were processed per second by the baseline algorithm). However, the best results (with the lowest $\delta = 0.04$) are offered by the version (11) of the algorithm (it is ca. 11% slower than (16)).

4   Conclusions and Future Work

In this paper, we presented a real-time algorithm for counting people in public vehicles. It is entirely based on the analysis of depth images acquired using the Kinect sensor placed above the bus door. It does not require any user interaction with moving people which is a main drawback of many real-life applications for analyzing and estimating people flow. An extensive experimental study performed on a number of depth image sequences captured in buses and reflecting many scenarios showed the efficacy of the proposed algorithm. Since it executes very fast (up to 16 frames per second), it can be applied in real-time applications to count people in vehicles. Finally, the algorithm may be executed to analyze very large transportation databases offline (in the batch processing mode).

Our ongoing research is aimed at incorporating new features extracted from images, including color [18], to increase the robustness of our algorithm [16]. We plan to compare the efficacy of the proposed algorithm with other state-of-the-art techniques, and run it in an embedded environment. This would enable determining the current number of passengers inside the moving bus. Finally, we plan to classify the extracted feature vectors using support vector machines [10].

References

1. Albiol, A., Mora, I., Naranjo, V.: Real-time high density people counter using morphological tools. IEEE Transactions on Intelligent Transportation Systems 2(4), 204–218 (2001)
2. Bernini, N., Bombini, L., Buzzoni, M., Cerri, P., Grisleri, P.: An embedded system for counting passengers in public transportation vehicles. In: Proc. IEEE ASME, pp. 1–6 (2014)
3. Chan, A.B., Liang, Z.S.J., Vasconcelos, N.: Privacy preserving crowd monitoring: Counting people without people models or tracking. In: IEEE Conference on Computer Vision and Pattern Recognition, CVPR 2008, pp. 1–7 (June 2008)
4. Chan, A.B., Vasconcelos, N.: Modeling, clustering, and segmenting video with mixtures of dynamic textures. IEEE TPAMI 30(5), 909–926 (2008)
5. Conte, D., Foggia, P., Percannella, G., Tufano, F., Vento, M.: A method for counting moving people in video surveillance videos. EURASIP Journal on Advances in Signal Processing 2010(1), 231–240 (2010),
   http://asp.eurasipjournals.com/content/2010/1/231240
6. Ferryman, J., Ellis, A.L.: Performance evaluation of crowd image analysis using the PETS2009 dataset. Patt. Recogn. Lett. 44(0), 3–15 (2014),
   http://www.sciencedirect.com/science/article/pii/S0167865514000191

7. Ge, W., Collins, R.T.: Crowd detection with a multiview sampler. In: Daniilidis, K., Maragos, P., Paragios, N. (eds.) ECCV 2010, Part V. LNCS, vol. 6315, pp. 324–337. Springer, Heidelberg (2010), http://dl.acm.org/citation.cfm?id=1888150.1888177

8. Gudyś, A., Rosner, J., Segen, J., Wojciechowski, K., Kulbacki, M.: Tracking people in video sequences by clustering feature motion paths. In: Chmielewski, L.J., Kozera, R., Shin, B.-S., Wojciechowski, K. (eds.) ICCVG 2014. LNCS, vol. 8671, pp. 236–245. Springer, Heidelberg (2014), http://dx.doi.org/10.1007/978-3-319-11331-9_29

9. Hsieh, J.W., Peng, C.S., Fan, K.C.: Grid-based template matching for people counting. In: IEEE 9th Workshop on Multimedia Signal Processing, MMSP 2007, pp. 316–319 (October 2007)

10. Kawulok, M., Nalepa, J.: Support vector machines training data selection using a genetic algorithm. In: Gimel'farb, G., et al. (eds.) SSPR & SPR 2012. LNCS, vol. 7626, pp. 557–565. Springer, Heidelberg (2012)

11. Kawulok, M., Szymanek, J.: Precise multi-level face detector for advanced analysis of facial images. IET Image Processing 6(2), 95–103 (2012)

12. Kawulok, M., Wu, J., Hancock, E.R.: Supervised relevance maps for increasing the distinctiveness of facial images. Pattern Recognition 44(4), 929–939 (2011), http://www.sciencedirect.com/science/article/pii/S0031320310004942

13. Lagodzinski, P., Smolka, B.: Application of the extended distance transformation in digital image colorization. Multimedia Tools and App. 69(1), 111–137 (2014), http://dx.doi.org/10.1007/s11042-012-1246-2

14. Maddalena, L., Petrosino, A., Russo, F.: People counting by learning their appearance in a multi-view camera environment. Patt. Recogn. Lett. 36, 125–134 (2014), http://www.sciencedirect.com/science/article/pii/S0167865513003796

15. Nalepa, J., Blocho, M.: Co-operation in the parallel memetic algorithm. International Journal of Parallel Programming, 1–28 (2014), http://dx.doi.org/10.1007/s10766-014-0343-4

16. Nalepa, J., Kawulok, M.: Fast and accurate hand shape classification. In: Kozielski, S., Mrozek, D., Kasprowski, P., Małysiak-Mrozek, B. z. (eds.) BDAS 2014. CCIS, vol. 424, pp. 364–373. Springer, Heidelberg (2014), http://dx.doi.org/10.1007/978-3-319-06932-6_35

17. Schofield, A.J., Mehta, P.A., Stonham, T.J.: A system for counting people in video images using neural networks to identify the background scene. Pattern Recognition 29(8), 1421–1428 (1996), http://www.sciencedirect.com/science/article/pii/0031320395001638

18. Starosolski, R.: New simple and efficient color space transformations for lossless image compression. J. of Vis. Commun. and Image Represent 25(5), 1056–1063 (2014)

19. Su, C.W., Liao, H.Y.M., Tyan, H.R.: A vision-based people counting approach based on the symmetry measure. In: IEEE International Symposium on Circuits and Systems, ISCAS 2009, pp. 2617–2620 (May 2009)

20. Viola, P., Jones, M.J., Snow, D.: Detecting pedestrians using patterns of motion and appearance. In: Proc IEEE Int. Conf. on Computer Vision, vol. 2, pp. 734–741 (2003)

21. Wu, B., Nevatia, R.: Detection and tracking of multiple, partially occluded humans by Bayesian combination of edgelet based part detectors. International Journal of Computer Vision 75(2), 247–266 (2007), http://dx.doi.org/10.1007/s11263-006-0027-7

22. Yahiaoui, T., Meurie, C., Khoudour, L., Cabestaing, F.: A people counting system based on dense and close stereovision. In: Elmoataz, A., Lezoray, O., Nouboud, F., Mammass, D. (eds.) ICISP 2008 2008. LNCS, vol. 5099, pp. 59–66. Springer, Heidelberg (2008), http://dx.doi.org/10.1007/978-3-540-69905-7_7

23. Zhao, T., Nevatia, R.: Bayesian human segmentation in crowded situations. In: Proceedings of the 2003 IEEE Computer Society Conference on Computer Vision and Pattern Recognition, vol. 2, pp. II–459–II–466 (June 2003)

24. Zhao, X., Delleandrea, E., Chen, L.: A people counting system based on face detection and tracking in a video. In: Sixth IEEE International Conference on Advanced Video and Signal Based Surveillance, AVSS 2009, pp. 67–72 (September 2009)

PCA Application in Classification
of Smiling and Neutral Facial Displays

Karolina Nurzynska[1][(✉)] and Bogdan Smolka[2]

[1] Institute of Informatics,
Silesian University of Technology, ul. Akademicka 16
44-100 Gliwice, Poland
karolina.nurzynska@polsl.pl
[2] Institute of Automatic Control,
Silesian University of Technology, ul. Akademicka 16
44-100 Gliwice, Poland
bogdan.smolka@polsl.pl

Abstract. Psychologists claim that the majority of inter-human communication is transferred non-verbally. The face and its expressions are the most significant channel of this kind of communication. In this research the problem of recognition between smiling and neutral facial display is investigated. The set-up consisting of proper local binary pattern operator supported with an image division schema and PCA for feature vector length diminishing is presented. The achieved results using k-nearest neighbourhood classifier are compared on a couple of image datasets to prove successful application of this approach.

1 Introduction

Emotions depicted on human face are the most important method of non-verbal communication between people. Regardless how well one can distinguish between different emotional states of others seeing only their facial display, it is rather easy to deceive the person. The consequences of this mistake might be unpleasant in personal life but also very severe when security measures are considered. Therefore, creation of automatic systems, which could deal with this issue, is an urgent research problem.

The research concentrating on facial expression recognition started in the last decade of XX century. First works used optical flow to follow the movement of muscles displayed in the image sequence [16]. Then this technique was supported with other means, e.g. holistic approach, explicit measurements, as is described in [5,9]. Next, another solution how to describe a subtle motion in image sequences is given in [6,7]. The last research concentrates on Facial Action Coding System [8] describing which muscles activate during a given emotional state. Finally, active shape models were exploited to describe the emotions in [13]. Although, eigenfaces introduced in [22] are commonly used for face recognition, more recently the local binary patterns [4] were applied to solve this problem [3,10]. Moreover, this texture operator is also exploited for emotion classification [21]. This technique gives very good outcome as reported in [14,19].

© Springer International Publishing Switzerland 2015
S. Kozielski et al. (Eds.): BDAS 2015, CCIS 521, pp. 398–407, 2015.
DOI: 10.1007/978-3-319-18422-7_35

In this research some existing algorithms for facial display classification into smiling and neutral expression are compared using similar experiment set-up and classification method. Moreover, the application of principal component analysis for the reduction of the feature vector length, which for considered methods is very long, is a novel contribution. Following section 2 describes the local binary patterns which are used for image description. Section 3 gives detailed information about the experiment set-up. Then, the results of performed experiments are described in section 4 and the conclusions are drawn in section 5.

2 Local Binary Patterns

Local binary patterns, (LBP), describe a texture in a form of a histogram. Each histogram bin represents a numerical value corresponding to a small texture area [20]. There are a couple of algorithms which might be used for the bin value calculation. Depending on the chosen approach, the distribution of the values in histogram changes as well as the length. In the following sections the two most popular versions of this texture operator are presented.

2.1 Basic LBP Operator

This operator was introduced by Ojala et al. [17] with assumption that the texture might be well described by its local pattern and its strength. Firstly, this approach assumed a 3×3 pixel neighbourhood in the monochromatic image I. Yet soon it was generalised to a circular neighbourhood, which radius R and number of sampling points P were parametrized [18]. Thus for each pixel (x, y) the neighbourhood is defined as follows:

$$
\begin{aligned}
g_p &= I(x_p, y_p), \\
p &= 0, ..., P - 1, \\
x_p &= x + R\cos(2\pi p/P), \\
y_p &= y - R\sin(2\pi p/P),
\end{aligned}
\tag{1}
$$

where g_p is a grey-scale value of the p-th sampled point. The gathered information from the local texture T, represents a joint distribution t, of the pixel: $T = t(g_c, g_0, g_1, ..., g_{P-1})$, for $g_c = I(x, y)$. Instead of storing the original pixel values, more information may be derived from the differences between the sampled points and the central value g_c. Moreover, since the central value is statistically independent, it might be omitted. Then the distribution is given as $t(g_0 - g_c, g_1 - g_c, ..., g_{P-1} - g_c)$. Still, there are many possibilities to code the information, therefore only the sign of the differences is recorded in the generic local binary patterns:

$$
LBP_{P,R}(x_c, y_c) = \sum_{p=0}^{P-1} s(g_p - g_c) \cdot 2^p,
\tag{2}
$$

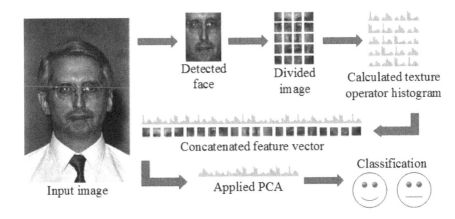

Detected face   Divided image   Calculated texture operator histogram

Concatenated feature vector

Input image   Applied PCA   Classification

Fig. 1. Experiment set-up

where

$$s(z) = \begin{cases} 1, & z \geq 0, \\ 0, & z < 0. \end{cases} \tag{3}$$

2.2 Uniform Patterns

The histogram generated with LBP contains 2^P different values (for $P = 8$ that gives 256 elements), hence as a feature vector is rather long. Moreover, it was observed that some patterns store more valid information than others. It was concluded [20] that the most descriptive labels are when there are up to two transition between 0 and 1 in the operator written as a series of bits, e.g. 01110000. This patterns were named a 'uniform' patterns, ($uLBP$) and have separate bins, while all other with higher number of transitions are considered non-uniform and are placed in one bin. This approach diminishes the number of histogram bins to $P(P - 1) + 3$ for P sampled points.

3   Experiment Set-up

In order to start classification between smiling and neutral facial displays it is necessary to perform some preliminary processing of the input image. The steps applied in presented framework are depicted in Fig. 1 and described below. Moreover, an overview of used images datasets is also presented.

3.1   Image Databases

The discussed approach for emotion classification was evaluated on three different databases (some exemplary images are presented in Fig. 2). First of them is the Cohn-Kanade AU-Coded Facial Expression Database [6,15] which consists

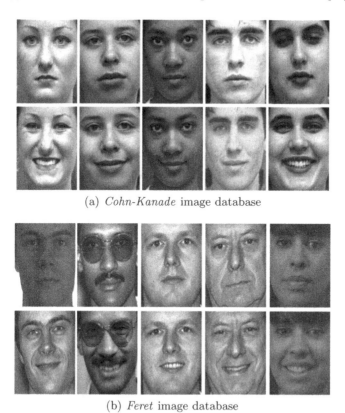

(a) *Cohn-Kanade* image database

(b) *Feret* image database

Fig. 2. Examples of images gathered in the databases. Neutral facial displays are in the top row, whereas smiling one in the bottom row.

of image sequences presenting the change of facial display from neutral to representing the basic emotion, like smile, anger, etc. From 82 sequences, images with neutral and smiling facial display were selected. These are monochromatic images with 640×480 pixel resolution. The lighting condition varies in the images, the subjects do not wear any covering elements, such as glasses or beards (see Fig. 2a).

Second database was created from images collected in *Feret* data set prepared for face recognition purposes [1]. Here, 62 images were selected to represent each emotional state. The subjects differ between the groups, however it is possible to have several pictures of one subject in the group. The images are also monochromatic with varying lighting condition. Some of the subjects are wearing glasses or have beards. Refer to Fig. 2b for some examples.

Due to a small number of facial expression in each of the accessed databases, it is difficult to generalise the obtained results, therefore a database containing images from several existing data sets was prepared. It contains data from previously described databases but also from *Iranian, Nottingham originals, Pain,* and *Utrecht* data sets [2]. There are 712 images in total for both expressions. Some of these images are in colour, other are monochromatic, the resolution also varies.

3.2   Face Image Acquisition

The first step of image preprocessing converts all images into grey-scale ones. Then, in order to describe emotion with texture operator, it is necessary to find and normalize the face region in the image. For this purpose a face detector described in [12] is applied. Since the original images are taken in various resolution and also the head size and distance from the camera changes between images, it was decided to resize the detected face into a constant resolution of 112×150 pixels for further processing.

3.3   Image Division

The emotions drawn on the face are related to its certain parts. According to research conducted by Ekman [8] it is possible to distinguish muscles which activate when displaying an emotion in the face. In this research the Face Action Coding System was developed, which binds the muscle movement with an emotion depicted in the face. Therefore, calculating a global *LBP* histogram for the whole image would be too general to describe precisely the visual information. In consequence, the image is divided into several sub-images, and the final feature vector is a concatenation of histograms calculated for each of them, as it is presented in Fig. 1. Moreover, the sub-images which present the region of muscle activation (in case of the smile: eyes and lips) convey more information, therefore it is possible to consider only these sub-images when building a feature vector by applying adequate masks.

In the presented work, three approaches for image division were considered. One of them divided the images into 20 sub-images using a mask with 4 columns

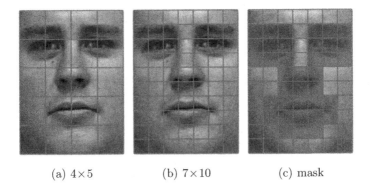

(a) 4×5   (b) 7×10   (c) mask

Fig. 3. Sub-images used to build a feature vector

and 5 rows. The second approach were more detailed and divided the images into 70 sub-images with 7 columns and 10 rows. Finally, in the second approach a mask was applied to analyse only the parts where the changes really took place as depicted in Fig. 3.

3.4 Principal Component Analysis

The length of the feature vector calculated to describe the image, depending on the chosen image division schema and texture operator applied, can result in hundreds up to thousands of elements. This severely complicates the classification process. Therefore, an attempt to diminish the number of elements by selecting those which convey the most descriptive information was undertaken.

In order to reduce the dimensionality of a large data set a principal component analysis is used [11], PCA. It is a mathematical projection, which transforms the data space description from the original one appointed by each feature in the feature vector into a new one, where the data variability is considered for space description. Then, the first principal component (eigenvector) is a vector corresponding to the direction in which the highest data variability is measured, next the second principal component reflects the less variable direction orthogonal to the previous one, and so on. Moreover, for each eigenvector, it is possible to define its importance (eigenvalue) with respect to whole data set. Choosing the most informative eigenvectors and summing up corresponding eigenvalues enables to claim which part of the initial data is described when only the picked up components are used. When this accuracy of information seems sufficient for the task solution, the data from eigenspace is transformed back using only the selected eigenvectors for calculation, what reduces the dimensionality of the original data space.

In presented research, the PCA parameters were obtained on the training set and then used to transform the feature vectors in testing data set. In experiments the results achieved for original features vectors are compared to those recorded when 97%, 98%, or 99% of initial information is exploited.

4 Results and Discussion

The experiments were performed on three previously described image data sets. The research were conveyed for two different image division schemas and the mask approach. Tables contain also information whether change of one of the texture operator parameters (radius) influences the results. The second parameter (number of sampling points) was constant because of its correlation with histogram length. The classification was performed using the k-nearest neighbourhood technique with cityblock distance metric and k equal to 9.

Tables 1 and 2 gather results achieved for the *Cohn-Kanade* database. The performance is comparable for all used texture operator. It is interesting to observe, that the radius does not influence the results when feature vector of full length is considered (last row of the tables). However, the results gathered with PCA application vary when the radius and applied image division

Table 1. *Cohn-Kanade* database - *LBP* texture operator

Information	$R = 1$			$R = 2$			$R = 3$		
	4×5	7×10	mask	4×5	7×10	mask	4×5	7×10	mask
97%	93.87	90.80	92.02	93.87	90.18	94.48	94.48	93.25	88.34
98%	92.02	94.48	90.80	92.64	90.80	96.32	93.25	93.87	87.73
99%	93.87	92.64	91.41	92.64	90.80	91.41	94.48	94.48	91.41
100%	95.09	94.48	93.87	96.93	95.09	94.48	96.32	95.09	94.48

Table 2. *Cohn-Kanade* database - *uLBP* texture operator

Information	$R = 1$			$R = 2$			$R = 3$		
	4×5	7×10	mask	4×5	7×10	mask	4×5	7×10	mask
97%	92.64	92.64	93.25	94.48	92.64	93.25	93.25	93.25	90.80
98%	92.02	91.41	90.80	94.48	88.96	92.64	95.71	91.41	90.80
99%	91.41	93.87	90.80	93.87	88.34	92.02	93.25	93.87	90.18
100%	94.48	95.09	94.48	96.93	95.09	95.71	96.93	96.32	96.32

schema changes. When considering *LBP*, the correct classification rate of 96% was recorded when 98% of original information was exploited with the set-up using mask and $R = 2$. On the other hand, for *uLBP* the performance is worse around 2% when compared to full feature vector used in case of set-ups using 4×5 image division and mask.

The situation is different when *Feret* (results presented in tables 3, 4) and *All* data sets are considered (outcomes gathered in tables 5, 6). First of all, since the images vary much more, the general accuracy of correct recognition rate for the original feature vector length is lower (refer to the last row in each of the tables). The best result for *Feret* database achieves 86% for *LBP* texture

Table 3. *Feret* database - *LBP* texture operator

Information	$R = 1$			$R = 2$			$R = 3$		
	4×5	7×10	mask	4×5	7×10	mask	4×5	7×10	mask
97%	63.71	67.74	74.19	66.94	58.06	65.32	70.97	69.35	66.13
98%	62.10	62.10	69.35	65.32	64.52	62.90	64.52	65.32	61.29
99%	71.77	64.52	72.58	70.16	67.74	62.90	63.71	68.55	62.90
100%	76.61	83.06	86.29	75.81	78.23	82.26	74.19	81.45	83.87

Table 4. *Feret* database - *uLBP* texture operator

Information	$R = 1$			$R = 2$			$R = 3$		
	4×5	7×10	mask	4×5	7×10	mask	4×5	7×10	mask
97%	64.52	70.16	69.35	75.00	64.52	67.74	61.29	61.29	65.32
98%	66.13	65.32	68.55	67.74	62.10	67.74	75.81	62.90	58.87
99%	65.32	65.32	73.39	69.35	62.90	74.19	67.74	62.90	61.29
100 %	78.23	75.81	79.03	76.61	82.26	81.45	77.42	83.06	84.68

Table 5. *All* database - *LBP* texture operator

Information	R = 1			R = 2			R = 3		
	4×5	7×10	mask	4×5	7×10	mask	4×5	7×10	mask
97%	73.35	68.02	74.05	74.33	61.01	69.00	74.61	59.19	64.52
98%	73.35	67.60	73.49	76.02	60.45	69.00	72.79	59.33	65.50
99%	75.32	69.57	73.63	76.30	61.85	71.81	74.61	57.92	65.22
100%	80.22	82.33	82.61	81.35	82.19	83.03	83.73	83.03	81.91

Table 6. *All* database - *uLBP* texture operator

Information	R = 1			R = 2			R = 3		
	4×5	7×10	mask	4×5	7×10	mask	4×5	7×10	mask
97%	74.19	66.34	76.44	77.14	59.05	69.28	76.72	58.91	67.60
98%	74.05	69.14	76.44	76.58	59.19	69.28	76.16	56.94	63.96
99%	74.61	70.27	75.32	76.58	61.85	69.85	77.00	59.61	62.55
100%	77.70	81.63	83.59	81.07	83.03	84.01	82.61	80.50	79.66

operator when mask is applied and $R = 1$, whereas for *All* database 83% score is recorded in several configurations. Next, application of PCA analysis diminishes the performance considerably in almost every case. There are, however, few exceptions. In *Feret* data set described with *uLBP* for image division 4×5 when $R = 2$ when 97% of original information is used and for the same image division schema when $R = 3$ and 98% of original information is applied.

The exploitation of PCA enables diminishing the feature vector length considerably. Table 7 presents the lengths of the feature vector when each of the texture operator is applied for each image description schema. Since the length of the feature vector after PCA application differs slightly in each experiment (even for similar set-up) instead of numeric length the part of the original feature vector length is given in percentage. One can see that in case of *LBP* the reduction is much higher when the other texture operator is applied. But when looking at the number of eigenvectors used in both cases they are rather similar.

From the performed experiments one can draw a conclusion that in case of *Cohn-Kanade* database the application of PCA might bring some benefits with

Table 7. Feature vector length

PCA%	LBP			uLBP		
	4×5	7×10	mask	4×5	7×10	mask
100%	5120	17920	9472	1180	4130	2183
Length is given as a percent of the original feature vector length						
99%	9	3	6	33	15	26
98%	7	3	6	25	13	23
97%	6	3	5	20	12	20

feature vector shortening without big impact on the correct classification accuracy. In other cases, however, application of PCA should be applied with care.

5 Conclusions

In this paper, a framework enabling classification between smiling and neutral facial displays was presented. The face was described using two texture operators: LBP and $uLBP$. Moreover, three different approaches to face description were given. Two of them consisted of image division schemas, whereas the third used specially designed mask. For each set-up the classification performance was calculated for the reference databases: *Cohn-Kanade*, *Feret*, and *All*. The presented results show that, although application of PCA reduced the feature vector length reasonably, it also in most cases is responsible for correct classification accuracy loss, therefore it should be used carefully.

Acknowledgement. This work has been supported by the Polish National Science Centre (NCN) under the Grant: DEC-2012/07/B/ST6/01227 and was performed using the infrastructure supported by POIG.02.03.01-24-099/13 grant: GCONiI - Upper-Silesian Center for Scientific Computation.

References

1. Feret database (July 2014), http://www.itl.nist.gov/iad/humanid/feret/feret_master.html
2. Smile databases (July 2014), http://pics.psych.stir.ac.uk
3. Ahonen, T., Hadid, A., Pietikainen, M.: Face description with local binary patterns: Application to face recognition. IEEE Trans. Pattern Anal. Mach. Intell. 28(12), 2037–2041 (2006), http://dx.doi.org/10.1109/TPAMI.2006.244
4. Ahonen, T., Matas, J., He, C., Pietikäinen, M.: Rotation invariant image description with local binary pattern histogram Fourier features. In: Salberg, A.-B., Hardeberg, J.Y., Jenssen, R. (eds.) SCIA 2009. LNCS, vol. 5575, pp. 61–70. Springer, Heidelberg (2009)
5. Bartlett, M.S., Hager, J.C., Ekman, P., Sejnowski, T.J.: Measuring facial expressions by computer image analysis. Psychophysiology 36(2), 253–263 (1999)
6. Cohn, J.F., Zlochower, A.J., Lien, J., Kanade, T.: Automated face analysis by feature point tracking has high concurrent validity with manual facs coding. Psychophysiology 36(2), 35–43 (1999)
7. Cohn, J.F., Zlochower, A.J., Lien, J.J., Kanade, T.: Feature-point tracking by optical flow discriminates subtle differences in facial expression. In: Proc. of the 3rd. Intern. Conf. on Face & Gesture Recog., Washington, DC, USA, pp. 396–401 (1998), http://dl.acm.org/citation.cfm?id=520809.796049
8. Ekman, P., Friesen, W.: Facial action coding system: A technique for the measurement of facial movement. Consulting Psychologists Press (1978)
9. Essa, I.A., Pentland, A.P.: Coding, analysis, interpretation, and recognition of facial expressions. IEEE Transactions on Pattern Analysis and Machine Intelligence 19(7), 757–763 (1997)

10. Heusch, G., Rodriguez, Y., Marcel, S.: Local binary patterns as an image preprocessing for face authentication. In: 7th International Conference on Automatic Face
and Gesture Recognition, FGR 2006, pp. 6–1 (April 2006)
11. Jolliffe, I.T.: Principal Component Analysis. Springer (2002)
12. Kawulok, M., Szymanek, J.: Precise multi-level face detector for advanced analysis
of facial images. IET Image Processing 6(2), 95–103 (2012)
13. Lanitis, A., Taylor, C.J., Cootes, T.F.: Automatic interpretation and coding of
face images using flexible models. IEEE Trans. on Pattern Analysis and Machine
Intelligence 19(7), 743–756 (1997)
14. Liao, S., Fan, W., Chung, A.C.S., Yeung, D.Y.: Facial expression recognition using
advanced local binary patterns, tsallis entropies and global appearance features.
In: 2006 IEEE International Conference on 2006 IEEE International Conference
on Image Processing, pp. 665–668 (October 2006)
15. Lien, J.J.J., Kanade, T., Cohn, J.Y., Li, C.: Detection, tracking, and classification of action units in facial expression. Journal of Robotics and Autonomous
Systems 31, 131–146 (1999)
16. Mase, K.: An application of optical flow – extraction of facial expression. In: IAPR
Workshop on Machine Vision Applications, pp. 195–198 (1990)
17. Ojala, T., Pietikäinen, M., Harwood, D.: A comparative study of texture measures
with classification based on featured distributions. Pattern Recognition 29(1), 51–59
(1996), http://www.sciencedirect.com/science/article/pii/0031320395000674
18. Ojala, T., Pietikäinen, M., Mäenpää, T.: A generalized local binary pattern operator for multiresolution gray scale and rotation invariant texture classification. In:
Singh, S., Murshed, N., Kropatsch, W.G. (eds.) ICAPR 2001. LNCS, vol. 2013,
pp. 397–406. Springer, Heidelberg (2001),
http://dl.acm.org/citation.cfm?id=646260.685274
19. Pantic, M., Rothkrantz, L.J.M.: Automatic analysis of facial expressions: The state
of the art. IEEE Trans. Pattern Anal. Mach. Intell. 22(12), 1424–1445 (2000),
http://dx.doi.org/10.1109/34.895976
20. Pietikäinen, M., Zhao, G., Hadid, A., Ahonen, T.: Computer vision using local
binary patterns 40, 13–9 (2011)
21. Shan, C., Gong, S., McOwan, P.W.: Facial expression recognition based on local
binary patterns: A comprehensive study. Image Vision Comput. 27(6), 803–816
(2009), http://dx.doi.org/10.1016/j.imavis.2008.08.005
22. Turk, M.A., Pentland, A.P.: Face recognition using eigenfaces. In: Proceedings of
IEEE Computer Society Conference on Computer Vision and Pattern Recognition,
CVPR 1991, pp. 586–591 (June 1991)

Detection of Tuberculosis Bacteria in Sputum Slide Image Using Morphological Features

Zahoor Jan[(✉)], Muhammad Rafiq, Hayat Muhammad, and Noor Zada

Department of Computer Science, Islamia College University
(Chartered University),
Peshawar, Pakistan
zahoor.jan@icp.edu.pk, rafiqmscs@hotmail.com,
{hayatvu,noorzadamohmand}@yahoo.com

Abstract. Automatic finding of tuberculosis bacteria is a medicinal imaging issue that includes the utilization of machine vision strategies. The manual technique for the detection of tuberculosis bacteria is costly, required much time and qualified persons to prevent errors. In this paper a new detection technique is presented. This technique is based on morphological features of bacteria object. The identification process is carried out using eccentricity, bounding box, area and aspect ratio *abstract* environment.

Keywords: Morphological features · Aspect ratio · HSV · TB bacteria detection

1 Introduction

In recent days Tuberculosis (TB) is the leading one of the infectious disease to cause death in developing countries. It influences fundamentally the lungs; this kind of tuberculosis is called pulmonary tuberculosis. As per WHO, TB is at number two after HIV/AIDS as the top murdering disease. More than 95% of TB death rate is in middle and low income countries and it is the top cause of death in women with age from 15 to 44. About sixty percent TB's patients are in Asia, and it is increasing rapidly [16].

The key difficulties are diagnosing TB and the introductory screening. The two principle techniques for screening of its bacteria (Mycobacterium) are fluorescent microscopy and bright field microscopy. In fluorescent microscopy the understand-ing's sputum smears specimens are stained utilizing Auramine.o, and in bright field microscopy they are stained with ZN (Ziehl-Neelsen) [1].

Fluorescent microscopes are costly and that is why they are used less. The most common and cheaper method of screening is bright field or ZN-staining technique which is the part of our research.

© Springer International Publishing Switzerland 2015
S. Kozielski et al. (Eds.): BDAS 2015, CCIS 521, pp. 408–414, 2015.
DOI: 10.1007/978-3-319-18422-7_36

2   Literature Review

Utilization of inadequate equipment, inexperienced staff and extra time has proved the manual method as non effective framework for diagnosing tuberculosis. A few techniques are utilized for the automation of the system which has increased the speed and sensitivity. M.K Osman et al. [4] attempted to solve the real challenges in image processing, which are bright and contrast stretching discussed by Jadhav Mukti, and Kale K.V [8]. Despite the fact that the improvement procedure enhances the visual appearance of the image however it has not finished up the detection of bacilli objects or the numbering of bacilli. Method [11] have combined the clustering and threshold-ing techniques which extends the previous work of Osman by studying and analyzing different color models such as HSI, RGB, and CY in clustering the bacilli objects. It shows a sensitivity ratio of 81.44%. R.A Raof et al. presented a method which examines the color attribute of the tubercle bacilli for the purpose to get accurate threshold value [6] and by using this threshold value the bacilli are filtered out from the image. Bayesian segmentation uses the probability of pixels as TB objects using prior knowledge of ZN stained colors and combine it with shape analysis [12].

The automatic identification of tuberculosis performed by Harshal Talati et al. focused around two stages, identification of objects by color, and identification by shape [14]. A Heuristic approach is used to classify the bacilli in a tree. Method presented in [13] is based on color segmentation, and the next step is the ex-traction of features that are movement invariant from sample images. Its recognition rate is 98.17%. Manuel G. Forero et al. technique is based on the heuristic acknowledge resulted from the bacilli shapes [2]. A classification tree was made using compactness, eccentricity, and the Hu's descriptors to identify an object as positive or negative tuberculosis.

The approach presented in [6] is based on Hue color component and it segments the TB bacilli by using an adaptive choice from the hue range. Thread length, width and area of the bacilli are the parameters that are used for identification of tuberculosis bacilli.

The analysis method of ZN stained images work on the images enhanced by contrast stretching and unsharp filtering [7]. By observing the statistical measures the quality of the images, the enhancement is carried out and the contrast stretching is proved as best enhancement technique for the purpose of segmentation and for identification.

The method presented by Osman et al. is based on the combination of thresholding and clustering process [5]. Global thresholding eliminates the unwanted pixels and K-mean clustering is used to classify the possible TB pixels. The CY model produced higher performance by achieving sensitivity, accuracy, specificity, and Jacquard coefficient rates up to 81.44%, 99.41%, 99.71%, and 0.641 % respectively.

An automatic system is developed by using neural networks in order to identify positive or negative TB objects [15]. Eccentricity and compactness features are obtained using segmentation and morphological operations. Two inputs, two outputs and 15 hidden layers neural network produces an accuracy rate of 88%.

The proposed method loads an RGB image of ZN stained sputum slide image and convert it into HSV color model. It consists of several steps which are explained in the Fig. 1.

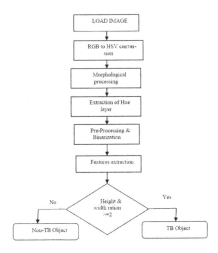

Fig. 1. Block Diagram of the Proposed Method

3   Methods

3.1   RGB to HSV Conversion

RGB color models consist of three colors, red, green and blue. While HSV or hue saturation model represents three values, i.e. The Hue represents the color range from 0° to 360° , saturation represents the strength of the color, and value means the bright-ness and darkness in the color. Conversion from RGB to HSV is explained by Gonzales and woods [3], as following.

$$
H = \begin{cases} \cos^{-1}\left[\dfrac{\frac{1}{2}[(R-G)+(R-B)]}{\sqrt{(R-G)^2+(R-B)(G-B)}}\right], if B \le G, \\[4ex] 2\pi - \cos^{-1}\left[\dfrac{\frac{1}{2}[(R-G)+(R-B)]}{\sqrt{(R-G)^2+(R-B)(G-B)}}\right], if B > G, \end{cases}
\tag{1}
$$

$$
S = 1 - \frac{3}{R+G+B}[min(R,G,B)].
\tag{2}
$$

$$
V = \frac{1}{3}(R+G+B).
\tag{3}
$$

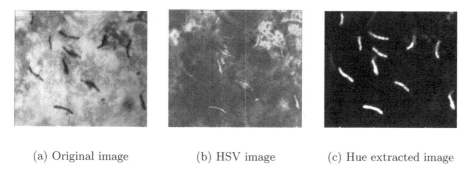

(a) Original image   (b) HSV image   (c) Hue extracted image

Fig. 2. RGB to HSV Conversion and Hue extraction

3.2 Morphological Processing and Extraction of Hue layer

During conversion the objects don't appear clearly visible, so morphological operations are applied on it to fill out the holes and smoothen the edges of the objects. As bacilli appear red in shape, therefore the main concerned in the HSV model is the hue layer. The hue layer is extracted from the HSV format of the image in order to identify the bacilli and red color lies at angle 0° to 30° .

3.3 Thresholding

Thresholding technique [9] is applied to convert the image into binary form, which contains only two values; 0 and 1. Value 0 represent black pixel and 1 represent white pixel. Threshold value that works better in this case is from 168 to 178. In the resultant image the TB bacilli appear in white color while the background appears in black color. Morphological operation closing is applied to the resultant image in order to join the broken objects.

Edge detection is used to minimize the quantity of data in an image, while maintaining the structural characteristics which can be used for further image processing. Several edge detection techniques where tested on sample image but Prewitt edge detection technique [10] proved best results. The image produced by canny edge detector is used as a separate image for further processing. This image and the previous binary image are added into a single image. In this process the pixel value of each input image is added to the corresponding value of each other input pixel and returns the sum in the corresponding pixel of the output image.

3.4 Features Extraction

This step involves the extraction of features which can be used for the further identification of tuberculosis bacteria. All the objects in the image are labeled in order to get its properties. These properties include eccentricity, bounding box, and area of the objects.

(a) Thresholded image   (b) Edge detection   (c) Addition of a and b

Fig. 3. Thresholding, Edge detection, and Addition

Area of an object is the product of its height and width. It is calculated to avoid small unwanted pixels. Eccentricity is defined as the ratio between minor axis and major axis of an elliptical shape [13].

To detect the rod shape like structure of TB bacteria eccentricity is calculated, if the eccentricity of an object is greater than eight then the object is considered as tuberculosis bacteria otherwise it is rejected. The value of eccentricity is calculated from the following equation [13].

$$e = \sqrt{1 - (\frac{b}{a})^2}. \tag{4}$$

Where e=eccentricity, a= length of major axis and b= length of minor axis. Bound-ing box is used to crop and select the region of interest. Hough transform is used to rotate the object at angle 90° . It detects the lines in the image and then returns the angles of those lines. Using these angles the objects are rotated in various directions.

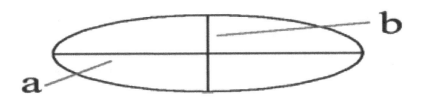

Fig. 4. Major and minor axis of an Ellipse

4   Experimental Results

Finally when the objects are rotated, the aspect ratio is calculated for each object in the sample image.

$$AR = \frac{Shape'sheight}{Shape'swidth}. \tag{5}$$

Based on various experiments a limit of acceptance is set for this ratio. If the ratio of is greater than or equal to 2 then the object is considered as TB bacteria otherwise it is rejected. Lists of possible of objects in a sample image are labeled by this ratio as shown in the Fig. 5.

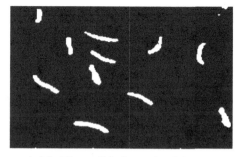

(a) All possible bacteria objects

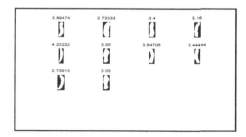

(b) Accepted objects labeled with their aspect ratios

Fig. 5. TB bacteria objects detection and Labeling

All the images were obtained from the Pathology Department, Hayatabad Medical Complex Peshawar. A total of 100 positive and 10 negative sample images were used for experimental purposes. The positive samples proved correct results as performed using manual technique. We have also used live images taken using a microscope having a camera called Olympus microscope. The propose method was implemented using a personal laptop and Matlab' Image Processing Toolbox software. The accuracy rate is calculated by the total number of objects in ZN stained image and the total number of objects detected in the final step, so it returns a value of about 90% which is more than I. Siena et al.[13].

$$Accuracy(\%) = \frac{Total\ No.\ of\ objects\ in\ sample\ image}{No.\ of\ detected\ TB\ bacteria} \times 100. \qquad (6)$$

References

1. Balakrishna, J., Shahapur, P.R., Chakradhar, P., Saheb, H.S.: Comparative study of different staining techniques – Ziehlneelsen stain, Gabbet's stain, fluorochrome stain for detecting of mycobacterium tuberculosis in the sputum (2013)
2. Forero, M., Cristobal, G., Alvarez-Borrego, J.: Automatic identification techniques of tuberculosis bacteria. In: Optical Science and Technology, SPIE's 48th Annual Meeting, pp. 71–81. International Society for Optics and Photonics (2003)
3. Gonzalez, R.C., Woods, R.E., Eddins, S.L.: Digital image processing using matlab. Pearson Prentice Hall, Upper Saddle River (2004)
4. Khusairi, O.M., Yusoff, M., Hasnan, J.: Colour image enhancement using bright and dark stretching techniques for tissue based tuberculosis bacilli detection (2009)
5. Khusairi, O.M., Yusoff, M., Hasnan, J.: Combining threshoding and clustering techniques for mycobacterium tuberculosis segmentation in tissue sections. Australian Journal of Basic and Applied Sciences 5(12) (2011)
6. Makkapati, V., Agrawal, R., Acharya, R.: Segmentation and classification of tuberculosis bacilli from ZN-stained sputum smear images. In: IEEE International Conference on Automation Science and Engineering, CASE 2009, pp. 217–220. IEEE (2009)
7. Mukti, J., Kale, K.V.: Analysis of zn-stained sputum smear enhanced images for identification of mycobacterium tuberculosis bacilli cells. Analysis 23(5) (2011)
8. Mukti, J., Mamta, B., Arjun, M., Kale, K.V.: Mycobacterium tuberculosis bacilli cells identification using moment invariant. International Journal of Machine Intelligence ISSN, 975–2927 (2011)
9. Otsu, N.: A threshold selection method from gray-level histograms. Automatica 11(285-296), 23–27 (1975)
10. Prewitt, J.M.S.: Object enhancement and extraction. Picture Processing and Psychopictorics 10(1), 15–19 (1970)
11. Raof, R.A.A., Mashor, M.Y., Ahmad, R.B., Noor, S.S.M.: Image segmentation of ziehl-neelsen sputum slide images for tubercle bacilli detection (2011)
12. Sadaphal, P., Rao, J., Comstock, G.W., Beg, M.F.: Image processing techniques for identifying mycobacterium tuberculosis in ziehl-neelsen stains (short communication). The International Journal of Tuberculosis and Lung Disease 12(5), 579–582 (2008)
13. Siena, I., Adi, K., Gernowo, R., Miransari, N.: Development of algorithm tuberculosis bacteria identification using color segmentation and neural networks. International Journal of Video and Image Processing and Network Security 12(4), 9–13 (2012)
14. Talathi, H., Patel, J., Laha, K., Kundargi, J.M.: Identification of tuberculosis bacilli using image processing (2012)
15. Veropoulos, K., Campbell, C., Learmonth, G., Knight, B., Simpson, J.: The automated identification of tubercle bacilli using image processing and neural computing techniques, pp. 797–802. Springer (1998)
16. WHO: WHO tb fact sheet (2014),
   http://www.who.int/mediacentre/factsheets/fs104/en/index.html

Automatic Medical Objects Classification Based on Data Sets and Domain Knowledge

Przemyslaw Wiktor Pardel[1(✉)], Jan G. Bazan[1], Jacek Zarychta[2], and Stanislawa Bazan-Socha[3]

[1] Interdisciplinary Centre for Computational Modelling, University of Rzeszów, Pigonia 1 Str., 35 - 310 Rzeszów, Poland
{ppardel,bazan}@ur.edu.pl
[2] Radiology Consultant, Department of Pulmonology, Pulmonary Hospital, Gladkie 1 Str., 34-500, Zakopane, Poland
jzar@mp.pl
[3] II Department of Internal Medicine, Jagiellonian University Medical College, Skawinska 8 Str., 31-066 Kraków, Poland
mmsocha@cyf-kr.edu.pl

Abstract. This paper describes the approach for automatic identifying organs from a medical CT imagery. Main assumption of this approach is the use of data sets and domain knowledge. We apply this approach to automatic classification of chest organs (trachea, lungs, bronchus) and present the results to demonstrate their usefulness and effectiveness. The paper includes the results of experiments that have been performed on medical data obtained from II Department of Internal Medicine, Jagiellonian University Medical College, Krakow, Poland. The experimental results showed that the approach is promising and can be used in the future to support solving more complex medical problems.

Keywords: CT images · Concept approximation · Classifiers · Decision trees · Medical object recognition · Object classification · Domain knowledge · Organs identifying

1 Introduction

An automatic identification of medical objects visualized by Computed Tomography (CT) imagery (e.g., organs, blood vessels, bones, etc.), without any doubt, could be useful, to support solving many complex medical problems using computer tools. This paper describes the approach for automatic identification of organs from medical CT imagery based on data sets and domain knowledge. Presented subject is concerned with substantial computation problem. It employs classifiers building for image data sets, where a *classifier* is an algorithm which enables us forecasting repeatedly on the basis of accumulated knowledge in new situations (see, *e.g.*, [2] for more details). Our approach is based on a two-level classifier. On the lower level, our approach uses a classical classifier based on a decision tree that is calculated on the basis of the local discretization (see,

© Springer International Publishing Switzerland 2015
S. Kozielski et al. (Eds.): BDAS 2015, CCIS 521, pp. 415–424, 2015.
DOI: 10.1007/978-3-319-18422-7_37

e.g., [11,3]). This classifier is constructed and based on the features extracted from images using methods known from literature (see [8,5] for more details). At a higher level of our two-level classifier, a collection of advisers works that is able to verify actions performed earlier by the lower-level classifier. This is possible by using domain knowledge injected to advisers. Each of the adviser is constructed as a simple algorithm based on a logical formula, that on input receives selected information extracted from a tested image and a decision returned by the lower-level classifier, and the output returns confirmation or negation for the suggestion generated by the lower-level classifier. It consists in the fact, that in a situation where the decision taken by the lower-level classifier, is clearly incompatible with domain knowledge, the adviser suggests to refrain from taking a decision. Thanks to this, increases the accuracy of such the two-level classifier, with a slight decrease in its coverage. To illustrate the method and to verify the effectiveness of presented classifiers, we have performed several experiments with the data sets obtained from Second Department of Internal Medicine, Collegium Medicum, Jagiellonian University, Krakow, Poland.

This paper is organized as follows. In the section 2, we present the problem of detecting organs from medical images. Some methods of features extraction that are used in order to construct attributes for the low-level classifier were presented in section 3. Next, in the section 4 we describe a low-level classifier used to the automatic medical object identification. Finally, we present the complete structure of two-level classifier and results of experiments performed on medical data sets for the automatic classification of chest organs (see section 5).

2   Organs Detection

The automatic detection, recognition and segmentation of anatomical objects in three-dimensional medical images, is an important and extremely challenging task, for a number of reasons, *e.g.*, objects may have no clear boundaries; there may be similar structures in close vicinity of an object; the constellation of objects and their neighborhoods can vary widely from individual to individual and some objects or parts of objects may be missing, and scans obtained in clinical practice often contain pathology (see, *e.g.*, [16,17] for more details).

In clinical applications, automatic medical object extraction is often a necessary preprocessing step for three-dimensional image reconstruction, quantitative analysis and computer-aided diagnosis. Due to the wide variety of shapes, the complexity of the topology, the presence of noise in a complicated background, and the diversity of imaging techniques, extracting medical objects automatically and accurately is a very difficult task [15].

Medical image analysis is one of the areas of computer vision where domain knowledge plays a very important role, because localized pixel information obtained from CT images is often ambiguous and unreliable [7]. The history of knowledge-based medical image analysis is older than the history of practical usage of CT imaging. One of the early studies in knowledge based medical image analysis was done by Harlow and Eisenbeisc [6] on radiographic image segmentation, when CT imaging was not yet available in hospitals. They proposed a top-down control system using a trees structured model description containing knowledge about locations and spatial relations of parts/organs of the human body. In his thesis work, Selfridgec [14] discussed image understanding systems in general and divided the causes of difficulties into problems of model selection, segmentation techniques, and parameter setting [7].

We conclude that the automatic detection of organs is the first step to understand medical images and it is necessary to begin the process of proper medical diagnosis support. To understand of the CT image correctly, a computer should detect and recognize all medical objects located on the image by using domain knowledge. The knowledge about objects located in the medical image, allows the correct identification of areas related to various medical problems.

In this work, we describe the approach to automatic classification of chest organs (trachea, lungs, bronchus) by using domain knowledge and a classifier.

3 Features Extraction

Making a detailed medical images analysis is necessary to move from the analysis of individual pixels to analyze objects located on the image. The image should be divided into homogeneous and disjoint areas that represent the objects or their parts. This process is called segmentation. Segmentation criteria proposed by Haralick and Shapiro [1] show that this not only an easy process but also very demanding. After completion of the segmentation process, may start the process of acquiring the features of the objects.

There is no "*ideal set of features*" which characterize the object. Features are selected individually depending on the recognized objects. In the computer analysis of the images, extracted features from the image, can be assigned to one of the categories, such as nontransformed structural characteristics (eg. moments, power, amplitude information, energy, etc.),transformed structural characteristics (eg. frequency and amplitude spectra, subspace transformation methods, etc.), structural descriptions (formal languages and their grammars, parsing techniques, and string matching techniques) and graph descriptors (eg. attributed graphs, relational graphs, and semantic networks) described in detail in [8] and [5]. In this publication we call these features as Low-Level Features (LLF). To prepare experiments presented in section 5 we use standard features such as surface area, width, height, cover regions and more advanced such as circuit of the object, the coefficient of thickness or shape. In total, for the purposes of the experiments we define 18 LLF features.

To understand the image, it is also necessary to define the additional features that will define the acquired domain knowledge from experts. We call these features Domain Knowledge Features (DKF). DKF can be assigned to one of the categories, such as:

- features used to describe domain knowledge about the number of objects that surround an analyzed object,
- features used to describe domain knowledge about the distance from analyzed object to surrounding objects,

– features used to describe domain knowledge about the size of objects that surround an analyzed object,
– features used to describe domain knowledge about position of an object.

To prepare experiments presented in section 5, we create features from each category eg. the number of objects on the right side of the analyzed object. In total, for the purposes of the experiments we define 13 DKF features.

The extracted features allow us to define properties of objects observed on medical images. Such properties are further represented by data sets and are used to create classifiers (see section 4). The typical model of data employed to classifier construction is a rectangular data table, which in the rough set theory [13] is called *a decision table*. The decision table consists of finite number of the rows called objects and the columns called the attributes. The rows represent the information about a single object instance and for that reason are called the objects (in our approach, any object represents a medical object observed on some image). The columns describe the features of the objects expressed as numerical or textual values, and are called the attributes. One of the columns (usually the last one) represents the decision class membership of an object and is named a decision attribute.

4   Low-Level Classifier

Classifiers also known in literature as *decision algorithms, classifying algorithms* or *learning algorithms* may be treated as constructive, approximate descriptions of concepts (decision classes). These algorithms constitute the kernel of *decision systems* that are widely applied in solving many problems occurring in such domains as *pattern recognition, machine learning, expert systems, data mining* and *knowledge discovery* (see, *e.g.*, [10,9,2] for more details).

In literature there can be found descriptions of numerous approaches to constructing classifiers, which are based on this way can often be not appropriate to classify unseen cases. For instance, if we have a decision table, where the number of values is high for some attributes, then there is a very low chance that a new object is recognized by rules generated directly from this table, because the attribute value vector of a new object will not match any of these rules. Therefore, for decision tables with such numeric attributes some discretization strategies are built to obtain a higher quality classifiers. This problem is intensively studied and we consider such paradigms of machine learning theory as *classical and modern statistical methods, neural networks, decision trees, decision rules*, and *inductive logic programming* (see, *e.g.*, [10,9,2] for more details).

In this section, we present a method of classifier construction that we have used for a decision table constructed on the basis of features defined in the previous section.

One of the most popular methods for classifiers construction is based on learning rules from examples (see, e.g., [13,2] for more details). Unfortunately, the decision rules constructed in discretization methods developed by Hung S. Nguyen (see [11,3]). In this paper we use local strategy of discretization (see [3]).

One of the most important notion of this strategy is the notion of *a cut*. Formally, the cut is a pair (a, c) defined for a given *decision table* $\mathbf{A} = (U, A \cup \{d\})$ in Pawlak's sense (see [13]), where $a \in A$ (A is a set of attributes or columns in the data set) and c, defines a partition of V_a into *left-hand-side* and *right-hand-side interval* (V_a is a set of values of the attribute a). In other words, any cut (a, c) is associated with a new binary attribute (feature) $f_{(a,c)} : U \rightarrow \{0, 1\}$ such that for any $u \in U$:

$$f_{(a,c)}(u) = \begin{cases} 0 & \text{if } a(u) < c \\ 1 & \text{otherwise} \end{cases} \tag{1}$$

Moreover, any cut (a, c) defines two templates, where a template we understand as a description of some set of objects. The first template defined by a cut (a, c) is a formula $T = (a(u) < c)$, while the second pattern defined by a cut (a, c) is a formula $\neg T = (a(u) \geq c)$.

In this paper, the quality of a given cut is computed as a number of objects pairs discerned by this cut and belonging to different decision classes. It is worth noticing that such quality can be computed for a given cut in $O(n)$ time, where n is the number of objects in the decision table (see, *e.g.*, [3]). The quality of cuts may be computed for any subset of a given set of objects.

In local strategy of discretization, after finding the best cut and dividing the object set into two subsets of objects (matching to both templates mentioned above for a given cut), this procedure is repeated for each object set separately until some stop condition holds. In this paper, we assume that the division stops when all objects from the current set of objects belong to the same decision class. Hence, the local strategy can be realized by using *decision tree* (see [3] and Fig. 1).

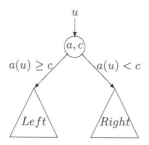

Fig. 1. The decision tree used in local discretization

The decision tree computed during local discretization can be treated as a classifier for the concept C represented by decision attribute from a given decision table $\mathbf{A}$. Let u be a new object and $\mathbf{A}(T)$ be a subtable containing all objects matching to template T defined by the cut from the current node of a given decision tree (at the beginning of algorithm work T is the template defined by the cut from the root). We classify object u starting from the root of the tree as follows:

Algorithm. *Classification by decision tree* (see [3])
Step 1. If u matches template T found for $\mathbf{A}$
     then: go to subtree related to $\mathbf{A}(T)$
     else: go to subtree related to $\mathbf{A}(\neg T)$.
Step 2. If u is at the leaf of the tree then go to 3
     else: repeat 1-2 substituting $\mathbf{A}(T)$ (or $\mathbf{A}(\neg T)$) for $\mathbf{A}$.
Step 3. Classify u using decision value attached to the leaf

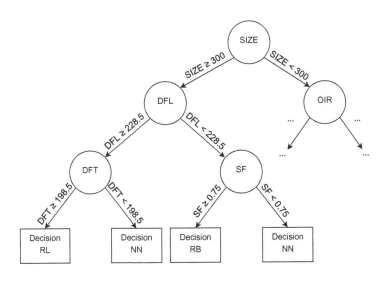

Fig. 2. The decision tree in organs detection

Figure 2 presents part of sample decision tree, computed for the problem of medical object classification. For example, for a medical object with size 560, distance from left edge (DFL) of CT image 200 and object shape factor (SF) equal 0.9, we follow from the root of the tree, down to the left subtree, as the object suits a template $SIZE \geq 300$. In the next step we tread a right tree $(DFL < 228.5)$ and again a left tree $(SF \geq 0.75)$, which consists of one node, called leaf, where we stop. The fitting path indicates that this object is classified as right main bronchi (Decision RB).

5   Two-Level Classifier and Experiments

To verify the effectiveness of classifiers based on our approach, we have implemented the two-level classifier in the IMPLA (Image Processing Laboratory), which is a continuation of the RSES-lib library (forming the kernel of the RSES system [4]), in the field of image processing. The IMPLA has been developed recently in Interdisciplinary Centre for Computational Modelling, University of Rzeszow, Poland.

Our experiments were carried out on the data obtained from the clinical hospital Jagiellonian University Medical College in Kraków (the patients were diagnosed with asthma). The entire data set counted 26 patients (19 woman, 7 man). The average age of patients was 58.12 years (st.dev. 6.78 years, age range from 47 to 70 years). In all patients, volumetric CT torso scans were performed at both full inspiration and expiration with using 16-channel multi-detector CT scanner Toshiba (manufacturer's model name: Aquilion). The acquired data were reconstructed using a kernel (FC86) with 1 mm increments. Images were stored in the Digital Imaging and Communications in Medicine (DICOM) format. For each patient was taken 300 to 400 images (full inspiration) with a resolution of 512x512 pixels. The total size of the data set for the experiment count 9655 CT images.

From all images we select every fifth image (20% of all images, 5mm increments) to pre-processing. As a result of the segmentation process, we acquired 7491 objects for experiments. For all the objects we set LLF and DKF features, further all objects are classified by an expert to one of the 7 classes (chest organs) presented in the table 1.

Table 1. Object classes

Class	Object	Number
TR	Trachea	671 (8,96%)
RL	Right lunge	1621 (21,64%)
LL	Left lunge	1616 (21,57%)
RB	Right main bronchi	190 (2,54%)
LB	Left main bronchi	211 (2,82%)
LL+RL	Object by gluing the left and right lungs	55 (0,73%)
OT	Other objects	3127 (41,74%)

The entire data set was divided 20 times randomly into two sets - a set with training data and a set with test data (around 70% of the data getting into a training set - 18 patients, other (around 30%) into test set - 8 patients). All the experiments we conducted on these datasets.

In the first experiment, we used only the LLF features described in the section 3. With using training data we built a classifier (widely described in section 4), which has been tested on test data. We designed a classifier to the automatic classification of chest organs.

In the second experiment (approach concept diagram presented on Fig. 3), we used the LLF features and in addition, we used an adviser working on DKF (Domain Knowledge Adviser, DKA). DKA suggest decisions based on domain knowledge eg. *"Left lung is located on the left side of medical image"*, *"Object located on the left side of medical image is probably not a right lunge"*. We prepare 15 DKA for all chest organs. Verification was followed on the basis of the DKF futures. Advisers are divided into two groups:

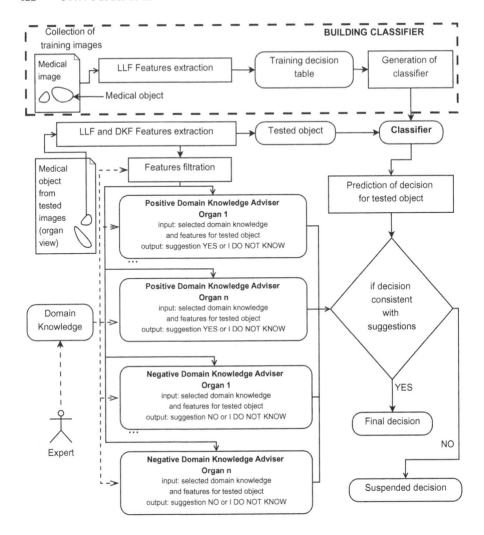

Fig. 3. Two-level classifier based on data sets and a domain knowledge

- Advisers to advise on YES eg. *"yes, this is probably the left lung"* (6 DKA),
- Advisers to advise on NO eg. *"no, this is probably not the left lung"* (9 DKA).

In this approach classification decision is dependent on suggestions of domain knowledge advisers. Advisers suggest what should be a decision (YES advisers) or suggested what should not be a decision (NO advisers). If any of the advisors suggested otherwise than the classifier (in some sense, the low-level classifier), decision was suspended.

We presented the results of the experiments in the table 2.

By using domain knowledge we have obtained an improvement in the automatic classification of each chest organ. The largest increase in the accuracy

Table 2. Experiments results

	Experiment 1 (LLF)			Experiment 2 (DKA)			Improvement	
Object	Accuracy	St.Dev.	Coverage	Accuracy	St.Dev.	Coverage	Accuracy	St.Dev.
TR	94,00%	3,61%	100%	**98,90%**	1,02%	94,58%	**+4,91%**	-2,59%
RL	97,64%	0,92%	100%	**98,32%**	0,61%	99,27%	**+0,68%**	-0,32%
LL	97,31%	0,98%	100%	**98,44%**	0,50%	98,82%	**+1,13%**	-0,48%
RB	76,77%	6,00%	100%	**88,19%**	4,06%	85,52%	**+11,41%**	-1,94%
LB	78,55%	6,04%	100%	**85,42%**	5,24%	88,35%	**+6,87%**	-0,80%
LL+RB	87,95%	24,01%	100%	**97,92%**	3,50%	94,09%	**+9,96%**	-20,51%
OT	94,73%	1,76%	100%	**96,80%**	1,16%	97,16%	**+2,07%**	-0,60%

(+11,41%) we have obtained on the right main bronchi objects. The smallest (+0,68%) on the right lunge. Noteworthy is the fact that all objects are classified with the accuracy more than 85%. In three cases the accuracy was greater than 98%. In addition, the use of domain knowledge improved the stability of the automatic classification. This can be seen by analyzing the standard deviation, for instance the stability of the the TR improved by 2,59%. In all tests DKA made the decision 15943 times - 15659 times it was the correct decision (98,22%).

All the decisions taken by the DKA pause the classifier decision where decisions are different. This is the direct reason for the decline coverage of the analyzed objects.

6 Conclusions and Further Works

The results of experiments performed on medical data sets indicate that the presented approach seems to be promising. The use of domain knowledge significantly improved the quality of the medical object identification. The next steps will focus on the use of time dependencies between medical images (object tracking in time) and the addition of a classifier resolving conflicts between advisers.

The presented approach can be used in the future to support solving more complex medical problems. We plan to use the results of research, among other things, to treatment of an asthmatic airway remodeling (see, e.g., [12] for more details).

Acknowledgement. This work was partially supported by the Polish National Science Centre grant DEC-2013/09/B/ST6/01568 and by the Centre for Innovation and Transfer of Natural Sciences and Engineering Knowledge of University of Rzeszów, Poland.

References

1. Image Segmentation Techniques, vol. 0548 (1985)
2. Bazan, J.: Hierarchical classifiers for complex spatio-temporal concepts. In: Peters, J.F., Skowron, A., Rybiński, H. (eds.) Transactions on Rough Sets IX. LNCS, vol. 5390, pp. 474–750. Springer, Heidelberg (2008)
3. Bazan, J.G., Nguyen, H.S., Nguyen, S.H., Synak, P., Wróblewski, J.: Rough set algorithms in classification problems. In: Polkowski, L., Lin, T.Y., Tsumoto, S. (eds.) Rough Set Methods and Applications: New Developments in Knowledge Discovery. STUDFUZZ, vol. 56, pp. 49–88. Springer, Heidelberg (2000)
4. Bazan, J.G., Szczuka, M.: The Rough Set Exploration System. Transactions on Rough Sets 3400(3), 37–56 (2005)
5. Cytowski, J., Gielecki, J., Gola, A.: Digital Medical Imaging. Theory. Algorithms. Applications. Problemy Współczesnej Nauki: Informatyka. Akademicka Oficyna Wydawnicza EXIT (2008) (in Polish)
6. Harlow, C.A., Eisenbeis, S.A.: The analysis of radiographic images. IEEE Transactions on Computers C-22(7), 678–689 (1973)
7. Kobashi, M., Shapiro, L.G.: Knowledge-based organ identification from ct images. Pattern Recognition 28(4), 475–491 (1995)
8. Meyer-Baese, A., Schmid, V.: Chapter 2 - feature selection and extraction. In: Meyer-Baese, A., Schmid, V. (eds.) Pattern Recognition and Signal Analysis in Medical Imaging, 2nd edn., pp. 21–69. Academic Press, Oxford (2014)
9. Michalski, R., et al. (eds.): Machine Learning, vol. I-IV. Morgan Kaufmann, Los Altos (1983, 1986, 1990, 1994)
10. Michie, D., Spiegelhalter, D.J., Taylor, C.C.: Machine learning, neural and statistical classification. Ellis Horwood Limited, England (1994)
11. Nguyen, H.S.: Approximate boolean reasoning: Foundations and applications in data mining. In: Peters, J.F., Skowron, A. (eds.) Transactions on Rough Sets V. LNCS, vol. 4100, pp. 334–506. Springer, Heidelberg (2006)
12. Niimi, A., Matsumoto, H., Takemura, M., Ueda, T., Nakano, Y., Mishima, M.: Clinical assessment of airway remodeling in asthma. Clinical Reviews in Allergy And Immunology 27(1), 45–57 (2004)
13. Pawlak, Z., Skowron, A.: Rudiments of rough sets. Information Sciences 177, 3–27 (2007)
14. Selfridge, P.G.: Reasoning about Success and Failure in Aerial Image Understanding. Reports // ROCHESTER UNIV NY. University of Rochester. Department of Computer Science (1981)
15. Shang, Y., Yang, X., Zhu, L., Deklerck, R., Nyssen, E.: Region competition based active contour for medical object extraction. Computerized Medical Imaging and Graphics 32(2), 109–117 (2008)
16. Staal, J., van Ginneken, B., Viergever, M.A.: Automatic rib segmentation and labeling in computed tomography scans using a general framework for detection, recognition and segmentation of objects in volumetric data. Medical Image Analysis 11(1), 35–46 (2007)
17. Zhou, S.K.: Discriminative anatomy detection: Classification vs regression. Pattern Recognition Letters 43(0), 25–38 (2014), (ICPR2012 Awarded Papers)

Spatial Data Analysis

Interpolation as a Bridge Between Table and Array Representations of Geofields in Spatial Databases

Piotr Bajerski[✉]

Institute of Informatics, Silesian University of Technology
Akademicka 16, 44-101 Gliwice, Poland
piotr.bajerski@polsl.pl

Abstract. Development of database technology facilitates wider integration of diverse data types, which in turn increases opportunities to ask ad hoc queries, and gives new possibilities of declarative queries optimization. For more than a decade, work on supporting multidimensional arrays in databases has been carried out, which led to such DBMSs as rasdaman, SciDB and SciQL. However, the DBMSs lack the ability to handle queries concerning geographic phenomena varying continuously over space (called geofields) which were measured in irregularly distributed nodes (e.g. air pollution). This paper addresses this issue by presenting an extension of SQL making possible to write declarative queries referencing geofields, called geofield queries. Geofield query optimization opportunities are also shortly discussed.

Keywords: Spatial databases · Array databases · SQL · GIS · Geofield · Coverage · Interpolation

1 Introduction

We can observe how functionality migrates from GIS applications to databases. This allows for better data integration, ACID guarantees, ad hoc querying, and automatic optimization of processing, when a declarative language, such as SQL, is used. Design and implementation of conventional relational database management systems (RDBMSs) have been driven by business requirements, and, as a consequence, they only partially meet scientific community requirements [16]. Especially during remote sensing and simulations, a lot of data is gathered that is naturally stored as multidimensional arrays [12,17]. However, storing, querying and modifying such multidimensional arrays are not supported by conventional RDBMSs. Such problems led to the development of array DBMSs. The appearance and development of array DBMSs have been driven by attempts to make multidimensional arrays the first class citizens in databases, and to integrate multidimensional arrays with relations. Currently, there is work in progress on adding multidimensional arrays support to ISO SQL standard [12].

Phenomena in space can be conceptualized as fields, varying continuously over space, and/or objects, being discrete spatial entities. To emphasize connections with geographical space in the paper, following [9], the terms *geofield*

© Springer International Publishing Switzerland 2015
S. Kozielski et al. (Eds.): BDAS 2015, CCIS 521, pp. 427–436, 2015.
DOI: 10.1007/978-3-319-18422-7_38

and *geoobject* are used in place of the terms *field* and *object*. In the OGC specifications the term *coverage* is used as a synonym of *geofield* [6]. In the paper queries concerning geofields will be called *geofield queries*. Geofields are divided into quantitative geofields and qualitative geofields. A *quantitative geofield* is a function assigning elements of its domain values measured on interval or ratio scale. A *qualitative geofield* is a function assigning elements of its domain values measured on nominal or ordinal scale, e.g. numbers of intervals of a geofield values. The domain of geofield may be 2D or 3D spatial, 3D or 4D spatio-temporal or of a higher dimension. As far as spatial data are concerned, the most common is 2D spatial, however, in the last years significant progress has been done in spatio-temporal prediction methods [8], which allows to generalize concepts presented in the previous work [4, 5].

Geofields are naturally represented by multidimensional arrays, when they are results of remote sensing or simulations on a regular grid – such a representation will be called *array representation*. However, geofields values are also often measured in monitoring networks with irregularly placed nodes, and such results are naturally stored in classical database tables – such a representation will be called *table representation*. Computing answers to geofield queries usually requires finding geofields values in places in which they had not been measured. In the case of the array representation simple and fast interpolation methods (e.g. developed for rasters processing) can be used. However, in the case of the table representation, more sophisticated and computationally intensive methods, such as Kriging, are needed [7]. A formula used for estimation of geofield values using its table representation will be called *geofield (mathematical) model*.

In the traditional approach to processing table representations of geofields, a database is only used for storing their point measurements. As a consequence, geofields processing leads to unloading the measurements and using external tools to compute array representations of the geofields. The result arrays are usually stored in a file system for further processing. Array DBMSs solve the problem partially by allowing to load the computed array representations back to the database. However, they miss the opportunity to include a conversion between table and array representations in the query optimization process.

The main idea of the presented work consists in adding operations on geofields to SQL and taking them into account during the query optimization phase. Of these operations, the conversion between table and raster representations of geofields is of key importance, as it can easily dominate the query execution time [3].

The remainder of the paper is organized as follows. section 2 outlines the Peano relation and PNR representation, as these concepts are fundamental to previous work on geofield queries optimization and influence the extension of SQL for geofield queries described in section 3. Section 3 contains the main contribution of the paper. Section 4 shortly describes how the presented SQL extension can help to speed up geofield query processing. It summarizes previously published geofield optimization rules and presents new ideas connected with optimization of quantitative geofields. Section 5 consists of a survey of related work. The paper is concluded in section 6.

2   Peano Relation and PNR Representation

The concepts of the Peano relation and the PNR representation are presented in [5], while [4] discusses efficiency of generating the PNR representation of a qualitative geofield based on the PNR representation and a geofield variability model. A thorough discussion of the concepts and exhaustive experimental results for 2D qualitative geofields are given in [2]. This section quotes the most important informations on the Peano relation and the PNR representation, which are used in the rest of the paper.

The PNR representation is based on *discrete coordinates*, in which sides lengths of the elementary quadrants are used as units [2]. This resembles approach used in array DBMSs, however, all geofields generated in a given query share the same discrete coordinates, which simplifies computations. The elementary quadrant can be treated as a synonym of the cell in array terminology. The area in which values of the given geofield must be computed to answer a query is called *geofield context region*. Elementary quadrants are ordered by a Peano N space-filling curve, and their ranges are stored in a relation valued attribute. Such a representation is called *PNR representation*. The PNR representation of a geofield consists of the PNR representation of zones in which geofield model values belong to the same intervals.

A quantitative geofield has predefined virtual attributes: `location` and `value`. The attribute `location` provides spatial coordinates of the center of an elementary quadrant (array cell), and the attribute `value` – the geofield value computed for this location. A qualitative geofield has predefined attributes `interval`, `lBound`, `uBound` and `shape`. The attribute `interval` stores the interval number described by the given tuple, the attributes `lBound` and `uBound` store lower and upper bounds of the interval respectively, and the attribute `shape` stores Peano keys of all elementary quadrants approximating fragments of the computational space in which values of the geofield model fall into this interval.

3   SQL Extension for Geofield Queries

In this section, an extension of SQL for geofield queries is presented. It consists of a data type, called `geofield`, the `create geofield` statement, and a short description how they can be integrated with SQL dialects used in array DBMSs. Quantitative geofields are created as multidimensional arrays, so the arrays operations can be used to process them. The PNR representation of qualitative geofields may be interpreted as a quadtree (or an octree, etc.) representation of a multidimensional array, so operations on such arrays may be easily adjusted to the PNR representation. The main difference between processing multidimensional arrays in array DBMSs (see section 5) and the approach presented in this paper is the possibility of generating on-the-fly only the needed parts of geofields during query processing in the presented approach.

Let us assume, that we have a table storing point measurements of ambient air pollution:

```
create table air_pollution (
    measurement_id integer primary key,
    pollutant_code varchar not null,
    measurement_location geometry not null,
    measurement_time timestamp not null,
    measured_value double not null,
    measurement_quality varchar not null,
    check (measurement_quality in ('ERROR', 'POOR', 'GOOD'))
);
```

3.1 Create Geofield Statement

The most basic and important part of the SQL extension is the create geofield statement:

```
create [virtual | materialized] [qualitative | quantitative]
            geofield <geofield_name> as
select <column_name | expression> as measurement_location,
       [<column_name | expression> as measurement_time,]
       <column_name | expression> as measurement_value
from <measurement_table_name>
where [<conditions_on_measurements> and]
      CRS = <CRS_code> and
      interpolation = <interpolation_parameters> and
      [square_classifier = <square_classifier_parameters> and]
      [intervals = <intervals_boundary_definion>] and
      [context_region = <geoobject | subquery | array_range> ]
      [resolution = <integer | double <unit>>];
```

The names measurement_location, measurement_time and measurement_value are treated as keywords, allowing to designate a column or an expression as, respectively, the locations of the measurement points, the time instants or the time periods of the measurements, and the measured values.

The create geofield statement can be used on its own or can be embedded into other SQL statements, such as: select, insert or update.

Geofields are by default virtual, which means that the create geofield statement only adds the definition of a geofield to the database dictionary. The materialized keyword has been added to improve the integration with multidimensional arrays (an example is given later), and as an optimization mechanism.

There is no clause for geofield domain definition in the create geofield statement, as the estimation of geofield values depends on the interpolation method and node search rule used. Instead, the user can define a geofield context region specifying an area on which geofield estimation can be undertaken. An attempt to estimate a geofield value outside its context region will return a special marker UGV (Unknown Geofield Value). The resolution column allows the user to determine the size of the elementary quadrants. It defaults to the array cell size used in a query.

All geofield properties definitions stored in the database dictionary can be overwritten when the geofield is used, e.g. in the select statement the default interpolation method or its default parameters can be changed, or in the insert statement the default context region can be redefined.

As in geostatistics usually Kriging is the best choice [7], to simplify geofield definition, a special registry of variograms models is added. Also, a registry for geofield variability models used by quandrant classifiers [4] is created.

The following statement creates an exemplary geofield of airborne suspended matter, called suspended_matter:

```
create virtual geofield suspended_matter as
select measurement_location, measurement_time,
       measurement_value
from air_pollution
where pollutant_code = 'SUSPENDED_MATTER' and
    measurement_quality = 'GOOD' and
    CRS = 'EPSG:4326' and
    interpolation = '{method: OK, semivariogramID: 71,
       nodeSearchType: QUADS, nodeSearchN: 6,
       nodeSearchR: 11 km}' and
    resolution = 5 m;
```

The geofield suspended_matter mathematical model uses measurements of airborne suspended matter from the table air_pollution (condition pollutant_code = 'SUSPENDED_MATTER') of good quality (condition measurement_quality = 'GOOD'), and is based on Ordinary Kriging (OK) with semivariogram with ID = 71. A quadrant search of measurement nodes, with six nodes for quadrant and the maximal distance of the search equal to 11 km, is used. Elementary quadrants (array cells) have side length of five meters. No context region is defined, so it must be explicitly provided in a query referencing the geofield, or it must be inferred from the query.

The definition of the geofield suspended_matter can be used to compute a quantitative geofield of the average-yearly distribution in 2013 of airborne suspended matter in the context region specified by coordinates given by a polygon:

```
select location, value
from suspended_matter
where year(measurement_time) = 2013 and
    context_region = polygon(49.95,18.38, 50.55,18.38,
       50.55,19.58, 49.95,19.58, 49.95,18.38);
```

A qualitative geofield, defined by four intervals: $(-\infty, 110]$, $(110, 165]$, $(165, 220]$, and $(220, \infty)$ $\mu g/m^3$, can be generated, using the BPO square classifier [4] (with: geofield variability model stored in the registry with ID = 10, and distortion coefficient = 0.8), by following query:

```
select interval, shape
from suspended_matter
where year(measurement_time) = 2013 and
    square_classifier = '{type: BPO, geofieldVariabilityModelID:
                          10, beta: 0.8}' and
    intervals = (110, 165, 220) and
    context_region = polygon(49.95,18.38, 50.55,18.38,
                             50.55,19.58, 49.95,19.58, 49.95,18.38);
```

3.2   Geofield as a Table Column

The following example illustrates how geofield can be used as a virtual column
in a table:

```
create table city (
    city_id integer primary key,
    city_name varchar not null,
    city_boundary geometry not null,
    sm_pollution geofield
);
```

Using the exemplary geofield suspended_matter, information about Gliwice
can be inserted by statement:

```
insert into city(city_id, city_name, city_boundary, sm_pollution)
values (1, 'Gliwice', <WKT_GEOM>,
        suspended_matter where context_region = city_boundary);
```

In the insert statement the value of the column context_region is set to
the boundary of Gliwice (<WKT_GEOM> is a placeholder for the polygon defining
boundary of Gliwice). Geofields can be defined inline, but defining them as separate statements (as in the running example) simplifies other statements, removes
redundancy, and makes the code more readable.

The following example shows how a geofield may be added as a virtual attribute of an array-typed column. The following example extends the example
from the listing 2 from [12], describing ISO SQL extension proposal.

```
create table landsat_scenes_air_pollution (
    id integer not null,
    acquired date not null,
    scene row (band1 integer,
              ...,
              band5 integer,
              sm_pollution geofield default suspended_matter
                                     where time = acquired
              )
              array [ x ( 1 : 5000 ) , y ( 1 : * ) ]
);
```

In the create table statement default geofield is provided. The condition time
= acquired imposes temporal restriction on the geofield. The geofield context
region is taken from the raster boundaries. Adding the materialized keyword

to the geofield definition would result in computation of its raster representation during a row insertion. If the user changes the interpolation method in a query, a new geofield raster representation will be computed using the modified geofield model.

4   Optimizability of Geofield Queries

This section shortly discuses opportunities given by using virtual (dynamically computed) geofields in declarative queries. The simplest and the most basic operation – in the context of this paper – is the conversion of a geofield from its table representation to an array representation. As the cost of the conversion is determined by interpolation, minimization of the number of places in which geofield values must be computed leads to significant query processing speedup [2, 4]. This can be achieved by integration of the conversion with an array constructor as shown in the previous section. If there are any conditions on geofield values in the query, passing them to the conversion called in an array constructor may significantly speedup query processing, as shown in [4] for qualitative geofields creation. This is especially profitable for qualitative geofields, but may be also used for quantitative geofields.

If a geofield query references more than one qualitative geofield, using the concepts presented in section 2 may significantly speedup query processing. The PNR representation may be used internally by a DBMS, and the intermediate result can be converted from the PNR representation to the array representation for further processing. The ideas presented in [5] and [4] may be easily generalized into more than 2D space, as far as an interpolation method is available for required dimensionality [8].

Let us assume, that a query states that values of a raster R must belong to an interval $i1$, and values of a geofield G must belong to an interval $i2$. We can distinguish three strategies for the query evaluation: 1) compute the geofield G for the intersection of the domains of the raster R and the geofield G, read the raster R, perform the selections on its values and find the cells which fulfill both conditions; 2) read the raster R, find the cells which values belong to the interval $i1$, compute the geofield G only for these cells and check if the computed values belong to the interval $i2$; 3) compute the geofield G and find its cells belonging to the interval $i2$, read only the chunks of the raster R which contain these cells, and check if they belong to the interval $i1$. As the interpolation is time expensive, usually the second strategy should be the best. However, if the selection on the geofield G values has high selectivity and the raster R is very large, then the third strategy may be the best. If both selectivities are low, then the first strategy may be the best, as anyway the whole geofield G must be computed and the whole raster R must be read. As experiments and theoretical analysis showed [2], usually qualitative geofield computation using quadrant classifiers [4] is much faster when the geofield context region consists of a small number of large quadrants, than when it consists of larger number of smaller quandrants. This can be explained by a rule of thumb telling that the quadrant classification cost is proportional to the number of its boundary elementary quadrants, while the raster computation cost is proportional to the number of its cells (elementary quadrants). As shortly discussed, even in such a simple case, the choice of the best strategy is not clear and a cost optimizer is needed [2,4]. The problem complicates significantly, when more geofields should be computed and combined by some operations, e.g. finding areas where some pollutants exceed given thresholds — this problem reminds the classical join order problem from relational databases [2].

5   Related Work

In [13] an idea and implementation of enriching the DBMS with interpolation were presented. The implementation was based on database procedures and temporary tables, not on an algebraic operation for geofield estimation, as is in this paper. The work was presented long before any significant development of array DBMSs took place, and included only the table representation of geofields, while this paper presents approach encompassing both representations. However, some ideas from [13] are still valid for array databases: geofield values interpolation during a query evaluation, choosing an interpolation method in a query, and specifying in a query which measurement values should be used in interpolation.

In [12] two approaches to the integration of multidimensional arrays with relations are distinguished: a) *array-as-attribute* – arrays are treated as the other data types, which allows to define columns of the array type, but prevents from usage of arrays without tables; and b) *array-as-table* – arrays are on the same level as tables, which allows to use arrays without tables, but does not allow to define columns of the array type.

In the work on the ISO SQL extension the array-as-attribute approach is followed [12], which is consistent with the support of one dimensional arrays in the current ISO SQL standard. Conversion between an array and a table is performed using the **unnest** and **nest** operators. However, during this conversion no geofield values can be estimated, what is possible in the presented SQL extension.

The rasdaman was the first array DBMS [12]. From the beginning it has been developed as a layer above conventional RDBMSs. The rasdaman follows the array-as-attribute approach. Each cell of an array may store one or more values of types specified in the array declaration. The coordinates of array cells are defined as bounded or unbounded integers. Only the simple nearest neighbor interpolation is supported for scaling [15].

SciDB is a DBMS designed and implemented to address scientific needs [17]. It implements only the array-as-table approach and does not support tables. SciDB provides linear algebra, but no spatial interpolation method (such as Kriging) is available. The package SciDBR provides integration of R with SciDB, which makes it possible to utilize sophisticated spatial interpolation methods, however, declarative geofield queries are not supported and their optimization is not possible.

SciQL is an array DBMS based on column-oriented DBMS MonetDB [10] and follows the array-as-table approach. In [10] only interpolation for time series is mentioned.

Some commercial and open source RDBMSs have been extended with raster data support, following the array-as-attribute approach. For example, Oracle RDBMS provides a set of functions and procedures for raster processing. As a consequence, usually procedural code in PL/SQL is needed, which limits the possibility of queries optimization. The current Oracle version 12c Release 1 supports only 2D rasters and provides only basic interpolation methods for rasters (near neighbor, bilinear, biquadratic, cubic and average) [14]. Also PostGIS provides support for 2D rasters with similar interpolation/resampling methods: nearest neighbor, bilinear, cubic, cubic spline, Lanczos, and inverse distance weighted [1].

To the best of the authors knowledge, none of the related DBMSs nor the ISO SQL extension under development contains SQL capabilities similar to the extension described in section 3 of the paper. As a consequence, optimizability discussed in section 4 is not possible in these DBMSs.

6   Conclusions

The presented extension of SQL for geofields processing widens data integration capabilities, gives new optimization possibilities, and eliminates the problem of unloading data from database for interpolation in external tools and loading the results back to the database for further processing. As discussed in the paper, adding the geofield support to the DBMS is more than just adding an interpolation function, as this extension significantly influences query optimization. Interpolation during query processing may be time consuming, but on the other hand, it eliminates the problems connected with measurement points selection based on the predicates included in the query, measurements updates, CRS conversions, as well as raster scaling, shifting and rotating. This can be especially valuable in a multi-resolution database [11], when an array representation of a geofield may be generated on-the-fly in the required resolution and CRS, instead of rescaling its previously generated and stored representation. The presented approach also allows to easily change the geofield model used (to change the interpolation method and/or adjust its parameters). The impact of such changes on the query answer can be easily evaluated as the difference may be computed by another SQL query.

Distinguishing between qualitative and quantitative geofields is not very important to the users writing queries, but has very significant impact on query optimizability.

References

1. PostGIS 2.1.4dev Manual. SVN Revision (12916)
2. Bajerski, P.: Using Peano Algebra for Optimization of Domain Data Sets Access during Analysis of Features Distribution in Space, PhD Dissertation. Poland, Gliwice (2006) (in Polish)
3. Bajerski, P.: Optimization of geofield queries. In: 1st International Conference on Information Technology, IT 2008, pp. 1–4 (2008)
4. Bajerski, P.: How to efficiently generate PNR representation of a qualitative geofield. In: Cyran, K.A., Kozielski, S., Peters, J.F., Stańczyk, U., Wakulicz-Deja, A. (eds.) Man-Machine Interactions. AISC, vol. 59, pp. 595–603. Springer, Heidelberg (2009)
5. Bajerski, P., Kozielski, S.: Computational model for efficient processing of geofield queries. In: Cyran, K.A., Kozielski, S., Peters, J.F., Stańczyk, U., Wakulicz-Deja, A. (eds.) Man-Machine Interactions. AISC, vol. 59, pp. 573–583. Springer, Heidelberg (2009)
6. Baumann, P.: OGC® GML Application Schema – Coverages, Open Geospatial Consortium, OGC 09-146r2, http://www.opengis.net/doc/GML/GMLCOV/1.0.1
7. Cressie, N.: Statistics for Spatial Data. Wiley (1995)
8. Cressie, N., Wikle, C.K.: Statistics for Spatio-Temporal Data. Wiley (2011)
9. Goodchild, M.F., Yuan, M., Cova, T.J.: Towards a general theory of geographic representation in GIS. International Journal of Geographical Information Science 21(3), 239–260 (2007)
10. Kersten, M., Zhang, Y., Ivanova, M., Nes, N.: SciQL, a query language for science applications. In: Proceedings of the 2011 EDBT/ICDT Workshop on Array Databases, Uppsala, Sweden, March 25, pp. 1–12 (2011)
11. Kozioł, K., Lupa, M., Krawczyk, A.: The extended structure of multi-resolution database. In: Kozielski, S., Mrozek, D., Kasprowski, P., Małysiak-Mrozek, B., Kostrzewa, D. (eds.) BDAS 2014. CCIS, vol. 424, pp. 435–443. Springer, Heidelberg (2014)
12. Misev, D., Baumann, P.: Extending the SQL array concept to support scientific analytics. In: Conference on Scientific and Statistical Database Management, SSDBM 2014, Aalborg, Denmark, June 30-July 02, p. 10 (2014)
13. Neugebauer, L.: Optimization and evaluation of database queries including embedded interpolation procedures. In: Proceedings of the 1991 ACM SIGMOD International Conference on Management of Data, Denver, Colorado, May 29-31, pp. 118–127 (1991)
14. Oracle and/or its affiliates: Oracle® Spatial and Graph. GeoRaster Developer's Guide, 12 c Release 1 (12.1) E49118-04 (November 2014)
15. rasdaman GmbH: rasdaman Query Language Guide, 9.0 edition (2014)
16. Stonebraker, M., Becla, J., DeWitt, D., Lim, K., Maier, D., Ratzesberger, O., Zdonik, S.: Requirements for science data bases and SciDB. In: Fourth Biennial Conference on Innovative Data Systems Research, CIDR 2009, January 4-7, Asilomar, CA, USA (2009) (Online Proceedings)
17. Stonebraker, M., Brown, P., Zhang, D., Becla, J.: Scidb: A database management system for applications with complex analytics. Computing in Science and Engineering 15(3), 54–62 (2013)

A Proposal of Hybrid Spatial Indexing for Addressing the Measurement Points in Monitoring Sensor Networks

Michał Lupa[✉], Monika Chuchro, Adam Piórkowski, Anna Pięta, and Andrzej Leśniak

Department of Geoinformatics and Applied Computer Science
AGH University of Science and Technology
al. Mickiewicza 30, 30-059 Cracow, Poland
{mlupa,chuchro,pioro,lesniak}@agh.edu.pl,
apieta@geol.agh.edu.pl
http://www.geoinf.agh.edu.pl

Abstract. One of the important features of data analysis methods in the area of continuous surveillance systems is a computation time. This article contains a research that is focused on improving the performance of processing by the most efficient possible indexation of spatial data. The authors proposed a structure of indexes implementation based on layered grouping of sensors, so as to reduce the amount of data in time windows. This allows to compare data at the layer-layer level, thereby reducing the problem of comparisons between all sensors.

Keywords: Spatial databases · Spatial indexing · R⁺-trees · R*-trees

1 Introduction

Floods are one of the largest natural hazards occurring in Poland and in the all central and east Europe. According to the National Water Management in Poland in the years 1950-2000 there were more than 600 floods of different coverage and different cause of the uprising. The most common cause of flooding is the spring thaw and heavy rain. Some of these phenomena are sudden and unpredictable, and some of them develops slowly and lasts a long time, depending on wheater condition. Predicting the occurrence of floods is a difficult and very complicated process. Analysis of the causes of flooding, flood assessment of the size and timing of individual movements of the flood wave is taken by a few research groups. More rarely considered issue, but also very important is assessment of physical state of the flood embankment during a flood, similar to the landslides monitoring systems [9,10] or other monitoring systems [8]. Knowledge about the current state of the flood embankment during both short-term, sudden and prolonged floods allows to make proper decisions related to the functioning of the flood embankment, to repair leaks and to evacuate of people from flood-threatened area. Systems of embankment monitoring are considered in several projects [25,5,17,19,21,23].

© Springer International Publishing Switzerland 2015
S. Kozielski et al. (Eds.): BDAS 2015, CCIS 521, pp. 437–447, 2015.
DOI: 10.1007/978-3-319-18422-7_39

In Poland, in cooperation with the AGH University of Science and Technology in Cracow and Cracow companies SWECO Hydroprojekt Ltd and Neosentio, project ISMOP (Computer Monitoring System for Flood Embankments) has been developed. This was founded under the NCBiR project (National Centre for Research and Development, Poland), which involves the creation of a real-time monitoring system of static and dynamic behavior of the embankments.

The monitoring systems constructions are differentiated. In most cases, these systems are data-driven dedicated, therefore creating software requires a complex solution. It is also possible to use existing solutions for part of the creating system. The ISMOP project is created as a data-driven system based on the analysis of data streams transmitted from the sensors in experimental flood embankment. This experimental flood embankment is constructed as a standard design in the region of the middle and upper river. A sensor networks are arranged inside the experimental flood embankment into the wells. The sensor network measures several key physical and geotechnical parameters (temperature, pressure, pore pressure, humidity, stress and strain, electral conductivity) of given time interval between one day (in stabile wheather condition) and max 15 minutes (during floods). Distance between two points measurenent is not more bigger than 5 meters and the sensors will be arranged in five layers.

The biggest problem is how to create a system that quickly, less than in 15 minutes, receives data sent from over 1000 sensors which measures seven parameters, processes the data, performs analysis and developes their visualization results. During the evaluation there are taken also into account hydrogeological and weather data. In the course of evaluation of the embankment several modules are used: time series, anomaly detection, prediction and expert systems (Fig. 1). Time series analysis module performs a comparison of the real data obtained from the sensors with data from numerical modeling performed in the Flac Software [15]. More information about the numerical models can be found in [19,20]. The next step is to compare the real data with previously obtained numerical models. For each sensor there are compared the real time data series with the numerical model obtained from Flac. Comparisons are performed simultaneously for all the sensors. The biggest challenge is to make the comparison for each single sensor which have a given length of time window with multiple numerical models. The number of numerical models for each sensor can be up to 1 thousand, also each numerical model could have a different length. In addition, with the passing of time observation of flood embankment state, analyzed the real time series are extended, causing an increase of time of analysis. By comparison of real data with numerical models is a select group of models with the highest value of similarity measure. For each numerical model is known the final state of flood embankment under the influence of certain preconceived parameters. Evaluation of the local flood embankment condition around a particular sensor is the calculated measure of the likelihood of maintaining the stability of the flood embankment. In contrast, the global assessment of the state of the flood embankment is average rating of probability for individual sensors. Only if the assessment given for the individual sensors indicates the stability of the

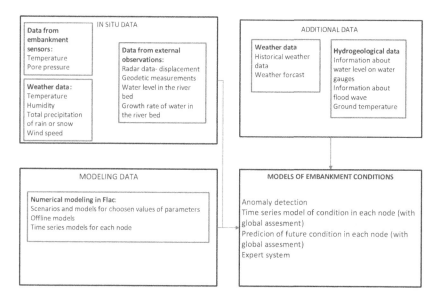

Fig. 1. The layered structure of the embankment measurement network

flood embankment disorder, the final assessment is the same as the lowest value of local assessment. Details of the analysis and comparison of real time series data from sensors inside the flood embankment and numerical modeling data from Flac can be found in [7] and in Fig. 2.

2 Spatial Data Indexation and Analysis

One of the main goals of the system is the division of the monitored embankments that run along the river segments. This procedure will maximize the accuracy of the predicted embankment behaviors that will be generated by the decision algorithms. The measurement network, depending on the selected section of the embankment, will consist of 500-1000 sensors placed at different depth levels (layers), which will generate data in a predetermined time interval. These data will be subject to continuous evaluation to determine whether the threat of flooding is real. Therefore, it becomes necessary to develop appropriate solutions for the analysis of large data volumes. These solutions should take into account the possibility of comparing time series, caused by incoming information in the form of streams, and the data stored statically (historical measurements and data obtained by numerical modeling). Moreover, these series, in addition to alphanumeric information, will also contain an attribute describing the geometry (position of the sensor in the embankment - the coordinates X, Y, Z), which greatly affect the reading, as well as processing efficiency. Therefore, having regard to the significance of the generated results, the comparison processes of static and stream data, and so the decision algorithms, should be devoid of any delays and procedures leading to a decline in efficiency of data processing.

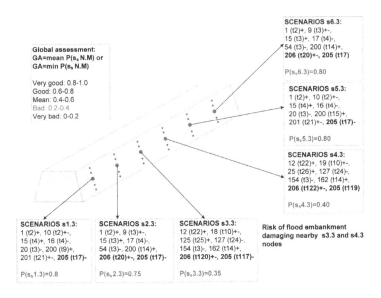

Fig. 2. The layered structure of the embankment measurement network

2.1   Indexation Methods

The performance of operations on spatial data can be improved in several ways. There can be a decomposition of query [18], changing the representation of data [3,4] or custom selectivity estimation [1,2] proposed. The most efficient method of improving query performance is using a proper index structures. The indexes are redundant structures, allowing a quick way to get to the sought data by some kind of arrangement and organization of their sets. In the greatest simplification this is a set of pairs:

$< key, value >$

where the key is an attribute (column or field) of searched data, and the value is a reference, or the physical address of the data (the record or object). This collection is organized in such a way (often radically different from the organization of the data in the source) to reduce the number of necessary IO operations to be performed to get to the key value. In the case of data with geometric attributes, which are oriented in the selected coordinate system (and therefore in some way already organized in space), the standard methods of indexing alphanumeric data fail [11,12]. This is particularly true in the context of spatial queries like: determine the objects intersection, return the values of all points inside the region. These queries are relate to the objects (and values assigned to them) with associated topological relationship or geometry. Therefore, a set of methods has been developed based on the indexing tree-structure (R^+-Tree, R^*-tree, etc.), which was adapted to the need for the organization and ordering of objects in space, as well as the space itself (2D, 3D). In the context of spatial indexes, the key in the set of pairs is the location, ie:

$< location, value >$

2.2   Spatial Data Used in Sensor Networks

The static spatial database (in this case a relational database, which has the ability to process spatial data) will be supplied with a set of numerical models and behavior scenarios of the embankment, generated during simulations conducted previously. This will create a reference point for the data stream, which will be compared "on the fly" with data stored in a relational database. As shown in chapter two, comparison will be based on the determination of the time series similarity measure, using sliding or agglomeration windows. This type of analysis, due to the need to sort through large data sets can result in a significant decrease in system performance. The risk of reduction in the data processing efficiency particularly affects agglomeration windows, which, together with the growth of the windows length, will cause an increase in the number of necessary comparison operations.

Thus, a necessary step is to develop appropriate indexing methods and grouping of data, in order to reduce the number of operations performed in the time windows. Monitoring will include data from sensors which are arranged in the space. The measuring grid is constructed in layers so as to uniformly cover the area of the monitored embankment, that is, three types of sensors are laid out alternately on one level (described in [20,7]). Depending on the height and shape of the embankments, there may be a few or several layers of sensors. This will allow one type of sensor to be placed every 1 meter at any point of the embankment crosssection. An examples of sensor arrangement is shown in Fig. 3. Given such a measurement structure, it was decided that the procedure which allows the greatest increase in efficiency of data mining is appropriate grouping of the input time series and layered indexing data from a spatial database. In this way, the analysis in time windows will operate in layers, as in the case of a considerable

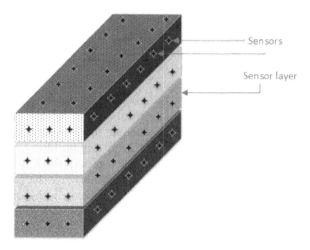

Fig. 3. The layered structure of the embankment measurement network

amount of compared series this can lead to a significant increase in productivity. This approach will simplify the process of comparing three-dimensional solid nodes to analyze of two-dimensional surfaces. Moreover, this method will allow for a better illustration of the measured parameters' variability, in relation to the height level from which they are monitored. This will allow it to become independent from the analysis of hydrogeological conditions, which assumes that the variability of certain parameters (such as humidity), depend on the depth of the sensor.

2.3   Choice of the Method

Spatial indexes are a common method for optimizing commercial databases, as well as an open source DMBS. Standard indexation mechanism is based on tree structures (R-tree, B-tree) and their modification (R*-tree, GiST). These mechanisms are often the only available method of indexing spatial data. Moreover, default spatial indexing structures are often dictated in advance, which does not always result in savings of processing time due to the type of indexed geometry. The R-trees space ordering is done by approximation based on Minimum Bounding Rectangles (MBR) methods and their optimization (R^+-trees, R*-trees) [13]. Therefore the choice of the appropriate indexation method for measuring network structure is a non-trivial task. First of all, there should be taken into account the nature of the data points (since each sensor and each measure will have a spatial attribute) and the type of queries. In this case, the standard R-tree structure (including the R^+-Tree and R*-tree) must be modified. One of them is the proposal of R^{++}-trees, contained in [24], where the authors make modifications to the R^+-trees for the purpose of indexing data points by using Double Surrounding Regions (Minimum and Maximum, to facilitate the insertion). This method is highly effective for top-N-queries or kNN-queries. The authors of article [16] described a new technique for indexing spatial points in N-dimensions called iDistance. This technique allows for increased efficiency of point queries, particularly in the case of real-time data. Indexation is done by choosing a certain number of reference points in space (the density of the point coverage in space is determined based on the given parameter). Then the length between each point and the nearest point of reference is calculated. This length is called iDistance and allows the mapping of a multi-dimensional area into a one-dimensional region value. The third step is indexing points by key, which is the calculated value iDistance. Among the solutions available in the literature there are also hybrid techniques that fit the problem. It is worth noting the method of QR*-tree [22], in which the authors made a hybrid of two tree structures, namely: Quad-tree and R*-trees. Analyzing the possibilities and advantages of the above methods, a hybrid indexation technique, which will combine the structures of PR-quadtree (PointRegion) and B-trees is proposed. This technique reminds the algorithm proposed in [14]. PR-quadtree works in such a way that each individual layer measurement is subjected to a recursive decomposition based on the coordinates of the points (sensors). Internal nodes can have a maximum of four children, each of which represents a different region obtained by the decomposition of the space coordinates (Fig. 4).

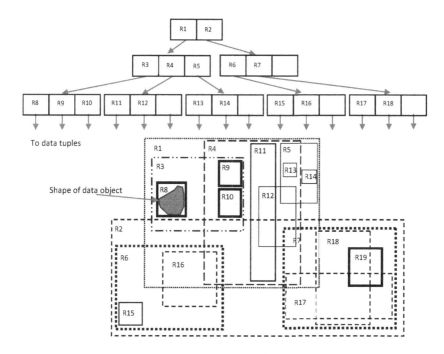

Fig. 4. Sample representation of points with PRegion-quadtree

Therefore each measurement layer will be represented in the form of a PR-quadtree (Fig. 5), which allows the reduction of a three-dimensional computational problem to a two-dimensional surface. In addition, each single sensor layer will be the first stage node in the B-tree. In each of the nodes PR-quadtrees are stored instead of record addresses. The resulting hybrid data structure adopts the form in which the vertex is defined earlier by measurement segment (embankments are divided into parts and each has 1000 sensors). The first level nodes are the various layers that contain sensors arranged at the same height level. Second-degree nodes will contain the addresses of sensors located in four major quadrants of the network, the nodes representing the third degree will contain the quadrants of each major quadrants, and so on. Individual sensors will form the leafs. Fig. 6 presents the data structure which is described above.

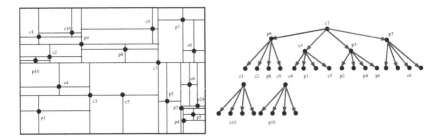

Fig. 5. The schema of a PR-quadtree index

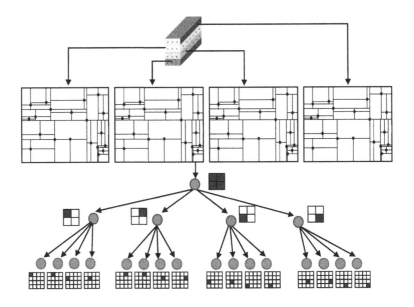

Fig. 6. The schema of hybrid index structure

2.4   Grouping and Data Analysis

In the earlier part of the paper, the authors presented the possibility of using spatial indexing for the purpose of increasing the efficiency of searching the modeled data. However, comparative analysis also includes information from the measurement network, which necessitates the development of mechanisms for integration of indexed data models and streams. What is more, the data stream cannot be indexed because of its unstable nature. Therefore, it was decided to divide the streams according to their spatial arrangement. The sensors are grouped by measuring layers, so that it imitates the spatial indexation models. This will enable layered analysis in time windows, which can significantly reduce the number of database readings. Grouping undoubtedly bring performance improvements. However, sometimes creating an index is redundant. Querying grouped data and utilizing the Cuboid-based architecture, as discussed in [6], has proven significant efficiency boost utilizing order dependency based optimization methods. This is important especially while considering chronological stream data.

3 Conclusions and Further Work

The article concerns very important issue, which is the effectiveness of the computation time in large data sets. This problem is especially evident in the context of spatial data processing that involves three-dimensional objects. Therefore, this article presents the possibility of a hybrid index construction which will allow for the computation reduction from three to two dimensions. This solution flattens the problem by grouping nodes in a layers (in the vertical or horizontal sections), which is particularly important advantage, for example, the case of the data from sensor networks and time window processing. The data will be compared at the layer-layer level, causing the reduction of comparisons between all sensors.

The future work involves carrying out the real tests of performance for the proposed method. The speedup of presented indexation will be described in comparison to traditional way of processing.

Acknowledgments. This work is financed by the National Centre for Research and Development (NCBiR), Poland, project PBS1/B9/18/2013 - (no 180535).

This work was co-financed by the AGH - University of Science and Technology, Faculty of Geology, Geophysics and Environmental Protection, as a part of statutory project.

References

1. Augustyn, D.R.: Applying advanced methods of query selectivity estimation in Oracle DBMS. In: Cyran, K.A., Kozielski, S., Peters, J.F., Stańczyk, U., Wakulicz-Deja, A. (eds.) Man-Machine Interactions. AISC, vol. 59, pp. 585–593. Springer, Heidelberg (2009)
2. Augustyn, D.R., Zederowski, S.: Applying CUDA technology in DCT-based method of query selectivity estimation. In: Pechenizkiy, M., Wojciechowski, M. (eds.) New Trends in Databases & Inform. Sys. AISC, vol. 185, pp. 3–12. Springer, Heidelberg (2012)
3. Bajerski, P.: Optimization of geofield queries. In: Proceedings of the 1st IEEE International Conference on Information Technology, pp. 1–4 (2008)
4. Bajerski, P., Kozielski, S.: Computational Model for Efficient Processing of Geofield Queries. In: Cyran, K.A., Kozielski, S., Peters, J.F., Stańczyk, U., Wakulicz-Deja, A. (eds.) Man-Machine Interactions. AISC, vol. 59, pp. 573–583. Springer, Heidelberg (2009)
5. Balis, B., Kasztelnik, M., Bubak, M., Bartynski, T., Gubała, T., Nowakowski, P., Broekhuijsen, J.: The urbanflood common information space for early warning systems. Procedia Computer Science 4, 96–105 (2011)

6. Chromiak, M., Wiśniewski, P., Stencel, K.: Exploiting Order Dependencies on Primary Keys for Optimization. In: Proceedings of the 23rd International Workshop on Concurrency, Specification and Programming, vol. 1269 (2014)
7. Chuchro, M., Lupa, M., Pięta, A., Piórkowski, A., Leśniak, A.: A Concept of Time Windows Length Selection in Stream Databases in the Context of Sensor Networks Monitoring. In: Bassiliades, N., Ivanovic, M., Kon-Popovska, M., Manolopoulos, Y., Palpanas, T., Trajcevski, G., Vakali, A. (eds.) New Trends in Database and Information Systems II. AISC, vol. 312, pp. 173–183. Springer, Heidelberg (2015)
8. Fidali, M., Jamrozik, W.: Concept of Database Architecture Dedicated to Data Fusion Based Condition Monitoring Systems. In: Kozielski, S., Mrozek, D., Kasprowski, P., Małysiak-Mrozek, B., Kostrzewa, D. (eds.) BDAS 2014. CCIS, vol. 424, pp. 515–526. Springer, Heidelberg (2014)
9. Flak, J., Gaj, P., Tokarz, K., Wideł, S., Ziębiński, A.: Remote Monitoring of Geological Activity of Inclined Regions – The Concept. In: Kwiecień, A., Gaj, P., Stera, P. (eds.) CN 2009. CCIS, vol. 39, pp. 292–301. Springer, Heidelberg (2009)
10. Gaj, P., Kwiecień, B.: The General Concept of a Distributed Computer System Designed for Monitoring Rock Movements. In: Kwiecień, A., Gaj, P., Stera, P. (eds.) CN 2009. CCIS, vol. 39, pp. 280–291. Springer, Heidelberg (2009)
11. Gorawski, M., Malczok, R.: Towards stream data parallel processing in spatial aggregating index. In: Wyrzykowski, R., Dongarra, J., Karczewski, K., Wasniewski, J. (eds.) PPAM 2007. LNCS, vol. 4967, pp. 209–218. Springer, Heidelberg (2008)
12. Gorawski, M., Malczok, R.: Indexing Spatial Objects in Stream Data Warehouse. In: Nguyen, N.T., Katarzyniak, R., Chen, S.-M. (eds.) Advances in Intelligent Information and Database Systems. SCI, vol. 283, pp. 53–65. Springer, Heidelberg (2010)
13. Guttman, A.: R-trees: A dynamic index structure for spatial searching, vol. 14. ACM (1984)
14. Hjaltason, G.R., Samet, H.: Improved bulk-loading algorithms for quadtrees. In: Proceedings of the 7th ACM International Symposium on Advances in Geographic Information Systems, pp. 110–115. ACM (1999)
15. Itasca Consulting Group, Inc.: FLAC Fast Lagrangian Analysis of Continua and FLAC/Slope – User's Manual (2008)
16. Jagadish, H.V., Ooi, B.C., Zhang, R.: iDistance Techniques. In: Encyclopedia of GIS, pp. 469–471 (2008)
17. Krzhizhanovskaya, V.V., Shirshov, G.S., Melnikova, N.B., Belleman, R.G., Rusadi, F.I., Broekhuijsen, B.J., Gouldby, B.P., Lhomme, J., Balis, B., Bubak, M., et al.: Flood early warning system: design, implementation and computational modules. Procedia Computer Science 4, 106–115 (2011)
18. Lupa, M., Piórkowski, A.: Spatial Query Optimization Based on Transformation of Constraints. In: Gruca, A., Czachórski, T., Kozielski, S. (eds.) Man-Machine Interactions 3. AISC, vol. 242, pp. 621–629. Springer, Heidelberg (2014)
19. Pieta, A., Bala, J., Dwornik, M., Krawiec, K.: Stability of the levees in case of high level of the water. In: 14th SGEM Geoconference On Informatics, Geoinformatics And Remote Sensing – Conference Proceedings, vol. 1, pp. 809–815 (2014)
20. Pięta, A., Lupa, M., Chuchro, M., Piórkowski, A., Leśniak, A.: A Model of a System for Stream Data Storage and Analysis Dedicated to Sensor Networks of Embankment Monitoring. In: Saeed, K., Snášel, V. (eds.) CISIM 2014. LNCS, vol. 8838, pp. 514–525. Springer, Heidelberg (2014)
21. Piorkowski, A., Lesniak, A.: Using Data Stream Management Systems in the Design of Monitoring System for Flood Embankments. Studia Informatica 35(2), 297–310 (2014)

22. Qiu, J., Guo, Q., Xiong, Y.: QR*-Tree: A New Hybird Spatial Database Index Structure. In: Qian, Z., Cao, L., Su, W., Wang, T., Yang, H. (eds.) Recent Advances in CSIE 2011. LNEE, vol. 126, pp. 795–802. Springer, Heidelberg (2012)

23. Stanisz, J., Borecka, A., Leśniak, A., Zieliński, K.: Selected levee monitoring systems. Przeglad Geologiczny 62(10/2), 699–703 (2014)

24. Šumák, M., Gurský, P.: R++-tree: An efficient spatial access method for highly redundant point data. In: Catania, B., et al. (eds.) New Trends in Databases and Information Systems. AISC, vol. 241, pp. 37–44. Springer, Heidelberg (2014)

25. Szydlo, T., Nawrocki, P., Brzoza-Woch, R., Zielinski, K.: Power aware MOM for telemetry-oriented applications using GPRS-enabled embedded devices – levee monitoring use case. In: Ganzha, M., Maciaszek, L., Paprzycki, M. (eds.) Proceedings of the 2014 Federated Conference on Computer Science and Information Systems. Annals of Computer Science and Information Systems, vol. 2, pp. 1059–1064. IEEE (2014), http://dx.doi.org/10.15439/2014F252

An Attempt to Automate the Simplification of Building Objects in Multiresolution Databases

Michał Lupa[1(✉)], Krystian Kozioł[2], and Andrzej Leśniak[1]

[1] Department of Geoinformatics and Applied Computer Science,
al. Mickiewicza 30, 30-059 Cracow, Poland
[2] Department of Geomatics, AGH University of Science and Technology,
al. Mickiewicza 30, 30-059 Cracow, Poland
{mlupa,krystian.koziol,lesniak}@agh.edu.pl
http://www.geoinf.agh.edu.pl

Abstract. The paper presents a method for the simplification of building objects in multiresolution databases. The authors present a theoretical foundation, practical ways to implement the method, examples of results, as well as a comparison with currently available generalization methods in commercial software. This algorithm allows the verifiability and reproducibility of results to be kept while minimising graphic conflicts, which are a major problem during the automatic generalisation process. These results are achieved by defining the shape of buildings, employing classification rules and adopting minimum measures of recognition on a digital map. Solutions included in this paper are universal and can successfully be used as a component in any automated cartographic generalisation process. Moreover, these methods will help to get closer to full automation of the data generalisation process and hence the automatic production of digital maps.

Keywords: Building generalisation · Cartographic generalisation · GIS · Spatial databasess

1 Introduction

Modern cartography is based on multiresolution spatial databases which are the source for derivative works, i.e. topographical, thematic or general geographic content maps that are created dynamically at smaller scales. These databases (MRDB) are created using cartographic generalization, allowing a certain level of generalisation of data depending on a certain scale to be achieved. Cartographic generalisation creates a caricature whose degree of similarity to the original is closely related to the accuracy of the scale. Moreover, the need for a smooth transition between the base reference and any different scale has resulted in the development of algorithms that allow the process of generalisation to be automatically carried out [6,12,13]. This is related to the need to develop a series of methods that, besides simplifying geometry, can identify and take into account the importance of objects in the surrounding space. The corresponding generalisation of data is also very important in the context of increasing the efficiency

© Springer International Publishing Switzerland 2015
S. Kozielski et al. (Eds.): BDAS 2015, CCIS 521, pp. 448–459, 2015.
DOI: 10.1007/978-3-319-18422-7_40

of searching large collections of information. As shown in [3–5, 11] optimisation of geospatial query processing time is possible, however, in the authors' opinion, this process should be associated with the simplification of objects due to of the linear relationship between the complexity of the geometry and query execution time, as demonstrated in the article [11]. These factors make the process of generalisation extremely problematic, because beyond the technical aspects of cartography, the process has some artistic and unique character. Olszewski [14] writes: cartographic generalisation is an individual and extremely complex process, so the challenge of modern cartography is to answer the question: is it possible to reconcile the subjectivity of the generalisation process with its automation? On the basis of the available literature and our own experience we can state that the complete automation of the generalisation process can be done only if the certain conditions are met. One such condition is the defining of a set of criteria by which the process of data generalisation is carried out. This will bring clarity, and thus, reproducible results. These criteria should be based on assumptions of cartographic semiotics, while bearing in mind the communication aspects, with particular emphasis on the perception of the audience. The method presented in this article is dedicated to data collections containing information on buildings which have characteristics that must be maintained. The procedures described allow for generalisation of the outlines of the foundations of buildings, based on assumptions about the recognition of objects on a digital map. A deciding factor during simplification is the minimum length of a façade recognized by scale operations [7, 8, 10, 15]. Simplification is performed based on an algorithm which refers to the theoretical assumptions presented in the paper [16], in which the author assumes that the vast majority of buildings have a rectangular shape which allows each edge of the building which is shorter than a certain accepted critical length (length is assumed empirically and thus causes significant limitations of the method) to be processed according to certain rules. Due to the brief description of the algorithm in paper [16], as well as the procedures performed in each of the three cases in the paper, the authors decided to develop the algorithm and enrich it with their own rules and mechanisms to identify cases and edge processing procedures. The aim of this work was to develop a method for simplifying the outlines of buildings which can generate unambiguous results without graphic conflicts. Complementing this method is, among others, the operator of rectangularisation operator [7–9], which can occur before, after, or repeatedly during generalisation. This work shows that there are many methods for assessing the generalisation of results, inter alia, by the number of changes and the value of these changes. Another measure of accuracy is the number of conflicts that arise as a result of the process. It has been shown that the correctness of results depends on the internal arrangement of data operators (the order of vertices) [8, 9] hierarchy, semantic classification (the significance of objects relative to the map type) and the context (the neighbourhood of generalised objects). This approach aims to maximise the accuracy of simplified data whose quality is often questionable due to a lack of fit to the shape or the wrong order of generalisation operators.

2   Simplification of Buildings

The algorithm proposed by [16] distinguishes three types of adjacent edge position, however none of them have mechanisms to increase the efficiency of locating and identifying these types, nor does the algorithm account for whether a building should remain in the operational scale or be removed. The algorithm also performs only modification of the identified edges' (short and adjacent edges) positions without any possibility of their reduction, resulting in an equal number of edges before and after simplification. The solutions proposed by the authors include mechanisms, which allow operator processing to be improved by appending additional preconditions related to the building's edge types. The described algorithm also takes into account all the aforementioned disadvantages. The operator can determine the visibility and degree of simplicity, which allow a building to be replaced by a minimum bounding rectangle (chapter 2.1). Moreover, the authors also propose a completely different approach to geometry processing, with an emphasis on maximum simplification by reducing the number of object edges, elimination or limitation of graphic conflicts generated by the algorithm, and the adaptation of results to the shape of the original building.

2.1   Identification

The first step of the algorithm is a function that allows identification of edges whose length does not meet the assumptions of the minimum recognition dimensions. The minimum dimensions are defined by Chrobak [10, 16] who described them with the following formula:

$$\epsilon_{03} = (0.3 + s) * 10^{-3} * M \, [m] \tag{1}$$

where:
s - line width (s >0,1)
M - scale denominator

Therefore the defined minimum dimensions can be a pattern from which the identification function can determine the number of "short" edges of the building. Next, the short edges are further modified. In the first iteration the function eliminates buildings that do not meet the assumptions of "visibility" in the target scale. These assumptions are described by the following formulas:

$$MBR_l \geq \epsilon_{03} \cap MBR_w \geq \epsilon_{03} \tag{2}$$

$$(MBR_l \geq 2 * \epsilon_{03} \cap MBR_w < \epsilon_{03}) \cup (MBR_l < \epsilon_{03} \cap MBR_w \geq \epsilon_{03}) \tag{3}$$

where:
MBR_w - width of building's minimal bounding rectangle
MBR_l - width of building's minimal bounding rectangle
ϵ_{03} - minimum recognition dimension for buildings (1)

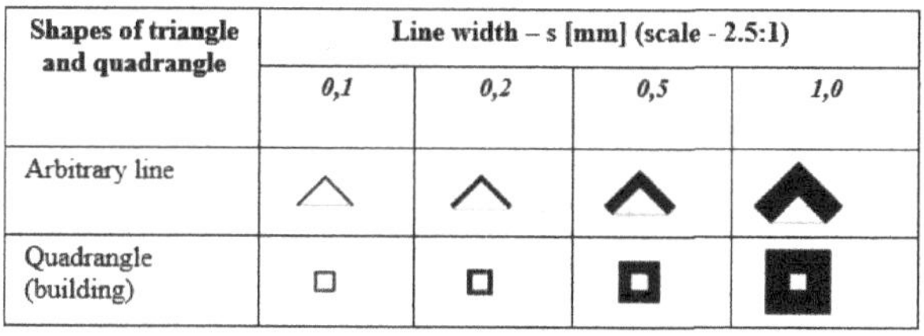

Shapes of triangle and quadrangle	Line width – s [mm] (scale - 2.5:1)			
	0,1	*0,2*	*0,5*	*1,0*
Arbitrary line	△	△	◭	▲
Quadrangle (building)	☐	◻	◼	◼

Fig. 1. Minimum dimensions for objects with varying edge widths

Figure 1 shows an example of the minimum recognition dimensions for buildings with different edge widths.

If the building does not meet the assumptions of the formula (2) an elimination operator is activated and the building is removed. If the assumptions (3) are fulfilled, the building is replaced by the MBR, otherwise it is removed. The veracity of the objectives set out in the formula (2) triggers the simplification operator. The object is then processed until all the edges are recognisable. Preserving the assumptions of building visibility and the use of minimum dimensions allows the unambiguity of the process to be achieved.

2.2   Ordering

Data must be organised before it is simplified. This treatment is necessary because it allows the subsequent identification and clear division of adjacent edges into "previous edge" and "next edge". A building with an ordered geometrical structure consists of vertices, edges and direction. So ordered geometry creates a clockwise directed graph . A pair of tangent edges next to the short edge will henceforth be denoted respectively as PREV_EDGE (the previous edge, index i-1) and NEXT_EDGE (the next edge, index i + 1). The short edge will adopt the designation SHORT_EDGE (current index i). Ordered geometric structure in the form of a directed graph will help prevent the formation of a so-called "envelope" while automatically creating polygons. When the algorithm has identified edges whose length does not meet the accepted criterion, its neighbourhood is considered in the form of two adjacent edges. Depending on their position relative to each other, the operator performs various geometric transformations on the object.

2.3   Case 0 Degrees

The first of the three considered cases is a situation in which the edges PREV_EDGE and NEXT_EDGE are positioned relative to each other at an angle close to 0 degrees. The authors allow the adoption of a certain angular tolerance, which is defined by Chrobak [8]. Degree of tolerance can also be adjusted manually by the user if there is a large variation in the geometrical structure of buildings. This situation is shown by the following figure (Fig. 2).

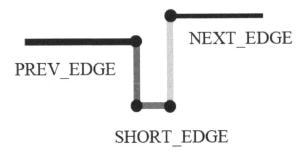

PREV_EDGE

NEXT_EDGE

SHORT_EDGE

Fig. 2. The previous and next edges in relation to each other at a 0 degree angle

The relative position of tangent edges is determined by calculating the azimuth of each edge, then the *Classifier* function identifies cases with an orientation 0 and 180 degrees. The classifier creates two auxiliary edges. The first is composed of the starting point of the previous edge (PREV_EDGE.firstPoint) and the center point of the next edge (NEXT_EDGE.Centroid). The second edge consists of the previous edge midpoint (PREV_EDGE.Centroid) and the end point of the next edge (NEXT_EDGE.lastPoint):

- *Edge1* - (PREV_EDGE.firstPoint, NEXT_EDGE.Centroid)
- *Edge2* - (PREV_EDGE.Centroid, NEXT_EDGE.lastPoint)

Edge2 INTERSECTS *Edge2* = TRUE

If *Edge1* and *Edge2* intersect and the absolute value of the difference of their azimuths oscillates around 0, the classifier identifies the situation as a "case 0 degrees" and modifies the geometry of the three edges (PREV_EDGE, SHORT_EDGE, NEXT_EDGE).

The algorithm then investigates the length of the adjacent edges PREV_EDGE and NEXT_EDGE to select the shortest of them. Next, from the chosen short adjacent edge the auxiliary edge is projected on a second edge at angle α_{se} (where α_{se} is an azimuth of SHORT_EDGE edge). Subsequently, an intersection point *IP* is designated. The next steps are as follows:

- removal of edge SHORT_EDGE

- if the short adjacent edge is a PREV_EDGE, its endpoint is changed, taking the coordinates of the intersection point *IP*:
PREV_EDGE'= (PREV_EDGE.firstPoint, *IP*); NEXT_EDGE takes the coordinates of the starting point of the *IP*: NEXT_EDGE' = (*IP*, NEXT_EDGE.lastPoint)

- if the short adjacent edge is a NEXT_EDGE, its starting point is modified, taking the coordinates of the point of intersection *IP*: NEXT_EDGE' = (*IP*, NEXT_EDGE.lastPoint); PREV_EDGE edge takes the coordinates of the end point *IP*: PREV_EDGE' = (PREV_EDGE.firstPoint, *IP*). This operation is illustrated in Fig. 3:

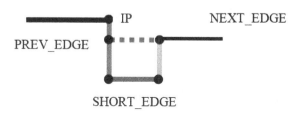

Fig. 3. Determination of an auxiliary edge and the point of intersectione

2.4 Case 90 Degrees

The second of these cases relates to situations in which adjacent edges are perpendicular. As in the case of 0 degrees, the first step is to analyse the position of NEXT_EDGE and PREV_EDGE in relation to their azimuth. Correct identification of the present case must involve both conditions. The first is the equation:

$$|\alpha_{PREV\_EDGE}| - |\alpha_{NEXT\_EDGE}| \approx \pm 90° \cup$$

$$|\alpha_{PREV\_EDGE}| - |\alpha_{NEXT\_EDGE}| \approx \pm 270° \cup$$

where:
α_{PREV_EDGE} - azimuth of previous edge
α_{NEXT_EDGE} - azimuth of next edge
azimuths have the same direction

The second condition, which was discussed in the case of 0 degrees in the previous section, is the intersection of the auxiliary edges:

- *Edge1* - (PREV_EDGE.firstPoint, NEXT_EDGE.Centroid)
- *Edge2* - (PREV_EDGE.Centroid, NEXT_EDGE.lastPoint)

Edge1 INTERSECTS *Edge2* = TRUE

Modification of the geometry in this case involves finding of an intersection point *IP* between adjacent edges. Therefore, two more auxiliary edges are constructed that are extensions respectively of PREV_EDGE and NEXT_EDGE.

Then, when the intersection point (*IP*) of the auxiliary edges has been determined the operator performs the following procedures:

- removal of SHORT_EDGE
- modification of an adjacent edge PREV_EDGE so that its end point is replaced with the coordinates of the *IP*: PREV_EDGE' = (PREV_EDGE.firstPoint, *IP*)
- modification of an adjacent edge NEXT_EDGE so that its end point is replaced with the coordinates of the *IP*: NEXT_EDGE' = (*IP*, NEXT_EDGE.lastPoint)

Figure 4 shows the geometry modification which is implemented in this case.

Fig. 4. On the left, the auxiliary edges and their intersection. On the right the result of edge simplification

2.5   Case 180 Degrees

The third case is associated with situations in which adjacent edges are positioned relative to each other at an angle of 180 degrees. As in the previous subsection, there are two factors determining the correct identification of the case. The first is described by the equation:

$$|\alpha_{PREV\_EDGE}| - |\alpha_{NEXT\_EDGE}| \approx \pm 0° \cup$$

$$|\alpha_{PREV\_EDGE}| - |\alpha_{NEXT\_EDGE}| \approx \pm 180° \cup$$

where:
α_{PREV_EDGE} - azimuth of previous edge
α_{NEXT_EDGE} - azimuth of next edge
azimuths have the same direction

- *Edge1* - (PREV_EDGE.firstPoint, NEXT_EDGE.Centroid)
- *Edge2* - (PREV_EDGE.Centroid, NEXT_EDGE.lastPoint)
Edge1 DISJOINT *Edge2* = TRUE

If these conditions are met geometry simplification procedures are launched for these three edges (PREV_EDGE, SHORT_EDGE, NEXT_EDGE). However, this case necessitates additional operations in order to minimise geometry errors (conflicts) arising from simplification. These operations are intended to be independent of the adjacent edge position, which in this case can take a variety of configurations. For this purpose, auxiliary edges are defined in points, which are the beginning and the end of the three edges, that is: the first point of the previous edge (PREV_EDGE.firstPoint) and the end point of the next edge (NEXT_EDGE.lastPoint). These points are the intersection of the two pairs of auxiliary edges, which are drawn perpendicular and parallel to the edge PREV_EDGE and NEXT_EDGE (Fig. 5).

The next step is to choose one of the two intersection points of the auxiliary edges by checking whether they are in the boundary contour of the building (defined by the OGC (Open Geospatial Consortium) function: Contains / Within) If so, the point is rejected, assuming that the point of intersection textitIP is disjointed from the boundary of building (located 'outside').

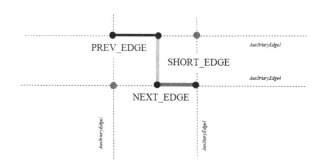

Fig. 5. Start and end points, auxiliary edges and intersection points of case 180 degrees

After determination of the intersection point textitIP, regardless of the configuration of the edge, geometry modification is performed:

- removal of SHORT_EDGE
- modification of an adjacent edge PREV_EDGE so that its end point is replaced with the coordinates of the *IP*: PREV_EDGE' = (PREV_EDGE. firstPoint, *IP*)
- modification of an adjacent edge NEXT_EDGE so that its end point is replaced with the coordinates of the *IP*: NEXT_EDGE' = (*IP*, NEXT_EDGE. lastPoint)

The results of this procedure are shown in Fig. 6.

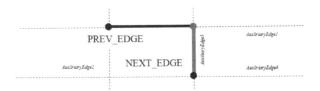

Fig. 6. The result of the geometry processing operation for 180 degrees

3 Example Results

For the purposes of this experiment it was assumed that both compact and loose data would be generalised as this is the only way to prove the quality of the generated results and hence the efficiency of the generalisation tool.

The experiment concerned building generalisation between an original scale of 1:500 and 1:5000 (BDOT500) [1] and a scale of 1:500 to 1:10000 (BDOT10K) [2]. These scales are representative of the Topographic Objects Database (BDOT) provided by law as a source of data for the development of standard cartographic studies as part of the infrastructure for spatial information in Poland (IIP) [1].

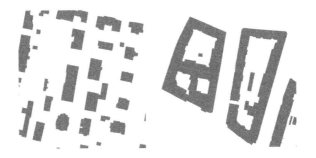

Fig. 7. Example data: loose buildings on the left, compact buildings on the right. The data were taken from the Topographic Objects Database (BDOT500, scale 1: 500).

The aim of the study was to compare the results of existing commercial solutions with the methods described in this article. For comparative studies the Simplify Buildings method was used, which is part of ArcGIS 10.2.2 software for generating a set of simplified buildings. The input tool parameters are minimum area and simplification tolerance. This method also performs a certain degree of building rectangularisation. Comparison of the two methods is shown in Fig. 8.

4   Methods Comparison

When the obtained results for sample scales and parameters are compared it can be noted that the ArcGIS algorithm does not entirely eliminate buildings (Fig. 7 and 8). This method causes ambiguity as it sometimes leaves or removes small objects and does not remove small (smaller than the tolerance) spaces between buildings. Methods statistics are compared in table 1.

Fig. 8. Example results showing on the left the original data,, in the middle the result of the ArcGIS tool for building simplification (parameters: tolerance: 5m and min. size: 25m2) and on the right the results obtained from the proposed algorithm. Operational scale: 1:10 000 and line width: 0.2 mm)

Table 1. Summary of total surface area and number of subjects before and after generalisation

Process	Total surface $[m^2]$	Number of objects	Total change of the size [%]	Total surface difference $[m^2]$
Input data	28383,72	143		
ArcGIS	28581,77	136	95%	198,05
Author's method	28424,36	123	86%	40,64

It is also worth noting that the input parameters are tolerance and minimum surface area, which can be problematic when results relevant to the adopted scale are desired (as in the proposed method, where scale is an input parameter). When the total surface area after processing with the ArcGIS tools are compared with results from the proposed solution we can observe that the difference is quite substantial. This is due to expansion of the objects during their merging and rectangularisation.

The main difference between the proposed algorithm and the existing solution is the lack of a surface objects elimination parameter and the use of minimum dimensions of drawing recognition. The authors present a solution that could also take into account the width of the line drawing, making possible its use in cartographic generalisation. A characteristic feature and a defect of both algorithms

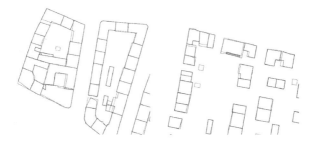

Fig. 9. The results of compact building simplification on the left and loose buildings on the right. Gray lines depict the ArcGIS tool results (5m tolerance and $25m^2$ min. size) and black depicts the proposed tool results (scale 1:10 000, line width: 0,2 mm))

is the possibility of the occurrence of topologically inconsistent objects, however, in the case of ArcGIS even self-intersections of buildings are produced (Fig. 9). This problem can be solved by implementing appropriate tools for detection of incorrect topological relations, based also on minimum recognition dimensions. This will be the subject of further study by the authors.

References

1. Dz. U. z 2013 r. poz. 383 Rozporządzenie ministra administracji i cyfryzacji z 12 lutego 2013 r. w sprawie bazy danych geodezyjnej ewidencji sieci uzbrojenia terenu, bazy danych obiektów topograficznych oraz mapy zasadniczej
2. Dz.U. z 2010 nr 76 poz. 489 Ustawa z dnia 4 marca 2010 r. o infrastrukturze informacji przestrzennej
3. Bajerski, P.: Optimization of geofield queries. In: Proceedings of the 1st IEEE International Conference on Information Technology, pp. 1–4 (2008)
4. Bajerski, P.: How to Efficiently Generate PNR Representation of a Qualitative Geofield. In: Cyran, K.A., Kozielski, S., Peters, J.F., Stańczyk, U., Wakulicz-Deja, A. (eds.) Man-Machine Interactions. AISC, vol. 59, pp. 595–603. Springer, Heidelberg (2009)
5. Bajerski, P., Kozielski, S.: Computational Model for Efficient Processing of Geofield Queries. In: Cyran, K.A., Kozielski, S., Peters, J.F., Stańczyk, U., Wakulicz-Deja, A. (eds.) Man-Machine Interactions. AISC, vol. 59, pp. 573–583. Springer, Heidelberg (2009)
6. Brassel, K.E., Weibel, R.: A review and conceptual framework of automated map generalization. International Journal of Geographical Information Science 2, 229–244 (1988)
7. Chrobak, T., Keller, S.K., Kozioł, K., Szostak, M., Zukowska, M.: The basics of digital cartographic generalization. Uczelniane Wydawnictwa Naukowo-Dydaktyczne AGH (2007)
8. Chrobak, T., Kozioł, K., Krawczyk, A., Lupa, M., Szombara, S.: Automatization of Generalization Process in Multiresolution Databases. Wydawnictwa AGH (2013)
9. Kozioł, K.: Operators for building generalization. Annals of Geomatics 10 (2000)

10. Kozioł, K., Lupa, M., Krawczyk, A.: The Extended Structure of Multi-resolution Database. In: Kozielski, S., Mrozek, D., Kasprowski, P., Małysiak-Mrozek, B., Kostrzewa, D. (eds.) BDAS 2014. CCIS, vol. 424, pp. 435–443. Springer, Heidelberg (2014)

11. Lupa, M., Piórkowski, A.: Spatial Query Optimization Based on Transformation of Constraints. In: Gruca, A., Czachórski, T., Kozielski, S. (eds.) Man-Machine Interactions 3. AISC, vol. 242, pp. 627–636. Springer, Heidelberg (2014)

12. Mackaness, W., Ruas, A., Sarjakoski, T.: Observation and Research Challenges in Map Generalisation and Multiple Representation. In: Generalisation of Geographic Information Cartographic Modelling and Applications, pp. 315–323. Elsevier (2007)

13. McMaster, R.B., Shea, S.K.: Generazization in Digital Cartography. In: Resource Publications for College Geography Resource Publications in Geography, pp. 1–67. Association of American Geographers (1992)

14. Olszewski, R.: Cartographic modelling of terrain relief with the use of computational intelligence methods (habilitation thesis). In: Prace Naukowe Politechniki Warszawskiej. seria Geodezja, Oficyna Wydawnicza Politechniki Warszawskiej (2009)

15. Saliszczew, K.A.: Kartografia Ogolna. Wydawnictwo Naukowe PWN (1998)

16. Sester, M.: Generalization based on least squares adjustment. In: International Archives of Photogrammetry and Remote Sensing, pp. 931–938. ISPRS Archives (2000)

Database Systems Development

Motivation Modeling and Metrics Evaluation of IT Architect Certification Programs

Michal Turek[✉] and Jan Werewka[✉]

Department of Applied Computer Science
AGH University of Science and Technology
al. Mickiewicza 30, 30-059 Krakow, Poland
{mitu,werewka}@agh.edu.pl
http://www.kis.agh.edu.pl

Abstract. The alignment of university curricula to the needs of the IT industry is a great challenge which needs analysis of various different aspects. IT architecture competencies and skills are very important to parties such as the IT industry, course providers, universities, and of course students. In this paper IT architect certification programs are analyzed as they need to be well-aligned to the needs of the industry. The range of IT architect certification programs on todays market is vast and rather complicated. This article describes a new lightweight method for quickly evaluating IT architect certifications using specially developed data collection methods. The method concentrates on non-domain certification features and introduces metrics which can be used to compare programs with each other. Broad research has been done to identify the most important domain-independent features of certificate programs for IT architects on the employee market today. These features have been selected, evaluated and combined into a metrics formula using a specially developed automated data mining process. The metrics can also be automatically updated in a process called "self-adaptation" after a specified period of time. The whole process assumes that the highest-ranking certificate programs from a previous time period can be used as a reference for establishing domain-independent features in the next period. Each certificate program can currently be evaluated only once per period based on the reference. The proposed solution will deliver a powerful tool for IT architect skill comparison, especially when there are many job candidates with different sets of certification documents to be assessed. The research results are currently being used to design architecture courses at the IT Architecture Academy at the AGH Univ. of Science and Technology.

Keywords: Data mining · IT architect · Software architect · Certificate programs

1 Introduction

IT industry, students and the universities themselves. However, these different parties view the problems differently. Alignment is essential for postgraduate students or training course participants, because they usually have experience

© Springer International Publishing Switzerland 2015
S. Kozielski et al. (Eds.): BDAS 2015, CCIS 521, pp. 463–472, 2015.
DOI: 10.1007/978-3-319-18422-7_41

and ideas that the industry needs. University educators are willing to deliver the best competences for the IT sector, but usually would like the competences to be independent of technology vendors. Preparing computer science (CS) and software engineering (SE) students for real-world jobs in the industry is a challenging task. To meet such demands, the inclusion of software engineering projects in computer science curricula is recommended. These project-based courses that can span the entire final year of a program are often referred to as capstone courses, and have been successfully used at many Universities. In paper [28] a company approach is proposed which can be treated as an adaptation of a capstone project with a larger number of students simulating company behavior. Paper [30] concludes that there is no reason to separate university teaching activities from R&D&I activities. As a result of their work in companies, professors have learned new concepts and methodologies that can help improve their teaching and their contact with companies, and subsequent reflection has enabled them to learn continuously and enhanced their motivation when working with students. Additionally, [29] includes an overview of various kinds of certification, but a knowledge- and experience-based certification was chosen. Finally, a practically-oriented approach was chosen in which training was combined with the real project work of participants and the content was explicitly oriented to the development of large, complex, software intensive systems using innovative software engineering techniques. The current article presents a selected systematic approach to designing training course curricula for IT sector architects and a new postgraduate study program to increase the competency of software architects. Different types of investigations need to be carried out to make proper decisions about setting up these curricula. The curricula are also intended as an educational offer to all students coming to AGH UST as part of a possible international exchange program. It is important that these programs educate IT architects according to their needs, as well as the needs of the industry. To resolve this problem, the following should be done:

- Identify work positions in IT companies associated with IT architects [33]
- Determine IT competencies and domain characteristics [32]
- Specify what non-domain features are of vital importance to the industry and the candidates (in this paper).

In considering the non-domain features of the courses it is difficult to evaluate and compare different solutions. The authors decided to perform such a study here based on:

- An analysis of training programs (for the determination of competencies).
- A review of certificates (for determining the characteristics of non-textual domain-specifics).
- An evaluation of the comparison of different solutions and developing practices concerning the setting up of new courses.

The approach will be evaluated at our Faculty of Electrical Engineering, Automatics, Computer Science and Biomedical Engineering, an institution that educates students at all levels of computer science (undergraduate, graduate, doctoral and postgraduate) and also delivers training courses.

2   Architecture Roles in IT Companies

The development and assessment of architect competencies in IT companies is a complex issue. A number of roles must be distinguished within an organization that differ in terms of scope of competencies, responsibilities, and position profile. As a rule, there are five categories of architects distinguished in companies that develop software [28,29,31,6]: software, infrastructure, information, and business and corporate architects managing all architecture initiatives within the organization. There is no common convention on the naming of IT architect roles. Fig. 1 shows the frequency of IT architect job titles found in the Internet.

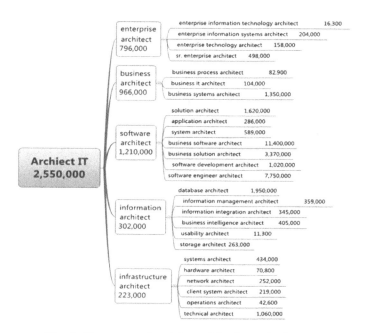

Fig. 1. Frequency of architect titles from the Internet

The correct assessment and development of IT architects within a company is one of the key elements in the process of developing competencies. Reliable assessment allows the current suitability of an employees competencies to be verified in terms of the positions that they currently hold, allowing employees to be placed on appropriate career paths. It is important that employees are provided with a clear message about the most desired competencies and that the company obtains a clearer picture of the level of qualifications of its employees.

3 Research Methodology

The goal of the project was to produce a coherent quality assessment system for certifying the skills and competencies of software architects. The resulting criteria should be the basis for assessing the weight and quality of a given certificate of competence, and above all, for obtaining a measurable comparative assessment of the competence of the holders of various certificates. Certificates were assessed not only in terms of their similarity to software architect competence profiles, but also other aspects. As a result, a solution needed to be created which could calculate the actual importance factor for any given certificate of competence. In the field of software architect competence, the competences division could simply be the results of mixing numerous competence profile proposals emerging from the Open CITS [13]. The aim of the proposal is naturally the creation of a more comprehensive certificate fitness evaluation procedure which leads to more precise assessment of a certified IT architect's competence. A typical and obvious outcome of the recruitment process in a Human Resources department is of course the choosing of the most competent person from the pool of available candidates to fill a given role. One of the criteria used might be a certification competence system. However, the problem is that the certificates market is vast and there is no way to effectively compare certificates. Alternatively (in a small pool of certificates), it is possible to research how popular or difficult to obtain certificates are, but without any significant comparative scale. The proposed solution introduces a universal procedure that defines a quantifiable metric for certificate comparison. The validation of personal competences confirmed by set of certificates held will rely on straight algebraic calculation and is a product of:

- the quality indicator of a certificate (focusing on all parameters not related to a certificate program),
- the level of certification fit to the required knowledge and skills.

Collecting atomic values of competence for the certificates held by a person will finally assess the personal competence documented by these certificates. A suitable certificate evaluator must interpret two groups of characteristics affecting the assessment the value of a certificate: referred to as the certification domain of knowledge and the domain-independent factors. There is no universal method of identifying a pool of components which is common to these groups, therefore a procedure was created for calculating an index based on a collection of time-varying components. A regular per-period indicator version revision is also necessary and appropriate due to rapid changes in the requirements of the industry. All "non-domain of knowledge" certificate characteristics (popularity, language, exam required etc.)will be placed in the non-program category of attributes. A set of such attributes will be the basis for calculating the index component. In addition, those individual components do not need to be construed as identically significant, which raises the problem of determining their actual meaning. Within the main solution path the following procedure will be defined and used during the research process:

- identify a list of the most significant market certificates of competence held by software architects,
- for each identified item determine the most important pool of non-program attributes and their meanings (evaluated in the form of percentage points),
- use this data to validate any certificate (including those outside of a predefined list) held by software architects.

A in-depth analysis of the certificate market and numerous case studies based on individual certification schemes will help us to choose non-program features and rank them (mainly in terms of the number of appearances in the description or reference to a particular certification system).A list of program-dependent features can be established in a classical way, thus these skills (possibly also linked to the possession of applied knowledge) determine the knowledge of a software architect. To maintain consistency throughout the process, each list will be limited to a flexible golden mean of twenty entities. It is assumed that a large number of entities could complicate the evaluation procedure. This limit has to be reasonably well maintained to ensure adequate analysis precision.

4   MC20 Certificate Quality Metrics

The proposed metrics is named "MC20" (Metrics of Competency, Twenty). Twenty is just a reasonable number which produces a high degree of precision from an acceptable amount of data each time a certificate is evaluated. The contents of the non-program features list were established as the result of a vast amount of real certificate program analysis (a set [16,21,8,19,23,17,11,3,10,26,9,12,4,5,7,1,2,20,15,24] was initially processed). The list was then reduced to 20 items. The analysis of certification list elements searches only for the weighting factors of twenty specific non-program attributes. Adjustable metrics calculations are not related to a specific certification system. The metrics calculation procedure for a particular certificate will therefore be independent of future market conditions associated with the certification system. Feature weights can be established when features have been sorted by popularity, which is the most significant determinant. The procedure determines the amount of references to the characteristics of certificates according to:

- certification systems official web sites (stressing the advantages of their own certification system),
- scientific publications, which usually refer to the official web sites of certifications,
- corporation web sites (those possible to reach) which accept particular certificates during the recruitment process,
- other web sites, containing references to the certification system (e.g. published training materials prior to the examination for the certificate)

A web text search engine has been involved into the process. Based on a standard Google search robot it was possible to prepare a cross-word automated search procedure. Data collection for the process started with the selection of twenty well-known certificate systems [16,21,8,19,23,17,11,3,10,26,9,12,4,5,7,1,2,20,15,24].

The question was: which non-domain certificate features are most important for certificate evaluation? However, this importance cannot be evaluated easily as it depends on a particular person's opinion, interpretation context and many other factors. Therefore, a huge amount of web references to all possible features was collected by an individual web search for each feature. In the course of this procedure each certification program feature (there were about 70 features defined in a preliminary test database) was assigned a dictionary of phrases considered as the feature's synonyms. For instance, "payment for training" is alternately described as a series of phrases such as "training price", "expensive", "certificate pricing", "cost", "prices" etc. If a feature was negative (and subsequently: bad for the certificate evaluation result), an additional intersection was performed with the words "not", "no", "avoid", "-less" etc. in the dictionary. The Google search engine was used for the Cartesian product search with all corresponding keywords. Well-known certification system names (as search keywords) were merged with all possible feature names and merged again with all the entries from a particular feature synonym dictionary. This resulted in a lot of search requests made with the Google search engine. A given certificate system was considered to have a feature whenever a search returned a result for an article, forum opinion, advertisement message etc. The number of hits assigned to the feature was then incremented, meaning "this feature is more important, because it was mentioned somewhere again". The histogram in Fig. 2 shows sample results listed with the hit count for each non-program feature (sorted by increasing hit-count). The sum of hit-counts for all the features in this case was 269.

Fig. 2. Histogram of certificate programs features

Let us now assume we are evaluating a real certificate held by a candidate. To do so, we have to establish the features that exist for every single entry in the list above for a given certificate system. If a feature exists, we just add a corresponding hit to the final certificate score. If no information about a feature is found, half a hit-count is added to the score. Finally, if a feature is definitely absent, no value is added to the score. The sum of hit-counts for all the features for the first sample process turned out to be 269. Having also just calculated a real certificate evaluation result we can generate a non-domain certificate value factor by dividing the sum of collected points by 269. This would be equivalent to simple value normalization into $< 0..1 >$ range. Finally, we can evaluate a popular non-program feature pack for each certificate system. We have to do this just once for each period of time. If another candidate comes with the same certificate there is no need to repeat the process. A certificate ranking database can be now built. This will enable an automatic "self-adaptation" feature for the MC20 metrics in future time periods. For this reason a special meta-model which describes evaluation data has been established. The model defines an entity for each certificate evaluation event, including certificate features possession (yes-or-no) for components. Non-domain feature rankings in the next period will be taken from a database as sums of all points collected during certificate evaluations in the past. The meta-model also introduces a timeline which allows for the storage of all the rankings from previously recorded periods of time. The certificate evaluation mechanism can now be improved with each subsequent period. This evaluation will also allow a strategy to be developed for the introduction of new non-domain features on a timeline. The self-adaptation process has been tested with an interactive software tool which was developed to calculate MC20 metrics and store information sent by users for each previously evaluated certificate. The tool can also perform a MCS20 "self-adaptation" procedure, generating new non-domain feature rankings for the next time period.

5   An Approach to Select the Best Certification Program

In the previous section the most important non-domain features were determined. Now, on the basis of the selected attributes it is important to compare and select the best certificate programs for a given scenario. This work will be continued with certificate program evaluation using the feature pairwise comparisons method [27] which delivers effective certificate comparison tools for IT-architects (Fig. 2). To apply the method in the first step, the attributes of certificate programs should be classified according to whether they are tangible or intangible. In the second step, certificate offers should be selected for the purpose of comparison. The offers are divided into reference offers (preselected) and submitted by other stakeholders at any time for comparison. For the reference offerings, the most well-known certification programs are selected. These certification programs are developed by the following institutions or organizations: SEI (Software Engineering Institute of Carnegie Mellon University), Open Group, IASA, and iSAQB. SEI (Software Engineering Institute of Carnegie Mellon University) offers a few certificate programs. SEI software architecture certifications [18] (Software Architecture Professional Certificate) are based on their own methodology using four main architecture development steps: quality attribute workshop, attribute driven design, architecture documentation, and evaluation of architecture solutions. Additionally there are other certificates related to software architecture like the SEI SOA Architect Professional Certificate.

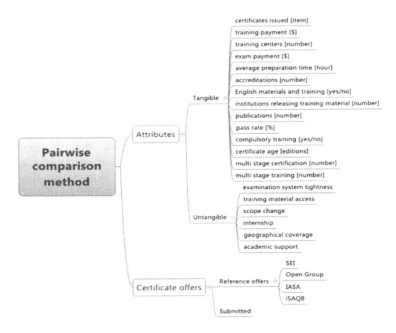

Fig. 3. Mind map of certificate program selection based on the pair wise comparison method

Open CA certification [25,22](Open Group Certified Architect) is primarily concerned with the competencies that an architect possesses. Applicants must prove that they have many years of real-life IT architecture design experience. IASA (International Association of Software Architects) [14] has developed a collection of architectural standards and best practices described in ITABOK (IT Architect Body of Knowledge). The architect certificate levels are distinguished associate, professional and masters. For the last two certificate levels some practical experience is needed. There is also a certificate program for software, information, infrastructure, business and enterprise architects. The iSAQB (international Software Architect Qualification Board) organization offers a certification scheme consisting of three levels: foundation, advanced and expert level, the latter of which is currently under development [13]. From an analysis of the four reference certification programs it is obvious that other attributes for comparison of certification programs should be selected. The non-domain attributes concern, for example, the independence of the examination provider, the type of examination (test, project work), whether practical experience is mandatory or not, and whether the certification program is based on their own methodology (SEI, Open Group) or on a body of knowledge (IASA, iSAQB). From the current analysis it may be concluded that the comparison of different certification programs is a considerably complicated task which should be thoroughly undertaken. However, the current results give very impressive insights which may be used to define courses better suited to IT sector employees.

6   Conclusions

Developing university courses that are best suited for industry is a difficult but important task. Domain and non-domain features must be considered when designing university level courses. The domain features not considered here are concerned with scope and competence levels. In determining domain features the authors propose the development of a software engineering landscape in which IT architecture characteristics will be included. The work described in this paper provides a reliable and universal tool for non-domain feature evaluation for IT architect certification programs. It can be easily extended to other domains, not only for software architect evaluation. The proposed metrics is also self-adaptive as it could be recalculated for new situations in the future IT architect certification program market. So, it is possible to store results and apply them for any newly discovered certification programs which need quick evaluation. The results of the certification program evaluations are being used for setting up courses by the IT Architecture Academy (http://www.it-architecture.agh.edu.pl).The authors also used their own experience from completing the SOA and Software Architecture Professional Certificate paths offered by SEI [18].

References

1. CCSA - check point certified security administrator, https://www.cpug.org
2. CCSP - cisco certified security professional,
   http://www.cisco.com/web/learning/certifications/professional/
   ccnp_security/index.html
3. CISCO training & certifications,
   http://www.cisco.com/web/learning/certifications
4. CISM - certified information security manager,
   http://www.isaca.org/Certification/
   CISM-Certified-Information-Security-Manager
5. CISSP - certified information systems security professional,
   https://www.isc2.org/cissp/
6. Common criteria for information technology security evaluation,
   http://www.commoncriteriaportal.org/files/ccfiles/CCPART3V3.1R4.pdf
7. CompTIA security+ group, http://certification.comptia.org
8. CompTIA Strata IT Fundamentals,
   http://certification.comptia.org/getCertified/certifications
9. CSSLP - certified secure software lifecycle professional,
   https://www.isc2.org/csslp-certification.aspx

10. HP expert one certification system, http://www.hp.com/certification/
11. IBM professional certification program, http://ibm.com/certify
12. IEEE computer society - premier organization of computer, http://www.computer.org/portal/web/guest/home
13. iSAQB certification, http://www.isaqb.org/certifications/
14. ITABOK - IT architect body of knowledge, iasa, http://www.iasahome.org/web/home/skillset
15. ITSEC - information technology security evaluation criteria, http://www.ssi.gouv.fr/site_documents/ITSEC/ITSEC-uk.pdf
16. Oracle partnerwork, http://www.oracle.com/partners/en/knowledge-zone
17. SANS computer security training, certification and research, http://www.sans.org/
18. SEI certification carnegie mellon, http://seicertification.clearmodel.com/get-certified/software-architecture
19. SOA certified professional, http://www.opengroup.org/architecture
20. TCSEC - trusted computer system evaluation criteria, https://www.ccn-cert.cni.es
21. The Open Group Architecture Framework (TOGAF), http://www.opengroup.org/architecture
22. The Open Group Certified IT Specialist (Open CITS) Program, https://www.opengroup.org/opencits/cert
23. TOGAF enterprise edition, http://www.opengroup.org/architecture
24. Zapthink SOA training & certification, http://www.zapthink.com/soa-training-certif
25. The Open Group Certified Architect (Open CA) Program: Conformance Requirements (Multi-Level), Document Number: X110 (2011)
26. Gallagher, P.R.: National Computer Security Center (1988), https://www.fas.org/irp/nsa/rainbow/tg009.htm
27. Kulakowski, K., Szybowski, J.: Tender with success – the pairwise comparisons approach. Procedia Computer Science 35, 1132–1137 (2014)
28. Meawad, F.: The virtual agile enterprise: Making the most of a software engineering course (2011)
29. Paulisch, F., Zimmerer, P.: A role-based qualification and certification program for software architects: An experience report from siemens (2010)
30. Plaza, I., Igual, R., Medrano, C., Rubio, M.: From companies to universities: Application of R&D&I concepts in higher education teaching (2013)
31. Steghuis, C., Voermans, K., Wieringa, R.: Interview report on the competences of the ICT-Architect (2005)
32. Werewka, J., Jamroz, K., Pitulej, D., Stepień, K.: A problem of development and assesment of architecture competences at it companies (in Polish). Studia Informatica, 311–312 (2014)
33. Werewka, J., Stepien, K.: A process of defining the it architects posts in software development companies. In: Organization Development Management in Multicultural Environment, pp. 241–252 (2014)

Supporting Code Review by Automatic Detection of Potentially Buggy Changes

Mikołaj Fejzer[(✉)], Michał Wojtyna, Marta Burzańska, Piotr Wiśniewski, and Krzysztof Stencel

Faculty of Mathematics and Computer Science
Nicolaus Copernicus University, Toruń, Poland
{mfejzer,goobar,quintria,pikonrad,stencel}@mat.umk.pl

Abstract. Code reviews constitute an important activity in software quality assurance. Although they are essentially based on human expertise and scrupulosity, they can also be supported by automated tools. In this paper we present such a solution integrated with code review tools. It is based on a SVM classifier that indicates potentially buggy changes. We train such a classifier on the history of a project. In order to construct a training set, we assume that a change/commit is buggy if its modifications has been later altered by a bug-fix commit. We evaluated our approach on 77 selected projects taken from GitHub and achieved promising results. We also assessed the quality of the resulting classifier depending on the size of a project and the fraction of the history of a project that have been used to build the training set.

Keywords: Bug detection · Code review · Github · Gerrit · SVM · Weka

1 Introduction

Contemporary projects use numerous tools to improve the quality of resulting software systems. Among them there are code review tools (e.g. Gerrit), builders (e.g. Maven) and testing environments (e.g. JUnit). Before a change is merged into the code repository, it is thoroughly proof-read and automatically checked if it breaks the building process or units tests. Unfortunately, only a fraction of all defects can be detected by reviewers. Prevalent software bugs are usually missed. In our opinion a proper automated support of the code review process can prevent a significant number of bugs that are often left unnoticed.

Inspiration and Related Work

In the literature there is number of approaches to the automatic verification of changes. The article [4] presents a method based on the extraction of knowledge from relations of data items; it is so called *multi-relational data mining*. It consists in building new relations in the form of "hypotheses" describing the given problem. They are implied by induction from known examples of existing relations. A training set is composed of positive and negative examples of the examined relation. They are described by existing relations and base knowledge. The article

© Springer International Publishing Switzerland 2015
S. Kozielski et al. (Eds.): BDAS 2015, CCIS 521, pp. 473–482, 2015.
DOI: 10.1007/978-3-319-18422-7_42

analyses C++ student projects and an automatically generated set. The goal of those analyses were to tell whether a given class contains errors or not.

On the other hand Jeong [8] struggles not only to recognize buggy changes but also tries to indicate the best person to review the code. His method is based on Bayesian networks and it was tested on Firefox's source code.

Ostrand [14] asked a questions whether adding the programmer to predictors would aid error detection based on quality metrics. He examined a dozen of projects using the negative binomial regression. The correlation of the identity of a programmer and bugs was confirmed. The presented tool was able to indicate 20% of the whole system that contained from 75% up to 90% of bugs.

We were especially interested in the findings of Sunghun Kim, who described his method in [11] based on SVM. Our approach to improve code review process was inspired by works of Stacy Nelson and Johann Schumann [13] who described how different approaches to code review might change the outcome of the reviews. The usage of parser combinators [12] for repository querying was influenced by [2]. The authors of this paper describe common pitfalls when mining Git repositories. The decision to use parser combinators allowed creating numerous parsers by combining reusable functions. However, it also might cause problems with memory management/garbage collection in case of large repositories.

Automatic analyses of the quality of changes have been also presented in [5,15,1,11,3,17].

Contribution

Recent studies on bug detection by Sunghun Kim [11,10] have inspired us to create a bug detection tool integrated with a code review system.

This automated tool identifies potential bugs in changes about to be merged into the repository of a version control system. It is integrated with Gerrit, a popular code review system. It is based on a classifier that takes changes and answers whether they are potentially buggy. This classifier is trained using the past of the project. We build the training set from past commits. We assume that a commit is clean if none of its changed lines has been later bug-fixed. All other commits are attributed to be buggy. The trained classifier is integrated into Gerrit and it suggests buggy changes. A reviewer is thus warned. However, he/she can ignore such a signal.

We prepared a proof-of-concept implementation of this classifier [6]. We tested it on 77 Github repositories and cross-validated it [11]. The results of these tests are promising. They also give insights how to create better tools in the future.

The contributions of this paper are as follows:

- A novel idea to build classifiers of changes submitted to version control repositories.
- A proof-of-concept implementation of the classifier, its thorough experimental evaluation on 77 Github projects and cross-validation. The implementation is avaliable on https://github.com/mfejzer/CommitClassification.
- An integration of the classifier into the code review system Gerrit.

The article is organized as follows. Section 2 presents the details of the creation of the training set and the build of the classifier. Section 3 shows the results of an experimental evaluation of the proposed method. Section 4 concludes and rolls out possible future directions of research.

2   The Method and Its Implementation Details

Our solution is based on an SVM (Support Vector Machine) classifier [21,9]. Such classifiers are hyperplanes in multidimensional spaces. Training such a classifier consists in selecting the parameters of a hyperplane so that it effectively separates the sets of negative and positive points.

In our method an SVM classifier tells clean and buggy commits apart. The training set is deduced from the development history of a project in terms of the prevalence of errors. In order to identify buggy commits we use the algorithm from [18] which discovers bugs based on later corresponding fixes. We assume that fixing commits are those that contain a term such as "fixes", "bug" etc. For each such a commit, we query the project repository for commits that added or altered the lines removed or modified in the fixing commit. Those commits are considered buggy, while all other commits are assumed clean. This creates the training set. It contains the content of commits and decisions whether they are clean or buggy. We use a specified number of initial commits to build the training set. We call this number the *history limit*.

Having the training set, we start building the classifier. Its multidimensional space is the set of vectors of all words that occur in commits. A single commit is interpreted as a bag of words in the space with discrete coordinates. A coordinate is the number of occurrences of the given word in a commit. This is the standard "bag of words" model.

In order to train the classifier (i.e. to find an appropriate hyperplane) we use a dedicated library from the Weka toolkit [7]. The resulting classifier is saved to be used in the future when new changes (candidate commits) arrive. The experiments discussed in section 3 show that the history limit of 100 commits is sufficient to build a good classifier.

The classification of a new change comprises its conversion to the "bag of words" model and then application of the trained classifier to decide if the change is buggy or clean. The resulting diagnosis is then communicated to the Gerrit code review system. This way the author and the reviewers are notified if the change seems dangerous or not.

The processing of commits begins by the execution of a series of git commands and parsing their results. We had chosen parser combinators [12] as a method to construct parsers. Our decision was motivated by the extensibility and easiness of unit testing enabled by this method. New and more complex parsers can be constructed from existing ones by the usage of special combining functions, provided by Scala language. This means that additional tools such as parser generators are not needed.

Table 1. The statistics of preliminary tests

Project	History limit	Training set size	Testing set size	Buggy	Clean	Correctly classified	Incorrectly classified	Revision
Gedit	500	400	100	16	484	96.00%	4.00%	99f6154
Egit	500	400	100	124	376	75.00%	25.00%	704f311
Svn	2000	1500	500	792	1208	78.25%	21.75%	840072

3 Experimental Evaluation

We divided our experimental evaluation into following three groups of tests:

1. preliminary tests, conducted on Egit, Gedit and Svn,
2. comparison tests, run on 77 Github repositories with the same history limit,
3. detailed tests, with different history limits on selected Github repositories.

The goal of the first group of tests was to check if our software worked for different kinds of version control repositories. We developed it to be used primarily with Git repositories, but we have also verified that it works fine with SVN (namely Apache SVN). We also wanted to examine differences in training the bug detection classifier when applied to various sizes of the project. We used a small project (Gedit), a medium project (Egit) and a large project (Svn). Our goal was to validate that our solution is general enough to be used for any size of a software project. Tests were run on projects cloned directly from Github, without any preprocessing.

The results of preliminary tests (the first group) are presented in table 1. They attest the efficiency of the implemented method. Those tests were conducted in a similar way as described in [11].

The results of preliminary tests inspired us to try our method on much broader class of projects.

As the second test group, we applied our solution to 77 project repositories. We partitioned them into four categories according to the project size as listed in table 2. The results show that the total number of commits determines the number of fixes that can be identified (see table 3). This has a significant impact on the process of classifier training.

The third group of tests was meant to determine how the history limit influences the accuracy of the classification. Tables 4 and 5 show that the history limit hardly influences the quality of the classification. However, we noted that the classifiers trained for very young projects are significantly worse than the classifiers obtained for older projects.

Another important question is how to set the history limit, i.e. how many commits use to train the classifier. The next test is devoted to answer this question. The training set comprised 80% of last commits selected randomly for given history limit. The remaining commits were used for evaluation, as shown in table 6. We obtained better results using only the last 500 commits (table 9) than 2500 (table 10) and 5000 (table 11) respectively. This shows that over-fitting could also be a concern in our solution.

Table 2. Groups of Github projects by repository size

Group name	Repository (MB) min	max	avg	Project names		
Small	0,5	6,30	4,01	beanstalkd devtools folly impress.js octopress Slim zipkin	chosen django-debug-toolbar gizzard jekyll paperclip stat-cookbook	devise flockdb httpie mosh shiny vinc.cc
Medium	6,80	19,00	11,79	android CraftBukkit html5-boilerplate libuv plupload scalatra tornado	boto facebook-android-sdk jquery memcached redis Sick-Beard	clojure flask knitr phpunit requests storm
Large	20,00	54,00	36,63	ActionBarSherlock ccv d3 Font-Awesome netty RestSharp ThinkUp	bitcoin CodeIgniter django-cms homebrew ProjectTemplate sbt	cakephp compass finagle libgit2 reddit SparkleShare
XL	57,00	757,00	178,00	akka elasticsearch hiphop-php mono rails SignalR TrinityCore	diaspora foundation kestrel node scala symfony zf2	django gitlabhq mongo phantomjs ServiceStack three.js

Table 3. The statistics of tests on Github projects by repository size

Group name	Correctly classified	%	Incorrectly classified	%
Small	72,95	82,28%	16,11	17,72%
Medium	81,74	82,29%	17,53	17,71%
Large	79,89	84,29%	15,26	15,71%
XL	84,40	84,40%	15,60	15,60%
On average	79.81	83,33%	16.12	16.67%

Table 4. Groups of Github projects by history length

Group name	History length			Project names		
	min	max	avg			
Small	1	985	607	beanstalkd facebook-android-sdk Font-Awesome memcached plupload shiny zipkin	ccv flockdb httpie mosh ProjectTemplate stat-cookbook	chosen folly impress.js octopress RestSharp vinc.cc
Medium	986	2693	1757	ActionBarSherlock CraftBukkit django-debug-toolbar html5-boilerplate paperclip Slim tornado	android devise flask kestrel phantomjs storm	clojure devtools gizzard libuv scalatra ThinkUp
Large	2694	5892	3859	bitcoin d3 jekyll netty redis ServiceStack SparkleShare	boto finagle jquery phpunit requests Sick-Beard	compass foundation knitr reddit sbt SignalR
XL	5893	94432	18965	akka diaspora elasticsearch homebrew mono scala TrinityCore	cakephp django gitlabhq libgit2 node symfony zf2	CodeIgniter django-cms hiphop-php mongo rails three.js

Table 5. The statistics of tests on Github projects by history length

Group name	Correctly classified	%	Incorrectly classified	%
Small	64.37	78.65%	19.11	21.35%
Medium	82.84	82.84%	17.16	17.16%
Large	84.32	84.32%	15.68	15.68%
XL	87.30	87.30%	12.70	12.70%
On average	79.81	83,33%	16.12	16.67%

Table 6. The relation between history limit and training/testing sets

History limit	Training set size	Testing set size
100	80	20
200	160	40
500	400	100
2500	2000	500
5000	4000	1000

Table 7. The statistics of tests with history limit set to 100

Project name	Correctly classified	%	Incorrectly classified	%	Bugs detected	False negative	False positive	Non bugs detected
akka	17	85.00%	3	15.00%	0	3	0	17
mongo	19	95.00%	1	5.00%	0	1	0	19
reddit	20	100.00%	0	0.00%	0	0	0	20
scala	19	95.00%	1	5.00%	0	1	0	19

Table 8. The statistics of tests with history limit set to 200

Project name	Correctly classified	%	Incorrectly classified	%	Bugs detected	False negative	False positive	Non bugs detected
akka	35	87.50%	5	12.50%	0	5	0	35
mongo	37	92.50%	3	7.50%	0	3	0	37
reddit	38	95.00%	2	5.00%	0	2	0	38
scala	36	90.00%	4	10.00%	0	4	0	36

Table 9. The statistics of tests with history limit set to 500

Project name	Correctly classified	%	Incorrectly classified	%	Bugs detected	False negative	False positive	Non bugs detected
akka	82	82.00%	18	18.00%	0	18	0	82
mongo	86	86.00%	14	14.00%	0	14	0	86
reddit	92	92.00%	8	8.00%	0	8	0	92
scala	82	82.00%	18	18.00%	0	17	1	82

Table 10. The statistics of tests with history limit set to 2500

Project name	Correctly classified	%	Incorrectly classified	%	Bugs detected	False negative	False positive	Non bugs detected
akka	386	77.20%	114	22.80%	29	95	19	357
mongo	389	77.80%	111	22.20%	7	103	8	382
reddit	402	80.40%	98	19.60%	6	92	6	396
scala	375	75.00%	125	25.00%	30	105	20	345

Table 11. The statistics of tests with history limit set to 5000

Project name	Correctly classified	%	Incorrectly classified	%	Bugs detected	False negative	False positive	Non bugs detected
akka	737	73.70%	263	26.30%	133	176	87	604
mongo	768	76.80%	232	23.20%	28	211	21	740
reddit	591	77.36%	173	22.64%	60	129	44	531
scala	732	73.20%	268	26.80%	158	185	83	574

These results prove that the history limit of 100 commits is sufficient to train a satisfactory classifier. This observation allows using our approach with daily training of the classifier from scratch. Then, the classification of incoming commits is performed using hundreds of most recent commits. This seems most appropriate since the classifier is continuously synchronized with the current (up to yesterday) composition and maturity of the development team.

4   Conclusions and Future Work

In this paper we have shown a method to detect possibly buggy changes. The method is based on a classifier trained using a project's history. The training set is built according to the assumption that clean changes are not later altered by bug fixing commits. We evaluated the method on a number of projects and achieved promising results.

We have observed that small projects are harder to analyze due to the low number of commits and thus fixes. However, code review is less likely to be used in such projects. Thus the benefits of potential error detection are marginal. Results of our evaluation show that only large projects with rich commit histories can benefit from our method of commit classification. Further research on small projects might possibly reveal a divergent assessment of quality than just fixes. Then, a completely different commit classifier may be prepared. Its goal will be similar, i.e. to identify whether a change is going to be reverted in the future. However, its training set will be based on possibly another assumption than taken in this paper. Another possible research option is to check how the number of developers corresponds to the size of a project and the error rate. It might allow preparing predefined default classification parameters to help easier integration with our tool.

We believe that the results and the algorithms from [16] can be used to enrich our solution and further increase the performance of the classifier. This effect can be obtained by taking into account additional information about a project such as the kind of a programming language (object oriented, functional, logic etc) and the type system used. It was shown in [16] that those factors can have significant impact on the prevalence of bugs, and thus also on our detection system.

We assume that better results can be obtained if developers tag bugfixes. Such tags [20] will eliminate the need to classify commits whether they are fixes or not. This can have a significant impact on the training quality. Furthermore, different bugfixing detection algorithms such as [19] can possibly also improve training quality.

References

1. Arisholm, E., Briand, L.C., Johannessen, E.B.: A systematic and comprehensive investigation of methods to build and evaluate fault prediction models. Journal of Systems and Software 83(1), 2–17 (2010), http://dx.doi.org/10.1016/j.jss.2009.06.055

2. Bird, C., Rigby, P.C., Barr, E.T., Hamilton, D.J., Germán, D.M., Devanbu, P.T.: The promises and perils of mining git. In: Proceedings of the 6th International Working Conference on Mining Software Repositories, MSR 2009 (Co-located with ICSE), Vancouver, BC, Canada,, May 16-17, pp. 1–10 (2009), http://dx.doi.org/10.1109/MSR.2009.5069475

3. Catal, C., Diri, B.: A systematic review of software fault prediction studies. Expert Systems with Applications 36(4), 7346–7354 (2009)

4. D'Ambros, M., Lanza, M., Robbes, R.: An extensive comparison of bug prediction approaches. In: Proceedings of the 7th International Working Conference on Mining Software Repositories, MSR 2010 (Co-located with ICSE), Cape Town, South Africa, May 2-3, pp. 31–41 (2010), http://dx.doi.org/10.1109/MSR.2010.5463279

5. D'Ambros, M., Lanza, M., Robbes, R.: An extensive comparison of bug prediction approaches. In: Proceedings of the 7th International Working Conference on Mining Software Repositories, MSR 2010 (Co-located with ICSE), Cape Town, South Africa, May 2-3, 2010, Proceedings. pp. 31–41 (2010), http://dx.doi.org/10.1109/MSR.2010.5463279

6. Fejzer, M.: Commit classification application. Project on Github code repository (2014), https://github.com/mfejzer/CommitClassification

7. Hall, M.A., Frank, E., Holmes, G., Pfahringer, B., Reutemann, P., Witten, I.H.: The WEKA data mining software: an update. SIGKDD Explorations 11(1), 10–18 (2009), http://doi.acm.org/10.1145/1656274.1656278

8. Jeong, G., Kim, S., Zimmermann, T., Yi, K.: Improving code review by predicting reviewers and acceptance of patches. Research on Software Analysis for Error-free Computing Center Tech-Memo (ROSAEC MEMO 2009-006) (2009)

9. Joachims, T.: Text categorization with support vector machines. In: Nédellec, C., Rouveirol, C. (eds.) ECML 1998. LNCS, vol. 1398, pp. 137–142. Springer, Heidelberg (1998)

10. Kim, S.: Adaptive bug prediction by analyzing project history. University of California, Santa Cruz (2006)

11. Kim, S., Whitehead Jr., E.J., Zhang, Y.: Classifying software changes: Clean or buggy? IEEE Trans. Software Eng. 34(2), 181–196 (2008), http://doi.ieeecomputersociety.org/10.1109/TSE.2007.70773

12. Moors, A., Piessens, F., Odersky, M.: Parser combinators in scala. CW Reports (2008)

13. Nelson, S.D., Schumann, J.: What makes a code review trustworthy? In: 37th Hawaii International Conference on System Sciences (HICSS-37 2004), CD-ROM / Abstracts Proceedings, Big Island, HI, USA, January 5-8 (2004), http://dx.doi.org/10.1109/HICSS.2004.1265711

14. Ostrand, T.J., Weyuker, E.J., Bell, R.M.: Programmer-based fault prediction. In: Proceedings of the 6th International Conference on Predictive Models in Software Engineering, PROMISE 2010, Timisoara, Romania, September 12-13, p. 19 (2010), http://doi.acm.org/10.1145/1868328.1868357

15. Radjenovic, D., Hericko, M., Torkar, R., Zivkovic, A.: Software fault prediction metrics: A systematic literature review. Information & Software Technology 55(8), 1397–1418 (2013), http://dx.doi.org/10.1016/j.infsof.2013.02.009

16. Ray, B., Posnett, D., Filkov, V., Devanbu, P.T.: A large scale study of programming languages and code quality in github. In: Proceedings of the 22nd ACM SIGSOFT International Symposium on Foundations of Software Engineering (FSE-22), Hong Kong, China, November 16 - 22, pp. 155–165 (2014), http://doi.acm.org/10.1145/2635868.2635922

17. Shivaji, S., Whitehead Jr., E.J., Akella, R., Kim, S.: Reducing features to improve code change-based bug prediction. IEEE Trans. Software Eng. 39(4), 552–569 (2013), http://doi.ieeecomputersociety.org/10.1109/TSE.2012.43

18. Sliwerski, J., Zimmermann, T., Zeller, A.: When do changes induce fixes? In: Proceedings of the 2005 International Workshop on Mining Software Repositories, MSR 2005, Saint Louis, Missouri, USA, May 17 (2005), http://doi.acm.org/10.1145/1083142.1083147

19. Tian, Y., Lawall, J.L., Lo, D.: Identifying linux bug fixing patches. In: 34th International Conference on Software Engineering, ICSE 2012, Zurich, Switzerland, June 2-9, pp. 386–396 (2012), http://dx.doi.org/10.1109/ICSE.2012.6227176

20. Treude, C., Storey, M.A.D.: Work item tagging: Communicating concerns in collaborative software development. IEEE Trans. Software Eng. 38(1), 19–34 (2012), http://doi.ieeecomputersociety.org/10.1109/TSE.2010.91

21. Vapnik, V.: Estimation of Dependences Based on Empirical Data: Springer Series in Statistics (Springer Series in Statistics). Springer-Verlag New York, Inc., Secaucus (1982)

Project Management in the Scrum Methodology

Maria Tytkowska[1], Aleksandra Werner[2]([✉]), and Małgorzata Bach[2]

[1] Predica Business Solutions, Warszawa, Poland
mtytkowska@gmail.com
http://predica.pl
[2] Silesian University of Technology, Gliwice, Poland
{aleksandra.werner,malgorzata.bach}@polsl.pl
http://www.polsl.pl

Abstract. Taking into account the current leading role of techniques based on the incremental-iterative programming, the system that allows the optimization of project teams developing software in the Agile methodology Scrum technique, is proposed in the paper. Presented tool automates the process of development project management. It is distinguished from the other tools of the same class, by the ability of creation the workflow tasks that allows to solve the problems associated with information and communication in the project team and the kanbanboard visibility that demarcates whether the task is in the process of implementation or in testing. In addition, it implements the algorithm of Sprints' generating and archiving.

Keywords: Project Management · Waterfall Agile · Scrum · Share-Point · Kanbanboard · Sprint · Backlog · Timer job · Project simulation

1 Introduction

The term 'management' refers to the set of actions aimed at achieving given objectives in an efficient and effective way [3]. Depending on the objectives of which group of people these activities are given, it can be said as a management of the organization, company, and also project. The activities included in the management process involve the resources of the organization, and this process is fulfilled with combining and coordinating various types of resources [12]. Therefore, the tasks of those who are responsible for the process realization, so-called managers, include: planning, decision making, organizing and controlling – both the budget and, in particular, the timetable of the work. Although the general rules for the management of various entities, regardless of their nature, are essentially unchanged. Many of them have their own characteristics, which cause that these entities become the subject of separate research and analysis. This is also observed in the case of the IT project and the software development process management.

The rapid expansion of computer science dealing with the IT projects' management can be seen in the recent years. The variety of options in software implementation and control models, as well as the growing number of tools supporting their examination in practice, is the visible evidence of this development.

© Springer International Publishing Switzerland 2015
S. Kozielski et al. (Eds.): BDAS 2015, CCIS 521, pp. 483–492, 2015.
DOI: 10.1007/978-3-319-18422-7_43

2   Software Development Models

The universal standard preparation for a software development process which might be applied in every IT project, has failed so far [9]. Each standard requires an individual approach, and the choice of the management technique must take into account many factors – namely the size of the developed system, used technology, the probability of requirements changeability during the project, the space of the project (a distributed or local team) or risks associated with the implementation. None of the previously developed models is excluded 'by definition' [10]. Additionally, when choosing or constructing the optimal software development process for a specific IT project, another criterion is very often taken into account. This is the 'success of the project resolution' for projects performed in a given technology, taking into consideration, inter alia, time and cost overruns and functional and technical requirements [15].

That complexity of processes involved in the software development is one of the reason why so many different models of software project management can be found in the literature [5]. The most popular account: Waterfall, iterative cascade model, and the Agile programming [10].

The Waterfall model assumes strict project division to the phases and their order, resulting in the inability of transition to the next phase before finishing the previous one [14]. It is one of the oldest models and now it is considered to be a classical software life cycle model.

As far as the Agile methodology is concerned it is based on iterative-incremental programming, which assumes the software implementation by the self-managing, closely cooperating teams [16]. The principles and the purpose of the Agile methodology are written in the Agile Manifesto [4] and its main feature is to create a product iteratively and to output the working software release after each iteration. In comparison with the cascade model, the strict sequential phase sequence is here abandoned.

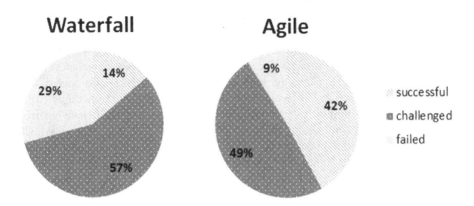

Fig. 1. The Waterfall and Agile methodologies comparison with respect to the project results [6]

Taking into account the criterion of the final success of the project being implemented in a given technique, a graphical comparison of the two above-mentioned models is shown in the Fig. 1.

The analysis of the pie charts (Fig. 1) and reported facts in literature [2,13] shows that the Agile methodology is better for IT projects than the cascade realization of different phases of the software product development. Therefore, the developed system presented in the article is just dedicated to the iterative-incremental approach, i.e. to the Scrum methodology.

During research regarding the current state of the solutions supporting the IT project management it was observed that there exist other tools, whose main objective is to efficiently manage IT projects according to the Scrum techniques. The examples of such systems, are: iceScrum, Banana Scrum or Agilefant [8,1]. The system presented in the article, is featured from other similar solutions by a customized approach to the problem of managing the process of software development. It is an attempt to develop a tool that supports the work of the entire project team and modernizes tasks' workflow between its members of the team, and is not just an assistance for the Product Owner or Scrum Master, as it is proposed in the mentioned solutions.

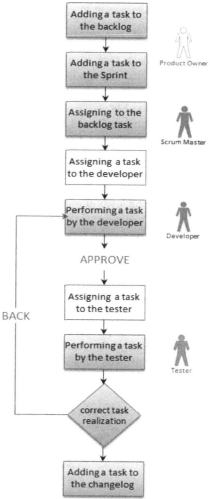

Fig. 2. Flow diagram of the teamwork performed in the task

3   System Assumptions and Architecture

The system was implemented by the usage of SharePoint 2013 [7] in the Visual Studio 2013 environment. It is worth pointing that thanks to the use of a WebPart components in the product, every project team member, being also a user of the system, can add other components to the default Scrum sites. It allowed to create a tool based on the default template, which in essence is a set of dedicated tools for each project participant.

A preliminary analysis which pre-determinate the final system requirements showed that each task in a single Sprint has its own life cycle. On this basis, it was found that the task and its life cycle should be the elements that will be managed by the proposed tool. The resulting task workflow are graphically presented in the Fig. 2.

It is worth emphasizing that this approach to the problem of software develop management is different from the offered one in the market-based solutions, as they usually focus only on the management of the people involved in the project (i.e. the project team).

Fig. 3 shows the implementation of the task workflow in the project team, implemented in the form of a state machine. It can be noted that depending on the stage of the implementation the task has various actions defined that can be executed. For example: the task may be ordered to implement by the group of developers by the Scrum Master. In the next step, after the programmer developed the assigned task, it can be moved for further verification to a group of testers. Depending the task on the results of testing, can either be accepted or sent back to the programmer with the recommendation to correct all detected errors. The task acceptance means state transition to the next one. In a defined task's workflow, during the consecutive steps, the appropriate statuses are assigned to the task like developing, testing, etc. In contrast, the final step only comprises the setting the project task status to *Closed*. Then the task workflow expires.

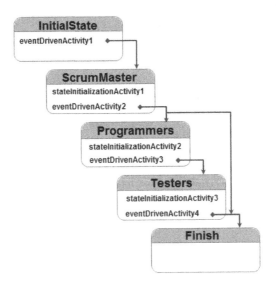

Fig. 3. State diagram of a task, implemented by a project team

The system was designed to enable tasks to be assigned to SharePoint groups and set their duration. Besides, it is possible to program private clock known as a timer job that will perform the established operations after fixed time (Fig. 4).

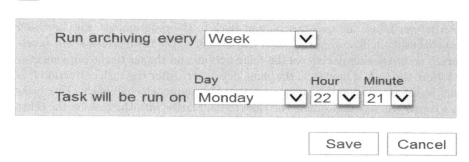

Fig. 4. The TimerJob configuration page

Taking into account the intuitiveness and clarity of the system, the feature allowing to assigning the user – user login, with the chosen color should be underlined. User's integration with the color facilitates the interpretation of the kanbanboard (Fig. 5).

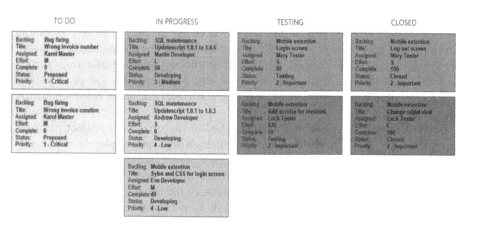

Fig. 5. The kanbanboard

It allows not only to read current stage of the work, but also to identify quickly those who are overworked and those who cannot cope with the assigned tasks.

While determining the system requirements special emphasis was put on the security of the workflow between the developer and the tester. To avoid the situation enabling the one tester verifying all tasks or the programmer choosing one tester he always works with, in the described product the developer can't send the task to the specific tester. After sending the task to the next stage, its field *Assigned To* is automatically completed with the group of all testers.

Every tester, after logging in, has two views: his own tasks and tasks in progress to take. Choosing from the set of tasks to take, after reading the contents of chosen task, the tester decides if to verify the task or not.

After verifying, the task can be sent back to the programmer, who has been implementing it. Knowing the particular author of the testing task is not necessary. The tester simply chooses the *undo* action, and thanks to the implemented solution, the task is moved to the same developer. After the task correction it is once again assigned to the previous tester. This approach enables checking tasks in a consistent way and saving the time to familiarize with the task by the other members of testers group.

4   Analysis of the Usefulness of the Developed System

In the IT projects, implementing all contracted functionality within the given time and budget is a very important aspect. In the Scrum methodology projects, budget and time are flexible, because after each iteration the Product Owner chooses the items that are to be completed in the coming Sprint. The described application allows to control the tasks and in particular the work efficiency of the project team. Therefore, after the Sprint completion, the individual team members project participation can be seen. It is visible who made the most of the tasks and what priorities they were, as well as what was their size. The another system advantage is the ability to check the efficiency of the testers and developers. If the tester often sends back tasks to the developers, it can be a sign of real efficacy of tests' performing, but on condition that the tasks are returned to different developers. The same situation can be a sign of a low productivity and poor quality of a developer's work, if the tasks are often returned to the same programmer.

Summarizing, in 9 areas that are identified in the process of project management (see pos. [11]), presented application improves or completely automates 5 of them. These are: integration management, cost management, communication, human and quality management. Other areas, i.e. delivery management, risk, scope and time, seemed not to be necessary for the daily work of the team producing the software. With a high probability it can be presumed that if they were implemented, it could either significantly make the use of the system difficult, or would be used only by a few users so rarely that others would regard it as the illegible redundant functionality.

4.1   Analysis on the Basis of the Project Simulations

One of the steps of testing the usefulness of the proposed system was the simulation of IT project development, modeling the actual environment of software team work. The part of this simulation was the estimation of wastes that could appear because of the developers and testers good communication shortage, and the lack of a fast response to errors.

For this purpose it was assumed that the Scrum Master, Product Owner, 10 Developers (marked with the symbols: P1 – P10) and 3 Testers participate in the project.

Table 1 presents an example of a set of tasks assigned to the developers for one Sprint with their size and the estimated time of execution (40h, 16h, 8h, 4h).

Table 1. The set of developers' tasks for one Sprint

Developer	Task					Time
	XL (40h)	L (16h)	M (8h)	S (4h)	Sum of tasks	
P1	1	2	0	1	4	76
P2	0	3	4	0	7	80
P3	1	0	1	7	9	76
P4	1	1	1	3	6	76
P5	2	0	0	0	2	80
P6	0	4	2	0	6	80
P7	0	0	4	10	14	72
P8	1	1	0	6	8	80
P9	1	0	4	1	6	76
P10	0	1	1	12	14	72
Sum of tasks	7	12	17	40	76	768

The next step of simulation was the prioritize the tasks. For the tasks with a given priority, the appropriate unit values were assigned to: high – 8u, important – 6u, medium – 4u, low – 2u.

The points assigned to the tasks are weighted values, which are calculated on the basis of the number of hours needed to implement the task, multiplied by the validity of it.

For example, if there are three tasks with size XL (40h) that have a high priority (8u), they obtain 3 * 40h * 8u = 960 points.

Table 2 contains the numbers of assigned points and rated losses resulting from errors during the acceptance and sending incorrect tasks to the customer. In this simulation, the individual tasks from each size class and validity were taken. There is not any XL task with low importance, because it seems to be logical from the project point of view.

In the next step, the percentage of the product quality degradation was calculated, as a percentage of the total error in the scoring scale project, according to the following formula:

$$quality\ degradation = \frac{cost\ of\ error * 100}{total\ number\ of\ points\ in\ the\ project}.$$

Table 2. Simulation of the weights of the individual errors

Developer	High (8u)				Important (6u)				Medium (4u)				Low (2u)				Sum
	XL	L	M	S	XL	L	M	S	XL	L	M	S	XL	L	M	S	
P1				1					1						1	1	4
P2		2				2								1	2		7
P3	1	2								1	2					3	9
P4			1				1		1		1					2	6
P5	1								1								2
P6			1		1	1					2				1		6
P7			1	2	1	2			2	2						4	14
P8		2					2				2			1			8
P9	1		1		1	1			1					1			6
P10	1		1								1	5				6	14
Sum of tasks	3	3	4	7	2	2	5	5	2	3	5	12	0	4	3	16	
Sum of points	960	384	256	224	480	192	240	120	320	192	160	192	0	128	48	128	4024
Errors	1	1	1	1	1	1	1	1	1	1	1	1	0	1	1	1	
Cost of error	320	128	64	32	240	96	48	24	160	64	32	16	0	32	16	8	1346
Degradation %	7,95	3,18	1,59	0,8	5,96	2,39	1,19	0,6	3,98	1,59	0,8	0,4	1,99	0,8	0,4	0,2	33,8

It can be noticed that making only one error of the described categories worsens the quality of the project at about 1/3. As a result of it, the raise of project costs and the reputation of the project team lowering can be seen. Making an error only once, depending on the validity of the task and of its size, has different impact on the deterioration of the quality of the released product. It can be simulated what number of the smaller error will be equivalent to making one error with the highest priority and XL size (table 3).

Table 3. Simulation of levelling serious, high-priority, error with an error of less priority

	High (8u)				Important (6u)				Medium (4u)				Low (2u)				Sum
	XL	L	M	S	XL	L	M	S	XL	L	M	S	XL	L	M	S	
Sum of tasks	3	3	4	7	2	2	5	5	2	3	5	12	0	4	3	16	76
Sum of points	960	384	256	224	480	192	240	120	320	192	160	192	0	128	48	128	4024
Errors	1											1		4	3	16	25
Cost of error	320	0	0	0	0	0	0	0	0	0	0	16	0	128	48	128	640
									Sum for tasks of Low priority:				320				
Quality degradation %	7,95	0	0	0	0	0	0	0	0	0	0	0,4	0	3,18	1,19	3,18	15,90
									Degradation for tasks of Low priority:				7,95				

Table 3 reveals one crucial error is equivalent to the failed acceptance of up to 24 tasks: 16 tasks with low priority and S size, 3 with M size, 4 with L size and 1 S-sized task with medium priority as well. Thus, the one XL task with the highest priority or a combination of the above-mentioned tasks worsens the quality of the project 7,95In the actual project, the number of undetected errors might be as follows (table 4).

It can be seen that errors made during the Sprint can deteriorate the quality of developing software even up to 43,53%. Thanks to the usage of the described

Table 4. The actual project errors' simulation

	High (8u)				Important (6u)				Medium (4u)				Low (2u)				Sum
	XL	L	M	S	XL	L	M	S	XL	L	M	S	XL	L	M	S	
Sum of tasks	3	3	4	7	2	2	5	5	2	3	5	12	0	4	3	16	76
Sum of points	960	384	256	224	480	192	240	120	320	192	160	192	0	128	48	128	4024
Errors	1	0	2	3	2	0	3	2	1	1	3	4	0	2	3	5	
Cost of error	320	0	128	96	480	0	144	48	160	64	96	64	0	64	48	40	1752
Quality degradation %	7,95	0	3,18	2,39	11,9	0	3,58	1,19	3,98	1,59	2,39	1,59	0	1,59	1,19	0,99	43,54

application, it is not possible to deliver the improper task to the customer. These tasks are verified and monitored all over the time. If the task is proved to be improper or incomplete during the verification it is returned to the developer immediately. In this case the task will be moved to another Sprint, if it is not able to be developed but for sure, it will not be delivered in an incorrect form to the customer. A better practice is definitely not providing assumed functionality than delivering it to the user in the useless form or with a fatal error which restricts the user.

5   Conclusions

The selected functionality of the system that supports the activity of the IT project teams was presented in the article. The described system is equipped with a number of options specific to this type of tool, but from a family of similar products it is distinguished by the following features: the list of color configuration entries [1], tasks' workflow between project team members and the ability to generate and archive Sprints.

Another important aspect that is not provided in the alternative solutions but included in the offered system is the automatic Sprint review. The decision of adding mentioned functionality to the system was made due to the principle of good programming practice. It states that at the end of the Sprint the list of changes implemented at this time should be sent to the Product Owner. Creating such a kind of a list can be really time consuming activity in situation where the summary is done only on the basis of the tasks' statuses or kanban-board. Therefore in the discussed system the Sprint summary, in the form of an Excel spreadsheet, is automatically generated and sent to the customer via an e-mail, without involving the Scrum Master or the Project Manager in it. Similarly, sending the unfinished tasks list with the justifications why they failed to finish during the Sprint, was implemented. Detailed simulations of the system usefulness and the conclusions after their completion are described in the point 4.1. The presented system can be used as a tool by the employees working in a variety of software projects. It can also be a basis for further research focused on the effective and efficient management of the software development process.

---

[1] Every team member has his own color that distinguishes him on the kanbanboard.

References

1. 20 Free Scrum Project Management Tools (January 30, 2015),
   http://www.mypmhome.com/20-free-scrum-project-management-tools/
2. Agile vs Waterfall: Comparing project management methods (January 30, 2015),
   http://manifesto.co.uk/agile-vs-waterfall-comparing-
   project-management-methodologies/
3. Encyclopedia of management (January 30, 2015),
   http://mfiles.pl/pl/index.php/Zarz
4. Manifesto for Agile Software Development (January 30, 2015),
   http://agilemanifesto.org
5. Project Management Methodologies (January 30, 2015),
   http://www.tutoriapoint.com/management_concepts/
   project_management_methodologies.htm
6. Raport CHAOS (January 30, 2015),
   http://www.onedesk.com/2013/01/waterfall-vs-agile/
7. SharePoint general development (January 30, 2015),
   http://msdn.microsoft.com/en-us/office/dn833469.aspx
8. Top Project Management Software Products (January 30, 2015),
   http://www.capterra.com/project-management-software/
9. Analysis and design of enterprise management systems. Mfiles.pl (2010),
   http://wydawnictwo.mfiles.pl/content/
   podglad-ksiazki-analiza-i-projektowanie-systemow-zarzadzania-przedsiebiorstwem
10. Chang, C.: Selecting an Appropriate Software Development Lifecycle (SDL) Model
    in an Agency Environment (January 30, 2015),
    http://www.metia.com/seattle/chong-chang/2012/08/
    sdl-model-in-an-agency-environment/
11. Duncan, W.R.: A Guide To The Project Management Body Of Knowledge
    (January 30, 2015), http://www-kiv.zcu.cz/~pergl/SAI/PMIBOK/project.htm
12. Griffin, R.W.: Fundamentals of Organization Management (January 30, 2015),
    http://pl.scribd.com/doc/119861242/R-W-Griffin-Podstawy-zarz
13. Mikoluk, K.: Agile Vs. Waterfall: Evaluating The Pros And Con (January 30,
    2015), https://www.udemy.com/blog/agile-vs-waterfall/
14. Royce, W.W.: Managing the Development of Large Software Systems (January 30,
    2015), http://www.cs.umd.edu/class/spring2003/cmsc838p/Process/waterfall.pdf
15. The Standish Group International Incorporated: CHAOS MANIFESTO 2013.
    Think Big, Act Small (January 30, 2015),
    http://www.versionone.com/assets/img/files/CHAOSManifesto2013.pdf
16. Wolf, H.: Agile projects in the classical Organization Scrum Kanban XP. Helion
    (2014)

Applications of Database Systems

Regression Rule Learning
for Methane Forecasting in Coal Mines

Michał Kozielski[1]([✉]), Adam Skowron[2], Łukasz Wróbel[3], and Marek Sikora[2]

[1] Institute of Electronics, Silesian University of Technology,
Akademicka 16, 44-100 Gliwice, Poland
[2] Institute of Informatics, Silesian University of Technology,
Akademicka 16, 44-100 Gliwice, Poland
[3] Institute of Innovative Technologies EMAG, Leopolda 31, 40-189 Katowice
{michal.kozielski,adam.skowron}@polsl.pl, lukasz.wrobel@emag.pl,
marek.sikora@polsl.pl
http://adaa.polsl.pl

Abstract. The rule-based approach to methane concentration prediction is presented in this paper. The applied solution is based on the modification called *fixed* of the separate-and-conquer rule induction approach. We also proposed the modification of a rule quality evaluation based on confidence intervals calculated for positive and negative examples covered by the rule. The characteristic feature of the considered methane forecasting model is that it omits the readings of the sensor being the subject of forecasting. The approach is evaluated on a real life data set acquired during a week in a coal mine. The results show the advantages of the introduced method (in terms of both the prediction accuracy and knowledge extraction) in comparison to the standard approaches typically implemented in the analytical tools.

Keywords: Prediction · Rule-based regression · Statistical rule quality evaluation

1 Introduction

The production processes monitoring systems have become a typical solution in the nowadays industry. In case of the safety parameters the proper monitoring becomes even more crucial.

One of the industry branches where safety monitoring systems play a very important role is coal mining. Miners in an underground coal mine can face several threats, like e.g. methane explosion or moze rock-burst. Therefore, the possibility of prediction of the future gas concentrations (methane in particular) that will be registered by sensors in the consecutive time intervals can be even more important then monitoring the current state of the process parameters.

The prognostic models of that type are scarce and further research and development in this area is still needed. Some research on the prediction of the methane concentrations was conducted in the recent years [8,9]. However, these

© Springer International Publishing Switzerland 2015
S. Kozielski et al. (Eds.): BDAS 2015, CCIS 521, pp. 495–504, 2015.
DOI: 10.1007/978-3-319-18422-7_44

approaches utilized also the readings of the sensor being the subject of forecasting. In such case the past indications of the sensor have obviously the strongest impact on the predicted future indications of this sensor, what makes it the most important attribute of the created prediction model.

The goal of this work is to analyze and evaluate different prediction models in application to the task of methane concentration prediction. The prediction will be delivered by a pair of rule-based methods (the method introduced in [10] and its modification described in this paper) which additionally deliver knowledge about the phenomenon in the form of rules. The results will be referred to the standard approaches typically implemented in the analytical tools.

The motivation of this work is fourfold. Firstly, it would be interesting to verify the possibility of methane concentration prediction in the location where the sensor is located only for a limited time (eg. a portable sensor). Secondly, our approach will enable the prediction even if the indications of the existing sensor are missing for some reasons (the previous approaches utilizing the prior measurements of the sensor being a subject of prediction are not able to perform this task). Therefore, the results of this work can be utilized in such tasks as automatic filling of long lasting missing sensor measurements or automatic identification of the sensor measurements manipulations. Thirdly, it would be interesting to find what are the dependencies in the analyzed data. For this purpose, we propose a modified version of the regression rule induction algorithm and we present a proposal of calculation of confidence intervals for the number of examples covered by the rule. The confidence intervals may then be used to the pessimistic and optimistic rule quality evaluation. Fourthly, the investigations undertaken are the component of a wider work, which is the decision support system to assist dispatchers and users of monitoring systems. The system is developed under a grant of the National Center for Research and Development.

Contribution of this work consists of two parts. At first, the work introduces the new *top-down fixed* approach to regression rule induction and a proposition of modification of the rule quality evaluation, which is based on confidence intervals for the examples covered by the rule. The work also presents the approach to methane concentration prediction, where the readings of the sensor being the subject of forecasting are not included into the model and the work contains the overall analysis and evaluation of the previously introduced and newly presented methods of regression rule induction in application to the given issue.

The paper is structured as follows. Section 2 presents the *top-down* (TD) approach to regression rule induction, its new modification called *fixed* and a new approach to rule quality evaluation. Section 3 presents experimental study covering a data set and reference methods description, and the results of the analysis. Section 4 summerizes the work.

2   Methods

The *top-down* (TD) approach to regression rule induction was utilized in the given work. This algorithm was introduced in [10]. In this work we introduce its modification (subsection 2.1), which is further referred as *fixed* (TDF), and a new approach to rule quality evaluation (section 2.2).

The *top-down* algorithm consists of two phases. During the growth phase a rule is built by the successive addition of elementary conditions. In the case of numerical attributes the elementary conditions are defined as the mean between two consecutive values of all the ordered values of the given attribute. Wheareas, in the case of the nominal attributes all possible values are selected. The condition for which the rule obtains the biggest quality improvement is added to the rule permanently. The process is then repeated. During the pruning phase the elementary conditions are deleted in order to further improve the rule according to the value of specified quality measure. The detailed description of the approach is presented in [10] and the general outline of the algorithm is presented by Algorithm 1.

Algorithm 1. Pseudocode of the *top-down* regression rule induction algorithm

---

   function RULEINDUCTION(*examples, ruleQualityMeasure*)
      ruleSet ← ∅
      uncoveredExamples ← *examples*
      while *uncoveredExamples* ≠ ∅ **do**
         rule ← *Grow*(*examples, uncoveredExamples, ruleQualityMeasure*)
         rule ← *Prune*(*rule, examples, ruleQualityMeasure*)
         covered ← *Covered*(*rule, examples*)
         uncoveredExamples ← *uncoveredExamples* \ *covered*
         ruleSet ← *ruleSet* ∪ {*rule*}
      end while
      return *ruleSet*
   end function

Next, the rules generated by means of the *top-down* method are postprocessed by means of the rule filtration algorithm in order to improve the clarity of the rule-based model and to speed up the prediction. Among several regression rule filtration methods that were analyzed and evaluated [1,14] the *Coverage* approach [14] was chosen in this work.

The quality measure utilized in this work is defined on the basis of the values p, n, P and N, where p denotes the number of positive examples covered by the rule, P is number of all positive examples, n stands for the number of negative examples covered by the rule and N is the number of all negative examples. These measures are designed for evaluation of decision rules, however dynamic reduction to classification [5] approach, which is used in our implementation of the rule induction algorithm, allows to apply them in induction of regression rules (what was analyzed in [10]). Thus, $C2$ quality measure was selected:

$$C2 = \frac{Np - Pn}{N(p + n)} \cdot \left(\frac{P + p}{2P} \right),$$ (1)

as it was shown in [10] that this measure is the most effective in terms of prediction precision.

2.1 The *Fixed* Algorithm

The regression value in the simplest form is a single value obtained from all examples covered by the fired rule. Such the simplest form of conclusion is primarily the most understandable and transparent way of presenting the prediction. This is also the fastest possible way of calculating the prediction. Alternatively, the implementation of linear model in a form of $w_0 + w_1 a_1 + ... + w_k a_k$ is based on the values of many of the attributes (k) multiplied by some weights (w) and usually allows to obtain a smaller prediction error [3]. However, it is evident that the clarity of rules with linear models decreases drastically. Therefore, in this work the rules are induced with a single numerical value as prediction. This value can be, however, identified by means of two different approaches.

In the first one we apply the median value of examples covered by the rule avoiding an issue of outliers in this way. Thus, the single numerical value is calculated as a rule conclusion after the rule is calculated.

In the second method, which is referred as *fixed*, the prediction of target value is slightly different and its principle is derived from classification rule induction algorithms. In classification approach, the rule induction is performed for each decision class separately. Examples belonging to one of the decision classes form a positive class while the remaining examples are included in the negative class. Then, the rules are successively built until all examples from the positive class are covered by at least one rule. After that, the next class is considered as positive and the process of single rule induction is repeated. The induction ends when all classes have been examined.

A key transferred idea that has been used in our algorithm is the consideration of the classes in turn. Thus, in regression rule induction algorithms each example and its target value is put down to the independent class. It means that the target value of a certain example (seed) is permanently assigned to the rule at the beginning and it does not change during the entire process of induction.

To permanently assign the target value to the rule we propose the method based on a centroid. At first, a centroid is calculated as a mean or mode value depending on the numerical or nominal attribute type. Next, the example which has the smallest euclidean distance to this centroid is identified by means of the k-dimensional tree algorithm. Finally, the example that was found is marked as the seed and its target value is assigned to the empty rule.

When the *fixed* approach is applied to the *top-down* rule induction algorithm, the rule that is being built has to cover the seed additionally. Such an assumption is necessary because otherwise, the rule could cover examples completely unrelated to the estimated target value what would impact the prediction error significantly.

The two approaches presented in this section are further referred as TD - the *top-down* algorithm with the median value as the prediction and TDF - the *top-down* algorithm with the *fixed* value as the prediction.

2.2 Confidence Intervals for Positive and Negative Examples Covered by a Rule

The quality of the rule, which is induced on the basis of the data sample, is subjected to a distribution of this sample. Therefore, in the perfect case the algorithm should consider entire population to prevent this problem. Otherwise, the statistical significance of the results should be assessed.

Application of confidence intervals to generation of association and decision rules was proposed in [12]. This approach can be partially adapted to induction of regression rules, what is presented in the following paragraphs as a contribution of this work.

As it was stated earlier the rule quality measure utilized in this work is defined, among others, on the basis of the values p and n, where p denotes the number of positive examples covered by the rule, n stands for the number of negative examples covered by the rule.

Each example from a data set U belongs to one of the two groups identified by p and n values or to the reminder group (not p and not n). The probability f of affiliation to the groups identified by p and n values can be estimated as:

$$f_x = \frac{x}{|U|}, \qquad x \in \{p, n\}, \tag{2}$$

where $|U|$ is the size of the data set.

Next, it is possible to calculate confidence intervals consisting of two values, which are lower and upper border of a given interval, what can be written as: $x_{low} \le x \le x_{high}, \quad x \in \{p, n\}$, and the probability of the occurrence of true value within this interval is equal to $1 - \alpha$ [12], where α is the confidence level typically having a value 1% or 5%.

It was chosen to apply the approach presented in [2] to calculate the borders of a given interval, and the value is rounded to the integer value of examples:

$$x_{low} = \left\lfloor \left(f_x - \chi \sqrt{\frac{f_x(1 - f_x)}{|U|}} \right) |U| + \frac{1}{2} \right\rfloor, \tag{3}$$

$$x_{high} = \left\lfloor \left(f_x + \chi \sqrt{\frac{f_x(1 - f_x)}{|U|}} \right) |U| + \frac{1}{2} \right\rfloor, \tag{4}$$

where $x \in \{p, n\}$ and χ is a square root of critical value from chi-square distribution for one degree of freedom and two estimated parameters. Assuming that $\alpha = 0.05$ the resulting square root of the critical value is $\chi = \sqrt{5.02}$.

Having the confidence intervals calculated it is possible to evaluate optimistic and pessimistic quality of the rules. The pessimistic approach utilizes the lowest value of p (p_{low}) and the highest value of n (n_{high}), whereas optimistic approach utilizes the highest value of p and the lowest value of n.

From the knowledge discovery perspective the best rules with respect to precision and coverage should be chosen as the most interesting. Therefore, minimal precision and minimal coverage threshold can be set to choose the best rules.

When pessimistic approach is utilized less rules will meet the threshold and the analyzed rule set will be smaller and therefore, easier interpretable. While in the case of resulting rule set being too small the optimistic approach can be utilized, what will result in a larger number of rules meeting the threshold.

3   Experimental Study

3.1   Data Set

In the given study the analysis of the measurements collected in a coal mine is considered. The methane concentration which is predicted can be dependent on a current mining activity, methane concentration in the other locations and ventilation measured in several ways.

The data set containing information about gases concentrations was registered on a longwall outlet (where the highest methane risk takes place). The topology of the coal mine part where the measurements were performed is presented in Fig. 1. The sensors indicated in Fig. 1 are explained in Table 1.

The measurement frequency of each sensor was 1 second. The data set contains measurements of 1 week aggregated at each 30 seconds. The aggregation functions applied to each sensor's data are presented in Table 1. The task was to predict the maximal value of MM116 for next 3 minutes. The other sensors described in Table 1 collected the values of the attributes utilized in a prediction model. Additional attribute (PD) of the model identifies if the combine

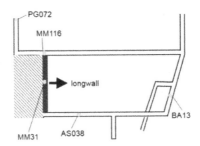

Fig. 1. Coal mine topology and sensors

Table 1. Description of the sensors marked in Fig. 1

Sensor	Sensor type	Description	Aggregation function
MM31	methanometer	methane concentration [%]	max
MM116	methanometer	methane concentration [%]	max
AS038	anemometer	air velocity [m/s]	min
PG072	airflow	airflow [m³/s]	min
BA13	barometer	pressure [hPa]	mean
PD	-	combain activity	dominant

works at a given time (dominant was applied as an aggregation function). The characteristics of the collected data are presented in Table 2.

Table 2. Data characteristics

Sensor	Min	Max	Median	Mean	Standard deviation
MM31	0.17	0.82	0.36	0.36	0.117
MM116	0.20	2.20	0.80	0.80	0.286
AS038	1.40	2.70	2.30	2.29	0.142
PG072	1.10	2.60	1.80	1.84	0.107
BA13	1067	1078	1075	1073	3.138

The entire data set was divided into two disjoint parts. First 70% of examples consisted of training set for model building, and last 30% of data created test set for model evaluation.

3.2 Other Tested Methods and Experimental Settings

The rule-based solution introduced in [10] and its modification presented in this paper were compered with several methods available in Weka 3.6.11 [13] and in R 3.1.2 environment [7]. We focus on methods which, similarly to rule-based ones, are able to express data model in comprehensible form. The methods that were applied are listed in Table 3. All algorithms were run in their default configuration, except the Cforest algorithm for which the number of trees was set to 1000. For the introduced TD and TDF methods the parameter defining the minimal number of examples that have to be covered by a rule was set to 3.

Table 3. Reference methods utilized in the analysis

Id	Description	Tool
Training mean	Mean of a target attribute on a training set	R
Linear regression	Linear regression	R
Ctree	Regression tree with statistical cut evaluation [4]	R
Cforest	Random forest utilizing Ctree method [11]	R
M5P	M5 trees [6]	Weka
M5RULES	M5 rules [3]	Weka

3.3 Results

The results of the analysis are presented in Table 4. Performance of each algorithm is described by root mean squared error (RMSE), maximal error on a test set, the percentage of a number of errors above the 0.3 threshold and size. The size is calculated as the number of rules for rule-based models, the number of leaves for trees and the number of trees for random forest. The results of the new rule-based methods introduced in this paper are highlighted in bold.

Table 4. Results of the analyzed methods

	RMSE	Max error	% Max error over 0.3	Size
TD C2	0.265	0.773	27.09	934
TDF C2	**0.181**	**0.611**	**10.82**	**100**
TD C2 filtered	0.249	0.757	23.88	510
TDF C2 filtered	**0.177**	**0.614**	**8.59**	**18**
Training mean	0.298	0.802	47.64	-
Linear regression	0.313	0.921	39.20	-
Ctree	0.216	0.600	16.88	137
Cforest	0.206	0.566	16.27	1000
M5P	0.205	0.761	15.81	195
M5RULES	0.220	0.700	18.80	53

The best results were achieved by the TDF C2 filtered method. Also the results before filtration (TDF C2) are characterized by a very low RMSE value, however the postprocessing reduces the number of rules significantly what makes them intelligible and what reduces the RMSE even more.

The histograms presenting error distributions (difference between the real value and the predicted methane concentration) for the TD and TDF methods are presented in Fig. 2. It can be noticed that larger number of the predictions performed by means of the TD method were underestimated and the maximal value of underestimation is higher. In case of TDF method the overall number of under- and overestimations seems to be similar and the maximal values are more similar. It shows that the results of TDF method are more balanced.

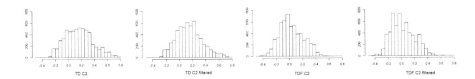

Fig. 2. The error distributions of the *top-down* rule-based method

The rules generated by means of the TDF method, such as two examples presented in Table 5, can be further statistically analyzed by calculation of confidence intervals and the pessimistic, optimistic or standard quality of the rules as it was discussed in section 2.2. Any quality measure can be calculated in this way, e.g., C2 measure (1), precision or coverage. Table 6 presents the pessimistic, standard and optimistic values of precision and coverage quality of the rules R1 and R2 presented in Table 5.

It can be noticed that rule R2 is more general what results in much higher standard coverage. However, the standard precision of the rules is very similar. Looking at the ranges derived by pessimistic and optimistic values we can also notice that the range of rule R1 precision is broader comparing to rule R2 precision and the pessimistic precision of rule R1 is significantly lower then the one

Table 5. Exemplary rules generated by means of the TDF method

	Rule R1			Rule R2
IF	PD = 1.0		IF	PD = 1.0
	BA13 ∈ [1072.867; 1076.287)			MM31 ≥ 0.37
	PG072 ≥ 1.75			BA13 ≥ 1068.088
	MM31 ∈ [0.405, 0.625)			
THEN	MM116 = 1.3 (0.200)		THEN	MM116 = 1.1 (0.201)

Table 6. Quality of the rules from Table 5

		R1		R2	
		Precision	Coverage	Precision	Coverage
	pessimistic	0.757	0.073	0.819	0.411
Rule quality	standard	0.831	0.087	0.839	0.428
	optimistic	0.895	0.100	0.858	0.445

of the rule R2. Summarizing, the pessimistic quality (values of precision and coverage) of rule R2 is significantly higher then the one of the rule R1.

The presented example of pessimistic and optimistic rule quality analysis shows that it can bring additional knowledge about the generated rules and also it allows to compare the rules more deeply.

4 Conclusions

An approach to methane concentration prediction, where the readings of the sensor being the subject of forecasting are not included into the model was presented in this paper. Several prediction methods were analyzed in this task, also the new method (TDF) of regression rule induction was proposed. The analysis was performed on a real life data consisting of a weekly measurements containing methane concentration collected at the coal mine.

The results showed that the best prediction quality was delivered by the TDF method which was introduced in this work. Some interesting characteristics of this method were also presented by means of the error distribution analysis.

The TDF method generates regression rules with a single value in a conclusion (calculated differently then in the works [5,10]). Such rules deliver easily interpretable knowledge about the analyzed phenomenon what is an additional advantage of this method. A new approach to pessimistic and optimistic evaluation of such rules was also presented in the work and evaluated on two exemplary rules generated for the given task.

The method presented in the paper will become a part of the decision support system assisting dispatchers and monitoring system users. Its effectiveness will be evaluated on numerous data sets analogously to the method presented in [10].

Acknowledgments. This research was supported by Polish National Centre for Research and Development (NCBiR) grant PBS2/B9/20/2013 in frame of Applied Research Programme. This work was also supported by the European Union from the European Social Fund (grant agreement number: UDA-POKL.04.01.01-00-106/09).

References

1. Amin, T., Chikalov, I., Moshkov, M., Zielosko, B.: Classifiers based on optimal decision rules. Fundam. Inform. 127(1-4), 151–160 (2013)
2. Gold, R.: Tests auxiliary to $\chi 2$ tests in a Markov chain. Annals of Mathematical Statistics, 56–74 (1963)
3. Holmes, G., Hall, M., Prank, E.: Generating rule sets from model trees. Springer (1999)
4. Hothorn, T., Hornik, K., Zeileis, A.: Unbiased recursive partitioning: A conditional inference framework. Journal of Computational and Graphical Statistics 15(3), 651–674 (2006)
5. Janssen, F., Fürnkranz, J.: Heuristic rule-based regression via dynamic reduction to classification. In: IJCAI Proceedings-International Joint Conference on Artificial Intelligence, vol. 22, p. 1330 (2011)
6. Quinlan, J.R., et al.: Learning with continuous classes. In: Proceedings of the 5th Australian Joint Conference on Artificial Intelligence, Singapore, vol. 92, pp. 343–348 (1992)
7. R Core Team: R: A Language and Environment for Statistical Computing. R Foundation for Statistical Computing, Vienna, Austria (2014)
8. Sikora, M., Sikora, B.: Improving prediction models applied in systems monitoring natural hazards and machinery. International Journal of Applied Mathematics and Computer Science 22(2), 477–491 (2012)
9. Sikora, M., Sikora, B.: Rough natural hazards monitoring. In: Rough Sets: Selected Methods and Applications in Management and Engineering, pp. 163–179. Springer (2012)
10. Sikora, M., Skowron, A., Wróbel, L.: Rule quality measure-based induction of unordered sets of regression rules. In: Ramsay, A., Agre, G. (eds.) AIMSA 2012. LNCS, vol. 7557, pp. 162–171. Springer, Heidelberg (2012)
11. Strobl, C., Boulesteix, A.L., Zeileis, A., Hothorn, T.: Bias in random forest variable importance measures: Illustrations, sources and a solution. BMC Bioinformatics 8(25) (2007)
12. Wieczorek, A., Słowiński, R.: Generating a set of association and decision rules with statistically representative support and anti-support. Information Sciences 277, 56–70 (2014)
13. Witten, I.H., Frank, E., Hall, M.A.: Data Mining: Practical Machine Learning Tools and Techniques, 3rd edn. Morgan Kaufmann, Amsterdam (2011)
14. Wróbel, L., Sikora, M., Skowron, A.: Algorithms for filtration of unordered sets of regression rules. In: Sombattheera, C., Loi, N.K., Wankar, R., Quan, T. (eds.) MIWAI 2012. LNCS, vol. 7694, pp. 284–295. Springer, Heidelberg (2012)

On Design of Domain-Specific Query Language for the Metallurgical Industry

Andrey Borodin[✉], Yuri Kiselev, Sergey Mirvoda, and Sergey Porshnev

Dept. of Radio Electronics for Information Systems,
Ural Federal University, Yekaterinburg, Russia
amborodin@acm.org, ykiselev.loky@gmail.com,
sergey@mirvoda.com, s.v.porhsnev@urfu.ru
http://urfu.ru

Abstract. Many systems have to choose between user-friendly visual query editor and textual querying language with industrial strength in order to deal with big amount of complex data. Some of them provide both ways of accessing data in a warehouse.

In this paper, we present key features of domain specific querying language, which we designed as a part of informational system of a steel production plant. This language aims to give an opportunity of easy data manipulation to those who know what the data actually is and to provide an easy way to discover what the data is for others. We also provide an evaluation of the designed language. The main focus of our estimation was to measure effort required for discovering dataset and deriving simple math expressions.

Although the paper overviews a data model for one specific domain, it can be easily applied for different domains.

Keywords: Query language · Domain-specific language · Data warehouse · Metallurgy · Steel plant

1 Introduction

Most of general-purpose database management systems (DBMS) have their own programming languages. Usually they are forms of structured query language (SQL), but also they can be a variation of DML, MDX, XQuery, etc. On top of these languages, software developers build all kinds of visual query construction toolkits, data browsers, and pivot tables. Specific data management systems operate on the top of predefined data types with their own unique features and requirements. Their developers can start directly from creating specific data visualization and querying tools, using standard DML as a media from user interface (UI) to DBMS or omitting this media at all in case of specific data warehousing techniques.

Our task was to create an automated analytical and modelling system for metallurgical production [1], which had contained a data warehouse and a query constructor. An existing system already had a visual query editor as a web

© Springer International Publishing Switzerland 2015
S. Kozielski et al. (Eds.): BDAS 2015, CCIS 521, pp. 505–515, 2015.
DOI: 10.1007/978-3-319-18422-7_45

service. Its users could define constraints and the form of the desired data. Then this visual query was compiled to Oracle PL\SQL query and executed on data warehouse.

After investigation of the system with metallurgical technical staff, we concluded that it was not the best fit for the plant. This was mainly because of different needs of the users. Some users were data scientists and they required specific features that were hard to implement with web UI. The other ones required a possibility of sharing significant parts of queries with their colleagues. Finally, some of them simply had no mouse devices or touch manipulators due to specific environmental restrictions, so all of the users expected all querying features to be available from a keyboard.

The initial move was to add keyboard shortcuts, and users actively welcomed this. A complication of the system leaded to a necessity of the query language as a main media between a query constructor and a data warehouse. This approach could have prevented the evolving complexity of a query editor, which can be used now as the only assistant tool in a query construction. We stated a list of the following advantages of domain-specific query language (DSQL):

1. Version control systems (VCS) out of the box. Software developers have been using safe source code tracking for many decades. Since queries are the main tools for metallurgical data analysts, they should be treated the same way, but the development of all VCS features in a custom query editor is not economically viable.

2. Portability. Text queries can be written even on a whiteboard and a notepad. One of the great advantages of text queries is that they are unambiguous: there are no hidden parts in a text query.

3. Detachable. Text queries can be run in different warehouses. They should not depend on identifiers, conditions, and environmental variables.

4. Fragmentation. An analyst can extract some feature from a query and partly pass it to his colleagues. E.g. from a query computing metallurgical length of all damaged slabs of specific operational brigade we can extract a part which searches all damaged slabs of the brigade.

5. Embedding. Software developers easily can embed queries into autotests, side subsystems can embed query parts into their code or resources.

6. Specification. DSQL can be a part of a systems applied programming interface (API) for third party systems if DSQL specification is precise enough.

This paper describes features of the DSQL we developed for our system (DSQLM M stands for metallurgy), although these six advantages are common for almost any DSQL, designed similarly. Term DSQL is generic and does not reflect its orientation towards steel production data [5]. DSQLM is used to point out our specific implementation. However, the variety of auxiliary data types is stored in systems data warehouse, which is not limited only to technical, chemical or signal data. Thus, most of design decisions of this DSQLM can be applied to any other domain.

2   Language Goals

Potential DSQLM users are steel plant employees; most of them are techno-logical data analysts who study properties of plant products in correlation with production process parameters. Some of them are support staff in finance, ITSM and management.

Most of the users have higher education; some of them hold PhD in tech-nology with works related to steel production. Most of them have neither deep knowledge of informational technologies nor experience in using SQL-like lan-guages. Most of the users know the programming data types while lack of the understanding of data processing algorithms.

In collaboration with the staff, we designed this set of goals:

1. DSQLM should be compatible with Google-style queries and document mod-els (bag of words - BoW [7]).
2. DSQLM should be declarative, i.e. to define not how to get the result but to declare the requirements for retrieving the result.
3. DSQLM should be as deterministic as possible. This means that two execu-tions of the same query should yield the same results. This goal is not strict, because overwhelming determinism could render impossible environmental variables usage, for example, use of the current time in a query would not meet such requirement.
4. DSQLM should aim to reduce the complexity of genealogy. In steel pro-duction, units are arranged in a genealogical graph; i.e. metal sheets are produced from different slabs and slabs are produced from different bars. An analyst should be able to examine properties of all bars while studying a certain sheet. This requires a language to have an easy way to define data aggregation over domain data model.
5. The result of DSQLM query is a data table, which can be exported to such tools like Excel, Statistica, and MATLAB.
6. The queries should be fragmental to extract features and combinable to develop query libraries. This point contradicts the 2nd advantage of text queries (portability), yet it can be handled with assistant tools extracting all-self-sufficient queries from query referencing libraries.
7. DSQLM is a tool to extract relevant data from a warehouse.

During language development, these were rather goals then requirements. Al-though users describe all of them as crucial, they cannot be achieved simultane-ously. Note that goals 4 and 5 are domain specific, yet all other goals are generic and can be applied to advance any DSQL

3   Domain Data Model

Although relational database management systems (RDBMS) suffer from the effect of object-relational impedance mismatch (ORIM) [2], such systems are most widely used for data warehousing. The ORIM is a set of conceptual and technical difficulties that are often encountered when a RDBMS is used by a program written in an object-oriented programming language or style.

While DSQL is not necessary object-oriented, ORIM is closely related to reducing of genealogy complexities. Using SQL as a query media, most genealogy manipulations are translated into JOIN constructions and non-obvious aggregation conditions that are hard to understand. We observed this in a visual query editor and made the following conclusion. Using SQL-like syntax and RDBMS data model (tuples and relations between them) would lead to unclear DSQL syntax and semantics.

An alternative for RDBMS would be object oriented and document [9] databases. In particular the system contains MongoDB warehouse as one of important storages. However, if DSQL were based on MongoDB query language it would require the understanding of JavaScript concepts that are hard to master for users. Moreover building DSQL that is strongly connected to data model would be restrictive in the future. The warehouse is designed to be easily adaptable to data of certain plant, whereas readjustment in language syntax, libraries and documentation could cause significant delays in the release of a system.

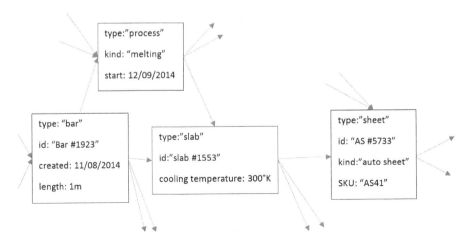

Fig. 1. Partial example of a dataset

Hence, we designed the following data model (Fig. 1 presents four objects of a possible dataset):

1. The main unit of data is an object ("slab", "sheet", "process", "firm", "user", "aggregate", etc.)
2. An object has different parameter values ("color", "weight", "length", "creation time", etc.)
3. An object has forward and backward references to objects. A set of forward objects is assumed as all objects dependent or created by referencing object. There is a corresponding backward reference for each forward reference.

4. Every object has a string parameter "type"; every type is an object itself with type "type". Type objects do not contain references.
5. Every parameter value contains a corresponding object with a type "parameter" and information related to its semantics, unit of measure and data type.

The data model is targeting the goal 4. Its also aims to reflect semantics of data. This means that a dataset neccesarily includes data about its structure as a first-class data.

Semantics of every reference depends on a target of this reference. However, the main purpose of the references is to reflect oriented genealogical graph.

This data model can be applied generically to any DSQL experiencing bottleneck in objects genealogical relations.

4   Language Features

Since the result of a DSQLM is a table, the core of textual query should be oriented towards table specification. A query includes parts, which can be essentially one of three kinds:

1. Column definition (Computable expression)
2. Row filter (Boolean expression)
3. Search filter (BoW)

If a table requires a Boolean column, this must be specified in column options discussed further. Query parts are delimited with commas and consist of an identifier, options, and a body. The body is surrounded with curly brackets; options are surrounded with square brackets. We chose an excessive amount of formatting characters to support an easy parsing and, if necessary, to develop clear formatting guidelines. Many languages like XML subsets tend to overwhelm their users with many strict control characters(i.e. the requirement of matching opening and closing tags), while languages like SQL tend to derive necessary information from the context (i.e. distinguish whether the inspected list of values are required for regular or merge insert clause). We have tried to achieve the balance between these two approaches by explicitly defining top-level delimiting information in many different characters.

Example 1. queryId1 *[options1]* {*body1*}, queryId2 {body2}, queryId3, queryId4 *[options4]*

Example 1 shows four different structures of a query part. Elements written in italic are not really compilable token streams. They are just placeholders, though body placeholders can be considered as BoW query part. The first query part contains identifier section, options section and body section. The second part contains no options. The third one contains only an identifier: such query part must be cross-compiled from query libraries. The last part is cross-compiled from a library with overridden options. A query part can also omit both identifier and options.

The options contain the information about column caption (if it is not equal to identifier), Boolean filter override, sorting order, sorting priority and sorting boost (applies only to BoW).

The result of a query is a table. Every row in it is based on some object from the data model. A row is created if it satisfies conditions of all row filters and search criteria. A query part is treated as a BoW if it cannot be parsed neither as a row filter, nor as a computable expression. An object meets search criteria if every word in BoW is contained in parameter name or parameter string value. We dont use stemming and lemmatization for now.

If a sorting option is applied to a search filter, objects ranking R is computed as follows:

$$R(query, object) = \sum_{pv} \frac{H(pv)}{N(pv)} \, ,$$

$$H(pv) = \sum_{i=1..N(query)} N_{gram}(query, pv, i) \, , \tag{1}$$

Where pv is selected from all values of object parameters, function H computes a corresponding rank for each parameter value. This rank is computed as follows. Every possible consecutive sequence of query words, founded in pv, increase rank by 1 (function $Ngram$ computes the number of occurrences of every n-gram from the query into pv). The calculated value are normalized by $N(pv)$ – the number of unique terms in the query. Then the values for every pv are summed up in the rank R of the object.

All words in a query are considered equally important, so weight coefficients like TF-IDF [11] are not applied. This helps better prediction a behavior of the ranking model. Some weight tuning can be achieved through manual manipulation with a boost coefficient in the options. Term exclusion is also implemented by assigning a negative boost coefficient to unwanted terms.

However, the actual sorting is based not on the computed rank R, but on the ordering value O computing as follows:

$$sign\,(boost) \cdot [log_2\,(|boost| \cdot R)] \, . \tag{2}$$

Where [ ] is a function of rounding to ceiling integer value no less than zero, sign($boost$) is a function that returns -1 for negative arguments and 1 otherwise, $boost$ is a boosting value, specified in column options. Boosts are especially convenient when we combine different queries (with different boosts).

Taking the logarithm of a rank is introduced to provide a meaningful BoW parts combination. The real values of R for two different objects are rarely equal. Which means that R can be used for sorting without any other sorting search clauses.

Boolean filters and computable expressions can access the parameter values by surrounding parameter names in double quotes.

Example 2. { "type" = 'slab'}, [desc]{ "creation time"}, { "Cr"}, { "V"}, { "Ni"}, { "Length" > "Width" / 2}, defective, today

Example 2 shows a query, which selects some chemical information for the newest slabs. Also slabs should be defective, created the same day when the query is executed and their length should be at least twice larger than width. These two restrictions of the query are extracted from libraries. The restriction on the creation date is context-dependent (addresses the current time). Cross-compilation can also create a context, i.e. a user can define his brigade number ("myBrigadeNumber{42}"), while common libraries use this number ("myBrigadeProduction{"operation brigade" = myBrigadeNumber }").

Parameters values can be extracted not only from the current row object, but also from referenced objects (with or without aggregation).

Example 3. Processes, melting, today, { "started"},
{ "Id".Next("type"='slab').SumStr[,]}

Example 3 shows a query that extracts an information with identifiers of all slabs grouped by processes started today and related to melting. Forward and backward references can be used in a chain and embedded into referencing conditions.

Example 4. Processes, melting, today, { "started"},
{ "Id".Next("type"='slab').Prev(type='bar' and "".Next.Next("state" = 'defective').Any).Count}

Example 4 shows the same set of processes as in Example 3, but calculates the total amount of different bars used in production of slabs of current process, which (bars) usage resulted in a defective sheet. This genealogical querying would result in an enormous amount of comboboxes and checkboxes in a visual query editor. If one would provide similar functionality.

5 Analysis

Designed DSQLM focuses mostly on data discovering and solving problems of metallurgical genealogy in data selection and preparation. Analytical capabilities (like SSA computation [6], regression and forecast analysis) are widely covered by a modeling system that is one of data consumers of this DSQLM. Although getting MDX-like pivot table using two sets of filter columns and an aggregation function could be implemented, this functionality was intentionally left undone. DSQLM was designed only to extract the relevant data.

DSQLM can aggregate data, but only by grouping on real objects in data warehouse. The language does not allow to set a limit on the rows in the result or to get a subset of retrieved objects for a query. These restrictions of DSQLM deters partial queries like "Top 5 selling mangers" in favor of queries like "All managers, ranged by sales" (financial information also is included in the system). However the system provides a viewer with the usual result paging functionality. Of course, a user can still achieve limit\offset constraints by tricking with context, but system does not encourage this programming style.

Creating a deterministic querying model with rich filtering possibilities is very challenging for indexing structures given that the estimated size of plant's raw operational data is measured in terabytes. Indexing questions will be covered in further papers.

The important thing is that ranking query part does not exclude objects from the result. To exclude objects not satisfying a query and order those by rank, the query should be included twice – with and without a sorting option. For example, a query "[desc]{ defective slab }", will be based only on the objects mentioning "defective" or "slab". While the result of a query "[desc]{ defective slab },[desc, boost 50]{ excess chromium }" will be based on the objects mentioning "defective" or "slab" or "excess" or "chromium". To exclude objects without mentioning certain terms one have to add an extra filter of Boolean type.

Ranking features of the system tends to favor consecutive occurrences of query words in the data. Lets look on the following example. A rank H for query "{steel plant data}" and parameter value "meaningful steel plant data" is 6, while for parameter value "steel plant collecting data" its only 4.

Note that the ranking model also takes into account the density of query words in parameter values, so same occurrences of query words would give lower rank for a large text, than for a small one. The default $boost = 100$ means that the minimum triggering term frequency is 1 term in 50 words of text. Ordering values O have large discreteness to avoid abuse of ranging features. Its not supposed to be used for retrieving just a few results satisfying queries. Instead of this the model should assist to retrieve the most relevant objects without limiting inspect window.

Sorting instructions can combine ranking criteria and computable functions. For example, there can be a query retrieving incidents sorted firstly by employees' last names, secondly by the rank of a query having a textual description of the incident, and finally by the amount of damage in monetary valuation.

6   Evaluation

A full-scale practical evaluation of DSQLM is yet to come along with the introduction of the next step of overall system in production. The language will be used to access systems data warehouse by plant chemical analysts. Their feedback will help us in further improvements of the DSQLM.

Nevertheless, we conducted two internal experiments. In our Department, we start to teach SQL and databases to sophomores. So we picked subjects from freshmen in order to simulate the plant employees who lack database knowledge but mastered a data types. We selected 10 such students.

The goals of the experiments were:

1. To evaluate an effort of dataset discovery with DSQLM
2. To evaluate an effort of deriving simple math expression with DSQLM

All subjects had been working separately and consequently. We explicitly asked them not to share any information about the experiment for the sake of evaluation clearance.

In the first experiment, the subjects were presented five rules defining data model (without Fig. 1 which contains significant details) and Examples 1-4 with notes. We asked them to discover dataset and state everything they understand about the data. The participants had 10 minutes to verbally describe their results. We expected them to mention all of the following concepts:

1. Steel, metal, melting or rolling
2. Production, process or output
3. Slab
4. Aggregate
5. Sheet

If a subject mentioned at least one word in each point we considered him mastered DSQLM. Nine of ten participants mastered DSQLM in allotted time and advanced next.

In the second experiment these nine subjects were given a formula and a text description of mean square deviation (MSD). We asked them to compute it for parameter value of a slab in each process. The subjects were forbidden to ask any additional information about DSQLM.

Five of nine subjects managed to derive required query.

The results of the first experiment showed that dataset discovery is effortless enough for production system. The results of the second experiment showed that expressiveness of DSQLM is acceptable for production system.

Subjects who mastered MSD query also noted that a sequence containing a parameter name, a reference path and aggregation function (*"length".Next.Avg(...)*) is not a very straightforward sequence of aggregation definition.

Our big concern regarding the upcoming real-world evaluations is an intentional high discreteness of the ranking. It was made to prevent overuse and provide deterministic, according-to-expectation ranging with use of multiple search clauses. However, this can spurn from usage those who are not familiar with information retrieval.

7 Related Work

One of the interesting works in the area of specialized query languages was published by Madaan and Bhalla [10]: they discuss difficulties of domain-specific querying of medical repositories. The key idea of replacing a visual query editor and a multistage text query with assisted input in our system is based on their work.

Tian et al. [15] presented NeuroQL a language for neuroscience data, which they compile to SQL and XQuery. This work contains a concept of query language compilation to SQL with segregation of DSQL data model from SQL data model. Our initial effort was to produce SQL tables reflecting domain model, but this work clearly stated that the scheme of database should not depend on changes in our domain specific data model.

Baratis et al. [3] presented Temporal Ontology Querying Language (TOQL) that aims to reduce the complexity of temporal ontology databases at least to the level of SQL. This work provided a fundamental background to address the main problem of metallurgical data. This data is aligned to oriented production graph, like TOQL is aligned to time. This particular feature of metallurgical data resembles a temporal database and is very familiar to every potential user of our DSQL.

BimQL [12] project is also of certain interest. They designed a language in order to help civil engineers with retrieving an information from construction models. The idea of querying objects of different types in the same table output was taken from BimQL. Also BimQL has an amazing concept of the result visualization that is natural for its domain. This particular feature is yet to be implemented in our DSQLM.

Development of DSQL would be much harder without LL-parsing by Parr [13] (also known as ANTLR). An alternative solution could be Fords packrat parsing [4] (also known as PEG). The initial DSQLM implementations used adaptive grammars, but this feature did not add clearance to DSQLM syntax. DSQLM grammar was developed with ANTRL 4, context free grammars shaped the way DSQLM is expressed.

8 Conclusion

In recent decades, querying languages and conventions have become widespread: search engines provide a wide range of search operators [11], IDEs have a lot of features available through built-in languages [8], genetic databases are even capable to help researchers to express which genome they are looking for [14]. There are numerous domains where specific querying languages can help to deal with specific problems. SQL, once created to rule them all, addresses generic problems, and it has been doing this well.

Maybe new kinds of structured user input superior to keyboard will end all text programming languages efforts at once. Currently it is certain: there is no better way to get a query than ask a user to type it. Neither gesture, nor speech recognition shows significant results in, for example, creating programs. Even if we would read mind somehow, there is no guarantee it will change a coding style.

In this article, we described the key ideas of our domain specific language that is not strongly connected with steel production. Described ranking model shows promising results for our domain, however it can be also applied for other domains as well.

Acknowledgment. Research is conducted under the terms of contract No. 02.G25.31.0055 (project 2012-218-03-167) with financial support by Ministry of Education and Science of the Russian Federation.

References

1. Aksyonov, K., Bykov, E., Aksyonova, O., Antonova, A.: Development of real-time simulation models: integration with enterprise information systems. In: ICCGI 2014, The Ninth International Multi-Conference on Computing in the Global Information Technology, pp. 45–50 (2014)

2. Ambler, S.W.: The object-relational impedance mismatch (update of February 15, 2006)

3. Baratis, E., Petrakis, E.G.M., Batsakis, S., Maris, N., Papadakis, N.: TOQL: Temporal ontology querying language. In: Mamoulis, N., Seidl, T., Pedersen, T.B., Torp, K., Assent, I. (eds.) SSTD 2009. LNCS, vol. 5644, pp. 338–354. Springer, Heidelberg (2009)

4. Ford, B.: Packrat parsing: simple, powerful, lazy, linear time, functional pearl. ACM SIGPLAN Notices 37(9), 36–47 (2002)

5. Fowler, M.: Language workbenches: The killer-app for domain specific languages (2005)

6. Golyandina, N., Osipov, E.: The "caterpillar" – ssa method for analysis of time series with missing values. Journal of Statistical Planning and Inference 137(8), 2642–2653 (2007)

7. Harris, Z.S.: Distributional structure. Word (1954)

8. JetBrains: Searching through the source code, http://www.jetbrains.com/idea/webhelp/searching-through-the-sourcecode.html

9. Kim, W.: Introduction to object-oriented databases, vol. 90. MIT Press, Cambridge (1990)

10. Madaan, A., Bhalla, S.: Domain specific multistage query language for medical document repositories. Proceedings of the VLDB Endowment 6(12), 1410–1415 (2013)

11. Manning, C.D., Raghavan, P., Schütze, H.: Scoring, term weighting and the vector space model. Introduction to Information Retrieval, pp. 109–133 (2008)

12. Mazairac, W., Beetz, J.: Towards a framework for a domain specific open query language for building information models. In: Proceedings of the 2012 eg-ice Workshop (2012)

13. Parr, T., Fisher, K.: Ll (*): the foundation of the antlr parser generator. ACM SIGPLAN Notices 46(6), 425–436 (2011)

14. Stypka, L., Kozielski, M.: Methods of gene ontology term similarity analysis in graph database environment. In: Kozielski, S., Mrozek, D., Kasprowski, P., Małysiak-Mrozek, B. z. (eds.) BDAS 2014. CCIS, vol. 424, pp. 345–354. Springer, Heidelberg (2014)

15. Tian, H., Sunderraman, R., Calin-Jageman, R.J., Yang, H., Zhu, Y., Katz, P.S.: NeuroQL: A domain-specific query language for neuroscience data. In: Grust, T., Höpfner, H., Illarramendi, A., Jablonski, S., Fischer, F., Müller, S., Patranjan, P.-L., Sattler, K.-U., Spiliopoulou, M., Wijsen, J. (eds.) EDBT 2006. LNCS, vol. 4254, pp. 613–624. Springer, Heidelberg (2006)

Approach to the Monitoring of Energy Consumption in Eco-grinder Based on ABC Optimization

Jacek Czerniak[1](✉), Dawid Ewald[1], Marek Macko[2], Grzegorz Śmigielski[2], and Krzysztof Tyszczuk[2]

[1] Institute of Technology, Kazimierz Wielki University in Bydgoszcz,
ul. Chodkiewicza 30, 85-064 Bydgoszcz, Poland
{jczerniak,dawidewald}@ukw.edu.pl
[2] Institute of Mechanics and Applied Computer Science,
Kazimierz Wielki University in Bydgoszcz, Bydgoszcz, Poland
mackomar@ukw.edu.pl

Abstract. This article is a part of the series dedicated to AI Methods Inspired by Nature and their implementation in the mechatronic systems. The Artificial Bee Colony (ABC) enables the optimization of the consumption of power supplied from photovoltaic cells. The paper includes a few implementations of ABC. Special emphasis was put on maintaining proper energetic balance, and on monitoring of power demand as well as energy sources used for the comminution. The ecological grinder was designed based on the autonomic unit. It can be called autonomous, because it does not need external control. The built-in computer system ensures monitoring and visualization of the current state of the energetic balance.

Keywords: Artificial Bee Colony · ABC · Eco-grinder · Optimisation

1 Introduction

Issues related to energy demand and acquisition in machinery systems are analyzed as part of the comprehensive assessment of the product which includes a number of processes. This relates for example to machining process, global assessment of energy resources and conversions, energy technologies, recycling technologies (as regards construction materials, polymers, food processing technologies, assessment of mechanical processes) [1,7,13,16,3,14,15]. Among those machines, one can distinguish grinders, where the unit energy demand for the comminution process is particularly important in mass processes and in very fine and colloidal grinding (comminution). The energy consumption aspects of the comminution technology are discussed broader because their aim is to minimize environmental impacts [18]. One of the assessment methods is the methodology of energy-related and environmental analyzes over the entire life cycle. In this perspective, the energy consumption of comminution is associated with the

© Springer International Publishing Switzerland 2015
S. Kozielski et al. (Eds.): BDAS 2015, CCIS 521, pp. 516–529, 2015.
DOI: 10.1007/978-3-319-18422-7_46

volume of the allowance (grinding product), which must be removed from the semi-finished product (the input material) in order to obtain the final product [15]. The modeling includes analysis of the analogies to manufacturing technologies, the most common of which in that group is machining based on the removal of the material allowance by means of the cutting edge of the tool, producing chips. Then comminution can be considered as machining with intermittent contact between a tool and material.

2   Energy Consumption of Comminution Process

Energy consumption of comminution process is defined as the demand for energy absorbed by the main drive of the grinder and used to overcome the material resistance during machining. This is energy value required to increase the specific surface area of the product [7,6,12]. In the simplified model of energy consumption for comminution process, it was proposed to determine the value of the energy using modification of the following formula:

$$E_c = k_c \cdot 10^{-3} \cdot \prod_{i=1}^{n} k_i \cdot V, J \tag{1}$$

where
E_c - energy consumption of comminution, J,
k_c - specific resistance of comminution, N/mm^2,
k_i - correction factors,
V - volume of the material layer to be removed, mm^3.
k_c values are calculated from the following formula:

$$k_c = d^{-m} \cdot k_{c1}, N/mm^2 \tag{2}$$

where
d - average grain size, mm,
k_{c1} - main value of the specific resistance of cutting, N/mm^2,
m - the factor dependent on the type of processed material.
k_{c1} and m values depend on the material type.

The correction factors define the influence of different process parameters (velocity, angles, gaps, radii etc.) on the cutting force and, as a result, on the value of required energy. Type and number of factors depends on the machining type. Number of factor ranges from zero (e.g. for grinding) to five (turning factors).

Removal of the material layer and comminution of the volume unit are characterized by processing efficiency - i.e. the ratio of the useful value of the selected physical input quantity to the corresponding value of the physical output quantity in the processing, which is usually expressed as percentage. Depending on the selected physical quantity, one can distinguish energy efficiency and material efficiency, while depending on the object, relative to which the efficiency is determined, there is: processing efficiency, processing machine efficiency, processing tool efficiency and processing instrumentation efficiency.

It is currently a global trend in manufacturing processes to put particular emphasis on environmental aspects. The key to environmental friendly design consists not only in observance of design rules, but first of all in minimization of environmental impact. Current tools: LCA (life cycle analysis) and DFE (design for environment) are strategies, where actions are focused on product and process.

Grindability factors are applied in practice by determining the unit energy demand for mass size reduction degree desired from the process point of view which is usually 1Mg (based on factors for specific material as well as input material and product grain size, and often process conditions). Hence, assuming required grinder capacity, the power necessary to execute the comminution process is determined from the following simple formula:

$$P_i = E_{op} \cdot Q \tag{3}$$

where

P_i - power of the grinder drive motor, kW,

Q - comminution capacity, Mg/h,

E_{op} - module, generalized unit energy demand, kWh/Mg.

In practice, the power of the grinder driving motor is determined based on the above formula, taking into account the efficiency of the motor, transmission gear, dynamic joints and relationships of the comminution elements. Having considered input components of the comminution process and having limited the capacity $Q_{f(2\div5)mm}$, depending on the product of specific grain size, e.g. $2 \div 5mm$, we finally reach the point where we can determine the module of disposed material susceptibility to desired comminution:

$$P_{op} = \frac{P_u + P_V + P_{d-d} + P_{ds} + P_{dp} + P_{dr} + P_{dq-s}}{Q_{f(2-5)mm}} \tag{4}$$

where

$Pd - d$ - power dissipated for internal accumulation (escape) in the grinder (input material), kW,

P_{ds} - power dissipated in the motor, kW,

P_{dp} - power dissipated in the transmission gear, kW,

P_{dr} - power dissipated in the dynamic joints of the quasi-cutting structure, kW,

P_{dq-s} - power dissipated in the grinding unit, during quasi-cutting of the material, kW.

Energy-related and environmental approach to the assessment of comminution processes and adjustment of the product to further processing reveals the problem of the input material grindability. The solution is the efficiency model, describing the relationship between the comminution energy during the phenomenon to the energy consumed by the machine for the purpose of comminution:

$$\eta_{q-\dot{s}} = \frac{E_{uq-\dot{s}(f)}}{E_{uq-\dot{s}(m)}} \tag{5}$$

where

$\eta_{q-\dot{s}}$ - efficiency of quasi cutting,

$E_{uq-\dot{s}(f)}$ - effective energy of the quasi-cutting under the phenomenon conditions (testing machine - static tests), kJ/kg,

$E_{uq-\dot{s}(m)}$ - effective energy of the quasi-cutting under actual conditions (testing machine - specific grinding unit), kJ/kg,

3 The Concept of the Polymer Waste Grinder

To enable an intelligent comminution system as a tool effectively control the process correctness, it is necessary to equip links of that system with appropriate interfaces and a human must be provided with proper communication media allowing supervision and control over the executed processes using that computer. Those goals have been accomplished using such techniques and technologies as: electronic data processing, image processing, construction and interpretation of databases and knowledge banks, modeling methods and special systems, simulation, visualization but also control, supervision and diagnostics etc. [10]. Consonance between the model and the original here means functional representativity i.e. that model behavior can be grounds for conclusions on the behavior of the original under specific conditions.

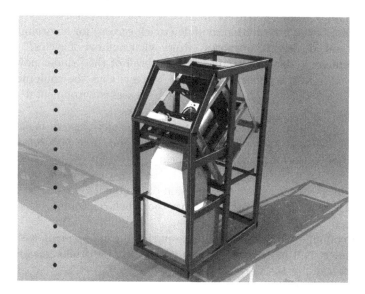

Fig. 1. View of the Waste-Sun-Grinding grinder concept

From the systemic point of view, the object as a model in the synthetic approach, due to the needs for: description, research, establishing, development and operation, should:

- enable presentation of the basic properties of the system,
- reflect structural relationships in the system, which are important due to functionality,
- reflect basic functional relationships in the system,
- enable presentation of the system development potential,
- allow to formulate decisions,
- allow assessment of the efficiency and destructiveness of the system.

4   Concept of the PET Bottles Grinder

The following conceptual stipulations were made based on preliminary assumptions:

- the 220 V (start-stop), 0,75 kW electric AC motor has been selected,
- a photovoltaic set has been selected as an alternative electric power source; the option of photovoltaic panel charging batteries connected to the power mains should be considered,
- power supply from the mains as the main source of energy with periodic power supply from the batteries,
- the design of the information module (touch screen) for environmental protection and educational purposes, "every what number of cycles?" a number of percentage is displayed indicating the level of the grinder power supply assistance from the photovoltaic cell / battery; thus the equipment can be regarded as educational and one that increases environmental protection awareness,
- safety - particular attention must be paid to safety, as we make available to public (without any preliminary training) a machine with rotating blade; necessary interlocking system preventing activation of the grinder with a limb inside must be provided as well as feeding with a push rod or level,
- operation periodical service, inspection of the blades' condition, emptying/ drying the tank, inspection of the drive, tooth belt tensioning, cells, batteries, permanent monitoring with a camera,
- size of comminuted bottles multiple-purpose grinding it is assumed that the grinder shall be fed with empty bottles from a drinks machine located in the corridor of the university,
- preliminary control of the input material quality should be ensured: e.g. full or empty bottle or partially empty bottle, glass bottle, with or without a cap, a can, a stone, a branch etc, as well as a preliminary identification system, which - in case of incorrect features of the input material - would reject it or eject it from the grinding space,

- photovoltaic cell exposure to sun light during the daytime must be ensured, potential locations are described in detail including installation place, PV panel orientation, detachment of the panel from the main part of the equipment as well as PV placement on the building roof,
- protection against theft once the equipment is made available in public place,
- changes in the preliminary design concept
- grinder drum of in-between bearings type,
- structure made of black or painted steel,
- Aluminium housing, painted
- recyclate container with drainage strainer,
- low tray located in the bottom part, designed to accommodate possible leakage and to keep the installation place clean,
- keeping high safety level,
- programmed continuation of the process following comminution of a bottle - i.e. several idle run revolutions in order to empty the grinder chamber,
- optional tilting of the axis of rotation to facilitate emptying of the grinder chamber.

5   Implementation of Artificial Intelligence (AI) in an Environmental Friendly Grinder

There are following three sources of inspiration for artificial intelligence as a computer science discipline:

- development of mathematics and computer science,
- observation of civilization phenomena,
- observation of nature.

An important place amongst many optimizing methods inspired by nature is occupied by so called Swarm Intelligence, which inspires researchers to create new metaheuristic optimization methods allowing to solve non algorithmic problems [2,4,5,19]. Those methods include the group of ant colony algorithms. Artificial bee colony algorithms are the collection of algorithms inspired by behavior of honey bees observed as early as in the middle of XX century by a German zoologist Martin Lindauer, who published his study in 1950, where he noticed that bees do so called "waggle dance" outside the swarm and carry little food to the hive at the same time. Thanks to further observations, Lindauer [11] concluded that only small percentage of bees participate in making decision on the new nest location. Most of bees do not participate in the foraging but they wait for the decision to be made by the "dancing" bees, who gradually reduce the number of nests. When finally a single nest is selected, all bees rise to fly towards the selected nest.

The behavior of bees when selecting a new nest can well be used for a decision made by any group. Thanks to three factors defined by scientists studying behavior of bees, such decision shall always be correct.

1. Organization allowing knowledge sharing by the entire group. It is much easier to make a proper and unanimous decision thanks to large number of objects taking part in the decision-making process and correct distribution of knowledge among the objects (UF).
2. The competition, thanks to which every object tries to improve at any time. The more good objects, the easier and quicker the decision-making process can be (EF1).
3. Balance, which is defined as making a decision on the basis of the opinion of many objects, keeping to the democracy rules at the same time. As a consequence, wrong decisions do not influence the final decision (EF2).

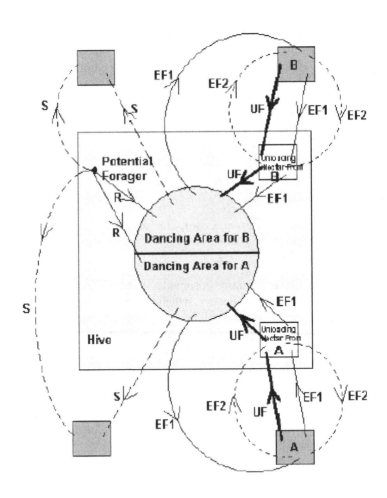

Fig. 2. Diagram of foraging behavior of bees

5.1   Implementation of the ABC Algorithm

Artificial bee colony (ABC) is a model proposed in 2005 by a Turkish scientist Dervis Karaboga [9,8]. Like other algorithms described herein, it is also based on herd behavior of honey bees. It differs from other algorithms in the application of higher number bee types in a swarm. After the initialization phase, the algorithm consists of the following four stages repeated by iteration until the number of repetitions specified by the used is completed:

- Employed Bees stage,
- Onlooker Bees stage,
- Scout Bees stage,
- storage of the best solution so far.

The algorithm starts with initialization of the food source vectors x_m, where $m = 1...SN$, while SN, is the population size. Each of those vectors stores n values $x_m, i = 1...n$, that shall be optimized during execution of that method. The vectors are initialized using the following formula:

$$x_{mi} = l_i + rand(0,1)x(u_i - l_i) \qquad (6)$$

where:

l_i-lower limit of the searched range,

u_i- upper limit of the searched range,

Bees adapted to different tasks participate in each stage of the algorithm operation. In case of ABC, there are 3 types of objects involved in searching:

- Employed Bees - bees searching points near points already stored in memory,
- Onlooker Bees - objects responsible for searching neighborhood of points deemed the most attractive,
- Scout Bees - (also referred to as scouts) this kind of bees explores random points not related in any way to those discovered earlier.

Once initialization is completed, Employed Bees start their work. They are sent to places in the neighborhood of already known food sources to determine the amount of nectar available there. Results of the Employed Bees work are used by Onlooker Bees. Onlooker Bees randomly select a potential food source using the following relationship:

$$v_i = x_{mi} + \varphi_{mi}(x_{mi} + x_{ki}) \qquad (7)$$

where:

v_i-vector of potential food sources,

x_{ki}- randomly selected food source,

φ_{mi}- random number from the range [-a,a]

Once the vector is determined its fitting is calculated based on the formula dependent on the problem being solved and the fitting v_m is compared with x_m. If the new vector fits better than the former one, then the new replaces the old one. Another phase of the algorithm operation is the Onlooker Bees stage.

Those bees are sent to food sources classified as the best ones and in those very points the amount of available nectar is determined. The probability of the x_m source selection is expressed with the formula:

$$p_m = \frac{fit_m(x_m)}{\sum_{m=1}^{SN} fit_m(x_m)} \tag{8}$$

where:
$fit_m(x_m)$- value of fitting functions for a given source.

Obviously, when onlooker bees gather information on the amount of nectar, such data is compared with results obtained so far and if the new food sources are better, they replace the old ones in the memory. The last phase of this algorithm operation is exploration by scouts. Bees of that type select random points from the search space and then check nectar volumes available there. If newly found volumes are higher than the volumes stored so far, they replace the old volumes. The activity of those bees makes it possible to explore the space unavailable for the remaining types of bees thus allowing to omit any extremes. Herd optimization algorithms, often combined with fuzzy logic [4], in particular ABC, have had many applications in mechanics, design optimization and mechatronics [2,5]. Based on the experiments [17], the algorithm developed by D. Karaboga provides satisfactory results both for linear and for square or non-linear functions. The tests were performed using the colony of 40 individuals, while the iteration limit amounted to 6000. The same test result was obtained every time for linear and square functions, while results for cubic and non-linear functions varied within very small range. Based on those tests, one can safely conclude that ABC is a perfect solution for optimization of functions. Application of ABC algorithm for optimization of energy demand and acquisition in an environmental friendly grinder is nothing but its implementation to solving of functions of many variables. Optimization of the demand for energy used in the comminution process, based on the efficiency model described in first paragraphs consists in fact in optimization of the relationship between energy consumption and efficiency. Although we deal with a difficult calculation problem which is also non-algorithmic one, the optimization using ABC yields good results at that stage of research.

5.2   Measurement of the Electric Power Quality

The electric power demand was measured integrated electric power meter manufactured by HIOKI company, model 3169-21.

For the needs of the performed tests, the HIOKI meter measurement capacity enables observation and analysis of the power consumption variations (its individual components, Fig. 7) depending on variable design and process-related conditions in grinders. Characteristic idle run Current [A] and Power [kW] curves which are shown as an example in Fig. 8 and Fig. 9, allow analysis of resistance to motion which is influenced by drive components and working components of

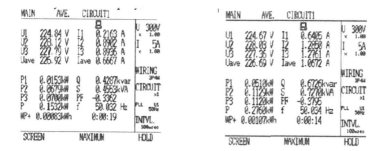

Fig. 3. View of the control panel of HIOKI meter, model 3169-21 and its indications during experiments

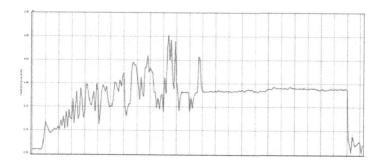

Fig. 4. Example of the power P [kW] curve of the grinder idle run, visualization prepared using Power measurement support software 9625 for HIOKI 3169-21 set

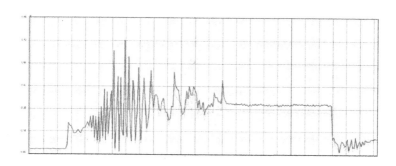

Fig. 5. Example of the power P [kW] curve of the grinder idle run after design changes of the drive system

the grinding space and others. Modernizations of selected components and assemblies of the drive system caused changes in the current and power curves, which was observed in other similar studies on environmental friendly grinders.

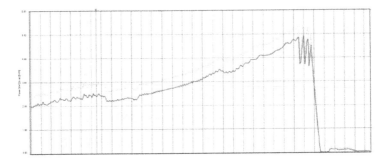

Fig. 6. Example of the power P [kW] curve of the grinder versus resistance increasing up to the motor stoppage

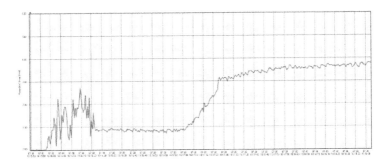

Fig. 7. Example of the power P [kW] curve of the grinder after design and process-related modifications and achievement of stable working process

Process record based on current and power curves of the grinder during idle run and grinding as well as during possible motor stoppage enable detailed identification of many complex factors. Sample tests of an industrial grinder are given in Fig. 10 and Fig. 11. Excessive cutting resistance cause motor stoppage (Fig. 10) while executed modifications allow relatively stable operation (Fig. 11) Those former test experiences with the HIOKI set now enable efficient formulation of the comminution process fundamentals as well as elementary rules of any mechanical components working in tandem. Target power value for a given process must be achieved within the above presented rotational speed range.

Motor rotational speed is measured using counter input PFI of USB 6008 card, to which signal from the induction sensor is connected. The sensor detects metal item placed on the pulley of the motor. Normally closed IME12-08NPOZW2K sensor of PNP type was used; detection of the metal item is indicated by a falling slope. The rotational speed can be determined by counting the number of slopes in a unit of time. The rotational speed is measured by means of a separate program thread. Applied sensors are characterized by power supply voltage range

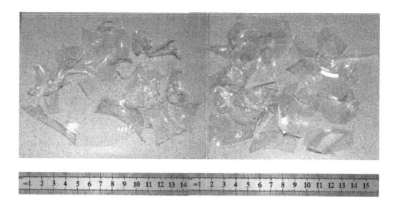

Fig. 8. View of shredded PET bottles

of 10 to 30 V. Therefore, in order to adapt USB card (which tolerates TTl levels) so as it supports the sensors, it was necessary to apply resistance divider (510 and 1 k). When sensors are supplied with 12 V voltage, the resultant voltage for high level amounts to 4.83 V. The form of the grinding product - PET bottles - was presented below. The form of the shredded material indicates to occurrence of shear and tension stresses. Size of shredded parts ranges from 2 to 8 mm.

6   Conclusion

The presented test station was built thanks to combination of knowledge on mechanical engineering, machine operation and electronic control and measurement instrumentation with artificial bee colony optimization method. Different versions of that method were applied many times to optimize mechanical designs. Optimization of the demand for energy used in the comminution (grinding) process, based on the efficiency model described at the beginning of the paper consists in fact in optimization of the relationship between energy consumption and efficiency. As demonstrated by tests, application of Artificial Bee Colony algorithm to optimize the above relationship improved energy balance of the object. Data acquired during measurements shall allow to determine optimum rotational speed and desired distance between cutting disks. As part of full automation of the comminution process, it is planned to use a sub-system designed to change the distance between cutting disks and enable analysis of the discharge product size during the comminution process. To do this, it is planned to apply video module of LabVIEW environment connected to an intelligent camera. The developed application enables visual monitoring of the comminution process. Visualization shall include states of sensors detecting material (both of inductive and capacitive type), position of the inlet flap, position of the grinding chamber flap, container overflow status as well as information on the voltage and current at photovoltaic cells, motor rotational speed and power consumption by the motor. Information on the input material weight, total weight of shredded material, number of grinding cycles and motor rotational speed since the application start-up is also displayed. An important element is the area of messages, where information on improper input material intended for grinding, on opened inlet flap etc. is displayed The application also provides the option to review stored important information as regards: date and precise time of the grinder start-up, duration of grinding process, input material weight as well as possible failures that could have occurred.

References

1. Alvarado, S., Algerno, J., Auracher, H., Casali, A.: Energy-exergy optimization of comminution. Energy 23(2), 153–158 (1998)
2. Brajevic, I., Tuba, M., Subotic, M.: Improved artificial bee colony algorithm for constrained problems. In: Proceedings of 11th WSEAS International Conference on Fuzzy Systems 2010, pp. 185–190 (2010)
3. Czerniak, J.: Evolutionary approach to data discretization for rough sets theory. Fundamenta Informaticae 92(1-2), 43–61 (2009)
4. Czerniak, J.M., Apiecionek, Ł., Zarzycki, H.: Application of ordered fuzzy numbers in a new oFNAnt algorithm based on ant colony optimization. In: Kozielski, S., Mrozek, D., Kasprowski, P., Małysiak-Mrozek, B. z. (eds.) BDAS 2014. CCIS, vol. 424, pp. 259–270. Springer, Heidelberg (2014)
5. Ewald, D., Czerniak, J.M., Zarzycki, H.: Approach to solve a criteria problem of the ABC algorithm used to the WBDP multicriteria optimization. In: Angelov, P., Atanassov, K.T., Doukovska, L., Hadjiski, M., Jotsov, V., Kacprzyk, J., Kasabov, N., Sotirov, S., Szmidt, E., Zadrożny, S. (eds.) Intelligent Systems'2014. AISC, vol. 322, pp. 129–137. Springer, Heidelberg (2015)
6. Flizikowski, J.: Micro- and Nano-energy grinding. Pan Stanford Publishing 2011 (2011)
7. Flizikowski, J.: Rozdrabnianie tworzyw sztucznych. ATR Publishing House, Bydgoszcz (1998)
8. Karaboga, D.: An idea based on honey bee swarm for numerical optimization. Technical report-tr06, Technical Report-TR06, Erciyes University, Engineering Faculty, Computer Engineering Department, Turkey (2005)
9. Karaboga, D., Basturk, B.: Artificial bee colony (ABC) optimization algorithm for solving constrained optimization problems. In: Melin, P., Castillo, O., Aguilar, L.T., Kacprzyk, J., Pedrycz, W. (eds.) IFSA 2007. LNCS (LNAI), vol. 4529, pp. 789–798. Springer, Heidelberg (2007)
10. Lesiak, P., Świsulski, D.: Komputerowa technika pomiarowa w przykadach (Examples of computerized measuring techniques). PAK, Warsaw (2002)
11. Lindauer, M.: Communication among social bees. Harvard books in biology (1971)
12. Macko, M.: Metoda doboru rozdrabniaczy wielokrawdziowych do przerbki materiaw polimerowych (Multi-edge grinders selection method for polymer materials processing), p. 177. Publishing house of UKW, Bydgoszcz (2011)
13. Marbac-Lourdelle, M.: Model-based clustering for categrocial and mixed data sets. Statitics, Universite de Lille 1, Ph.D. Thesis (2014)
14. Pahl, M.H.: Zerkleinerungstechnik. Praxiswissen Verfahrenstechnik. Mechanische Verfahrenstechnik. Auflage, Leipzig (1993)
15. Pieńkowski, G.: Ocena energochonnoci operacji obrbki ubytkowej (energy consumption assessment of material removal processes). Thesis, Wrocaw University of Technology (2006)

16. Sameon, D.F., Shamsuddin, S.M., Sallehuddin, R., Zainal, A.: Compact classification of optimized boolean, reasoning with particle swarm optimization. Intelligent Data Analysis 16, 915–931 (2012)
17. Stanarevic, N., Tuba, M., Bacanin, N.: Enhanced artificial bee colony algorithm performance. Latest Trends on Computers 2, 443–444 (2010)
18. Tanaka, K.: Assessment of energy efficiency performance measures in industry and their application for policy. Energy Policy 36, 2877–2892 (2008)
19. Weger, O., Henseler, P., Bongulielmi, L., Birkhofer, H., Meier, M.: Sustainability in the information society. In: 15th Int. Symp. Inf. for Environ. Protection ETH Zürich, Switzerland, Umwelt. aktuell (2001)

Mercurydb - A Database System Supporting Management and Limitation of Mercury Content in Fossil Fuels

Sebastian Iwaszenko[1(✉)], Karolina Nurzynska[2], and Barbara Białecka[1]

[1] Central Mining Institute,
Pl. Gwarków 1, 40-166 Katowice, Poland
siwaszenko@gig.eu
[2] Insitute of Informatics, Silesian University of Technology,
ul. Akademicka 16, 44-100 Gliwice, Poland

Abstract. Mercury is commonly known as a harmful element for environment and human beings. Due to the high mobility and lack of degradation in environment, reduction of mercury emission became one of the main fields of limitation in its adverse impact on environment and health. The database systems, addressed in gathering, transforming and disseminating information about mercury sources, ways of transportation in environment and technologies used to limit its influence, can be a useful management supporting tool. It can help scientists, stakeholders and decision makers in straggle to restrict mercury influence on people's lives and environment.

In the article a concept and design of database system dedicated to support management of mercury content in fossil fuels and energy sector wastes was presented. The solutions chosen and being implemented to provide users with ability to store measurements of mercury and accompanying species as well as sample material characteristic were described. The structure of object data model and its mapping into relational database structure is given. Design of software the developed is described and detailed solutions improving flexibility in data management and exploration are depicted. The article summarizes the use of database system as a support in toxic substances content management.

1 Introduction

Mercury is well known from its diverse impact on human health and animals. Reduction of its emission to environment and particularly atmosphere becomes the aim of activities in global scale. It is especially important, because of high toxicity of mercury species, accompanied with their mobility and capability to easily cross different compartments of natural environment. Moreover, substances containing mercury are hardly biodegradable and can be transported to very long distances [15,2]. There is no little wonder that the United Nations, European Commission and national organizations have started preparation of legal solutions and recommendations to reduce the undesirable emission of mercury to the environment [3].

© Springer International Publishing Switzerland 2015
S. Kozielski et al. (Eds.): BDAS 2015, CCIS 521, pp. 530–539, 2015.
DOI: 10.1007/978-3-319-18422-7_47

The fossil fuel combustion along with mining and metallurgy is recognized as a main source of mercury release into the environment. Despite steps taken to address this issue, the knowledge of mercury content in the fuels as well as its release rate to environment and residual in industrial, post combustion wastes are uncertain and very limited. Therefore a wide range of research activities has been started to measure the mercury content in coal and coal originating fuels, to determine mercury emission to the atmosphere as a result of fossil fuel combustion. Research aimed at elaboration of mercury balance in every stage of technologies used in energy sector are also carried out. However, to make the research results really impact the level of mercury released from anthropogenic sources, the management supporting tool is necessary. Not only should it gather the data about the contaminant, but also support decision makers with transformed and aggregated information to help them in management process. There is a research project started recently in Central Mining Institute (GIG) in Katowice, Poland to take an inventory of mercury content in fuels and its circulation. One of the project deliverable is a database system supporting management of mercury content in fossil fuels. In the article, the scope and design of the developed software is presented along with its potential applicability to mercury content management.

2   System Scope

A functionality of the database system, supporting management and limitation of mercury content in fossil fuels, include few areas:

- gathering, storing, and disseminating data obtained from measurements and supporting measurements processes,
- gathering, storing, and disseminating data describing selected additional data (coal production, fuel production, waste generation, coal origin etc.),
- information about technologies used in mining and energy sector,
- assessment of selected fuels and technology impact on mercury content in materials and its emission into environment.

System is responsible for collecting data from experimental tests and measurements. It is crucial, that not only mercury content should be saved. As the behavior of contaminant is determined by its form and accompanying chemical compounds, several parameters for each of tested sample is usually measured. Gathering the whole information is necessary for further data mining. Upon gathered data, the correlation between toxic species release and fossil fuel characteristic should be determined. It is also important that exact place of sampling and sampled material is saved. The analytic laboratories are the data sources for that kind of information. Apart from testing samples, the laboratories perform quality control tests, based on reference material. The system should be also capable of tracking the quality control tests. Moreover, it is expected that database would monitor entered sample test results and identify the values which are inconsistent with reference test results.

Appropriate support for management requires that decision makers are supplied with information of current fossil fuels market and available technological solutions. It is important to access the information of amount of coal being excavated, the coal processing (converting raw coal into fuel of required parameters) parameters, combustion technology used and so on. It is also important to track amount and parameters of wastes generated and its storage location. This kind of data is provided by coal mines, power plants, logistic companies, and wastes operators. The system provides decision makers with complete information of possible mercury (and to some extent other chemical species) emission at every technological step. The mercury release models (mostly statistical) allow estimation of expected environmental impact for chosen fuel and technology. The system forms foundation for development of decision models in the future. A decision trees were chosen as a basic support tool [11,12].

During conversion of fuel into energy, the technologies used play important role in mercury content management. Some of them can increase the contaminant release, while others can stabilize it in fuel structure. It is also important to keep track of energy requirements of given technology and amount of wastes it generates. The data is gathered by surveys and provided technology suppliers.

One of the most important aspects of developed solution is estimation how different scenarios of fuel usage in energy sector impact the mercury release to the environment. It is assumed that system should implement a set of algorithms helping users in determination of expected contaminant emission depending on submitted data on planned technology usage, fuel mass streams and fuel parameters.

The database is a heart of the system. The stored data form basis for decision and mercury emission models development. The database is not only responsible for tracking data consistency, but also identifying the exceptional values in them. Its structure and functionality is vital for optimization of energy production processes.

The system is expected to be used by three groups of users:

- decision makers
- scientists
- laboratory staff

The user interface is different for each users' group and focus on its interests. For example, for laboratory employees only forms allowing management of measurements data are accessible. The whole system user interface is developed using web technologies.

3   Data Model and Database Structure

The object data model was designed for developed solution. It was a basis for future development of relational database structure. The development the data model was based on object-oriented approach [4,5]. This allows the representation of both entities and complicated relationships that exist between them in real world. A problem of samples and sampling places can be an example of such relationship. On the one hand, each sample has to be associated with an object indicating its origin. On the other hand, this origin (place of sampling) can be very diverse - ranging from the coal seam, the semiproducts to products or substances taken from technological precesses. Each of these entities should also be present in the system even if it is not used as a sampling place. The inheritance was used to represent this kind of relationship. The object representing the sample can manage the information (and representing objects) on the place of its origin in a uniform way .

All classes in data model have a common ancestor, a *BaseObject* class. The *BaseObject* class implements identity control, support for lazy data loading and bulk operations on data objects' properties. Identity control is supported by identity property, which value is unique for each object. It was decided that identity is read only and can be set during object creation only. The value of identity field is controlled by data repository (RDMS).

The data objects are composed of other data objects and their collections. It was decided that lazy data loading would be implemented. Though the lazy component creation is implemented in a few known libraries (e.g. Managed Extensibility Framework [1]), dedicated implementation was developed. The mentioned library is more oriented in object creation than data loading from database. Therefore development of our own solution allowed better code optimization. During object creation, the contained objects are created as a placeholders for real objects. They are not filled with data, the only meaningful field they have is identity. Once the contained object is accessed, an event is fired signaling, that object have to be filled with real data. It is up to components responsible for communications with data repository to fulfill this requirements. Similarly a contained collections of data objects are filled. The base functionality for lazy loading is implemented in BaseObject.

For most of designed data model classes composition was sufficient to model relations between objects. However, there are a few cases, where inheritance has to be used. As an example, a set of classes representing places from which samples for tests were taken. The samples can be virtually taken from anywhere. It can be taken from product, coal bed in mine, waste, semiproduct and so on. Therefore it was proposed that every data object, that could possibly represent the sampling place should be inherited from *SamplingPlace* abstract class. The class covers all common functionality required, e.g. location information. The proposed object model is shown in the Fig. 1.

The most difficult aspect of transforming the object model into relational database structure was connected with representation of inheritance relationship. The method used compromises the flexibility and efficiency, and is similar to the solution presented in [13] and [14]. In the database structure it is represented by additional table, *SamplingPlaces* (Fig. 2). The table links the information from *Samples* table with appropriate record, representing the place where the sample was taken. When sample data are read from repository, the *Sampling-Places* is checked to distinguish which, of inherited objects, should be created.

Then, dummy object (allowing lazy data loading) is created. When the object is accessed, it is filled with data as in any other data object. Though the chosen relational representation requires using additional joins it still prefers data loading efficiency over flexibility - for example, adding another *SamplingPlace* specialization would require modification of *SamplingPlaces* table structure. Each table in relational has five diagnostic fields. For each record a date and time of creation and last modification is stored. The information of user, who created the record, and the user who made the last modification is also remembered. The information is used for diagnostic and administration purposes (e.g. solving users problems with database operation). The records are marked for deletion using *IsDeleted* field.

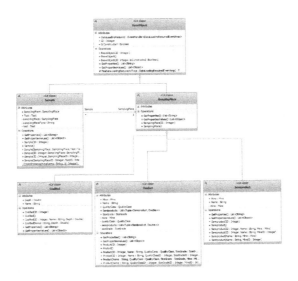

Fig. 1. Data object model representing sample and a place at which it was taken

Collecting measurements data is always connected with appropriate units handling. The problem was previously addressed both theoretically [8] and practically [9]. The solution chosen for developed system assumed that:

- the measurements can be loaded into system in different units for the same quantity,
- there are commonly used, preferable units for every measured quantity,
- the units for the same quantity can be recalculated using linear function.

The proposed solution link each of quantities (a list of quantities is stored in system's repository) with set of possible units (Fig. 3). *Quantity* represents any physical or chemical property that can be measured. The measurement is performed according to formal procedure represented by *AnalyticMethod*. Collection of measurements is stored in *Test* object. *Quantity* has a circular reference with

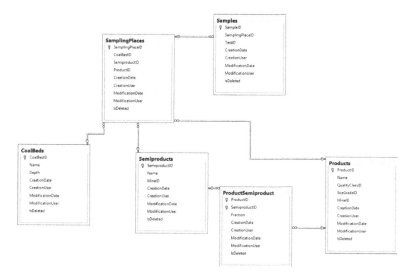

Fig. 2. Inheritance mapping to relational model. The *Users*, *Mine*, *QualityClass* and *SizeGrade* tables are omitted for clarity.

Unit object. The set of *Unit* objects referencing one *Quantity* object represents all units which can be used for the quantity. Among them one unit is marked as 'Base', which is used in units conversions. It is assumed that all units can be converted to each other using linear function. Parameters of the function (slope and intercept) are stored for each *Unit* object. Their values are set to perform conversion between given unit and the 'Base' unit. Therefore, the values for 'Base' unit are 1 for slope and 0 for intercept. It is assumed, that the values of measurements saved in repository are formerly converted to 'Base' units.

Similarly, one of the units is referenced as default for *Quantity*. If not specified otherwise, the data, before presentation to the user, are converted to that default unit. Needless to say, there is only one 'Base' and one default unit allowed for given *Quantity*, though it can be the same.

4   System Architecture

The system is designed as a multicomponent one with roughly defined multi-layer architecture (Fig. 4). It was assumed that the business layer is composed with three components responsible for basic data transformation, modeling, and registering measurements data. All components cooperate with each other and transmit the data to data access component and user interface. They are supported with *Object Data Model* and *Reasoning Rules* components. *Object Data Model* defines all objects which represents domain data. It is used for all data transformation and transmission. *Reasoning Rules* component is responsible for storing modeling and data analysis algorithms. It is used mainly by *Reasoning/Modeling* component. Its role is to provide a ways of sophisticated data

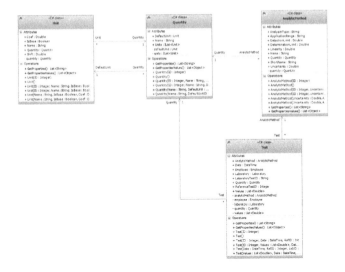

Fig. 3. Data object model representing units of measure management

analysis, required during 'what if' scenarios analysis. Data access component is an abstraction layer between data repository (a relational database management system) and the system. It converts data retrieved from repository into data model objects set. The business layer perform all kinds of data transformations necessary to fulfill user requirements. It translates command issued by UI layer into calls to data access layer and its internal functions. The business layer forward completes data into UI layer using data model objects and standard .NET collections. The UI components implement set of views (which include also forms) for data presentation and manipulation. Internally it takes advantage of MVVM design pattern [7], which is a Microsoft's modification of commonly known MVC pattern [6]. The data model objects and collections are transformed to fulfill visualization requirements.

5 System Implementation

The system was developed in C# with extensive use of Microsoft Visual Studio 2013. Microsoft SQL Server 2014 was chosen as a data repository. Because most of interaction with database system is limited to reading information, it was assumed, that the data access layer will be implemented from scratch. Use of Entity Framework and NHibernate was considered but finally rejected. Both frameworks force the specific model of interactions between data repository and object data model, which seemed not suited for developed system. Moreover, there were some efficiency problems observed when using of Entity Framework library in earlier works [10].

The most of data access layer functionality ware implemented with use of generics. Dictionaries were used for mapping between data model properties and

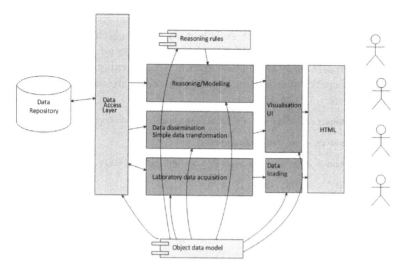

Fig. 4. Data object model representing units of measure management

relation attributes names. A double dispatch mechanism was used for appropriate control flow in hierarchical object structure, and in fact is a base mechanism for lazy data loading. As was mentioned before, data access layer creates and fills with data taken from repository the top most data object. All internal objects are created as surrogates, and are used only for holding proper identity of related records in database. When the first access to the property occurs, the event is raised, signaling, that internal object has to be filled with real data. As an example lets consider code excerpt from Waste data object:

```
public class Waste : BaseObject
{
    protected Mine mine;
    ...
    public Mine Mine
    {
        get
            {
            if (!mine.IsConstructed)
            {
                DataLoadingRequiredEventArgs e;

                e = new DataLoadingRequiredEventArgs(mine);

                mine = FireDataLoadingRequired<Mine>(e);
            }

            return mine;
        }

        set
        {
            mine = value;
        }
    }
    ...
}
```

If Mine object is not constructed, an event is raised. To properly load the data from the repository, it is not sufficient to know which object should be constructed (*Mine*). The context of the object construction (the *Mine* object contained inside *Waste* object) must be known also. To call appropriate loading method, double dispatch mechanism is used. The event procedure just calls the dispatch mechanism:

```
void obj_DataLoadingRequired(object sender, DataLoadingRequiredEventArgs e)
{
    DataAccDispatcher disp = new DataAccDispatcher();

    e.RequestedComponent = disp.Dispatch(sender, e.RequestedComponent);

    e.Fulfilled = true;
}
```

In the *Dispatch* method, upon types of two objects – a component raising the event and the object being constructed appropriate method responsible for data loading is called:

```
internal BaseObject Dispatch(object sender, object p)
{
    return DispatchSpec(sender as dynamic, p as dynamic);
}

private BaseObject DispatchSpec(object entity, object component)
{
    throw new ArgumentException();
}

private BaseObject DispatchSpec(BaseObject entity, IList component)
{
...
}

private BaseObject DispatchSpec(BaseObject entity, List<double> component)
{
...
}

private BaseObject DispatchSpec(BaseObject entity, BaseObject component)
{
...
}
```

In the case of *Waste* and *Mine* object, the last method is called, causing *Mine* object filling with data and using Waste object as a parent. If the list of data objects should be created (representing e.g side 'one' in one to many relation) the *DispatchSpec(BaseObject entity, IList component)* is called. The mechanism proved to be reliable and efficient.

The components forming each layer are created using Class Factory design pattern. All components are expected to run in server environment. The system is currently tested by selected group of end users.

6   Conclusions

The concept of a new database system dedicated as a support for limitation of mercury release to environment was presented. The system was designed using multilayer architecture, with Web based user interface. The object domain data model has been presented as well as its mapping to relational database structure. The selected software solutions were described. The developed system proved its robustness during internal tests and is currently tested by group of selected end users.

Acknowledgments. The research presented in this work is funded by the National Research and Development Centre in project titled: 'Opracowanie bazy danych zawartości rtęci w krajowych węglach, wytycznych technologicznych jej dalszej redukcji wraz ze zdefiniowaniem benchmarków dla krajowych wskaźników emisji rtęci', agreement no. PBS2/A2/14/2013.

References

1. Microsoft Corporation: Managed Extensibility Framework homepage, http://mef.codeplex.com/
2. United Nations Environment Programme (ENAP). The Global Atmospheric Mercury Assessment: Sources, Emissions and Transport. Geneva, Switzerland (2008)
3. Industrial Emissions Directive, 2010/75/EU (2010)
4. Coad, P., Yourdon, E.: Object-oriented analysis, vol. 2. Yourdon Press, Englewood Cliffs (1991)
5. Coad, P., Yourdon, E.: Object-oriented design. Yourdon Press (1991)
6. Gamma, E., Helm, R., Johnson, R., Vlissides, J.: Design patterns: elements of reusable object-oriented software. Pearson Education (1994)
7. Ghoda, A.: HTML5, JavaScript, and Windows 8 Applications. In: Windows 8 MVVM Patterns Revealed, pp. 99–107. Springer (2012)
8. Gruber, T.: Toward principles for the design of ontologies used for knowledge sharing. International Journal of Human-Computer Studies (43), 907–928 (1993)
9. Łączny, J., Iwaszenko, S., Michalak, M.: System wspomagający ocenę emisji co_2 ze zwałowisk odpadów powęglowych. Studia Informatica 32(2B(97)), 373–386 (2011)
10. Nurzyńska, K., Iwaszenko, S., Choroba, T.: Database application in visualization of process data. In: Kozielski, S., Mrozek, D., Kasprowski, P., Małysiak-Mrozek, B., Kostrzewa, D. (eds.) BDAS 2014. CCIS, vol. 424, pp. 537–546. Springer, Heidelberg (2014)
11. Quinlan, J.R.: Induction of decision trees. Machine Learning 1(1), 81–106 (1986)
12. Quinlan, J.R.: Simplifying decision trees. International Journal of Man-Machine Studies 27(3), 221–234 (1987), http://www.sciencedirect.com/science/article/pii/S0020737387800536
13. Rahayu, J.W., Chang, E., Dillon, T.S., Taniar, D.: A methodology for transforming inheritance relationships in an object-oriented conceptual model to relational tables. Information and Software Technology 42(8), 571–592 (2000)
14. Rumbaugh, J., Blaha, M., Premerlani, W., Eddy, F., Lorensen, W.E.: Object-oriented modeling and design, vol. 199. Prentice-Hall, Englewood Cliffs (1991)
15. Sloss, L.L.: Economics of Mercury control. IEA Claen Coal Centre, CCC/134 (2008)

Liquefied Petroleum Storage and Distribution Problems and Research Thesis

Marcin Gorawski[1,2](✉), Anna Gorawska[1,2], and Krzysztof Pasterak[1,2]

[1] Institute of Computer Science, Silesian University of Technology,
Akademicka 16, 44-100 Gliwice, Poland
{Marcin.Gorawski,Anna.Gorawska,Krzysztof.Pasterak}@polsl.pl
[2] Department of Data Spaces and Algorithms, AIUT Ltd.,
Wyczolkowskiego 113, 44-100 Gliwice, Poland
{mgorawski,agorawska,kpasterak}@aiut.com

Abstract. The greatest threat to the environment and aquatic life is an uncontrolled fuel leakage, which is also extremely hazardous to health and safety of people. Guaranteeing the reliability of a leak detection system is probably the ultimate purpose of fuel management systems. However, there are more problems that ought to be solved before or simultaneously with detecting possible outflows of fuel products. In this paper we highlight major research opportunities consistent with wetstock management and statistical inventory reconciliation. The main goal is to outline thesis on the nature and impact of numerous phenomena on the inventory reconciliation methods. Issues considered in this paper include but are not limited to sensor miscalibration, data acquisition, and transmission problems as well as leak detection from both, tanks and connected pipeline.

Keywords: Fuel leak detection · Sensor miscalibration · Petrol station · Inventory reconciliation · Statistical inventory reconciliation · Leak from pipeline · Data science · Stream data warehouse

1 Introduction

Even a small inflow of liquid oil products or its water mixtures to the ground or aquifers, can easily damage the ecosystem in a manner that cannot be completely reversed [6]. In the event of an uncontrolled outflow, fuel can travel through underground rivers injuring the environment and aquatic life over a wide area. Moreover, it can jeopardize health and safety of people by contaminating drinking water supplies or through risks arising from its highly flammable nature. Therefore, to minimize the adverse effects of loss of containment, service stations must comply with numerous legislations governing operation and maintenance of each and every element in a facility infrastructure [2,3,17]. With costly clean-ups followed by legal penalties, the more accurate wetstock management and leak detection systems are, the more profits can be made. Consequently, earliest possible detection of fuel loss became central to successful storage and distribution of liquid fuels.

© Springer International Publishing Switzerland 2015
S. Kozielski et al. (Eds.): BDAS 2015, CCIS 521, pp. 540–550, 2015.
DOI: 10.1007/978-3-319-18422-7_48

In modern service stations the problem of detecting an unaccounted fuel can be handled by built-in monitoring wells or sensors. Although purely mechanical solutions may indicate leak before any liquid enters the ecosystem [2], they are not risk-free. Another solution to the problem is performing manual leak tests periodically, which guarantee almost zero probability of leakage detection before an uncontrolled outflow occurs. Moreover, manually derived data is relatively imprecise due to human errors. Aforementioned approaches rely on mechanical or testing methods, while it is feasible to provide an efficient wetstock management system by means of constant statistical analysis of delivery, storage, and dispensing data. This base knowledge is a prerequisite for the inventory reconciliation [17], which this paper will be concentrated on.

The detection of an unaccounted fuel is not a trivial task – apart from abnormal fuel storage behaviours (e.g. leaks, sensors-related issues) there are several factors that have to be taken into account. Natural phenomena like evaporation or fuel thermal expansion may result in discrepancies between measured stock and expected values, which do not necessarily indicate leakage. Furthermore, numerous factors commonly overlap forming inventories of utterly complex data. Profound knowledge of all phenomena nature and impact on mentioned datasets is the key in distinguishing problems possibly resulting in disastrous effects from ones induced by the natural fuel behaviour.

All vessels storing liquids and pipelines may suffer from an uncontrolled outflow. The problem of anomaly detection was raised by researchers in the context of storage and distribution of liquids e.g. water, fuel, gas. In [15] classification and inventory reconciliation techniques were applied to leak control, while in [13] authors have used plastic optical fibre sensing. There is a group of pipe-leak oriented solutions [7], where e.g. fault detection clustering [14], fuzzy system classification [5], pattern recognition [20], support vector machine learning [4] or a frequency response diagram [12] were used to detect and locate leakage.

The issues of storage and distribution of liquid fuels are key challenges both from financial and environmental reasons. In this paper, anomaly detection challenges will be demonstrated with regard to operation of a petrol station infrastructure. However, the thesis and remarks in this paper can be easily transferred to all types of liquid fuel storing and/or dispensing facilities. Works carried out in this area are the subject of a project implemented by AIUT Ltd. in collaboration with the Institute of Theoretical and Applied Informatics of the Polish Academy of Sciences. The project is founded by the Polish Council of the National Centre for Research and Development within the *DEMONSTRATOR+* program [1].

Service Station Infrastructure

Each petrol station consists of the same elements: fuel tanks, pumping and distributing utilities, and dispensers with nozzles. During normal operation, when no malfunctions are considered, fuel is operating according to a following schema: delivery – storing – pumping – dispensing.

There are many types of connections between tanks and dispensers, more specifically – between tanks and individual nozzles. The basic criterion for division in accordance with [2] is how fuel is being pumped between infrastructure components. Vacuum or pressure systems cause suction or injection of fuel into the pipeline.

A pumping system, as well as the type of a storage, seemingly has no identifiable impact on inventory data. However, a pumping system reflects on the amount of fuel drawn from the tank into connected pipework. While station infrastructure schema available for the software vendor consists only of details on tanks and dispensers, length and configuration of pipework itself are not available. Therefore, piping should be considered a *black box*, with only inputs and outputs known. As a consequence, approximation of the amount of fuel present in piping is extremely difficult not only due to lack of aforementioned information, but because of the fuel thermal expansion phenomenon changing product volume, which was unknown from the very beginning.

Nevertheless, for a service station it is not always feasible to introduce modern, mechanical methods of fuel loss detection [17]. The adaptation to comply with legislations must be made without any changes to the petrol station infrastructure. Therefore, inventory reconciliation methods are popular alternative for the most common storage systems with underground, single-skin tanks and piping.

2   Data Analysis and Mining

As it was stated in the introduction, for most petrol stations it is not feasible to introduce modern, mechanical means of fuel management. Thus, measurements of stored, delivered, and dispensed product must become basis for surely statistical analysis. In this scenario, two main sources of data can be distinguished: tanks and dispensers, i.e. nozzles, which produce measurements carrying information about stored and sold fuel respectively. As a consequence, data representing a current state of mentioned facilities is commonly supplying a dedicated software application by two types of data streams. Third type of inventory records represents the amount of delivered petroleum. However, deliveries can be detected through analysis of tank volume – sudden and relatively large increase in value may indicate a delivery. Thus, such inventories are not shared with a wetstock management system vendor.

2.1   Tank Measurements

In modern stations Automatic Tank Gauges (ATG) [3] are increasingly used, thus, basic information on stored fuel ought to be drawn from product height. Then it is possible to transform measured fuel height to volume using an appropriate function $f: h \to v$, where h denotes height and v stands for volume.

Table 1. Simulated tank measurements

Timestamp	TankId	Fuel Volume	Temperature	Water Valume
2014-01-01 07:00:00	3	28358,8912789761	15	0
2014-01-01 07:00:00	4	37536,4535629835	15	0
2014-01-01 07:05:00	1	9129,33206202403	15	0
2014-01-01 07:05:00	2	18850,602708773	15	0
2014-01-01 07:05:00	3	28309,2665066639	15	0

The function is called *calibration curve* and it strongly depends on a shape and position of a particular tank and may change over time due to tank malformations. Therefore, from the analysis point of view, more unified data lies in the volume values.

The simulated measurements presented in Table 1 contain inventory data from four tanks. Fuel height was omitted in presentation, but temperature and water volume were taken into account. Obviously water presence in A tank is not desirable; therefore, commonly there are additional water level probes installed in tanks. Measuring water level is based on differences in liquids density. For simulation purposes a reference temperature of 15 Celsius degrees was used.

2.2   Nozzle Measurements

Records of dispensed product are represented mainly by the volume measurements with regard to a specific type of product, tank, and nozzles. Unlike in case of tanks, nozzle measurements made at a given moment of time are not simple to define. It has to be distinguished that values can be expressed as summarized volume of fuel dispensed through a nozzle (i.e. total counter) or information regarding current transaction only (i.e. transaction counter). Table 2 presents records obtained from the same simulation as mentioned in the previous section.

Table 2. Simulated nozzle measurements

Timestamp	NozzleId	TankId	Transaction Counter	Total Counter
2014-01-01 07:02:00	24	4	0	207,638426564652
2014-01-01 07:03:00	13	1	0	223,688673885022
2014-01-01 07:03:00	14	2	0	64,6584545488741
2014-01-01 07:03:00	15	3	12,4583333333333	60,0984271005254
2014-01-01 07:03:00	16	4	0	132,636625676526

2.3   Inventory Reconciliation

The single measurement in terms of fuel station management are not reliable sources of information. Only when a greater dataset is created, complex trends and phenomena may be identified. As a consequence, measurements are aggregated over fixed time windows and each represent the total amount of stored, dispensed or delivered fuel within single time period.

In best case scenario, fuel volumes associated with current stock and purchased volume are equal. Unfortunately, it is purely hypothetical situation that serves as a reference for real data analysis. Achieving the balance between delivered and sold fuel can be handled by means of the inventory reconciliation techniques [17,18,19]. It can be performed monthly, daily or every few minutes, where each mode designates an adequate aggregation time. In literature the result of the inventory reconciliation is called *variance* or *error* [15,17,18]:

$$Var = V_s - V_p - V_d \qquad (1)$$

where V_s stands for the volume of sold product and V_p for the volume of fuel pumped from the tank during sale transactions. The last symbol, V_d, represents delivered fuel, i.e. the amount of product poured into the tank during a delivery. In such case, when delivery occurs the volume of fuel pumped out from the tank is less than 0. In the ideal scenario variance equals 0.

The value of a single error represents state of a service station in a given period of time. Long term analysis requires unified presentation of summarized variance values, which is served by a *cumulative variance* (CV). The CV is the basis for detecting many problems, e.g. sensors miscalibration and leakage. Analysis of a single phenomenon occurrence has to be held under the assumption that other are absent. Specific problems will be described later in this paper.

3   Sensor-Related Issues

The accuracy of measuring devices affects measurement correctness and, as a consequence, quality of data. Each petrol station is equipped with a massive range of sensors, which can produce corrupted datasets which are subsequently processed giving invalid results, e.g. in reconciliation. In this section we will focus primarily on the volume sensors installed in tanks and dispensers.

Tank Sensors Miscalibration. The visible symptom of the fuel sensors miscalibration is either over- or underevaluation of real fuel volume which in consequence leads to emergence of virtual surpluses or leaks. Moreover, situation is further complicated by a partial distortion of the calibration curve. The curve usually changes only in some particular segments, thus causing *the height-to-volume* transformation incorrect only within certain ranges. Therefore, the result of reconciliation may be correct only in some volume ranges.

An original form of the calibration curve is created during an initial tank calibration – starting with an empty tank, the tank is successively filled with exactly known amounts (volumes) of fuel and every time product height is being measured with a dipstick or an ATG. While manual measurements made with a dipstick is performed by a petrol station worker, accuracy is vulnerable to a human error. Moreover, during the initial calibration the tank should not be used for any other purpose than the calibration. Later, when tank position or shape changes, it is essential to perform the recalibration without pausing any system – preferably by relying solely on data and software solutions.

As described in the previous sections, reconciliation of a tight and well-calibrated tank should result in an error equal to 0. When the probability of any leakage is known to be 0 too, it is possible to utilize the CV (cumulative variance) [15,17,18] as a calibration quality criterion.

Nozzle Sensors Miscalibration. Measurements of dispensed fuel are made on the dispensers site, thus, the amount of sold product is expressed in volume for each nozzle separately. Unlike with tanks, where height is a basic criterion, we can omit the height-to-volume transformation as well as recalculation of calibration curve. However, measuring devices installed in nozzles, i.e. fuel-flow sensors, can also introduce errors to measurements. Generally sensors placed in nozzles ought to be calibrated every six months to assure measurements correctness. Practice shows that, mostly due to financial reasons, it is performed less frequently or only when major problems occur. Nozzle sensor miscalibration may lead to misrepresentation of returned volume values. In order to eliminate that pejorative impact a *calibration coefficient* is included in all calculations involving sold fuel volume. Usually it is a factor that the volume value is multiplied by. The calibration coefficient must be calculated for each nozzle separately and should be actualized, while sensor miscalibration may worsen with time.

A Probe Hang. Another issue is mechanical suspension of a probe floating on a fuel surface. In such condition, the fuel probe can be located either below or above the fuel level. In terms of data analysis, it results in production of invariable measurements, even when stored fuel volume is changing due to deliveries or sales. When reconciled, such data creates virtual surplus – despite the normal sale, measurement from probes shows that fuel level tends to remain constant.

A Fuel Thermal Expansion. A natural factor complicating data analysis is a temperature. Given the same mass of stored fuel, its volume may differ due to temperature discrepancies. With the rise of a fuel temperature, the volume increases, thus, comparison of fuel volumes in different locations or time periods must be associated with a temperature compensation. A fuel expansion phenomenon is even more significant when liquids are mixing during a delivery – the product temperature changes rapidly before it compensates. The thermal expansion should also be considered in accordance with seasonal weather changes.

4 Leak Detection

According to norms [2] and [3] five classes of leak prevention systems can be distinguished, where each concerns leaks from tanks and connected piping. Solutions consistent with classes from I to III and V use physical devices and a special infrastructure to detect anomalies consistent with an uncontrolled fuel outflow, whereas class IV systems rely sorely on the data analysis. That is why in older petrol stations sensors are used only to gather data during normal operation and adequate statistical methods are detecting leak-related incidents. Although our primary focus is set on class IV systems, as most petrol stations cannot be enhanced by an automatic leak detection, remarks on the leakage nature are present for any means of wetstock control.

Leaks from Tank and Piping. The difference between an outflow from a tank and a leak from the connected pipework is not only based on the physical localization of the leak, but it is also noticeable in the datasets collected from the station. When a leak from a tank is considered, the measured fuel volume decreases gradually during the whole time frame the leak was presented. A leak from piping results in an observable loss only in fuel sale intervals – such operation involves pumping and thus the fuel is present in the pipeline only when sale transactions occur.

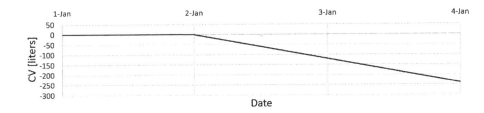

Fig. 1. Cumulative variance for leak introduced to the tank

Figure 1 presents a cumulative variance chart for a well-calibrated tank with a tightness problem. A leak of 5 litres per hour was introduced on the 2nd of January and was present till the very end of the simulation. The cumulative variance chart is constant until that specified day, after which linear decrease in value is observed. In inventory reconciliation techniques negative balance between sold, dispensed, and delivered fuel may indicate leakage.

Leaks from piping are more difficult to detect – fuel loss is observable only when product is being sold. Therefore, on Figure 2 from the 3rd of January, when leak (5 liters per hour) was applied, decrease in value is observable but it is not linear. Periods when cumulative variance is stable are consistent with no fuel being dispensed from connected tank, while decreasing trend denotes fuel loss during dispensing.

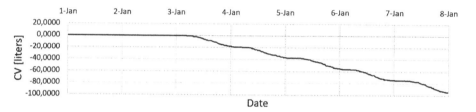

Fig. 2. Cumulative variance for leak from pipeline

More importantly, when intensity of customer arrivals is relatively low, leak from pipeline can remain undetected for a longer period of time. Moreover, even if such leakage is detected, it is not possible to specify its exact location. Structure and length of connected pipework is unknown, therefore with only statistical data it can be only indicated whether it occurs or not.

The other criterion of leak division is associated with an intensity of a leak. Two cases can be distinguished: constant and variable leaks. The former occurs when a leak rate is constant over time, while the latter is characterized by a variable rate. Usually it is a decreasing function of fuel volume as the amount of fuel stored in a tank has a direct impact on the pressure causing a leak. It is worth noticing that leaks with variable rate can only occur in tank because in pipes there is a constant pressure forced by a pump.

5 Data Acquisition and Transmission Problems

Besides previously mentioned problems, associated with different phenomena and anomalies that can occur on petrol stations, data can be affected with variety of distortions and malformations created during transmission or acquisition of the data. Most common issues of that type, which increase analysis complexity, are:

- missing data,
- data anachronism,
- data corruption.

Aforementioned issues are consequences of linking data sources with proper outputs through a transmission medium. In addition, all devices that generate data can also suffer from various disorders and produce corrupted data.

The first problem to consider is lack of certain data packages, which can be caused by sensor-related or networking issues. Such phenomenon can be quite effortlessly repaired, using interpolation techniques. It is applicable only when the missing part of data is relatively small.

The next problem is consistent with data packets being sent via multiple channels. It is possible for some of these packages to experience serious delays whereas the other (transmitted via a different channel) are delivered on time. This leads to a data anachronism. To neutralize potentially unsafe effects of this anomaly, two different approaches can be considered: wrong measurements can be rejected or input module can wait for a certain time interval to ensure that all delayed data has arrived. In the second method all measurements suffer delay equal to this interval.

The last phenomenon described in this section is data corruption. Due to different disturbances some bits of particular data packages can be changed causing malformation of the whole measurement. It can happen either on the data source site or during transmission. Nevertheless, wetstock management systems must be able to detect corrupted data and exclude it from its computations while they can violate further results significantly.

6 Proposed Solution's Architecture

The data from fuel stations are usually collected every few seconds and their amount depends on the number of tanks and nozzles. Complex wetstock analysis is often performed on data gathered from many stations at once, even the whole country area can be potentially considered. The stream nature [9,16] of fuel station data sources implies requirement of special tools capable of native processing of data streams. In terms of wetstock management following features are the most important when managing the stream source data:

- *real-time processing* – analysis oriented on the detection of mentioned storage and distribution problems ought to be performed as fast as possible for environmental and financial reasons,
- *the ability to process large data sets* – storing historical data along with relatively high frequency of generation; however, causes fast growth of size, it enables to perform advanced analysis concerning already determined issues as well as detecting periodic, subtle procedures, e.g. theft during deliveries,
- *handling data sources located within a large geographical area* – collecting data streams from a distributed measurement system [11],
- *quick access to data* – an accurate analysis may require data from any of time periods and on different aggregation levels.

When big datasets and complex analysis are considered, a Data Warehouse system is an obvious choice. Therefore, we propose stream-dedicated solution – *Stream Data Warehouse* (StrDW) [8,10]. An architecture of this specific system fully satisfies aforementioned requirements and adds many features supporting processing and maintaining of stream data, e.g. a stream ETL [8], a specialized processing engine StrSOLAP. Moreover, data streams supplied by the service station sensors have to be stored in a specialized data structure adapted to their continuous arrival specification. An StrMAL [10] engine is intended for fast access to current and historical data with regard to a selected aggregation level. This has a significant importance during data analysis involving trend and feature mining, as in the leak detection case.

7 Summary

Data discrepancies presented in the fuel volume inventories may be an indication of serious problems. Possible disastrous effects of issues highlighted in this paper can cause variety of economical and environmental losses. Discussed phenomena have either direct or indirect influence on the quality of data. The situation is further complicated by the fact that these factors are frequently independent from each other and may overlap. Therefore, the biggest challenge is the determination of problems source by a proper classification of the mentioned data discrepancies nature.

In this paper variety of research areas were discussed, i.e. tank and nozzles miscalibration, fuel probe hang, the thermal expansion impact on the volume measurements, the calibration curve malformation, leak detection as well as determination of water occurrence in a tank. The nature of each phenomena was presented with possible impact on the inventory reconciliation. This paper serves a purpose of a state-of-art support, while presented pool of knowledge is a very basis for successful management of storage and distribution of variety of liquids, especially fuels. To the best of our knowledge, there is no commercial solution providing unified environment for wetstock management with leakage detection and complex trend analysis that links current and historical data. Our future goal is to provide such unified system in a form of the Stream Data Warehouse with highly specified decision support and reporting services.

References

1. DEMONSTRATOR+ program, The Polish Council of the National Centre for Research and Development, http://www.ncbir.pl/en/domestic-programmes/demonstrator/
2. EN 13160-1. Leak Detection Systems - Part 1: General Principles (2003)
3. EN 13160-5. Leak Detection Systems - Part 5: Tank Gauge Leak Detection Systems (2005)
4. Chen, H., Ye, H., Lv, C., Su, H.: Application of support vector machine learning to leak detection and location in pipelines. In: Proceedings of the 21st IEEE Instrumentation and Measurement Technology Conference, IMTC 2004, vol. 3, pp. 2273–2277. IEEE (2004)
5. Da Silva, H.V., Morooka, C.K., Guilherme, I.R., da Fonseca, T.C., Mendes, J.R.P.: Leak detection in petroleum pipelines using a fuzzy system. Journal of Petroleum Science and Engineering 49(3), 223–238 (2005)
6. Erkman, S.: Industrial ecology: an historical view. Journal of Cleaner Production 5(1), 1–10 (1997)
7. Ferrante, M., Brunone, B., Meniconi, S., Karney, B.W., Massari, C.: Leak size, detectability and test conditions in pressurized pipe systems. Water Resources Management 28, 4583–4598 (2014)
8. Gorawski, M., Gorawska, A.: Research on the Stream ETL Process. In: Kozielski, S., Mrozek, D., Kasprowski, P., Małysiak-Mrozek, B., Kostrzewa, D. (eds.) BDAS 2014. CCIS, vol. 424, pp. 61–71. Springer, Heidelberg (2014)

9. Gorawski, M., Gorawska, A., Pasterak, K.: A survey of data stream processing tools. In: Information Sciences and Systems, pp. 295–303. Springer International Publishing (2014)

10. Gorawski, M., Malczok, R.: On efficient storing and processing of long aggregate lists. In: Tjoa, A.M., Trujillo, J. (eds.) DaWaK 2005. LNCS, vol. 3589, pp. 190–199. Springer, Heidelberg (2005)

11. Gorawski, M., Marks, P., Gorawski, M.: Collecting data streams from a distributed radio-based measurement system. In: Haritsa, J.R., Kotagiri, R., Pudi, V. (eds.) DASFAA 2008. LNCS, vol. 4947, pp. 702–705. Springer, Heidelberg (2008)

12. Lee, P.J., Vítkovský, J.P., Lambert, M.F., Simpson, A.R., Liggett, J.A.: Leak location using the pattern of the frequency response diagram in pipelines: a numerical study. Journal of Sound and Vibration 284(3-5), 1051–1073 (2005)

13. Morisawa, M., Muto, S.: Plastic optical fibre sensing of fuel leakage in soil. Journal of Sensors 2012 (2012)

14. Murvay, P.S., Silea, I.: A survey on gas leak detection and localization techniques. Journal of Loss Prevention in the Process Industries 25(6), 966–973 (2012)

15. Sigut, M., Alayón, S., Hernández, E.: Applying pattern classification techniques to the early detection of fuel leaks in petrol stations. Journal of Cleaner Production 80, 262–270 (2014)

16. Stonebraker, M., Çetintemel, U., Zdonik, S.: The 8 requirements of real-time stream processing. SIGMOD Rec. 34(4), 42–47 (2005)

17. United States Environmental Protection Agency: Standard Test Procedures For Evaluating Leak Detection Methods: Statistical Inventory Reconciliation Methods. Final Report (1990)

18. United States Environmental Protection Agency: Introduction to Statistical Inventory Reconciliation for Underground Storage Tanks (1995), http://www.epa.gov/oust/pubs/sir.pdf

19. United States Environmental Protection Agency: Straight Talk on Tanks - Leak Detection Methods for Petroleum Underground Storage Tanks and Piping (2005)

20. Zhang, J.: Designing a cost effective and reliable pipeline leak detection system. Pipes and Pipelines International 42(1), 20–26 (1997)

Idea of Impact of ERP-APS-MES Systems Integration on the Effectiveness of Decision Making Process in Manufacturing Companies

Edyta Kucharska[1], Katarzyna Grobler-Dębska[1(✉)], Jarosław Gracel[1], and Mieczysław Jagodziński[2]

[1] Department of Automatics and Biomedical Engineering,
AGH University of Science and Technology,
30 Mickiewicza Av, 30-059 Krakow, Poland
{edyta,grobler,gracel}@agh.edu.pl

[2] Department of Automatics, Silesian University of Technology,
Akademicka 2A, 44-100 Gliwice, Poland
mieczyslaw.jagodzinski@polsl.pl

Abstract. The aim of this article is to present an analysis of the impact of ERP-APS-MES systems integration on decision-making process in the manufacturing company of global nature. As a part of the article the flow of data in the integrated ERP, APS, and MES systems and including of these data in Business Intelligence systems were analyzed. Two types of BI systems have been considered, the first classic, based on the search technologies and processing of OLAP data, the second type are the systems of "In-Memory" class. There has been also described the impact of ERP APS and MES systems integration, on the efficiency of business management at all levels: strategic, tactical and operational. The approach proposed by the authors aims to improve significantly the reliability of data. As a consequence this will reduce the risk of erroneous data used in planning and evaluation of implementation processes. According to the authors the integration of these systems will significantly increase the effectiveness of management in the manufacturing company.

Keywords: ERP · APS · MES · Business Intelligence · Decision Making Process · Integration of systems · Information system

1 Introduction

Due to the increasing market demands, the basis of manufacturing companies thrive is their ability to efficient and effective management. In order to increase competitiveness, continuous analysis and optimization of processes at the strategic, tactical and operational levels is required. It is extremely important that the data were collected in real time. Furthermore, an important factor is the way of data processing, i.e. an efficient extraction of reliable knowledge of the collected data. Such management methods require adequate support systems. Strategic management support systems are ERP systems [12], which are

S. Kozielski et al. (Eds.): BDAS 2015, CCIS 521, pp. 551–564, 2015.
DOI: 10.1007/978-3-319-18422-7_49

designed for detailed calculation and monitoring costs across the enterprise and have modules for management reporting (Business Intelligence), or very often are integrated with external BI tools. The disadvantage of these systems is the manual reporting of data from the production, which may interfere with the analysis. For this reason, these systems need to be supplemented in the area of improving the efficiency of production, to ensure the stability and security of supply and quality of production. Thus, in order to support the implementation of the strategy in large manufacturing companies, there should be introduced the integration of ERP-level system with systems that include its functionality, implementation planning and execution of production, APS and MES-level systems.

The aim of this article is to present the idea of the use of integrated ERP, APS and MES systems to automatic reporting of data from production, which will improve the accuracy of automatic analysis in Business Intelligence module. As part of the article the flow of information between the integrated ERP-APS-MES systems were analyzed. The article analyzes the impact of automatic downloading of reliable data in real-time from the integrated ERP-APS-MES systems to report in BI systems. This approach will allow to conduct accurate analysis and to raise the effectiveness of the strategy described by Balanced Score Card. This paper is a continuation of the work presented in [4,5].

2   Company Management

Enterprise management, especially manufacturing, requires planning and decision-making at many levels. Due to the increasing market demands, continual analysis and optimization of processes at the strategic, tactical and operational levels is required. It is important that decisions are taken on the basis of reliable data, preferably collected in real time. Such methods require the support of management information systems. To management at the strategic level initially used methods concerned mainly plans for the budget of the company and allowed only for short-term planning cycles (annual). These methods were based on data on the tangible assets of the company, depreciating impact on the value of intangible assets of the enterprise. Recognizing the importance of intangible assets in strategic planning methods have been developed, taking into account not yet taken into account factors (eg. a employees skills development and building relationships with the customer).

One of the tools supporting this approach to management is developed by Harvard Business School and is called balanced scorecard [10]. Its main aim is to look at all the resources in the company (human resources, business units, the management, information technology, budget and investment) through the prism of the strategy. Set targets should take account of the current state of the company's resources, at the same time defining the appropriate management of these resources in order to implement the strategy. The essence of the Balanced Scorecard is the presentation and analysis of the achievements of the company simultaneously in four perspectives: Financial (including revenue growth, profitability and risk from the point of view of shareholders), Client (improving the quality of customer service and creating value of products and services), Internal Processes (management and optimization of business and production processes in order to raise the value of the company in terms of customers and shareholders), Learning and growth (creating a culture leading to the creation of innovation, development of employee competencies and implementation of optimization initiatives). As part of the balanced scorecard strategy maps templates have been developed for companies operating in different business models and providing various types of products and services for the four strategies: Low Total Cost, Product Leadership, Complete Customer Solutions and Lock-in [9].

The directions of the company set out in the strategic plan determine an action plan at the tactical level and consequently determine the tasks to be performed at the operational level. On the basis of a long-term strategic plan objectives, tactical plans of the company, including a period of 2-3 years on average, are built. Tactical plans based on a certain amount of different types of required resources (human, material, financial) and the desired deadlines determine how to implement. Whereas on the basis of tactical plans are developed operational plans. For the purposes of the business plan objectives specific tasks to carry out are determined. In turn, to achieve the strategic objectives there is a necessity of the appropriate implementation of tactical plans, and operational activities for specific business units, teams and even individual employees. In addition, it is necessary to provide employees with knowledge on the impact of their work on the implementation of strategic objectives and the implementation of processes of continuous strategy updating based on experience from the performance of current tasks. The proper use of resources which the company possess, a good tactical decision making and well thought-out operational movements thanks to which the proper tasks will be carried out in a proper time affect the ultimate success of strategic plans. It is therefore necessary at every stage of reliable and up to date knowledge, what provides the proposed ERP-MES-APS integration.

3 Supporting the Company Management Processes with Information Systems

Supporting processes with information systems is extremely important because of the ability to perform effectively current task and develop the company. On the market there are many different classes of such systems and smaller, usually specialized dedicated applications. Depending on the scope of implemented functionality we divide them into systems of different class. In the manufacturing companies IT support is provided mostly by the systems of ERP, APS, MES class. Standards of these systems have been developed in [1,7,13] and the article uses the standard definitions and concepts. It is worth remembering that the particular systems are based on these standards, but they include additional functionalities to increase their competitiveness in the market for this type of software and they use their specific vocabulary.

Enterprise Resource Planning systems are the largest and most advanced group of systems supporting the management. All company data are stored and processed in one database with appropriately controlled access. The most important areas of business activities such as distribution and logistics are supported. Some of the larger ERP systems have the functionality for planning, but use besides the MRP method simple methods of scheduling.

Advanced Planning and Scheduling Systems contain the functions of planning and short-term scheduling, using advanced algorithms, which are often lacking in ERP systems. APS systems generate optimal (suboptimal) plans and schedules taking into account the operational capabilities of the company and the availability of materials while considering business objectives. In some systems, apart from schedule creation on the basis of known limitations it is possible to create schedules taking into account different criteria (e.g. not only time but also a cost criterion) [6]. APS in comparison to ERP have more analytical capabilities, use algorithms based on linear programming, and genetic algorithms.

The basic tasks of the Manufacturing Execution System is the production process management and reporting the current status of orders in real-time. Efficient and effective management of the production process is carried out in this system, based on accurate and currant production coming from control systems and data acquisition systems.

Business Intelligence systems allow to create multi-dimensional analysis and are an important tool for managers and specialists in making business decisions and strategy of the company. Visualization of data and reports in the form of management dashboards allows to avoid viewing large amounts of data, rapid signaling in case of adverse events and presentation customized according to the needs of the business user.In each company data are collected in different information systems (ERP, CRM, SCM, Excel, Access files, or e-mail), which is a consequence of the intensive development of information technologies. Transactional systems to support daily business activities are made in different technologies and are not with them related in any way. Currently, BI systems are divided into two classes [2]: classic systems based on OLAP data processing technologies and the Class "In-Memory" systems. Classical solutions using OLAP are based on a data warehouse, which stores all the information needed to carry out multi-dimensional analysis of management. The data from transactional systems are subject to downloading, transformation and loading into a properly designed data warehouse. The data can then be visualized in presentation systems or advanced mechanisms are used for data mining and visualize the processed data. "In-Memory" systems allows to process data in memory of personal computers. There is no need to place into the workstations memory of all data from all over the enterprise. Sufficient are only source data of the areas of interest of individuals, eg. sales, inventory, customers, etc. Choosing BI system class depends on the purpose of the system, the number and location of users, infrastructure, business processes, organization, and the quantity and quality of the source data.

4 Analysis of Impact of Commonly Found Integration Scenarios on BI

In the recent period in the leading ERP systems the Balanced Scorecard application has been implemented as one of the modules of Business Intelligence [8]. This application supports the implementation of the concept in corporate strategy in the framework of the Balanced Scorecard methodology and its dissemination throughout the enterprise. In addition, apart from the financial factors that have been the genesis of the action of ERP so far, the management has an access to the indicators showing the state of customer service, internal processes and the skill level of employees. Moreover, there is the possibility of simultaneous presentation of strategies and indicators what facilitate analysis of the impact of specific projects and activities on the company's performance in the long term.

The concept of BI systems development assumes the simplification of the analitical software service, so as to eliminate the human factor in decision-making is largely replacing employees with expertise knowledge. Implementation of such a comprehensive BI approach is a multistep process consisting of [8]: Step 1 - reporting from production systems (operating) and statistics; Step 2 - inquiries processing OLAP and data mining; Step 3 - Business Intelligence packages and analytic applications; Step 4 - automation of decision-making processes; Step 5 - automation of "smart" process.

Innovative companies are currently using the tools defined in step 3 and 4. According to the authors disadvantage of these tools is the lack of reliability of the data collected in step 1. For example, the lack of efficient scheduling of tasks has a negative impact on the level of utilization of the existing machine park. The lack of automatic tracking of workload tasks may adversely affect the impossibility of determining the real time of delivery. These problems can be eliminated when the manufacturing company at the same time complies with the integration of the three ERP, APS, and MES systems.

In the manufacturing companies common ones are systems integration in pairs: ERP with MES systems [3], as well as ERP with the systems of APS class. It is unlikely to meet cooperation between MES and APS [11]. Even if there are all three systems in the company, according to the authors the flow of information is wrongly overlooked from the MES system to APS whereas APS system data are usually transmitted to the MES through ERP (for example as production orders scheduled for implementation).

In the ERP systems planning and scheduling tools of production (CRP, RRP, MPS) are based on methods that use stepper procedures. In developed production orders planned material availability and known limitations of production capacity are taken into consideration. This approach is simple and can be used in the case of a stable demand for the products of the company. However it is not enough in complex planning situations. In order to solve these problems ERP system integration with the APS is applied, which generates plans and production schedules based on required data from the ERP system using more sophisticated methods and algorithms to optimize production [14]. The figure (Fig. 1) presents an example of planning process using the ERP system. In this

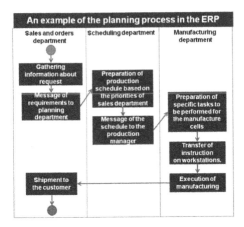

Fig. 1. An example of the planning process in the ERP

case, on the basis of information from the sales and forecasts the information about the demand is collected and the demand plan is created. This plan is often created in a spreadsheet instead of in the ERP system and transmitted (eg. sent by e-mail) to the planning department. Then, based on the priorities from sales department, inventory data from the ERP system and conversation with the production department on the possible execution time the production plan in the planning department is created. This plan is sent to the production manager. Then, based on the knowledge of ongoing assignments and discussions with masters and production leaders a detailed task instruction is prepared to complete for manufacturing cells. Cards are printed and quantity of work performed is reported by filling out timesheets manually and sending it to the master after the end of the shift. Then the data are entered into the ERP system. Thus, using only basic modules of the ERP causes that information between sales and orders, planning and production are transferred with the delay on paper sheets or using spreadsheets. Similarly, the reporting process from production is performed. However, in manufacturing companies, there is a need for the automatic exchange of these and many different kinds of information. The significant problem in the case of using only an ERP system is the lack of reliability of data caused by the fact that a lot of important data on the level of production is transferred and reported by employees in a manual way, after some time. For example, from the ERP system in manual way (in the form of paper timesheets) the detailed information on the work schedule is transmitted to the line workers (operators). In addition, feedback on the progress of implementation of the order is completed by operators card once per shift or once a day, and then manually entered into the ERP system. Then often delay in reporting generates subsequent delays in the delivery of the product to the customer. An additional disadvantage of the ERP system in the company's independent operation, are deficiencies in access to real-time information on the operational level, ie. in the management and execution processes of production. These deficiencies translate into a negative impact on strategic objectives. The solution to information flow delay issues in the flow of information and errors in data entry is ERP system integration with MES system, which monitors production in real time.

In this integration scenario in which systems of ERP class are integrated only with APS class systems, scheduling production orders is significantly hampered by the lack of accurate information on the current status of the work centers and their load, and the current state of operational production warehouses. Most often this data is updated daily or weekly during manual inventory. In addition, after the execution of the allocation of tasks to particular working outlets APS system does not get real time information about the progress of ongoing work, failures and stoppages and shortages of materials. This information must be refilled manually by workers directly in the APS system or indirectly in the ERP system. Then usually the rebuilding of schedule can be started manually by the decision of the planner.

In case of the systems integration of ERP and MES systems the key role in the planning process plays experience and knowledge of a planner. Giving priorities to the orders and assigning them to specific workstations depends on him. In such scenario, it is very difficult to integrate multi-criteria analysis, simulation and optimization of the schedule according to the selected output variables. Technologist should have the tools for the development of many concepts and options of designed products, performing a detailed analysis of these models and data preparation for the technical preparation of production, so that on these basis the planner could create full production orders, optimal production plan and schedule execution and coordinator supervise time delivery.

To improve efficiency of the production process, operators should automatically receive information about execution of current tasks on the visual panels at the workstations. Planners should automatically receive the necessary information for effective planning of the availability of the machinery, breakdowns and stoppages, availability of materials in manufacturing warehouses, and current state of execution. In addition, planners could get real information about the workload and cost of each operation. Sales staff could have real-time information on the status of orders placed for production and verify potential threat of failure to meet the delivery deadlines. The authors believe that this will happen if you use the simultaneous integration of all three ERP-APS-MES systems instead pair integration. In summary, the current approach can cause that for the automation of analysis and decision-making processes (step 3 and 4) incorrect assumptions are accepted. According to the authors, it is possible to achieve greater efficiency of BI tools and thereby improving the implementation of the strategy through the use of ERP-APS-MES systems integration.

5   Idea of ERP-APS-MES Integration in Manufacturing Company

Integration of information systems in a manufacturing company is not an easy task, both for technical reasons, as well as functional. Technical problems may include the diversity of network standards and communication protocols and interfaces between applications and databases. We can speak about the methods of solving them only in the specific case of integration, where the systems and environment in which they are going to operate are already defined. In contrast to functional problems first of all belong an appropriate combination of modules integrated applications, and optimal selection of messages sent between systems. Their solution is usually at the implementation stage and causes a significant and costly modifications. The integration of these systems in terms of functionality will look significantly different depending on the type of production company. The integration of ERP-MES APS systems in independent factories and factories with distributed networks of production resources will look differently, not to mention the development of solutions tailored to the unique production engineering processes within various manufacturing industries.

For these reasons, the authors in this section present a general scenario of ERP APS-MES systems integration which are in accordance with functional standards of data systems and the impact of this solution on the functioning of BI systems. This scenario is only a sketch to develop optimal integration scenarios taking into account the needs of specific companies.

5.1   Integration Scenario of ERP, APS, and MES Systems

In the case of the ERP-APS-MES integration systems much information can be transferred automatically. The figure (Fig. 2) shows examples of messages exchanged between specific modules of ERP-APS-MES integration systems.

Automatic download from ERP and MES systems of information set required to calculate the schedule causes that the plans created in the APS system are based on reliable and currant data and not planned data, often entered with errors. The production schedule is created based on the information on the demand for orders from the main production plan, the planned availability and cost of material and human resources, information about the current financial capabilities from the ERP system and the current information on the status of work in progress from the MES system. Choosing and confirming a specific timetable by plannist causes automatic transformation into specific job (task) production and display them on the panels (MES interface) on the working cells. Then MES integration system with programmable controllers of the machines causes automatic track of order execution and recording of all emergency events, stoppages, shortages of raw materials and lack of quality of manufactured products, which may affect the postponement of the completion of the order. Moreover, based on these data, you can take immediate intervention actions involving the creation of a new schedule by the APS system and/or negotiations with the customer in order to decide jointly on the date of execution of the order. On the other hand, after the completion of the order immediately logistics and sales department will be informed from the MES system and can realize the shipment to the customer. Moreover, in the MES system are continuously recorded (and updated in the ERP system) all consumptions and material movements on production and, in addition, important from the perspective of financial management module,

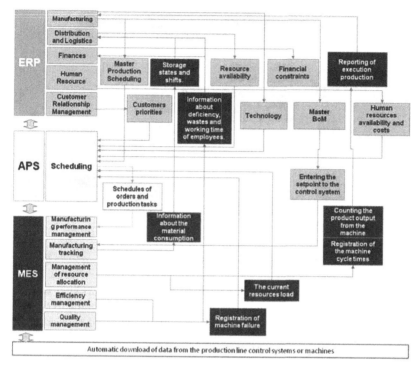

Fig. 2. Examples of messages exchanged between ERP, APS, MES based on [4]

the amount of generated waste. The ERP system receives from the MES system also real data on production costs (ie. the cost of materials, waste, energy and the media, human resources) in real time, allowing you to control the current profitability in the context of specific contracts, product groups, groups of machines or human teams. Furthermore, automatically reported progress of individual employees enable fast and reliable settlement in HR and payroll module. In addition, the integration of ERP-MES-APS systems allows to send automatically technological recipes to production lines with specific control instructions eliminating the possibility of errors during pre-production. The figure (Fig. 3) presents an example of planning process using the ERP-APS-MES.

Integration of all three systems at the same time has a big impact on the quality of analysis BI systems. Firstly, automatically downloaded directly from the control systems in real-time detailed data from the production can be used for current analysis in "In-Memory" systems, supporting management at the operational level. Integration is a save of time required for preparation of analyzes. Data can be processed and analyzed without delay. Secondly, automatically collected data are more reliable, and thus prepared analysis are more accurate. This applies both simple current reports as well as more complex analyzes performed on the basis of data collected in the data warehouse. It should be note that in addition to data from the integrated ERP-MES-APS systems data collected in the warehouse can come from other sources, eg. from the Internet.

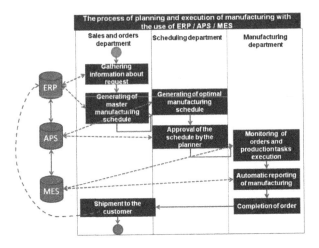

Fig. 3. An example of the planning process in the ERP-APS-MES systems integration

Moreover, in the case of integration would be needed less workload associated with ETL processes, because integration forces prior arrangement types and data formats. As a consequence integration can be a source of consistent input to support systems and automation decisions.

5.2   Idea of Systems Integration in Furniture Manufacturing

To illustrate the proposed idea of all three systems integration in this section we present our analysis of the support mattresses production problem in furniture manufacturing. Currently, ERP system - IFS Application is applied in this manufacturing company and integration with APS-MES i.e. Wonderware MES (with custom production order scheduling) system is planned.

The mattress is a final product of the production process. It consists of foam block, bonnel springs, frame and mattress cover. This production is performed in batches. To produce mattress the foam should be properly cut, bonnel springs should be prepared, frame should be made earlier. These semi-finished products are assembled together. Then the core mattress is encased by cover and final product is packed in foil. Because of a foam cutting, the springs and frames fabrication processes are independent, and for that reason they can be executed in parallel.

To illustrate the idea of ERP-APS-MES systems integration and its benefits we limit our example to a part of the production e.i. mattresses foam templates preparation. Currently, main process phases consist of the following ones: 1. Foam long-blocks cutting with result: foam short-blocks, 2. Foam Short-blocks warehouse placement, 3. Foam Short-blocks dispatching to detailed cutting, 4. Detailed cutting of short-blocks into individualized foam templates (according to mattresses specification). Sample visualization of the information flow in conjunction with production process is presented in the figure (Fig. 4).

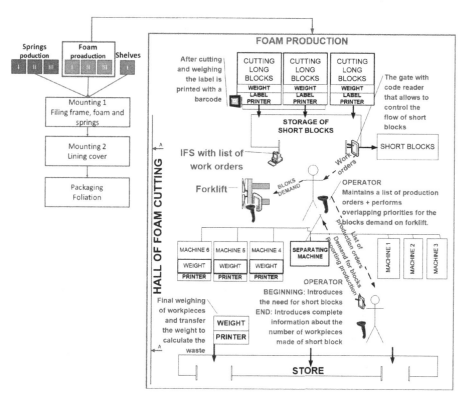

Fig. 4. An example of the flow of information in the mattress production without ERP-APS-MES integration

As a result of the analysis carried out in has been evaluated company that the total production efficiency losses are estimated at approximately 20%. This includes 10% of problems with raw materials. It result from the fact that:

- raw materials (foam short blocks) do not arrive on time to an operator of cutting foam machine. This is due to the fact that the demand for short block (a specific type, quantity, required delivery date) is transmitted from the operator of cutting foam machine to a forklift operator on the paper and with delay. The forklift operator, providing the foam block, has no current information about where, when and what type of the raw materials/semi-finished products should be delivered to the production line. It is a logistical problem.
- there is a large amount of waste after the foam cutting in some types of products. This waste is not processed, but it could be used with a profit (e.g., some pieces of the waste foam can be used for other production orders). The IFS Application reports only an indication of the percentage of waste, but there is no detailed information about which types of products, which machines or which operators generate the greatest amount of waste. This is a quality problem.

The authors propose to apply the full integration of the threeIFS Application - Wonderware MES with AP systems. We focus here on the logistic problem. As a result of the systems integration the following flow of information should be taken into account:

- Wonderware MES (with APS) gets information from IFS about the planned production orders on a given date
- in Wonderware MES (with APS) the priorities of the order will be assigned, the order will be split into a suborders and transferred to the operator terminals on specific machines. Thereby an operator knows immediately what to do
- after downloading the foam short blocks from the warehouse, a forklift operator will scan bar code block and information about it will be recorded in the storage module MES. At the same time it will be synchronized as a storage transaction in the ERP (IFS) system
- through(thanks to) the information on the terminal, the foam blocks will go immediately to the correct cutting position (machine)
- after the order execution, the operator will save this fact on a control panel (in MES) and information about order completion will be synchronized online with the ERP system as the completed production order - if the quality lacks occurs, such information shall be reported in the MES and included in the APS. Then APS has the ability to modify the scheduling of production. Moreover this information is synchronized with the IFS system that can generate a request to the supplier for additional resources (because of a super standard of the quality lack number).

According to the authors, among others, the improvements in the area of logistics (i.e., current information about the state of the short blocks magazine, including the costs of materials) and in the production area (i.e., production schedule optimization in the APS, covering the current information about lack of quality or the machine failure and the ability to take into account the additional operations on the same day) are the expected profit from the integration ERP-APS-MES. In addition, such a flow of information allows in the Business Intelligence analyzing in real time the exact time of the individual production orders implementation. The authors believe that it is needed to calculate the actual level of machine utilization and cost-effectiveness of production orders. Moreover, the current BI analysis for the amount of the defects occurrence on the change and comparing it with existing trends may lead to a decision to stop the cutting machine and overshoot its parameters. Such information on the unavailability of the machine can be sent to the APS and the current production schedule may be modified (for the next day or the next shift). Systems integration presented in the furniture manufacturing is planned and in the future it will be possible to investigate the improvement of production efficiency.

6 Expected Business and Management Effects of ERP-APS-MES Systems Integration

Each of the ERP, APS, and MES systems working independently affect the achievement of the objectives established in different strategic perspectives. Each of these systems due to their specialization and available functions supports the strategic perspective only to a limited extent. Lack of access to information on the operational level, ie. in the management and implementation processes of production translates into a negative impact on strategic objectives. For example, the lack of effective task scheduling has a negative impact on the level of use of existing machinery (internal processes perspective). Lack of automatic tracking of workload tasks may adversely affect the impossibility of determining the real time of delivery (customer perspective). In addition, the lack of control of the material consumption may adversely affect the margin earned on the product or product line (internal processes perspective).

Integration of ERP-APS-MES systems has an immediate and undeniable impact on operating the business. Operational business benefits of integration is to shorten the duration of the whole process of production (Lead Time), as well as fast access to reliable information, which become the basis for decision making (Expected Value of Information). By integrating ERP-APS-MES systems there will be possibility to provide, apart from improving operational and tactical, comprehensive support for all strategic perspectives. Automation of business processes in manufacturing companies assisted by the action of integrated ERP-APS-MES systems leads to-time optimization executed operations and the reduction of production costs, improve reliability and punctuality in business, and that translates into better financial result, a better image to customers and increase goodwill in the eyes of shareholders.

In addition, worth emphasizing is the fact of impact of ERP-APS-MES systems integration on the effectiveness of BI systems. Companies which would introduce this integration can not only minimize the time of the preparation of reports, but are provided with reliable data. Reports prepared on the basis of data collected automatically instead of manually reporting production will provide accurate information while minimizing the time of analysis. In the case of the warehouses data, it is important that the analyzed data will be good quality and reliable. Shorter time of processing data is desirable even for analysis prepared "for the next day". In that case analyzes "In-Memory" BI systems, may be prepared for the specific workstations in real time without significant delays. It is very important for the work-in- progress performing, especially when production items are often changing. Preparation of rapid and dedicated to workstations analysis is currently of the great importance, because the employees are cluttered an excess, often unnecessary, information. Then they may not notice the individual factors that could have negatively affect for the execution of production schedule. Properly prepared reports could help in faster identification of such potential risks (before disturbing tendency is noticed by operator) and much earlier indicate the need to convert the production plan by the APS system.

7 Conclusions

In this article we have shown the impact of integrated ERP-APS-MES systems to improve the efficiency of the manufacturing enterprise management at all levels: strategic, tactical and operational. Article also allows to put forward the thesis that the integration of ERP-APS-MES systems automates in considerable extent the flow of information and shortens the process of planning and scheduling. Moreover, it allows to reorganize sub-optimal processes in sales, planning and production. By integrating these systems, apart from improvement of operational and tactical actions, comprehensive support for all strategic perspectives can be also provided. Aggregated accurate and updated data from the ERP-APS-MES systems, analyzed in BI systems allow to proper reasoning and to make proper decisions. They can also provide a basis for the creation of a knowledge base, and the "best" practices within the organization. In addition, continuous monitoring of the current level of achieving the strategic objectives enables companies to early diagnosis of risks in achieving the goals and modify strategies.

References

1. APICS: Supply chain and operations management, http://www.apics.org
2. Chaudhuri, S., Dayal, U., Narasayya, V.: An overview of business intelligence technology. Communications of the ACM 54, 71–100 (2011)
3. Gao, X., Zhang, R., Zhang, Y., Jing, S.: Research focus on MES oriented communication among enterprise informatization system. In: IEEE Asia-Pacific Services Computing Conference, pp. 365–369 (2012)
4. Gracel, J., Grobler-Dębska, K., Dutkiewicz, L.: Analysis of the flow of information between ERP, APS and MES systems in production management. In: Research Group Pro Futuro, pp. 825–838 (2011)
5. Grobler-Dębska, K., Gracel, J.: Intelligence tools by integrated erp, aps and mes systems. PAR Pomiary Automatyka Robotyka 15, 183–187 (2011)
6. Hvolby, H.H., Steger-Jensen, K.: Technical and industrial issues of advanced planning and scheduling (APS) systems. Computers in Industry 61, 845–851 (2010)
7. ISA: International society of automation, http://www.isa.org
8. Januszewski, A.: The functionality of Management Information Systems. Scientific Publishing PWN (2008)
9. Kaplan, R., Norton, D.: The strategy-focused organization: how balanced scorecard companies thrive in the new business environment. Harvard Business School (2001)
10. Kaplan, R., Norton, D.: Strategy maps: converting intangible assets into tangible outcomes. Harvard Business School (2004)
11. Liu, W., Chua, T.J., Larn, J., Wang, F.Y., Cai, T.X., Yin, X.F.: APS, ERP and MES systems integration for semiconductor backend assembly. In: 7th International Conference on Control, Automation, Robotics and Vision, pp. 1403–1408 (2002)
12. Madapusi, A., D'Souz, D.: The influence of ERP system implementation on the operational performance of an organization. Int J. Information Management 32(1), 24–34 (2012)
13. MESA: Manufacturing enterprise solutions association, http://www.mesa.org
14. Van Nieuwenhuyse, I., De Boeck, L., Lambrecht, M., Vandaele, N.: Advanced resource planning as a decision support module for ERP. Computers in Industry 62(1), 1–8 (2011)

A Concept of Decision Support in Supply Chain Management – A Hybrid Approach

Paweł Sitek[(✉)]

Department of Control and Management Systems,
Kielce University of Technology, Poland
sitek@tu.kielce.pl

Abstract. This paper describes the hybrid approach to optimization of decision problems in supply chain management (SCM). The hybrid approach proposed here combines the strengths of mathematical programming (MP) and constraint logic programming (CLP), which leads to a significant reduction in the search time necessary to find the optimal solution, and allows solving larger problems. The proposed hybrid approach is presented as a concept of an additional layer of decision-making in integrated systems, for example ERP, DRP, etc. This solution allows the implementation of complete decision-making models, additional constraints as well as a set of questions for these models. The illustrative model presented in the article illustrate the advantages of the approach.

Keywords: Constraint Logic Programming · Mathematical Programming · Optimization · Supply Chain Management · Decision Support

1 Introduction

The supply chain is commonly seen as a collection of various types of companies (raw materials, production, trade, logistics, transport, etc.) working together to improve the flow of products, information and finance. As the words in the term indicate, the supply chain is a combination of its individual links in the process of supplying products (material/products and services) to the market [10]. A comprehensive review of models and methods in Supply Chain Management has been presented in [7,10].

Problems related to the design, integration and management of the supply chain affect many aspects of production, distribution, warehouse management, supply chain structure, transport modes etc. Those problems are usually closely related to each other, some may influence one another to a greater or lesser extent. There are interconnectedness and a very large number of different constraints: resource, time, technological, and financial, etc [7].

For the reasons named above, a need arises for tools and solutions to support the work of managers. The tools must allow making optimal decisions by getting quick answers to the questions asked in the management process.

Unfortunately, most enterprise management systems, such as ERP, DRP, etc. are designed to enable everyday operation of the company by collecting current

© Springer International Publishing Switzerland 2015
S. Kozielski et al. (Eds.): BDAS 2015, CCIS 521, pp. 565–574, 2015.
DOI: 10.1007/978-3-319-18422-7_50

information and supporting core processes. Certainly they can be great support for managers but such functionalities as optimization, asked questions etc. are not provided. Specialized modules for optimization or analyses based on data mining are available for purchase. However, they are costly and refer to selected areas such as scheduling or planning [1].

This paper deals with a problem of supply chain management modeling, optimization and analysis. An important contribution of the presented hybrid approach is to propose a Modeling and Decision Support Layer (MDSL) that supports the modeling, optimization and analysis of decision problems in the supply chain. In MDSL two environments, mathematical programming (MP) and constraint programming (CP), in which constraints are treated in different ways and different methods are implemented, were combined to use the strengths of both in the presented layer.

The rest of the paper is organized as follows: section 2 describes our motivation and methodology. Section 3 gives the concept of the novel hybrid CP/MP approach and MDSL. The optimization model as an illustrative example is described in section 4. Computational examples and tests of the implemented model are presented in section 4.4. The discussion on possible extensions of the proposed approach and conclusions is included in section 5.

2   Methodology and Motivation

Based on numerous studies and our own experience, the constraint-based environment [2,3,4,5] is believed to offer a very good framework for representing the knowledge, information and methods needed for the decision support. The central issue for a constraint-based environment is a constraint satisfaction problem (CSP), which is a mathematical problem defined as a set of elements whose states must satisfy a number of constraints. CSP represents the entities in a problem as a homogeneous collection of finite constraints over variables, which is solved using constraint satisfaction methods. Constraint satisfaction problems on finite domains are typically solved using a form of search. The most widely used techniques include variants of backtracking, constraint propagation, and local search. Constraint propagation embeds any reasoning that consists in explicitly forbidding values or combinations of values for some variables of a problem because a given subset of its constraints cannot be satisfied otherwise [2]. Effective search for the solution in CSP problems depends considerably on the effective constraint propagation, which makes it a key method of the constraint-based approach. Constraint logic programming (CLP) is a form of constraint programming (CP), in which logic programming is extended to include concepts from constraint satisfaction. A constraint logic program contains constraints in the body of clauses (predicates).

Based on [2,7,10,11,13,14] and previous work [4,5,6,20], some advantages and disadvantages of these environments can be observed. An integrated approach of constraint programming (CP) and mixed integer programming (MIP) can help to solve optimization problems that are intractable with either of the two methods alone [8].

Both MIP and finite domain CP/CLP involve variables and constraints. However, the types of the variables and constraints that are used, and the ways the constraints are solved are different in the two approaches [8,9].

In both MIP and CP/CLP, there is a group of constraints that can be solved with ease and a group of constraints that are difficult to solve. The easily solved constraints in MIP are linear equations and inequalities over rational numbers.

Integrity constraints are difficult to solve using mathematical programming methods and often the real problems of MIP make them NP-hard. In CP/CLP, domain constraints with integers and equations between variables are easy to solve. The system of such constraints can be solved over integer variables in polynomial time. The inequalities between variables, general linear constraints, and symbolic constraints are difficult to solve in CP/CLP (NP-hard). This type of constraints reduces the strength of constraint propagation. In addition, the CLP as declarative environment provides a simple way to formulate queries.

Observations above together with the knowledge of the properties of CLP and MP systems enforce the integration. The integration called hybridization consists in the combination of both systems and the transformation of the modeled problem.

The motivation underlying this research was to implement this approach as a MDSL to support managers in the modeling and optimization of decision problems in SCM. This solution is better than using MP or CLP separately. What is difficult to solve in one environment can be easy to solve in the other.

3   Modeling and Decision Support Layer (MDSL) - Concept and Implementation

The hybrid approach to modeling and optimization of decision problems in SCM is able to bridge the gaps and eliminate the drawbacks that occur in both MP and CLP approaches. To support this concept, we propose the Modeling and Decision Support layer (MDSL), where:

- All types of questions can be asked: general questions: Is it possible ...? Is it feasible ...? and specific questions: What is the minimum ...? What is the number ...? (the list of example questions in the MDSL is shown in table 1);
- knowledge related to the sustainable supply chain can be expressed as linear and logical constraints;
- the novel method of constraint propagation is introduced (obtained by transforming the decision model to explore its structure);
- constrained domains of decision variables (domain solution), new constraints and values for some variables are transferred from the CLP into the MP system in order to optimize;
- efficiency of finding solutions to larger size problems is increased.

Linking various types of constraints in one environment and using the best and already proved problem optimization and transformation methods constitute the base of the MDSL architecture. The concept and architecture of this platform with its CLP-predicates and MP-procedures is presented in Fig. 1.

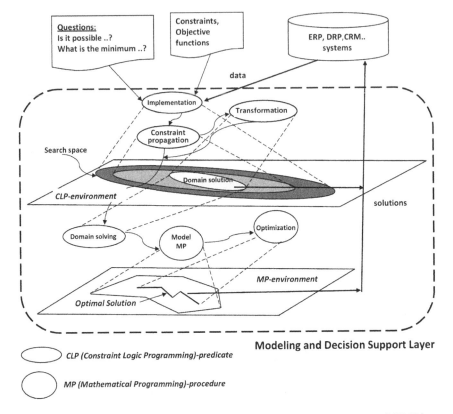

Fig. 1. Detailed scheme of the Modeling and Decision Support Layer (MDSL)

Table 1. A set of sample questions for the MSDL

Question	Description
Q1	Is timely execution of orders possible with existing capacity distributors V?
Q2	Is execution of orders possible within a given environmental cost K in time T?
Q3	Is timely execution of orders possible with the specified number of means of transport d in time T?
Q4	What is the minimum number of each type of transport means necessary for the timely execution of customer orders?
Q5	What is the minimum cost of timely execution of orders?
Q6	What is the minimum capacity of distribution centres for the execution of customer orders in time T?

From a variety of tools for the implementation of CLP techniques in the hybrid solution platform, ECL^iPS^e software [15] was selected. ECL^iPS^e is an open source software system for the cost-effective development and deployment of constraint programming applications. Environment for the implementation of MP in MDSL was LINGO by LINDO Systems [16].

4   Illustrative Example

The proposed approach was verified and tested for the illustrative example, which is the authors' original model of cost optimization for the supply chain with multimodal transportation. The proposed model is the cost model taking into account different types of parameters, i.e., space (area/volume occupied by the product, distributor capacity and capacity of transportation unit, recycling centre capacity), time (duration of delivery and service by distributor, etc.) and a transportation mode. Multimodality in this example is understood as the possibility of using different modes of transportation: railway, commercial vehicles, heavy trucks, etc. Illustrative example is a decision model from the perspective of 3PL (third-party logistics) provider.

4.1   Objective Function

The objective function (1) defines the aggregate costs of the entire chain and consists of five elements. The first element comprises the fixed costs associated with the operation of the distributor involved in the delivery (e.g. distribution centre, warehouse, etc.). The second element corresponds to environmental costs of using various means of transport. The third component determines the cost of the delivery from the manufacturer to the distributor. Another component is responsible for the costs of the delivery from the distributor to the end user (the store, the individual client, etc.). The last component of the objective function determines the cost of manufacturing the product by the given manufacturer.

$$
\sum_{s=1}^{E} F_s Tc_s + \sum_{d=1}^{L} Od_d \left(\sum_{i=1}^{N} \sum_{s=1}^{E} Xb_{i,s,d} + \sum_{s=1}^{E} \sum_{j=1}^{M} Yb_{j,s,d} \right) +
$$
$$
\sum_{i=1}^{N} \sum_{s=1}^{E} \sum_{d=1}^{L} Koa_{i,s,d} + \sum_{s=1}^{E} \sum_{j=1}^{M} \sum_{d=1}^{L} Kog_{s,j,d} + \sum_{i=1}^{N} \sum_{k=1}^{O} \left(C_{i,k} \sum_{s=1}^{E} \sum_{d=1}^{L} X_{i,s,k,d} \right) \tag{1}
$$

4.2   Constraints

The model was based on constraints (2)..(23) Constraint (2) specifies that all deliveries of product k produced by the manufacturer i and delivered to all distributors s using mode of transport d do not exceed the manufacturer's production capacity.

Constraint (3) covers all customer j demands for product k ($Z_{j,k}$) through the implementation of delivery by distributors s (the values of decision variables $Y_{i,s,k,d}$). The flow balance of each distributor s corresponds to constraint (4). The possibility of delivery is dependent on the distributor's technical capabilities - constraint (5). Time constraint (6) ensures the terms of delivery are met. Constraints (7a), (7b), (8) guarantee deliveries with available transport taken into account. Constraints (9), (10), (11) set values of decision variables based on binary variables Tc_s, $Xa_{i,s,d}$, $Ya_{s,j,d}$. Dependencies (12) and (13) represent the relationship based on which total costs are calculated. In general, these may be any linear functions.

The remaining constraints (14)..(23) arise from the nature of the model (MILP). The most important variables, parameters, and model indices are shown in table 2. Detailed description of the model and constraints is shown in [6].

Table 2. Indices, parameters and decision variables

Symbol	Description
	Indices
k	product type ($k = 1..O$)
j	delivery point/customer/city ($j = 1..M$)
i	manufacturer/factory ($i = 1..N$)
s	distributor /distribution center ($s = 1..E$)
d	mode of transport ($d = 1..L$)
N	number of manufacturers/factories
M	number of delivery points/customers
E	number of distributors
O	number of product types
L	number of mode of transport
	Input parameters
F_s	the fixed cost of distributor/distribution center s
$C_{i,k}$	the cost of product k at factory i
$Koa_{i,s,d}$	the total cost of delivery from manufacturer i to distributor s using mode of transport d
$Kog_{s,j,d}$	the total cost of delivery from distributor s to customer j using mode of transport d
Odd	the environmental cost of using mode of transport d
$Z_{j,k}$	the customer demand/order r j for product k
	Decision variables
$X_{i,s,k,d}$	delivery quantity of product k from manufacturer i to distributor s using mode of transport d
$Xa_{i,s,d}$	if delivery is from manufacturer i to distributor s using mode of transport d then $Xa_{i,s,d} = 1$, otherwise $Xa_{i,s,d} = 0$
$Xb_{i,s,d}$	the number of courses from manufacturer i to distributor s using mode of transport d
$Y_{s,j,k,d}$	delivery quantity of product k from distributor s to customer j using mode of transport d
$Ya_{s,j,d}$	if delivery is from distributor s to customer j using mode of transport d then $Ya_{s,j,d} = 1$, otherwise $Ya_{s,j,d} = 0$
$Yb_{s,j,d}$	the number of courses from distributor s to customer j using mode of transport d
Tc_s	if distributor s participates in deliveries, then $Tc_s = 1$, otherwise $Tc_s = 0$
CW	Arbitrarily large constant

4.3   Model Transformation

The transformation is an important and inseparable part of the hybrid approach (see Fig. 1). This is what proposed transformation of the problem distinguishes hybrid approach from the other known from the literature [8,9]. The idea of

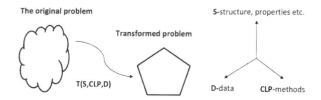

Fig. 2. Multidimensional transformation of the decision problem in the hybrid approach

Table 3. Answers for question Q1 $(V_1, V_2, V_2$-distributors capacity, T-time, Fc-objective function)

V_1	V_2	V_3	T	Answer	Fc
6000	6000	6000	14	Yes	6411
1500	1500	1500	14	Yes	6411
1000	1000	1000	14	Yes	7416
6000	6000	6000	12	Yes	6834
1500	1500	1500	12	Yes	6834
1000	1000	1000	12	Yes	7932
6000	6000	6000	10	No	—
1500	1500	1500	10	No	—
1000	1000	1000	10	No	—

multidimensional transformation is shown in Fig. 2. Due to the nature of the decision and optimization problems in SCM (adding up decision variables and constraints involving a lot of variables), the constraint propagation efficiency decreases dramatically. The idea was to transform the problem by changing its representation without changing the very problem. At the stage of transformation, the structure of, and maximum knowledge about the problem have to be used, including the knowledge about the orders, technical capacity of the producers, distributors and recycling centers. All permissible routes were first generated based on the fixed data and a set of orders, then the specific values of parameters i, s, k, d, were assigned to each of the routes. In this way, only decision variables X (deliveries) had to be specified. This transformation fundamentally improved the efficiency of the constraint propagation and reduced the number of backtracks and decision variables. This is due to the simple fact that it should be set-values for the one decision variable instead of five.

4.4 Numerical Experiments

In order to verify and evaluate the proposed approach, many numerical experiments were performed for the illustrative example. All the experiments relate to the supply chain with seven manufacturers *(i=1..7)*, three distributors *(s=1..3)*, ten customers *(j=1..10)*, three modes of transportation *(d=1..3)*, twenty types of products *(k=1..20)*, and fife sets of orders *(P1(10), P2(20), P3(40), P4(60),*

Table 4. Answers for question Q2 (K-environmental cost, T-time, Fc-objective function)

K	T	Answer	Fc
5675	14	Yes	6411
5600	14	Yes	6535
4600	14	No	—

Table 5. Answers for question Q5 (both approaches)

	Hybrid Approach MDSL			Integer Programming LINGO		
P(No)	Fc	Time V	C	Fc	Time V	C
P5(120)	14627 235	2047(2001)	556	15729* 900**	6881(6429)	27439
P4(60)	6411 234	1037(993)	564	7023* 900**	6881(6429)	15721
P3(40)	4572 34	783(741)	562	5028* 900**	6881(6429)	11941
P2(20)	2920 23	487(455)	552	3059* 900**	6881(6429)	8161
P1(10)	1766 4	299(277)	542	1766 743	6881(6429)	6271

Fc	the value of the objective function
Time	time of finding solution (in seconds)
V(V)	the number ofdecision variables (integer decision variables)
C	the number of constraints
*	the feasible value of the objective function after the time T
**	the calculation was stopped after 900 s

Table 6. Answers for question Q3 (d_1-mean of transport, T-time, Fc-objective function)

d_1	T	Answer	Fc	d_1	T	Answer	Fc
10	14	Yes	6507	10	12	No	—
1	14	Yes	6526	1	12	No	—

Table 7. Answers for question Q4 (d_1,d_2,d_3-means of transport)

d_1	d_2	d_3	T	Answer	Fc
unlimited	0	0	14	No	—
0	35	0	14	Yes	7119
0	0	unlimited	14	No	—
unlimited	0	0	12	No	—
0	35	0	12	Yes	7234
0	0	unlimited	12	No	—

P5(120), -(n)-the number of orders in set P). Computational experiments consisted in asking questions (table 1) for the model (section 4.1, 4.2), [6] implemented in the MDSL. Each question was asked many times, for set of different parameters. In order to compare the results and effectiveness of the MDSL, the model (section 4.1, 4.2) was also implemented in mathematical programming environment (LINGO) for question Q4. The results of these experiments are shown in the tables 3..8.

Table 8. Answers for question Q6 (V_1,V_2,V_3-distributors capacity)

V_1	V_2	V_3	T	Answer	Fc
2628	0	0	14	Yes	7060
0	2628	0	14	Yes	6503
0	0	2628	14	Yes	7496
2628	0	0	12	Yes	7060
0	2628	0	12	Yes	6503
0	0	2628	12	Yes	7496
unlimited	unlimited	unlimited	10	No	—

As you can see, the use of MDSL can also ask questions by the manager which is extremely useful in the supply chain management. You can obtain information about the possibilities of timely execution of orders, demand for storage capacity, transport, etc. Only the use of MDSL was able to finding optimal answers within acceptable time (below 900s) for Q5.

5 Conclusion

This paper provides a robust and effective hybrid approach to modeling and optimization of SCM problems, implemented with the hybrid approach which incorporates two environments (i) mathematical programming (LINGO) and (ii) constraint logic programming (ECL^iPS^e). The models presented as examples are transformed using the MDSL, which results in the new representation of the problem. The application of this hybrid method leads to a substantial reduction in (i) number of decision variables (up to twenty three times), (ii) number of constraints (up to fifty times) (iii) computing time (more than one hundred times faster). The proposed solution method with MDSL is recommended for decision-making problems whose structure is similar to that of the presented models (section 4). This structure is characterized by (i) constraints and the objective function in which decision variables are added up and (ii) logical constraints which are difficult to implement in mathematical programming-based models. In the versions to follow, implementation is planned of other supply chain layers such as remanufacturing, reverse logistic, etc. [12,17] and for automotive supply chains [18]. In contrast, the very idea of a hybrid approach and methodology can be used in many new areas such as cloud computing and applications [19] and job-shop scheduling [21].

References

1. SAP Advanced Planning and Optimization 7.0 - SAP Help Portal Page (2014), http://help.sap.com/apo (accessed October 12)
2. Apt, K., Wallace, M.: Constraint Logic Programming using Eclipse. Cambridge University Press, Cambridge (2006)
3. Bocewicz, G., Nielsen, I., Banaszak, Z.: Iterative multimodal processes scheduling. Annual Reviews in Control 38(1), 113–132 (2014)

4. Sitek, P., Wikarek, J.: A Hybrid Approach to the Optimization of Multiechelon Systems. Mathematical Problems in Engineering 2015, Article ID 925675, 12 pages (2015), doi:10.1155/2015/925675

5. Sitek, P., Wikarek, J.: Hybrid solution framework for supply chain problems. In: Omatu, S., Bersini, H., Corchado Rodríguez, J.M., González, S.R., Pawlewski, P., Bucciarelli, E. (eds.) DCAI 2014. AISC, vol. 290, pp. 11–18. Springer, Heidelberg (2014)

6. Sitek, P., Wikarek, J.: A hybrid approach to supply chain modeling and optimization. In: Federated Conference on Computer Science and Information Systems (FedCSIS), pp. 1223–1230 (2013)

7. Christou IT.: Quantitative Methods in Supply Chain Management, Models and Algorithms (2012), doi 10.1007/978-0-85729-766-2

8. Bockmayr, A., Kasper, T.: A Framework for Combining CP and IP. Branch-and-Infer, Constraint and Integer Programming: Toward a Unified Methodology Operations Re-search/Computer Science Interfaces 27, 59–87 (2014)

9. Milano, M., Wallace, M.: Integrating Operations Research in Constraint Programming. Annals of Operations Research 175(1), 37–76 (2010)

10. Ramanathan, U., Ramanathan, R.: Supply Chain Strategies, Issues and Models (2014), doi:10.1007/978-1-4471-5352-8

11. Lang, J.C.: Production and Operations Management: Models and Algorithms. Production and Inventory Management with Substitutions. Lecture Notes in Economics and Mathematical Systems, vol. 636, pp. 9–79 (2010)

12. Grzybowska, K.: Supply Chain Sustainability - analysing the enablers. In: Golinska, P., Romano, C.A. (eds.) Environmental Issues in Supply Chain Management - New Trends and Applications, pp. 5–40. Springer (2012)

13. Schrijver, A.: Theory of Linear and Integer Programming. John Wiley & sons (1998) ISBN 0-471-98232-6

14. Rossi, F., Van Beek, P., Walsh, T.: Handbook of Constraint Programming (Foundations of Artificial Intelligence). Elsevier Science Inc., New York (2006)

15. Eclipse (2014), www.eclipse.org (accessed January 20)

16. Lindo Systems INC. (2014), http://www.lindo.com (accessed)

17. Seuring, S., Müller, M.: From a Literature Review to a Conceptual Framework for Sus-tainable Supply Chain Management. Journal of Cleaner Production 16, 1699–1710 (2008)

18. Golinska, P., Fertsch, M., Pawlewski, P.: Production ow control in the automotive industry - quick scan approach. International Journal of Production Research 49(14), 4335–4351 (2011)

19. Bąk, S., Czarnecki R., Deniziak S.: Synthesis of Real-Time Cloud Applications for Internet of Things. Turkish Journal of Electrical Engineering & Computer Sciences (2013), doi: 10.3906/elk-1302-178

20. Sitek, P., Wikarek J.: A hybrid framework for the modelling and optimisation of decision problems in sustainable supply chain management. International Journal of Production Research (2015), doi: 10.1080/00207543.2015.1005762

21. Korytkowski, P., Rymaszewski, S., Wisniewski, T.: Ant colony optimization for job shop scheduling using multi-attribute dispatching rules. The International Journal of Advanced Manufacturing Technology 67, 231–241 (2013)

ePros – A Bioinformatics Knowledgebase, Toolbox and Database for Characterizing Protein Function

Florian Heinke[✉], Daniel Stockmann, Stefan Schildbach, Mathias Langer, and Dirk Labudde

University of Applied Sciences Mittweida
Technikumplatz 17, 09648 Mittweida, Germany
{fheinke,stockman,schildba,mlanger2,labudde}@hs-mittweida.de
http://www.bioforscher.de/

Abstract. Proteins are macromolecules that facilitate virtually every biological process. Information on functional and structural characteristics of proteins is invaluable in life sciences, but remain difficult to obtain, both computationally and experimentally.

In recent work, we have introduced a novel method for functional characterization, which we refer to as protein energy profiling. The ePros (energy profile suite) is an online knowledgebase, toolbox and database that provides a webspace for protein energy profile analyses to the scientific community. The objective of ePros is to offer a free-for-all repository of energy profile data, annotations, visualizations, as well as tools that can aid in deducing relations complementing and supporting findings made by traditional bioinformatics methods.

In this paper, we discuss the underlying biological and theoretical backgrounds used by implemented methods and tools, and also introduce recent enhancements and developments. ePros is available at http://bioservices.hs-mittweida.de/Epros.

Keywords: Bioinformatics · Protein structure · Energy profiling · Protein function

1 Introduction

Over the last decades, bioinformatics has become the backbone of modern life sciences. Today, the utilization and development of algorithms, databases, visualization techniques and tools form a major part in current research processes. Due to the rapid improvements of experimental techniques (e.g. high-throughput screening and sequencing), scientists have to cope with an exponentially growing amount of data. Thus, modern bioinformatics software has to be intuitive, easy-to-use, and present results in a comprehensible and most-informative manner; but it is also required to be fast enough to scan through millions of DNA sequences and thousands of protein structures within only a few minutes. On the

© Springer International Publishing Switzerland 2015
S. Kozielski et al. (Eds.): BDAS 2015, CCIS 521, pp. 575–584, 2015.
DOI: 10.1007/978-3-319-18422-7_51

other hand, underlying methods must perform reliably and show an optimal trade-off between biological sensitivity and speed.

In our research, we are interested in the functional characterization of proteins. Proteins are macromolecules that facilitate virtually every biological process. Although tremendous scientific efforts have been made, important questions on the general nature of proteins remain unanswered: What are the molecular and structural properties that ensure a protein's function? How do proteins interact with their environment? Given the sequence of a protein, how does the corresponding three-dimensional protein structure look like? Finding answers to these questions is a major goal in molecular biology and life sciences - and of great importance for the researcher who aims at characterizing the functionality of a given protein of interest.

Recent studies have demonstrated that sequences of energy values, referred to as energy profiles, obtained from protein structures using fine and coarse-grained energy models can be used to characterize protein functionality and structural features [10,11,7,9]. Mrozek et al. had introduced a dynamic programming approach for generating energy profile alignments [10,11]. Furthermore, they had shown that energy profile alignment analyses can be employed to investigate the changes of conformational energy distributions that can be caused by structural rearrangements as results of f.e. protein-protein interactions or ligand binding [10]. Using fine-grained energy models based on all-atom energy breakdowns to obtain energy profiles, as proposed by Mrozek et al., considers a variety of physio-chemical information (van der Waals and electrostatic interactions as well as chemical bond potentials). However, this approach comes along with drawbacks that are common in working with all-atom energy models, such as high computational demands, long processing times, and ambiguities in interpreting resulting potentials. To implement an automated energy profile generation pipeline for large-scale analyses thus becomes challenging. However, such analyses can be of interest in structural biology, as entire protein folds and families can be investigated in the focus of characterizing and understanding the energetic features that determine protein structure, stability and function [10,11,7,9].

In an effort to circumvent the aforementioned drawbacks in energy profile computation, a straightforward coarse-grained energy model had been developed, which considers local residue packing characteristics by means of statistical physics approaches (see section 3 for a brief overview). As shown by Heinke et al. [9,7,8] this model is capable to cope with the task of characterizing protein structure changes caused by protein-environment interactions. In these studies energy profiling was employed to investigate energetic changes in models of mutant aquaporin 2 structures — a key membrane protein in maintaining the body's water balance. Nephrogentic diabetes insipidus, a rare disease in which patients suffer from excessive urine excretion, is linked to mutations in the aquaporin 2 gene. Energy profiling of aquaporin 2 models gave insights into the correlations between structural changes caused by mutations and water flux. Besides this particular example, energy profiling is applicable to a wide range of questions.

The energy profile suite (eProS) provides a web-based solution to address such tasks. Implemented tools available via eProS include energy profile calculation from protein structure, visualization, exhaustive and heuristic large-scale database searching, energy profile comparison, protein family information, browsable biological annotations from various sources, and energy profile prediction from protein sequence. Derived information can support findings made by means of classic bioinformatic methods, and thus can aid in deducing enigmatic functional characteristics.

In this work we briefly present novel implementations and enhancements of eProS as well as currently available methods. Furthermore, a short summary of theoretical backgrounds is given.

2   Components and Application

2.1   Aligning Energy Profiles and Large-Scale Database Searching

As main applications in protein energy profiling, eProS implements global and local energy profile alignment algorithms as well as alignment visualization and scoring. In general, computing an energy profile alignment from two profiles aims at finding the optimal profile overlap and quantifying biological significance of the alignment by means of applying an appropriate scoring function. Hence, implemented alignment algorithms supply an automated methodology for deducing overlapping and mismatching regions. Such information can be valuable in understanding functional similarities and dissimilarities [9]. Energy profile and alignment visualizations provide an intuitive presentation for working with the protein data of interest, which can be uploaded as PDB (Protein Data Bank [13]) files or as energy profile files (downloaded from eProS after previous calculations), specified by PDB IDs, or as protein sequences. Here, an underlying database of pre-calculated energy profiles ensures fast access to the data. With this database it is also possible to search for similar energy profiles — referred to as search hits — and thus investigate family or fold-specific energetic fingerprints. Thus, searching for energy profiles similar to a query protein will result to a ranked list (referred to as 'hit list').

Currently, there are two search strategies available. The first conducts alignment computations and alignment significance testing to each entry in the database. This exhaustive strategy is guaranteed to find the best global or local overlap and thus performs best with respect to sensitivity [9]. However, required search times are generally long, due to the time complexity of $\mathcal{O}(pnD)$, where n is the length of the query energy profile (equals the size of the corresponding protein chain), D is the total number of energy profiles in the database, and p is a query-dependent variable used for testing alignment significance. For example, processing an energy profile with $n = 500$ residues takes about five hours.

To cope with this problem, we developed a heuristic approach, which is essentially based on correlating obtained alignment scores with alignment consistencies and alignment lengths. To evaluate biological sensitivity, hit lists obtained using this heuristic method were compared to hit lists obtained via exhaustive searching by means of Webbers's rank-based distances (RBDs) [16]. In summary, the heuristic method performs in $\mathcal{O}(nD)$, however utilizes the exhaustive strategy for realigning and re-testing k best hits (120 proteins by default) of the initial hit list. It has shown to be 25 - 37 times faster. The average observed RBD ($p = 0.95$) is 0.36, indicating a good agreement with the exhaustive search strategy.

2.2   Enhancements of User Interfaces and Data Presentations

In addition, compared to the methodology elucidated in [9], the current implementation and database layout of eProS have been redesigned entirely. Using the BioJava framework [12] has increased software flexibilities and capabilities, since well-defined data structures and implementations of protein structure file parsers provided by this framework allow reliable processing. In addition, the user interface has been revised completely, which is now entirely based on HTML5 technologies. In previous eProS versions, visualizations were entirely based on Applets and communications between them, leading to increasing discrepancies with respect to browser compatibility and security aspects over the last years.

The current revised version of the eProS database stores about 270,000 and 7,000 energy profiles from globular and α-helical membrane protein chains, respectively. In addition to the database, various annotations on protein functionality (Gene Ontology (GO) terms [2] and E.C. numbers) as well as structure/domain classifications from SCOP [1] and CATH [3] are supplied. External links to corresponding databases (such as Pfam [5], BRENDA [14], and UniProtKB [15]) as well as to corresponding literature yield a wide range of information to the user. Mappings between annotations and energy profile data provided via direct links afford a key feature of eProS, which we refer to as 'reverse annotation lookup', where knowledgebase-like browsing and analyzing energy profile data starting from annotation (f.e. a Pfam ID) become possible. Reverse annotation lookup facilitates top-down approaches in accessing energy profile data, in which the data can be quickly obtained based on a given annotation. This energy profile data can be downloaded from the database and used for further analyses, such as detecting energetic properties which can aid in drawing conclusions about protein family or fold-specific characteristics.

Working with eProS can be roughly separated in two primary approaches. First, data can be accessed directly by following database links, leading to single energy profile files and text files that contain direct download links for automated download and subsequent offline energy profile processing. Second, eProS provides various tools for computing, accessing and working with energy profile data (see table 1 for detailed descriptions). These tools also offer graphical, human-understandable representations to the user, such as colorized plotting and energy-to-structure mappings (see Fig. 1A).

2.3   ePfam: Browsing Energy Annotated Protein Families

Furthermore, we recently introduced the protein energy profile family viewer, which helps to understand energy conservations that characterize protein families.

This tool provides visualizations of aligned sequences with known energy profile entries directly extracted from Pfam sequence alignments (see Fig. 1B for an example output). Pre-processing presented data essentially included filtering available Pfam alignments (v 27.0) for sequences with known structures, followed by mapping aligned sequence data to computed energy profiles in our database, leading to a set of high-quality, family specific multiple energy profile alignments. Discretization of aligned energy profiles using 4-quantiles of observed energy values led to a 4-state alphabet, which was further used to represent energetic states of all residues in the filtered Pfam alignment, as well as generating family specific energy logos. These logos present a convenient way to illustrate the degree of energetic conservation per alignment column and which energy states contribute to conservation. The protein energy profile family viewer allows users to access and visualize about 5,000 pre-processed energy-profile enriched Pfam alignments, mapped energetic states, and logos. Finally, underlying data is free available for download in image and raw text formats.

2.4   Utilizing ePros

As depicted in 2, the modular design allows to combine available tools in numerous ways, which ensures a large degree in usage flexibility. Thus, ePros provides a framework of methods that can be employed in a variety of energy profiling strategies. However, the kinds of application differ with respect to the biological problem addressed. Starting from calculated, predicted or retrieved energy profile data, the user can freely access visualizations and annotations, download the data, and combine the individual modules as required.

Table 1. The following tools and accepted input formats are currently available at ePros

Tool	Description	Id[1]	struc[2]	EP[3]	seq[4]
eCalc	calculates an energy profile from globular or α-helical membrane protein structure	✓	✓	✓	
eGOR	predicts an energy profile from sequence				✓
eAlign	performs global and local energy profile alignments	✓	✓	✓	✓
eSearch	runs a database-wide search for similar energy profiles, bests hits are shown in a list ordered ascending by score	✓	✓	✓	✓
eMut	visualizes ΔE values of two proteins of the same length	✓	✓	✓	✓

[1] RCSB Protein Data Bank (PDB) structure identifier [13]
[2] PDB structure file
[3] Energy Profile file
[4] Protein sequence from which the energy profile is predicted using the eGOR algorithm

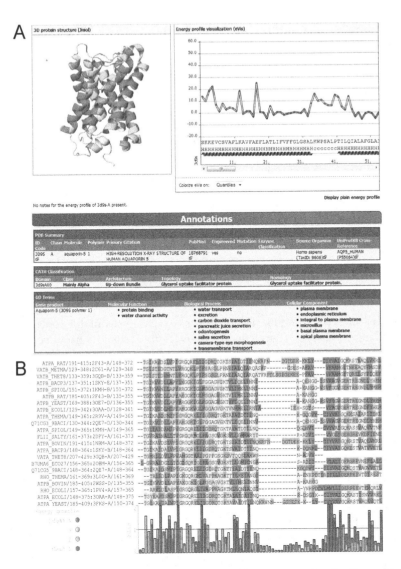

Fig. 1. A: eCalc output for human aquaporin 5 (PDB ID 3d9s). eProS provides energy profile plots, energy-to-structure visualization, and various browsable sources of annotation. Beside links to external databases, eProS offers annotation mappings that leads the user to an overview of annotation-specific energy profiles. By that, energy profiles of proteins sharing the same function or fold can be accessed and analyzed easily. B: Output of the protein energy profile family viewer. Protein family (Pfam, [5]) alignments of structures belonging to the same family are enriched with energy profile information. Color-representations illustrate energetic states of single residues. In addition, these energetic states are used to compute energetic conservation per alignment column and their degrees of contribution. This results to so-called energy profile logos, as shown on the bottom. These logos provide an intuitive illustration of energetic conservations as well as family-specific energy conservation patterns.

3   Theoretical Background: Energy Model and Energy Profile Prediction

To understand biological and biophysical significance of the energy model employed by ePros, we give a brief overview of the theoretical background in this section. For more information, please refer to the discussions in references [7,4,9].

The underlying coarse-grained energy model applied by ePros services utilizes concepts of statistical physics. In essence, residue potentials E_i are approximated

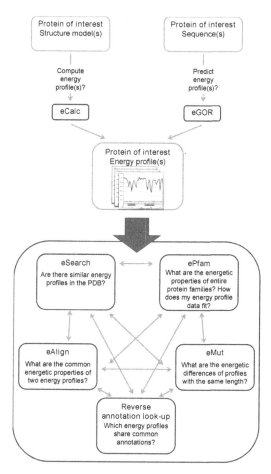

Fig. 2. Overview of ePros components. This schematic illustrates individual available components and tools (highlighted by blue boxes), their relations to each other, and which biological questions can be addressed by utilizing them. Besides starting from computing or predicting energy profiles from structures and sequences respectively, the user can also access energy profile data by querying the database with a PDB identifier of interest. Having energy profile data at hand, the ePros components can be employed to assemble working pipelines that are specifically suited for the underlying biological question.

by means of buriedness/exposed preferences for each corresponding amino acid a_i of the 20 canonical amino acids. Employing the inverse Boltzmann's principle to ratios of buried/exposed occurrence frequencies (denoted as f_{bur,a_i} and f_{exp,a_i} in the following) gives a knowledge-based potential formula for two interacting amino acids of type a_i and a_j:

$$E_i = \sum_{j \in Protein, i \neq j} g(i,j) \left[\underbrace{-\ln\left(\frac{f_{bur,a_i}}{f_{exp,a_i}}\right)}_{e_i} \underbrace{-\ln\left(\frac{f_{bur,a_j}}{f_{exp,a_j}}\right)}_{e_j} \right] \tag{1}$$

$g(i,j)$ denotes a contact function with a cutoff distance of 8Å. Here, C_β atom coordinates (or C_α atom coordinates in case of glycine) are considered as spatial residue-reference points. The frequencies f_{bur,a_i} and f_{exp,a_i} have been obtained from about 2,700 non-redundant globular protein structures. In case of α-helical membrane proteins, each log term in Equation 1 is extended by an additional term k_a, which is a measure of amino acid-specific hydrophobicity. k_a contributes to 1 as an additional energy term that is needed to approximate the change in free energy observed in residues located in the membrane bilayer relative to residues exposed to the solvent.

$$e_i = -\ln\left(\frac{f_{bur,a_i}}{f_{exp,a_i}}\right) + k_{a_i} \tag{2}$$

The sequence of energy potentials $E_1, E_2, \ldots, E_n$ of a given protein structure with length n corresponds to its energy profile.

Energy profile prediction from sequence is facilitated by the eGOR tool. Analogous to [6], the underlying algorithm analyses the amino acid composition in a sequence frame of seven residues with respect to observed amino acid occurrences and co-occurrences, including underlying discretized energetic states. The eGOR algorithm estimates the mutual information content in the given sequence frame and, based on this, predicts the energetic state of the residue located in the middle of the sequence frame. Iterating over the sequence results to the predicted energy profile. The accuracy of predicting the correct energetic state is 70%.

4   Conclusions

Inferring protein function and structural properties of proteins is a major task in bioinformatics. Protein energy profiling can afford valuable data on sequence-structure-function relationships. The study of common energy profile pattern in protein families and fold classes can reveal characteristics that establish a protein's function or fold. Including data of predicted energy profiles can aid in expanding this knowledge on protein sequence level and might be useful in drawing conclusion on uncharacterized proteins which can be rather difficult by classic approaches.

The ePros provides a web-based solution to address such tasks. Due to its modular architecture and broad spectrum of cross-linked annotations, the ePros is also applicable to various tasks in structure biology. Energy profile analyses provide orthogonal information to data obtained by classic approaches and, thus, can contribute to protein function prediction and characterization.

Acknowledgments. The authors thank the European Social Fund (ESF), Sächsisches Staatsministerium für Wissenschaft und Kunst (SMWK), the Free State of Saxony, and the University of Applied Sciences Mittweida for support and funding. Additional thanks go to Silvio Oswald for aiding in processing Pfam alignment information and energy profile mappings. Finally, we would like to thank Karolin Wiedemann for contributing valuable ideas that greatly helped developing heuristic energy profile search strategies.

References

1. Andreeva, A., et al.: Data growth and its impact on the scop database: new developments. Nucleic Acids Res. 36(Database issue), D419–D425 (2008), http://dx.doi.org/10.1093/nar/gkm993
2. Blake, J.A., et al.: Gene ontology annotations and resources. Nucleic Acids Res. 41(Database issue), D530–D535 (2013), http://dx.doi.org/10.1093/nar/gks1050
3. Cuff, A.L., Sillitoe, I., Lewis, T., Clegg, A.B., Rentzsch, R., Furnham, N., Pellegrini-Calace, M., Jones, D., Thornton, J., Orengo, C.A.: Extending cath: increasing coverage of the protein structure universe and linking structure with function. Nucleic Acids Res. 39(Database issue), D420–D426 (2011), http://dx.doi.org/10.1093/nar/gkq1001
4. Dressel, F., Marsico, A., Tuukkanen, A., Schroeder, M., Labudde, D.: Understanding of SMFS barriers by means of energy profiles. In: Proceedings of German Conference on Bioinformatics, pp. 90–99 (2007)
5. Finn, R.D., et al.: Pfam: the protein families database. Nucleic Acids Res. 42(Database issue), D222–D230 (2014), http://dx.doi.org/10.1093/nar/gkt1223
6. Garnier, J., Gibrat, J.F., Robson, B.: Gor method for predicting protein secondary structure from amino acid sequence. Methods Enzymol. 266, 540–553 (1996)
7. Heinke, F., Labudde, D.: Membrane protein stability analyses by means of protein energy profiles in case of nephrogenic diabetes insipidus. Comput. Math. Methods Med. 2012, 790281 (2012), http://dx.doi.org/10.1155/2012/790281
8. Heinke, F., Labudde, D.: Functional Analyses of Membrane Protein Mutants involved in Nephrogenic Diabetes insipidus: An Energy-based Approach. In: Research on Diabetes, 1st edn. iConcept Press Ltd. (2013)
9. Heinke, F., Schildbach, S., Stockmann, D., Labudde, D.: epros–a database and toolbox for investigating protein sequence-structure-function relationships through energy profiles. Nucleic Acids Res. 41(Database issue), D320–D326 (2013), http://dx.doi.org/10.1093/nar/gks1079
10. Mrozek, D., Malysiak, B., Kozielski, S.: An optimal alignment of proteins energy characteristics with crisp and fuzzy similarity awards. In: FUZZ-IEEE 2007, pp. 1–6 (2007)
11. Mrozek, D., Malysiak-Mrozek, B., Kozielski, S.: Alignment of protein structure energy patterns represented as sequences of fuzzy numbers. In: Annual Meeting of the North American Fuzzy Information Processing Society, NAFIPS 2009 (2009)

12. Prlić, A., et al.: Biojava: an open-source framework for bioinformatics in 2012. Bioinformatics 28(20), 2693–2695 (2012), http://dx.doi.org/10.1093/bioinformatics/bts494

13. Rose, P.W., et al.: The rcsb protein data bank: new resources for research and education. Nucleic Acids Res. 41(Database issue), D475–D482 (2013), http://dx.doi.org/10.1093/nar/gks1200

14. Schomburg, I., et al.: Brenda in 2013: integrated reactions, kinetic data, enzyme function data, improved disease classification: new options and contents in brenda. Nucleic Acids Res. 41(Database issue), D764–D772 (2013), http://dx.doi.org/10.1093/nar/gks1049

15. UniProt Consortium: Activities at the universal protein resource (uniprot). Nucleic Acids Res. 42(Database issue), D191–D198 (2014), http://dx.doi.org/10.1093/nar/gkt1140

16. Webber, W., Moffat, A., Zobel, J.: A similarity measure for indefinite rankings. ACM Transactions on Information Systems 28 (2010)

Evaluation Criteria
for Affect-Annotated Databases

Agata Kolakowska, Agnieszka Landowska, Mariusz Szwoch,
Wioleta Szwoch[✉], and Michal R. Wrobel

Faculty of Electronics, Telecommunications and Informatics,
Gdansk University of Technology, Poland
szwoch@eti.pg.gda.pl

Abstract. In this paper a set of comprehensive evaluation criteria for affect-annotated databases is proposed. These criteria can be used for evaluation of the quality of a database on the stage of its creation as well as for evaluation and comparison of existing databases. The usefulness of these criteria is demonstrated on several databases selected from affect computing domain. The databases contain different kind of data: video or still images presenting facial expressions, speech recordings and affect-annotated words.[1]

Keywords: Affective computing · Database quality · Affect-annotated databases

1 Introduction

Emotions play an important role in human interaction, often influencing our behavior in a more or less predictable way. This fact is usually neglected by standard human-computer interfaces, which are based on sets of rigid rules. However they might be effective and user-friendly in the sense of easiness of use, they do not resemble human interaction in any way. On contrary, affect-aware software may not only recognize and interpret users emotions, but also adapt its behavior and functionality to users needs according to the detected states. Such systems may be applied in many fields such as healthcare [29], entertainment [15], education [17], software engineering [30], etc.

A great many methods have been developed in order to identify users emotional states. They are based on different data sources such as visual (facial expression, gestures) [14], audio (speech, voice) [24], textual (semantics) [13], physiological (heart rate, temperature, skin conductance etc.) [28], input devices (keyboard, mouse, touch-screen) [16,11] and on a combination of them, what can significantly improve emotion recognition abilities [24,32]. Most of

---

[1] The research leading to these results has received funding from the Polish-Norwegian Research Programme operated by the National Centre for Research and Development under the Norwegian Financial Mechanism 2009-2014 in the frame of Project Contract No Pol-Nor/210629/51/2013.

© Springer International Publishing Switzerland 2015
S. Kozielski et al. (Eds.): BDAS 2015, CCIS 521, pp. 585–597, 2015.
DOI: 10.1007/978-3-319-18422-7_52

these approaches rely on supervised training to create a classifier of emotions. Its effectiveness and generalization ability highly depends on a set of data used for training. Depending on the data source, the datasets may differ significantly. For visual recognition it would be videos or images, and for the audio voice samples. Furthermore sentiment analysis technique would utilize dictionaries of affect-annotated words. The quality of these data has a great impact on the effectiveness of the recognition algorithms.

There are many databases and datasets that can be useful in the process of creating a specific affect recognition system. Many of them are publicly available for non-commercial purposes, such as research or education. Unfortunately, they differ significantly in terms of size, content and quality. Therefore algorithms that use different datasets for training are difficult to compare. Moreover datasets often consider narrow sets of recognized emotions, acquire training data in perfect conditions or from experiment participants showing very homogeneous characteristics. All these factors limit the usefulness of these datasets in different real-life applications.

The problems mentioned before may cause a real dilemma while choosing a proper dataset for an application, so there is a need for a set of reliable criteria to facilitate making the right choice. Some evaluations of the available datasets can be found in the literature [32,22], but none of them proposes a set of general factors to be taken into account during the process of designing any affect recognition system. In this paper, seven universal criteria are proposed to evaluate the quality and usefulness of affect-annotated databases regardless their data type and field of application.

The paper is organized as follows: section 2 presents the database evaluation criteria, section 3 describes the proposed criteria for databases containing data of different types and gives some examples of available datasets, section 4 provides final conclusions.

2   Evaluation Criteria for Affect-Annotated Databases

Affect-annotated databases are created for different purposes. Therefore, they can differ significantly in quality, size, completeness, formats, etc. Unfortunately, there are no commonly accepted standards for creation and evaluation of such database. In this section we propose seven general criteria that cover most aspects of affect-annotated database quality and usefulness of their content for different applications. Verifying any database against these criteria can give an answer about its generality and completeness that are required for its common and widely usage.

When creating or evaluating a set of emotion-labeled data several important questions must be answered taking into account different objectives that should be fulfilled, at least in some extent, to be useful for classifiers training and validation. These aspects cover such topics as:

1. *Representativeness* is the key attribute of a training set in each supervised training process. Showing emotions differs strongly among people, so it is

essential to gather as much diverse data as possible. In general, a database should contain samples from a large set of people who differ in age, gender, race, culture, occupation, and other aspects. Though it is possible to narrow any aspect of participants due to the assumed field of application, such as children emotion, received results could not be generalized on other cases, limiting the database usefulness.

2. *Granularity* of an affective database informs about the set of emotions or affect states that are distinguished and represented. This criterion seems crucial from its usability point of view. Unfortunately, there are no commonly accepted set of expressions that could be used in all applications. In fact, the choice of an interesting set of expressions depends on the field of applications and on the assumed model of emotions. When the discrete model is used, the granularity criterion is represented by the number of different labels assigned. In the case of dimensional model the number of possible values of its components determines the number of distinguishable affective states.

3. *Size* and *completeness* inform about the total number of fully or partially assembled samples for each human object and each class of emotions. To train a classifier, it is essential to have large enough number of examples per class (emotion). In general, datasets with greater number of data/items are more useful for training and testing purposes. Size and completeness criteria of a database are partially related to the representativeness factor, however, a database might be small and yet representative and complete.

4. *Labeling quality* is an important usability factor that can be considered on different levels of details. At minimum level it is required to label the emotional state of each sample and basic metadata about the human object. Such labeling can be done as simple naming of an affective state or by expressing it according to one or more commonly accepted affect models, such as PAD [23]. No matter if the gathered data are artificial or natural, an independent label evaluation made by a number of judges should be done. The second description level concerns additional data labeling that could facilitate data preprocessing and creation of recognition algorithms.

5. *Procedural quality* describes a set of factors that assure proper conditions of data acquisition. These factors include, among others, the way of emotion expression, that can be spontaneous or posed where the latter can be performed by amateurs or professional actors. The second important factor refers to acquisition conditions which can be more or less natural. The common problem is that many collections of affective data contain posed expressions that come from studio recordings, which can highly differ from those observed in everyday life. Unfortunately, it is not easy to create a database of well indexed, spontaneous data from natural environment.

6. *Data-level quality* refers to all technical aspects of data acquisition such as number of channels, sampling rate, resolution quantization error, data representation model, and others. On one hand, assuring high quality data is very important for data processing and further reasoning stages, but on the other hand, data gathered with a top-quality professional hardware, may need

additional procedures for quality degradation to adapt them for consumer applications.

7. Database *popularity* and its *availability* are important factors that cannot be neglected in the process of database choosing for algorithm validation and evaluation. The popularity level, in the case of scientific purposes, can be evaluated by the number of citations of the first article that introduced the data set. In many cases the database *popularity* is highly related to its quality expressed by the previously given criteria. Database *availability* determines its usefulness for other researchers and contains the license and publication policy, standard conformance, and additional supporting tools for database exploration, complement and extension. The databases that are paid, or use custom propriety formats with no additional access tools cannot become popular and are practically useless.

Almost all of the proposed criteria are general and should be adjusted according to the specific aspects of a database in a particular research field. In the following chapter we describe three important types of affect-annotated databases, give specific descriptions of the evaluation criteria and use them to evaluate some of publicly available databases.

3   Evaluation of Representative Affect-Annotated Databases

Affect recognition methods can use different information channels, such as video [29,14], audio [15,31], standard input devices [16,12], physiological signals [28,3], depth information [6], and others. These channels can be also used in different combinations in multimodal systems that typically achieve higher recognition rate due to fusion of different information. For each information channel a specific database should be used for supervised training algorithms as well as for validation of a developed approach. In this section three different kinds of databases, i.e. visual, audio and text, are considered according to the evaluation criteria proposed in the previous section. Three affect-annotated databases of each kind were chosen to be evaluated in details with the predefined criteria. Their choice was justified by the datasets diversity in order to depict, how the proposed criteria apply to various databases' characteristics.

3.1   Still Images and Video Databases of Facial Expressions and Emotions

Visual channel carries most information about the world surrounding humans. That is why it plays so important role in interpersonal communication. People send a lot of visual (non-verbal) information, especially using facial expressions, but also with gestures, body movements, and others. Analysis of information from a video camera is also the most popular source of non-invasive and non-intrusive information about human emotions. There are a number of algorithms based on these data, which can provide users emotions recognition. Some of these algorithms concentrate only on facial expression recognition (FER) as it is usually done by humans while analyzing affect of other people [8].

Still images is the most popular way to represent visual information. However, as facial expressions and human emotions have temporal character, video databases containing short recordings can be more useful. Visual databases contain usually two-dimensional images (2D) or video sequences. Unfortunately, algorithms using video camera are very sensitive to face illumination conditions, causing great problems with dark or unevenly illuminated scenes. Possible solutions include storing pairs of stereoscopy images or data with an additional channel for scene depth information. As depth sensors use additional light source (e.g. infrared) the information is generally insensitive to different ambient light conditions. Rapid development and increasing availability of relatively cheap RGB-D consumer sensors allows for creation of on-line systems for the recognition of facial expressions and, going further, human emotions and moods [27].

Evaluation of affect-annotated databases containing visual information requires taking into account some additional aspects of criteria given in the previous section. Because most of the image and video databases features are very similar, we provide common evaluation criteria.

The *representativeness* criterion should reflect the diversity of people participating in the sessions in the aspect of human gender, race, age, hairstyle, and others. The diversity requirements strongly depend on the purpose the images are to be used for. Other features worth considering are wardrobe accessories such as glasses or hats. In addition to the diversity it may be important to consider whether expressions were spontaneous or posed. In the later case, the professional actors participation may affect the assessment of representativeness.

Size and *completeness* in the case of facial expression database are mainly determined by the number of video or image files and the number of participants. Another important factors are the total number of expressions recorded and the number of recordings per participant.

For *labeling quality* two additional indexing levels should be considered for facial expressions. The first is Facial Action Coding System (FACS), which is a human-observer based system using so called action units (AU) to describe even subtle changes in facial features [10]. Apart from AU labeling, images or recording of faces should also include additional meta-data such as acquisition parameters, scene environment, and also location of characteristic facial points (landmarks).

Procedural quality criterion covers both proper conditions of image or video acquiring as well as conditions for proper emotion expression. Scene composition describes face location and orientation in an image and plays an important role in face recognition as well as in recognition of facial expressions and emotions. In the facial expression recognition field it is often assumed that a face stays oriented towards the camera. However, it may not be true especially when expressing natural emotions. In that case human pose often changes dynamically what should be reflected in a database. Other important factors, that can influence face location or expression recognition, are: scene illumination, background and the presence of other people. Generally, recordings of human facial expressions and emotions may be divided into two categories: posed and spontaneous sessions. Posing allows for recording extreme expressions like human cry, envy, and so on. The drawbacks of posed recordings are human problems to act such emotions and their artificial character. On the other hand recording of spontaneous emotions is much harder and requires a lot of time to catch the desired emotion or expression. Moreover, it is almost impossible to record extreme emotions mentioned earlier. That is why databases of facial expressions and emotions should contain images or recordings of both types, possibly played by professional actors or actresses.

Table 1. Selected image and video databases described with the proposed criteria

Criteria \ Database	Extended Kanade [19]	Cohn- MMI Facial Expression Database [21]	FEEDB [26]
Representativeness	Diversity of races and ages	White, in the age of 20 - 33 years	European people, in the age of 19-26 years
Granularity	6 basic emotions + 1 neutral	8 emotions	30 emotions and facial expressions
Size and completeness	128 persons, 593 expressions, 10708 images	5 persons, 493 expressions	50 persons, 1550 recordings
Labeling quality	FACS, Action Units, Emotions (only for 327 expressions)	Action Units	Action Units, SAM
Procedural quality	Posed expressions, professional lighting and background	Posed expressions, professional lighting and background	Posed and natural expressions, standard lighting conditions, people in the background
Data-level quality	PNG files: 104 files in BW, 640x480, poor quality, 13 files in color, 640x480, poor quality, 6 files in color, 720x480, good quality	JPEG files: 1200x1600, in good quality	color, XED and AVI video very files, BMP frames, standard VGA resolution
Popularity	High (345 citations in Google Scholar)	High (334 citations in Google Scholar)	Low

Data-level quality strongly depends on the hardware used for data acquisition and the technical properties of files, such as resolution, number of colors, and sampling rate in the case of video recordings. The effective resolution of a face in an image is an important factor of the image quality and can influence recognition efficiency. For three-dimensional representation depth resolution (Z axis) should correspond to scene resolution in an image plane (X and Y axis). It means that the face should be located at certain optimal distance from the camera. Additionally such aspects as light conditions, and scene exposure as well as any auto-adjustment functions should be considered, such as auto-illumination level, automatic white balance etc.

Table 1 contains detailed description of the selected image and/or video datasets with the use of the proposed criteria. It presents great variety of sample databases according to particular criteria allowing to construct own preferences for choosing learning and testing datasets.

3.2   Speech Databases (Audio)

Emotions affect our utterances. Not only is it important what the speaker says, but also how he or she says that. Features such as pitch and volume may be used do discriminate between some emotional states [22,31]. The description below contains issues which should be taken into account while choosing or designing a speech database.

The *representativeness* of a speech database is determined by the number of speakers, their profile and the type of texts uttered. The essential parameter of the database is the recordings language and the type of utterances. They can be either natural, taken for example from a TV program, or artificial, taken from actors in a studio. In the case of artificial utterance, it is important to know whether or not the experiment participants are professional actors, whose reactions in speech may be exaggerated [22]. The artificial texts may or may not have an emotional context. The latter option might entail difficulties with acted uttering, especially for non-professionals. The advantage of artificial utterances is that it is possible to record the same sentences from all actors. Moreover, from the point of view of future data processing, it is also important whether these are short utterances containing single sentences or words or long speeches, which require searching through to match specified moments. Other relevant features, which may affect the degree of manifesting emotions in speech, are speakers sex, age and nationality.

Size and *completeness* are determined by the number of samples per speaker, which depend on the number of represented emotions. In the case of speech databases their size may be also measured in the number of hours recorded.

Labeling quality is one of the main factors determining the quality of training data. The simplest, though quite subjective, way of assigning labels may be done in the case of acted utterances when the actors are asked to demonstrate a specified emotional state. Usually the labels are judged depending on their recognizability or naturalness in order to reject the samples which do not fulfill prespecified requirements.

Procedural quality depends on the way an experiment is designed and then conducted. However professional actors are able to act without special stimulus, some elicitation is usually performed. Sometimes short descriptions of situations are provided. A speech may be also guided by the experimenter. It is important

Table 2. Selected speech databases described with the proposed criteria

Criteria ＼ Database	Emotional prosody Speech and Transcripts [18]	Berlin Database of Emotional Speech[7]	Belfast Naturalistic Database [9]
Representativeness	8 professional actors (5 females, 3 males); short (2 or 3 words) semantically neutral utterances (dates and numbers) in English	10 non-professional actors (5 females, 5 males) in the age of 21-35; sentences in German from everyday life, semantically neutral	125 speakers (31 males, 94 females); natural (clips from TV programs) and artificial (studio interactions) recordings
Granularity	15 emotional states: disgust, panic, anxiety, hot anger, cold anger, despair, sadness, elation, happiness, interest, boredom, shame, pride, contempt, neutral	7 emotional states: anger, boredom, disgust, anxiety/fear, happiness, sadness, neutral	dimensional model (two dimensions: activation, evaluation)
Size and completeness	over 9 hours recorded in 15 files; 15-30 samples per emotion for each user	about 500 utterances; 40.5 MB; each sentence read by every user with each of the 7 emotional states	298 clips lasting from 10 to 60 s
Labeling quality	labels assigned according to the emotional states acted by the actors after being provided with the situational context	500 out of 800 selected by 20 judges according to their recognizability and naturalness	2D labelling: activation + evaluation; rated by 5 judges in regular intervals
Procedural quality	emotions induced by short descriptions of situational context; possibility of repeating the utterances	10 from 40 people selected by three judges; emotions induced by short descriptions of example situations; possibility of repeating sentences	clips selected to cover a wide range of emotional states; each speaker recorded in both an emotional and a neutral state (according to judges)
Technical quality	sampling rate 22.05 kHz; 16 bits per sample; two channels; WAV files; some background noise between utterances; transcripts available	sampling rate 48 kHz, downsampled to 16 kHz; WAV files; the recording level adjusted between the loudest and the most quiet utterances	clips saved in MPEG files, audio data extracted in WAV files
Popularity	high	very high, easy access	high

whether or not the speaker is allowed to repeat the sentences he is not satisfied with. Finally the database may contain either all recordings or only the ones selected by judges.

Data-level quality of recordings is determined by factors such as the number of channels, sampling rate, coding resolution. Moreover any background noise may affect the eventual analysis, which may be especially troublesome in the case of natural recordings. Another unfavorable attribute are long recordings in contrast to separated utterances, as they require manual splitting. File format is also an important parameter. Finally some databases contain only speech, whereas others are multimodal ones usually containing video as well.

Table 2 contains detailed description of three selected speech datasets with the use of the proposed criteria that confirm usefulness of the proposed criteria in the task of speech database evaluation.

3.3   Affect-Annotated Dictionaries

Sentiment analysis of textual information derived from websites, forums or dialogues, that is frequently used in opinion mining, requires a dictionary of affect-annotated words. There are several dictionaries, that differ in size, method of construction and labeling technique, including: Orthony's Affective Lexicon [20], General Inquirer[1], SentiWordNet [2], WordNet-Affect [25], Subjectivity Lexicon [4], Affective Norms for English Words (ANEW) [5] to name just a few. The lexicons of affect-annotated words can be evaluated using the proposed criteria, however, detailed definition and metrics must be provided.

Representativeness factor in evaluation of a dictionary is an important issue, as a selected word subset might be useful in some application areas only. To evaluate a dictionary the following characteristics should be considered: language, number of words and the inclusion criteria, especially inclusion of the most commonly used words, colloquialisms and perhaps offensive phrases. It's important to emphasis, that the lexicon word set should be matched to characteristics of the application area.

Labeling quality of a lexicon is the most important criterion, as people may differ significantly in meanings assigned to words, and some of the words are memes, that might have a special meaning in a specific culture. Representativeness of manually created dictionaries can be defined as factor of the size and diversity of the group of people, that annotated words. Evaluation of automatically generated databases representativeness depends both on the quality of the source material and the generation method and in some cases it might be difficult to evaluate the factor of automatically created lexicons due to lack of data regarding representativeness of the source. Although representativeness and labeling quality are the most important characteristics and should be evaluated separately, they are strongly correlated with the procedural quality of a lexicon. Procedural quality of a manually annotated lexicon depends on the repeatability of the construction process - how many people annotated one word, was there any kind of reliability check performed. For automatically annotated databases the algorithm of annotation should also be checked against repeatability - would

Table 3. Selected dictionaries described with the proposed criteria

Criteria \ Database	ANEW [5]	WordNet Affect [25]	SentiWordNet [2]
Representativeness	English,+multiple national, the most commonly used words; no colloquialisms; no swears	English, only terms related to affect	English,(almost) all words
Granularity	Annotation based on PAD (continuous 3D scale)	Labels: emotion (positive, negative, ambiguous, neutral), attitude, feeling	Labels: positive, nega-tive, neutral (of sum 1)
Size and completeness	1040, the most frequently used words only	Manual (1903) Automatic(4787)	115.000
Labeling quality	Manual, based on referential SAM questionnaire	Manual-automatic	Automatic (semi-supervised learning and random-walk process)
Procedural quality	Each word annotated many times by different people, no reliability check, one combination of word sets only	For manual part no information on labeling process is provided. Automatically annotated words were derived from manual core using WordNet	Derived automatically from Word-Net, Post-checked with manual results (with p-normalized Kendall T-distance of 0,23-0,28)
Technical Quality	Single download-able file: word-annotation: P,A,D: means, SD	Multiple files, downloadable, Version 1.0: information was pro-vided on manual or automatic creation, Version 1.1: no information	Single download-able file, no information on reliability
Popularity	very high - referential lexicon	moderate	high

we obtain the same labels if considering different order of annotation or dissimilar starting point. Affective lexicons are usually distributed as dictionary files (one or multiple) containing word-label pairs. Technical quality of a dictionary can be decomposed into the following criteria:

- even/uneven labeling for all words - are all words labeled with the same precision and in the same way or is there uneven distribution of labels (are there 'empty' labels),

– group/subgroup annotation - some dictionaries contain overall as well as annotations that were provided by subgroups (e.g. male/female),
– the existence of some process metrics (e.g. number of evaluations, standard deviation), that allows to estimate the uncertainty related to the actual word-label assignment,
– application support.

Table 3 contains detailed description of the selected lexicons with the use of the proposed criteria. The most important criteria for evaluation of affect-annotated lexicons include: representativeness, granularity and labeling quality. Manually created lexicons contain a limited number of words, however they are the most reliable and popular. Automatically annotated lexicons, although containing a large number of words, are usually assigned with a limited set of labels, that does not conform to some applications requirements.

4 Conclusions and Future Works

For many emotion recognition algorithms, the selection of a suitable training set is the key factor for proper operation. This paper presents criteria, which may be used in evaluation of the affect-annotated databases.

Some of the seven defined criteria are general, some have to be specified in detail for different databases. The paper describes all the criteria from the perspective of three different database types: visual, speech, and dictionaries. The evaluation performed for three databases of each type has proved suitability of the developed criteria.

The development of comprehensive and reliable affect-annotated databases is a difficult task. Therefore provided criteria may be treated as guidelines in the process of their developments. None of the proposed factors may be regarded as the primary one. Assigning weights to the criteria is the systems designer task and it is closely related to a given application.

References

1. General Inquirer (February 21, 2013), http://www.wjh.harvard.edu/~inquirer/
2. Baccianella, S., Esuli, A., Sebastiani, F.: SentiWordNet 3.0: An enhanced lexical resource for sentiment analysis and opinion mining. In: Conference on Language Resources and Evaluation, pp. 2200–2204 (2010)
3. Bailenson, J., Pontikakis, E., Mauss, I., Gross, J., Jabon, M., Hutcherson, C., Nass, C., John, O.: Real-time classification of evoked emotions using facial feature tracking and physiological responses. Int. J. Human-Computer Studies 66(5), 303–317 (2008)
4. Banea, C., Mihalcea, R., Wiebe, J.: A bootstrapping method for building subjectivity lexicons for languages with scarce resources. In: 6th International Conference on Language Resources and Evaluation, Marrakech (2008)
5. Bradley, M.M., Lang, P.J.: Affective norms for English words (ANEW): Instruction manual and affective ratings. Technical Report C-1, The Center for research in Psychophysiology, University of Florida (1999)

6. Burgin, W., Pantofaru, C., Smart, W.D.: Using depth information to improve face detection. In: Proc. of the 6th Int. Conf. on Human-Robot Interaction, pp. 119–120 (2011)
7. Burkhardt, F., Paechke, A., Rolfes, M., Sendlmeier, W., Weiss, B.: A database of German emotional speech. In: 9th European Conference on Speech Communication and Technology (2005)
8. Castrillon-Santana, M., Deniz-Suarez, O., Anton-Canalis, L., Lorenzo-Navarro, J.: Face and facial feature detection evaluation - performance evaluation of public domain Haar detectors for face and facial feature detection. In: Third Int. Conf. on Computer Vision Theory and Applications, VISAPP 2008, pp. 167–172 (2008)
9. Douglas-Cowie, E., Cowie, R., Schroeder, M.: A database of German emotional speech. In: 15th International Congress of Phonetic Sciences, Barcelona (2003)
10. Ekman, P., Friesen, W.V.: Facial Action Coding System. Consulting Psychologist Press (1978)
11. El Menshawy, D., Mokhtar, H.M.O., Hegazy, O.: A keystroke dynamics based approach for continuous authentication. In: Kozielski, S., Mrozek, D., Kasprowski, P., Małysiak-Mrozek, B., Kostrzewa, D. (eds.) BDAS 2014. CCIS, vol. 424, pp. 415–424. Springer, Heidelberg (2014)
12. Epp, C., Lippold, M., Mandryk, R.L.: Identifying emotional states using keystroke dynamics. In: Conf. on Human Factors in Computing Systems, pp. 715–724 (2011)
13. Gill, A.J., French, R.M., Gergle, D., Oberlander, J.: Identifying emotional characteristics from short blog texts. In: 30th Annual Conference of the Cognitive Science Society, pp. 2237–2242 (2008)
14. Gunes, H., Piccardi, M.: Affect recognition from face and body: Early fusion vs. late fusion. In: IEEE International Conference on Systems, Man and Cybernetics, vol. 4, pp. 3437–3443 (2005)
15. Jones, C., Sutherland, J.: Acoustic emotion recognition for affective computer gaming. In: Peter, C., Beale, R. (eds.) Affect and Emotion in Human-Computer Interaction. LNCS, vol. 4868, pp. 209–219. Springer, Heidelberg (2008)
16. Kołakowska, A.: A review of emotion recognition methods based on keystroke dynamics and mouse movements. In: Proc. of the 6th Int. Conf. on Human System Interaction (2013)
17. Landowska, A.: Affect-awareness framework for intelligent tutoring systems. In: 6th International Conference on Human System Interaction (2013)
18. Liberman, M., et al: Emotional prosody speech and transcripts LDC2002S28. philadelphia, linguistic data consortium (2002), https://catalog.ldc.upenn.edu/LDC2002S28
19. Lucey, P., Cohn, J.F., Kanade, T., Saragih, J., Ambadar, Z., Matthews, I.: The extended Cohn-Kanade (CK+): A complete dataset for action unit and emotion-specified expression. In: IEEE Computer Society Conference on Computer Vision and Pattern Recognition (Workshops), pp. 94–101 (2010)
20. Ortony, A., Clore, G.L., Foss, M.A.: The referential structure of the affective lexicon. In: Cognitive Science, pp. 341–364 (1987)
21. Pantic, M., Valstar, M.F., Rademaker, R., Maat, L.: Web-based database for facial expression analysis. In: IEEE International Conf. on Multimedia and Expo, Amsterdam (2005)
22. Pittermann, J., Pittermann, A., Minker, W.: Handling emotions in human computer dialog. Springer (2010)
23. Russell, J., Mehrabian, A.: Evidence for a three-factor theory of emotions. J. Research in Personality 11, 273–294 (1977)

24. Schuller, B., Lang, M., Rigoll, G.: Multimodal emotion recognition in audiovisual communication. In: IEEE International Conference on Multimedia and Expo (2002)

25. Strapparava, C., Valitutti, A.: Wordnet-affect: an affective extension of WordNet. In: 4th International Conference on Language Resources and Evaluation, Lisbon, pp. 1083–1086 (2004)

26. Szwoch, M.: FEEDB: a multimodal database of facial expressions and emotions. In: Proc. of the 6th Int. Conf. on Human System Interaction, pp. 524–531 (2013)

27. Szwoch, M.: On facial expressions and emotions RGB-D database. In: Kozielski, S., Mrozek, D., Kasprowski, P., Małysiak-Mrozek, B., Kostrzewa, D. (eds.) BDAS 2014. CCIS, vol. 424, pp. 384–394. Springer, Heidelberg (2014)

28. Szwoch, W.: Using physiological signals for emotion recognition. In: Proc. of the 6th Int. Conf. on Human System Interaction, pp. 556–561 (2013)

29. Tivatansakul, S., Chalumporn, G., Puangpontip, S., Kankanokkul, Y., Achalaku, T., Ohku-ra, M.: Healthcare system focusing on emotional aspect using augmented reality: Emotion detection by facial expression. In: Advances in Human Aspects of Healthcare, pp. 375–384 (2014)

30. Wróbel, M.R.: Emotions in the software development process. In: 6th International Conference on Human System Interaction (2013)

31. Yacoub, S., Simske, S., Lin, X., Burns, J.: Recognition of emotions in interactive voice response system. In: 8th European Conference on Speech Communication and Technology, Geneva (2003)

32. Zeng, Z., Pantic, M., Roisman, G., Huang, T.: A survey of affect recognition methods: Audio, visual, and spontaneous expressions. IEEE Transactions on Pattern Analysis and Machine Intelligence 31, 39–58 (2009)

The Mask: A Face Network System for Bell's Palsy Recovery Surveillance

Hanen Bouali[✉] and Jalel Akaichi

BESTMOD Laboratory,
ISG Tunis, University of Tunis, Tunis, Tunisia
hanene.bouali@gmail.com

Abstract. Bell's palsy is a sudden disease and very complex to diagnose. Doctors and physicians are required to be more agile and to be able to adapt quickly and efficiently to such disease. Mostly, doctors opt for the electromyography technique but the low quality services makes its realization difficult. To cope with this major problem, we provide a home nursing system. Our solution consists of modeling the EDA of Bells palsy system and proposing a conceptual framework that can guide research by providing a visual representation of theoretical aspects. However, a model should be good enough to be useful. To reach that, we present a mask for the face network system to detect and supervise Bells palsy.

1 Introduction

System modeling is an essential technology for managing interactions that occur across complex system components in a network. It helps in improving system understanding through visual analysis. In addition to that, system models are abstract description of systems whose requirements are being analyzed. It represents the system from different perspectives:

- External perspectives: showing the system context, interactions and information exchanged.
- Behavioral perspectives.
- Structural perspectives: showing the system architecture.

To take advantages from these benefits, researchers use system modeling to model or represent how real world works or to forecast future behavior in many systems such as weather prediction, economic models, natural sciences, engineering discipline (biology, physics)... Besides, the human body is a complex system susceptible to rapidly change and such instability may give risk to severe disease. Among this sub-systems, the facial nerve which his failure causes facial paralysis; that effective treatments are not well defined. As physicians are required to be more agile to respond to the rapidity and to be able to adapt quickly and efficiently to such disease, we need to introduce primarily an event driven architecture (EDA). An EDA is a software architecture pattern promoting the production, detection, consumption and reaction to events which is a significant change in state. Among key characteristics of EDA we adopt:

© Springer International Publishing Switzerland 2015
S. Kozielski et al. (Eds.): BDAS 2015, CCIS 521, pp. 598–609, 2015.
DOI: 10.1007/978-3-319-18422-7_53

- Timeliness: systems may publish events as they occur and not only storing the events locally and process them in package.
- Asynchrony: the publishing system does not wait for the receiving systems to receive and process events.
- Complex Event Processing CEP: the discipline of processing events to create complex events. CEP is a technology capable to match events with predefined contexts and patterns between events.

The main objective of this paper is to detect Bells palsy. To do that, physicians opt for the technique of Electromyography (EMG). The EMG send signals of intensities (events streams) recorded in facial nerves endings (Muscles and Glands). But, with the increasing cost of services and the patients inability to access to diverse services, the physicians technique is not always available. Hence, we present in this paper a nursing home technique in the form of a wearable ubiquitous system for detecting Bells palsy. To fulfill our solution, we represent the EDA of Bells palsy system and we propose a conceptual framework that can guide research by providing a visual representation of theoretical aspects. We opt for data processing model to show how the data is processed at different stages and the flow of information from one process to another. This model is simple and has intuitive notation which show end to end processing of data. But, in reality this model is very complicated to use. To cope with this major problem we present a more suitable solution for the face network system for the Bells palsy surveillance. The system consists of a wireless body area network that is deployed on the face of the patient. This system includes all the wearable sensors and a PDA as wearable computer. The system proposed manage the patient statues anytime anywhere hence the appointment of ubiquitous smart wearable systems (USWS). The remainder of this paper is as follows: in section 2, we model event by modeling the EPN components and show relationships between them and we provide a public health scenario for our use case: Bells palsy system. In section 3, we propose a facial nerve modeling which are the event producers of our EPN. In section 4, we present wearable systems developed in the literature to cope with diverse disease and. In section 5, we present our main contribution of this paper by modeling events deriving from face network system. And finally, in section 6, we summarize the work and propose new perspectives to be done in the future.

2   Bell's Palsy Modeling

Facial Paralysis is a disease affecting the facial nerve that, despite the techniques that are used to accelerate recovery, effective treatment is not yet well defined. In our search we opts the electromyography. Motor neurons transmit electrical signals that cause muscles to contract. An EMG translates these signals into graphs, sounds or numerical values that a specialist interprets. These numerical values allow a comparative analysis. Examination is done by setting up a needle electrode successively in the eyebrow muscles, orbicularis oculi, orbicularis oris and in the cover of the chin on the healthy side and the paralyzed side.

EMG activity was recorded at rest and during a voluntary contraction. By referring to the values calculated by the EMG on the healthy side is performed comparative analysis. These values are called threshold. This threshold is used later in our visualization algorithm to compare them to values calculated on the paralyzed side. On each patients visit, doctors apply EMG until patients completely recovering. This code is comprehensible by specialists and physicians but most of times patient doesnt understand specialized language concerned with medicines which is characterized by pretentious syntax, vocabulary, meaning or graphics. Understanding facial nerve anatomy permits us to better understand the Bells palsy disease. Once facial nerve structure is established and facial nerve structure is analyzed. We note that to better understand this disease a graphical modeling of the structure can be the efficient way. A useful way of representing the knowledge is by using graphs. They have been proved as an effective way of representing objects [7]. There are two kinds of graphical models; those based on undirected graphs and those based on directed graphs. Our main focus will be directed graphs. Graphical modeling presents several advantages:

– It shows the relationships between facial nerve components.
– It clearly illustrates the complexity of facial nerve bifurcations.
– It helps to understand the general idea by providing a global view of design, the text will be minimized.

In addition to the advantages of graphical modeling, the nature of facial nerve is composed of facial nerve bifurcations and connection between those bifurcations; this is the basic structure of a graph: nodes and edges. Hence, to benefit from all this advantages and the schematic nature of facial nerve, we modeled the facial nerve components bifurcation using an oriented graph. Once the facial nerve components are studied, bifurcation of the facial nerve was observed in the four components. Recall that the facial nerve stream is considered as a moving object circulating through a defined network, the facial nerve in our case of studies. A moving object is characterized by the trajectory in which circulates the facial nerve stream which is facial nerve [2]. In Fig. 1, we simplified the bifurcation of the facial nerve, where their end-points innerve the facial muscles. For each bifurcation, we place an intersection point to acquire a new portion. For the bronchial motor components, nodes represent intersection points or muscles and arcs represent the connections between them. The start node of the graph describes the beginning of the facial nerve of one side (the left side or the right side) of the face, the end-nodes represent facial muscles, and arcs describe connections between nodes. The choice of the oriented graph is explained by the fact that the facial nerve stream is unidirectional. Then, we identify each portion with its start-point and its end-point. We can generate the measures of amplitude and frequency of the stream nerve using electrical tests (e. g. electrodes) applied to each portion. We can measure also the intensity of the facial nerve stream crossing one specific portion. An intensity superior to a threshold means that the stream nerve is crossing a portion, otherwise is not.

For the visceral motor components, nodes represent intersection glands and edges represent the connections between them. The start node of the graph

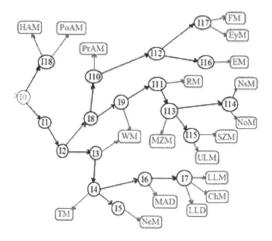

Fig. 1. Facial Nerve Muscle Graph

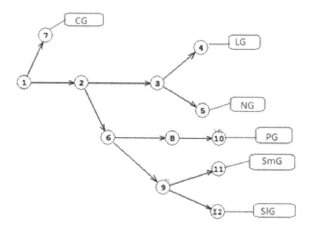

Fig. 2. Facial Nerve Glands Graph

describes the beginning of the facial nerve of one side (the left side or the right side) of the face, the end-nodes represent facial glands, and edges describe connections between nodes 2.

3   A Face Network Wearable System

3.1   Introduction

Real systems for disease prevention, symptom detection and diagnosis are required to prevent from different disease. To reach those efforts smart wearable systems (SWS) have been increasing in the last years. This development is due to the increasing cost of services, inability to access diverse services especially for elderly and the low quality services in some medical institutions. SWS is a kind of nursing home and a permanent admission to a core home. An SWS include a wide range of wearable implantable devices, sensors, wireless communication networks, user interface, algorithms for data capture and decision support. These systems are able to measure vital signs such as temperature, health rate, oxygen level, blood pressure... These measurements are forwarded via a wireless sensor network either to a central connection node or directly to a medical center. A subject can wear the device during normal daily life while medical professionals monitor the patient state in real time.

3.2 Related Works

We present in this section different wearable systems developed by researchers to monitor and manage the health patient statues. These sensing system scan be divided into:

- Worn by an individual as an accessory [13,14,15,12]
- Implantable in vivo [9,3,10,18]
- Portable
- Embedded in the users outfit as part of clothing [9,8]
- Embedded in pieces of object, fourniture [21,4,5]

Authors in [19] develop a wearable multi-channel fNIRS system for brain imaging in freely moving subjects during an outdoor activity in real life environments. It is employed in neuroscientific research. Moreover, the work proposed in [8]presents a wearable ECG monitoring device based on Electro-optic acquisition technology which a network of optical fiber and components placed at specefic recording location. This system represents a T-shirt with embedded optical sensors-optical module. Since the health statues have to be managed anytime and anywhere, authors in [1] develop a wearable system for ubiquitous healthcare service (UHS) provisionning. This system uses a wireless biomedical sensor. It monitor the oxygen situation level, heart rate, activity and locationof the elderly. The physiological data collected are uploaded to a health sensor via GPRS/internet for analysis. The UHS ensure privacy, integrity and authentication protocols. Otherwise, in order to assess cervical spine mobility, authors in [6] develop a wearable inertial system(IS). The IS includes two inertial sensors linked to be a light data logger worm at the waist. A new omnidirectionnal camera based scene labeling approach for augmented indoor topological mapping [20]. A semantic labels of the different types of transitions between places using a wearable cartadioptric vision system. Later on, a wearable line-of-sight detection system was developed by authors in [17] using micro fabricated transparent optical sensor or eyeglasses. The sensor detects the difference in the intensity of hight reflected from the purpil and the white of the eye and then determines the view angle. The system is also equipped with a CDD camera to acquire the users field of view to deduce the line-of-sight. A cable free network system on conductive fabric was used for wearable EMG measurement system [1] for power supply and electric shield for noise reduction is proposed. Tghis system is cable-free, confortable wear,free installation on the wear by pins, high communication ability and high power supply ability. For the safety and the well-being of children, authors propose in [11] a vest which automatically gather and provide information about the location and well-being of children. In this system, teachers and parents are able to receive alerts and notification when a child moves accross certain retricted area. These information are available through the web using sensor web enablement.

3.3   A Wearable Mask for Bells Palsy Detection

Bells palsy is a disease affecting the facial nerve that despite the techniques used to detect and used to accelerate recovery, effective treatments are not yet well defined. Besides, causes of Bells palsy are not clear and it develops suddenly. A person might have Bell's palsy first thing in the morning - they wake up and find that one side of the face does not move. Most people who suddenly experience symptoms believed to be temporary and can be confused with other facial paralysis disease. To avoid those problems, patients and doctors needs to detect Bells palsy instantly by receiving an alert. This can be reached by placing sensors in Facial Nerve ending which sends signals to an EPA to detect whether Bells palsy exists or not. In theory this method seems to be excellent, practical and easy to use. But once we go to practice it is very difficult for the patient to be comfortable with both sensor and cables. Our idea is to use this same technique of sensor and alert notification while offering patients a more comfortable and easy to use solution. The idea is to place the sensors in the form of a mask that a patient suspecting having a facial paralysis can wear it. Each component of the mask has a sensor connected to facial nerve endings (muscle and gland). Subsequently these sensors send the EMG signals in the form of event streams to the EPN for detecting facial paralysis. The patient receives a notification sent by a doctor or nurse on his device. These alert notifications are in the form of an event. In the Fig. 3 we present in details the solution mechanism.

4   Event Modeling

4.1   Event Processing Network Components

A conceptual model presents view of event processing systems, aimed at describing the important concepts and their relationships at an abstract level, removed from any technical details. The conceptual model we introduce in this paper provides unified and well defined concepts of event processing network (EPN) describing the scenario of Bells palsys wearable system. An EPN models consists of four components[16]:

– Event Producer: known also as event source or event emitter. An event producer might for example be an application, data store, service, business process, transmitter or sensor.

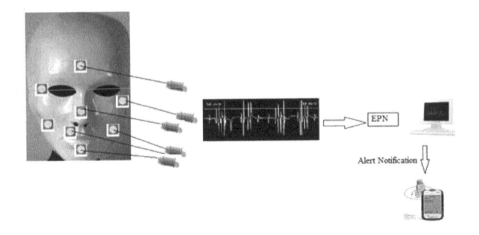

Fig. 3. The mask: A face Network System

- Event Consumer: also known as an event sink, is interested in events to perform its responsibilities.
- Event Processing Agent (EPA) : also known as event mediators, EPA is needed to detect patterns on raw events, then to process these events through enrichment, transformation, validation and finally to derive new events and publish them. EPA gives a sense to the event. EPA is made up of three stages:
  - Pattern Detection: selecting events for processing according a specified pattern
  - Processing: applying the processing function on the selected events satisfying the pattern, resulting in derived pattern
  - Emission: deciding where and how to publish the derived events.
- Event Channel: a mechanism for delivering events streams from event producers and event processing agents to event consumer and events processing agents. Event channel create an ordered and combined event stream from the sources and provide this stream to each sink.

Figure 4 shows the relationships between events channel, producer, consumers and processing agent composing Bells palsy system to be studied in details in the rest of the paper.

4.2 Public Health Scenario

An EPN describes how events received from event producers are directed to the consumers through agents that process these events by performing transformation, validation or enrichment. Any event flowing from one component to another must flow through an event channel connection component. The primary responsibility of an EPN is to receive events from producers, pass them on

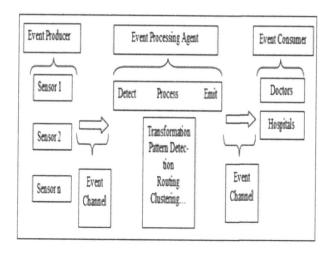

Fig. 4. Event Processing Network

the combination of EPA to process the events and deliver the events to the right consumer. Combining EPN and Facial Nerve Modeling, physicians can easily detect Bells palsy immediately by receiving an alert notification (Agent A3) and see the recovery progress (Agent A4) of the disease once detected. The Bells palsy Alert Scenario describes the alerting system in cases that are indication of real words or potential outbreaks that pose a risk to patients. For the scenario, two stages are primarily considered:

- No detection of Bells Palsy
- Detection of Bells palsy

For the detection of Bells palsy is decomposed under different stages to represent the gravity of Bells palsy by computing a score of recovery. A physician, comparing the threshold to EMG intensities received and detects Bells palsy. The detection of Bells palsy is published as an event. The event receiver normalizes the event (EMG signals) and performs some basic quality and origin checks before passing it to the EPA where signals are transformed, ordered, clustered and validates to check whether the Bells palsy id detected. In the case of Bells palsy detection notification are sent as corresponding events to the event consumer. Different types of events are sent through event channel. These system process many different types of events, including events that represent:

- EMG signals and intensities
- Detected faults in the network
- Changes in the state of network elements
- Noisy events
- Actions of human operators that respond to network events.

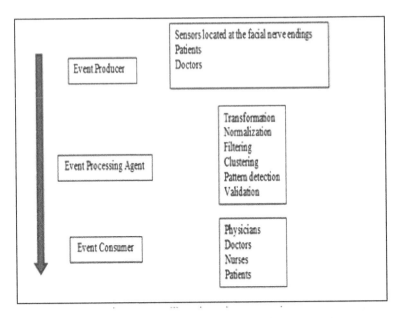

Fig. 5. Bell's palsy Alert Overview

Table 1. Event Producer

Event Producer	Description
EMG Sensors	EMG sensors send intensities:emit events
Physicians	The physician may detect Bells palsy and emit a detection event
Patient	The patient may emit extra information to doctors

Access to this event stream allows physicians to:

– Normalize network event to a common format for consistent visualization
  and processing
– Automatically discover and maintain up-to-date models
– Provide dashboards to physicians or nurses for the purpose of problem

The following tables list EPN components of the Bells Palsy Alert Scenario
described before.

After defining the various actors, we provide a mapping of the Bells palsy
alert scenario to the event processing conceptual architecture. Figure 6 provides
a graphical depiction of the part of the EPN used in the scenario, up to the point
that the technician is dispatched.

Table 2. Event Consumer

Event Consumer	Description
Physicians	Physicians subscribe to Bells palsy notification system
Nurses	Nurses subscribe to Bells palsy notification system
Patients	Patients subscribe to Bells palsy notification system
Dashboards	Interactions view of significant events including diagnostics and disease gravity

Table 3. Event Types

Event Type ID	Event Type	Attributes
E1	Bells Palsy Detection	ID, PID, Details
E2	Bells Palsy Gravity	ID, PID, Score
E3	Muscle intensity	ID, PID, Mintensity, MLocation, MThreshold
E4	Gland intensity	ID, PID, Gintensity, GLocation, GThreshold
E5	Patient Information	PID, gender, age
E6	Extracted Event	ID, details

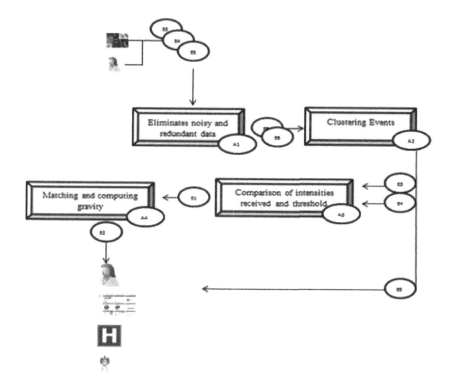

Fig. 6. EPN for Bell's palsy Alert Scenario

4.3 Example

Table 4. Event Types

Agent Properties	Specification
ID	A2
Name	Agent Clustering
Type	Clustering
Input	E6
Output	E3, E4, E5
Specification	Receives event streams and cluster them according to their types
Comments	Agent uses an pipelined On-line Off-line clustering algorithm based on support vector machine and data representative points techniques in order to keep an up-to-date model

5 Conclusion and Future Works

Bells palsy is a sudden and mysterious disease. For this, researchers offer solutions to detect the disease and helping patient and doctors in diagnostics. To do that, we opt in this paper for modeling a face network system for Bells palsy recovery surveillance. We provide a mapping of the Bells palsy alert scenario to the event processing conceptual architecture. To make this solution practical, we propose a wearable mask for patient to wear it in case of doubt. The mask is composed by EMG placed at facial nerve endings. Those EMG sends events to the EPN model to detect Bells palsy.

References

1. Akita, J., Shinmura, T., Sakurazawa, S., Yanagihara, K., Kunita, M., Toda, M., Iwata, K.: Wearable electromyography measurement system using cable-free network system on conductive fabric. Artificial Intelligence in Medicine 42(2), 99–108 (2008)
2. Bouali, H., Akaichi, J.: Mobile data modeling in human body network "bell's palsy case study". In: GLOBAL HEALTH 2012, The First International Conference on Global Health Challenges, pp. 33–40 (2012)
3. Chaudhary, A., McShane, M.J., Srivastava, R.: Glucose response of dissolved-core alginate microspheres: towards a continuous glucose biosensor. Analyst 135(10), 2620–2628 (2010)
4. Cook, D.J., Augusto, J.C., Jakkula, V.R.: Ambient intelligence: Technologies, applications, and opportunities. Pervasive and Mobile Computing 5(4), 277–298 (2009)
5. Coughlin, J.F., Pope, J.: Innovations in health, wellness, and aging-in-place. IEEE Engineering in Medicine and Biology 27(4), 47–52 (2008)

6. Duc, C., Salvia, P., Lubansu, A., Feipel, V., Aminian, K.: A wearable inertial system to assess the cervical spine mobility: comparison with an optoelectronic-based motion capture evaluation. Medical Engineering & Physics 36(1), 49–56 (2014)

7. Eshera, M.A., Fu, K.S.: An image understanding system using attributed symbolic representation and inexact graph-matching. IEEE Transactions on Pattern Analysis and Machine Intelligence 5, 604–618 (1986)

8. Fernandes, M.S., Correia, J.H., Mendes, P.M.: Electro-optic acquisition system for ecg wearable sensor applications. Sensors and Actuators A: Physical 203, 316–323 (2013)

9. Giorgino, T., Tormene, P., Lorussi, F., De Rossi, D., Quaglini, S.: Sensor evaluation for wearable strain gauges in neurological rehabilitation. IEEE Transactions on Neural Systems and Rehabilitation Engineering 17(4), 409–415 (2009)

10. Halliday, A.J., Moulton, S.E., Wallace, G.G., Cook, M.J.: Novel methods of antiepileptic drug delivery – polymer-based implants. Advanced Drug Delivery Reviews 64(10), 953–964 (2012)

11. Jutila, M., Rivas, H., Karhula, P., Pantsar-Syväniemi, S.: Implementation of a wearable sensor vest for the safety and well-being of children. Procedia Computer Science 32, 888–893 (2014)

12. Lanata, A., Scilingo, E.P., De Rossi, D.: A multimodal transducer for cardiopulmonary activity monitoring in emergency. IEEE Transactions on Information Technology in Biomedicine 14(3), 817–825 (2010)

13. Loseu, V., Ghasemzadeh, H., Ostadabbas, S., Raveendranathan, N., Malan, J., Jafari, R.: Applications of sensing platforms with wearable computers. In: Proceedings of the 3rd International Conference on PErvasive Technologies Related to Assistive Environments, p. 53. ACM (2010)

14. Ma, Z.: An electronic second skin. Science 333(6044), 830–831 (2011)

15. Miwa, H., Sasahara, S.I., Matsui, T.: Roll-over detection and sleep quality measurement using a wearable sensor. In: 29th Annual International Conference of the IEEE Engineering in Medicine and Biology Society, EMBS 2007, pp. 1507–1510. IEEE (2007)

16. Moxey, C., Edwards, M., Etzion, O., Ibrahim, M., Iyer, S., Lalanne, H., Monze, M., Peters, M., Rabinovich, Y., Sharon, G., et al.: A conceptual model for event processing systems. IBM Redguide publication (2010)

17. Ozawa, M., Sampei, K., Cortes, C., Ogawa, M., Oikawa, A., Miki, N.: Wearable line-of-sight detection system using micro-fabricated transparent optical sensors on eyeglasses. Sensors and Actuators A: Physical 205, 208–214 (2014)

18. Pathan, S.A., Jain, G.K., Akhter, S., Vohora, D., Ahmad, F.J., Khar, R.K.: Insights into the novel three 'd's of epilepsy treatment: drugs, delivery systems and devices. Drug Discovery Today 15(17), 717–732 (2010)

19. Piper, S.K., Krueger, A., Koch, S.P., Mehnert, J., Habermehl, C., Steinbrink, J., Obrig, H., Schmitz, C.H.: A wearable multi-channel fnirs system for brain imaging in freely moving subjects. NeuroImage 85, 64–71 (2014)

20. Rituerto, A., Murillo, A.C., Guerrero, J.J.: Semantic labeling for indoor topological mapping using a wearable catadioptric system. Robotics and Autonomous Systems (2012)

21. Schmidt, A., Strohbach, M., Van Laerhoven, K., Friday, A., Gellersen, H.-W.: Context acquisition based on load sensing. In: Borriello, G., Holmquist, L.E. (eds.) UbiComp 2002. LNCS, vol. 2498, pp. 333–350. Springer, Heidelberg (2002)

Author Index